STUDENT SOLUTIONS
to Accompany

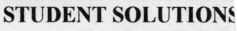

SALAS AND HILLE'S
CALCULUS
ONE VARIABLE
EARLY TRANSCENDENTALS

Seventh Edition

REVISED BY
GARRET J. ETGEN

John Wiley & Sons, Inc.
New York • Chichester • Brisbane • Toronto • Singapore

ISBN 0-471-13377-9

Printed in the United States of America

10 9 8 7 6 5 4 3 2 1

CONTENTS

CHAPTER 1

SECTION 1.2

1. rational, complex   **3.** rational, complex   **5.** integer, rational, complex

7. integer, rational, complex   **9.** integer, rational, complex   **11.** irrational, complex

13. $x = 13.201201\cdots,\quad 1000x = 13201.201201\cdots.$ Therefore, $999x = 13188$ and $x = \dfrac{13188}{999}$.

15. $x = 0.2323\cdots,\quad 100x = 23.2323\cdots.$ Therefore, $99x = 23$ and $x = \dfrac{23}{99}$.

17. $x = 5.252252\cdots,\quad 1000x = 5252.252252\cdots.$ Therefore, $999x = 5247$ and $x = \dfrac{5247}{999}$.

19. $\dfrac{3}{4} = 0.75$ **21.** $\sqrt{2} > 1.414$ **23.** $-\dfrac{2}{7} < -0.28517$

25. $|6| = 6$ **27.** $|3 - 7| = 4$ **29.** $|-5| + |-8| = 13$

31. $|5 - \sqrt{5}| = 5 - \sqrt{5}$ **33.**

35.   **37.**

39.   **41.**

43.   **45.**

47.   **49.**

51. bounded, lower bound 0, upper bound 4   **53.** not bounded

55. not bounded   **57.** bounded above, upper bound $\sqrt{2}$

59. $x^2 - 10x + 25 = (x - 5)^2$ **61.** $8x^6 + 64 = 8(x^2 + 2)(x^4 - 2x^2 + 4)$

63. $4x^2 + 12x + 9 = (2x + 3)^2$ **65.** $x^2 - x - 2 = (x - 2)(x + 1) = 0;\quad x = 2, -1$

67. $x^2 - 6x + 9 = (x - 3)^2;\quad x = 3$ **69.** $x^2 - 2x + 2 = 0;\quad$ no real zeros

71. $R(x) = \dfrac{x-1}{x^2}$; R is not defined at $x=0$; $R(x)=0$ at $x=1$.

73. $R(x) = \dfrac{x^2-1}{x^2+1}$; R is defined for all real numbers $(x^2+1>0$ for all $x)$;

     $R(x)=0$ at $x=1,-1$.

75. $R(x) = \dfrac{2x+3}{x^2+2x+5}$; R is defined for all real numbers $(x^2+2x+5>0$ for all $x)$.

     $R(x)=0$ at $x=-3/2$.

77. $\dfrac{5!}{8!} = \dfrac{5\cdot4\cdot3\cdot2\cdot1}{8\cdot7\cdot6\cdot5\cdot4\cdot3\cdot2\cdot1} = \dfrac{1}{336}$ **79.** $\dfrac{9!}{3!6!} = \dfrac{9\cdot8\cdot7\cdot6\cdot5\cdot4\cdot3\cdot2\cdot1}{3\cdot2\cdot1\cdot6\cdot5\cdot4\cdot3\cdot2\cdot1} = 84$

81. Let r be a rational number and s an irrational number. Suppose $r+s$ is rational. Then $(r+s)-r=s$ is rational which contradicts the fact that s is irrational.

83. The product of a rational and an irrational number may either be rational or irrational; $0\cdot\sqrt{2}=0$ is rational, $1\cdot\sqrt{2}=\sqrt{2}$ is irrational.

85. Suppose that $\sqrt{2}=p/q$ where p and q are integers and $q\neq0$. Assume that p and q have no common factors (other than ±1). Then $p^2=2q^2$ and p^2 is even. This implies that $p=2r$ is even. Thus $2q^2=4r^2$ which implies that q^2 is even, and hence q is even. It now follows that p and q are both even and contradicts the assumption that p and q have no common factors.

87. Let x be the length of a rectangle that has perimeter P. Then the width y of the rectangle is given by $y=\frac{1}{2}(P-x)$ and the area is

$$A = x\left(\frac{1}{2}P-x\right) = \left(\frac{P}{4}\right)^2 - \left(x-\frac{P}{4}\right)^2.$$

It now follows that the area is a maximum when $x=P/4$. Since $y=P/4$ when $x=P/4$, the rectangle of perimeter P having the largest area is a square.

SECTION 1.3

1. $2+3x<5$

     $3x<3$

     $x<1$

     Ans: $(-\infty,1)$

3. $16x+64\leq16$

     $16x\leq-48$

     $x\leq-3$

     Ans: $(-\infty,-3]$

5. $\frac{1}{2}(1+x)<\frac{1}{3}(1-x)$

     $3(1+x)<2(1-x)$

     $3+3x<2-2x$

     $5x<-1$

     $x<-\frac{1}{5}$

     Ans: $(-\infty,-\frac{1}{5})$

7. $x^2-1<0$

     $(x+1)(x-1)<0$

```
+++ 0 ----- 0 +++++
    -1        1
```

     Ans: $(-1,1)$

9. $x(x-1)(x-2)>0$

```
-- 0 ++ 0 -- 0 ++
   0    1    2
```

     Ans: $(0,1)\cup(2,\infty)$

11. $x^3-2x^2+x\geq0$

     $x(x-1)^2\geq0$

```
---- 0 ++++ 0 ++++
     0      1
```

     Ans: $[0,\infty)$

13.
$$x^2 + 1 > 4x$$
$$x^2 - 4x + 1 > 0$$
$$x^2 - 4x + 4 > 3$$
$$(x-2)^2 > 3$$
$$x - 2 > \sqrt{3} \ \text{ or } \ x - 2 < -\sqrt{3}$$

Ans: $(-\infty, 2-\sqrt{3}) \cup (2+\sqrt{3}, \infty)$

15.
$$1 - 3x^2 < \tfrac{1}{2}(2 - x^2)$$
$$2 - 6x^2 < 2 - x^2$$
$$0 < 5x^2$$

True if $x \neq 0$.

Ans: $(-\infty, 0) \cup (0, \infty)$

17.
$$\frac{1}{x} < x$$
$$x - \frac{1}{x} > 0$$
$$\frac{x^2 - 1}{x} > 0$$
$$x(x-1)(x+1) > 0 \qquad \text{(by 1.3.1)}$$
$$(x+1)\,x\,(x-1) > 0$$

Ans: $(-1, 0) \cup (1, \infty)$

19.
$$\frac{x}{x-5} \geq 0$$
$$x(x-5) > 0 \ \text{ or } \ x = 0 \quad \text{(by 1.3.1)}$$

Ans: $(-\infty, 0] \cup (5, \infty)$

21.
$$\frac{x}{x-5} > \frac{1}{4}$$
$$\frac{x}{x-5} - \frac{1}{4} > 0$$
$$\frac{4x - (x-5)}{4(x-5)} > 0$$
$$\frac{3x + 5}{4(x-5)} > 0$$
$$4(x-5)(3x+5) > 0 \quad \text{(by 1.3.1)}$$
$$(3x+5)(x-5) > 0$$

Ans: $\left(-\infty, -\tfrac{5}{3}\right) \cup (5, \infty)$

23.
$$\frac{x^2 - 9}{x+1} > 0$$
$$(x+1)(x-3)(x+3) > 0 \quad \text{(by 1.4.1)}$$
$$(x+3)(x+1)(x-3) > 0$$

Ans: $(-3, -1) \cup (3, \infty)$

25. $x^3(x-2)(x+3)^2 < 0$

$(x+3)^2 x(x-2) < 0$

```
+++ 0 +++++++ 0 ------ 0 +++
    -3          0         2
```

Ans: $(0,2)$

27. $x^2(x-2)(x+6) > 0$

$(x+6)\,x^2(x-2) > 0$

```
+++ 0 ---------- 0 --- 0 +++
   -6            0      2
```

Ans: $(-\infty,-6) \cup (2,\infty)$

29. $\dfrac{2x}{x^2-4} > 0$

$2x(x+2)(x-2) > 0$

$(x+2)\,x\,(x-2) > 0$

```
---- 0 +++++ 0 ------ 0 ++++
   -2        0        2
```

Ans: $(-2,0) \cup (2,\infty)$

31. $\dfrac{1}{x-1} + \dfrac{4}{x-6} > 0$

$\dfrac{x-6+4(x-1)}{(x-1)(x-6)} > 0$

$\dfrac{5x-10}{(x-1)(x-6)} > 0$

$5(x-2)(x-1)(x-6) > 0$

$(x-1)(x-2)(x-6) > 0$

```
--- 0 +++ 0 ----------- 0 +++
   1      2             6
```

Ans: $(1,2) \cup (6,\infty)$

33. $\dfrac{2x-6}{x^2-6x+5} < 0$

$2(x-3)(x-1)(x-5) < 0$

$(x-1)(x-3)(x-5) < 0$

```
---- 0 +++++ 0 ------ 0 ++++
   1         3        5
```

Ans: $(-\infty,1) \cup (3,5)$

35. $\dfrac{x^2-4x+3}{x^2} > 0$

$x^2(x-1)(x-3) > 0$

```
+++ 0 ++++ 0 -------- 0 ++++
   0       1          3
```

Ans: $(-\infty,0) \cup (0,1) \cup (3,\infty)$

37. $|x| < 2$

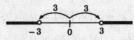

Ans: $(-2,2)$

39. $|x| > 3$

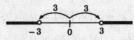

Ans: $(-\infty,-3) \cup (3,\infty)$

41. $|x - 2| < \frac{1}{2}$

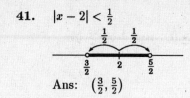

Ans: $\left(\frac{3}{2}, \frac{5}{2}\right)$

43. $0 < |x| < 1$

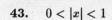

Ans: $(-1, 0) \cup (0, 1)$

45. $0 < |x - 2| < \frac{1}{2}$

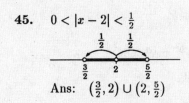

Ans: $\left(\frac{3}{2}, 2\right) \cup \left(2, \frac{5}{2}\right)$

47. $0 < |x - 3| < 8$

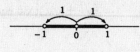

Ans: $(-5, 3) \cup (3, 11)$

49. $|2x + 1| < \frac{1}{4}$

$\left|x - \left(-\frac{1}{2}\right)\right| < \frac{1}{8}$

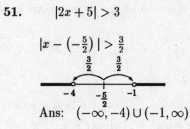

Ans: $\left(-\frac{5}{8}, -\frac{3}{8}\right)$

51. $|2x + 5| > 3$

$\left|x - \left(-\frac{5}{2}\right)\right| > \frac{3}{2}$

Ans: $(-\infty, -4) \cup (-1, \infty)$

53. $|5x - 1| > 9$

$\left|x - \frac{1}{5}\right| > \frac{9}{5}$

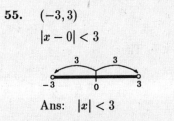

Ans: $\left(-\infty, -\frac{8}{5}\right) \cup (2, \infty)$

55. $(-3, 3)$

$|x - 0| < 3$

Ans: $|x| < 3$

57. $(-3, 7)$

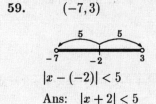

Ans: $|x - 2| < 5$

59. $(-7, 3)$

$|x - (-2)| < 5$

Ans: $|x + 2| < 5$

61. $|x - 2| < A \implies 2|x - 2| = |2x - 4| < 2A \implies |2x - 4| < 3$
             provided that $0 < A \le \frac{3}{2}$

63. $|x + 1| < 2 \implies 3|x + 1| = |3x + 3| < 6 \implies |3x + 3| < A$
             provided that $A \ge 6$

65. $x < \sqrt{x} < 1 < \frac{1}{\sqrt{x}} < \frac{1}{x}$

67. If a and b have the same sign, then $ab > 0$. Suppose that $a < b$. Then $\quad a - b < 0 \quad$ and

$$\frac{1}{b} - \frac{1}{a} = \frac{a - b}{ab} < 0.$$

Thus, $\quad (1/b) < (1/a)$.

69.   With $a \geq 0$ and $b \geq 0$

$$b \geq a \implies b - a = (\sqrt{b} + \sqrt{a})(\sqrt{b} - \sqrt{a}) \geq 0 \implies \sqrt{b} - \sqrt{a} \geq 0 \implies \sqrt{b} \geq \sqrt{a}.$$

71.   By the hint

$$\left|\, |a| - |b| \,\right|^2 = (|a| - |b|)^2 = |a|^2 - 2|a|\,|b| + |b|^2 = a^2 - 2|ab| + b^2$$

$$\leq a^2 - 2ab + b^2 = (a - b)^2.$$

$$\overset{\displaystyle\llcorner}{\quad (ab \leq |ab|)}$$

Taking the square root of the extremes, we have

$$\left|\, |a| - |b| \,\right| \leq \sqrt{(a - b)^2} = |a - b|.$$

73.   With $0 \leq a \leq b$

$$a(1 + b) = a + ab \leq b + ab = b(1 + a).$$

Division by $(1 + a)(1 + b)$ gives

$$\frac{a}{1 + a} \leq \frac{b}{1 + b}.$$

75.   Suppose that $a < b$. Then

$$a = \frac{a + a}{2} \leq \frac{a + b}{2} \leq \frac{b + b}{2} = b.$$

$\dfrac{a + b}{2}$ is the midpoint of the line segment $\overline{ab}$.

SECTION 1.4

1. $d(P_0, P_1) = \sqrt{(6 - 5)^2 + (-3 - 0)^2} = \sqrt{1 + 9} = \sqrt{10}$

3. $d(P_0, P_1) = \sqrt{[5 - (-3)]^2 + (-2 - 2)^2} = \sqrt{64 + 16} = 4\sqrt{5}$

5. $d(P_0, P_1) = \sqrt{(2 - 2)^2 + (-2 - 4)^2} = \sqrt{0 + 36} = 6$

7. $\left(\dfrac{2 + 6}{2}, \dfrac{4 + 8}{2}\right) = (4, 6)$ **9.** $\left(\dfrac{2 + 7}{2}, \dfrac{-3 - 3}{2}\right) = \left(\dfrac{9}{2}, -3\right)$

11. $\left(\dfrac{\sqrt{3} + 0}{2}, \dfrac{0 + \sqrt{3}}{2}\right) = \frac{1}{2}(\sqrt{3}, \sqrt{3})$

13. $m = \dfrac{5 - 1}{(-2) - 4} = \dfrac{4}{-6} = -\dfrac{2}{3}$ **15.** $m = \dfrac{2 - 2}{4 - (-3)} = \dfrac{0}{7} = 0$

17. $m = \dfrac{b - a}{a - b} = -1$ **19.** $m = \dfrac{0 - y_0}{x_0 - 0} = -\dfrac{y_0}{x_0}$

21.   Equation is in the form $y = mx + b$.   Slope is 2; y-intercept is -4.

23. Write equation as $y = \frac{1}{3}x + 2$. Slope is $\frac{1}{3}$; y-intercept is 2.

25. Line is vertical: $x = \frac{1}{4}$. Thus slope is undefined and there is no y-intercept.

27. Write equation as $y = \frac{7}{3}x + \frac{4}{3}$. Slope is $\frac{7}{3}$; y-intercept is $\frac{4}{3}$.

29. $y = 5x + 2$ **31.** $y = -5x + 2$ **33.** $y = 3$ **35.** $x = -3$

37. Every line parallel to the x-axis has an equation of the form $y = a$ constant. In this case $y = 7$.

39. The line $3y - 2x + 6 = 0$ has slope $\frac{2}{3}$. Every line parallel to it has that same slope. The line through $P(2, 7)$ with slope $\frac{2}{3}$ has equation $y - 7 = \frac{2}{3}(x - 2)$, which reduces to $3y - 2x - 17 = 0$.

41. The line $3y - 2x + 6 = 0$ has slope $\frac{2}{3}$. Every line perpendicular to it has slope $-\frac{3}{2}$. The line through $P(2, 7)$ with slope $-\frac{3}{2}$ has equation $y - 7 = -\frac{3}{2}(x - 2)$, which reduces to $2y + 3x - 20 = 0$.

43. $m = 1 = \tan\theta$; $\theta = 45°$ **45.** line is vertical: $x = \frac{3}{2}$; $\theta = 90°$

47. $m = -\frac{3}{4} = \tan\theta$; $\theta \cong 143°$ **49.** $m = \tan 30° = \frac{1}{3}\sqrt{3}$;   line: $y = \frac{1}{3}\sqrt{3}\,x + 2$

51. $m = \tan 120° = -\sqrt{3}$;   line: $y = -\sqrt{3}\,x + 3$

53. $\left(\frac{1}{2}\sqrt{2}, \frac{1}{2}\sqrt{2}\right), \left(-\frac{1}{2}\sqrt{2}, -\frac{1}{2}\sqrt{2}\right)$ [Substitute $y = x$ into $x^2 + y^2 = 1$.]

55. $(3, 4)$ [Write $4x + 3y = 24$ as $y = \frac{4}{3}(6 - x)$ and substitute into $x^2 + y^2 = 25$.]

57. $(1, 1)$; $\alpha \cong 39°$ [$m_1 = 4 = \tan\theta_1$, $\theta_1 \cong 76°$; $m_2 = \frac{3}{4} = \tan\theta_2$, $\theta_2 \cong 37°$]

59. $\left(-\frac{2}{23}, \frac{38}{23}\right)$; $\alpha \cong 17°$ [$m_1 = 4 = \tan\theta_1$, $\theta_1 \cong 76°$; $m_2 = 19 = \tan\theta_2$, $\theta_2 \cong 87°$]

61. Using the formula: (a) $\frac{2}{13}$ (b) $\frac{29}{13}$

63. $d((0, 1), l) = 1/\sqrt{113}$, $d((1, 0), l) = 2/\sqrt{113}$, $d((-1, 1), l) = 7\sqrt{113}$

The closest point is $(0, 1)$;   the point $(-1, 1)$ is farthest away.

65. We select the side joining $A(1, -2)$ and $B(-1, 3)$ as the base of the triangle.

length of side AB: $\sqrt{29}$

equation of line through A and B: $5x + 2y - 1 = 0$

length of altitude from vertex $C(2, 4)$ to side AB: $\dfrac{|5(2) + 2(4) - 1|}{\sqrt{29}} = \dfrac{17}{\sqrt{29}}$

area of triangle: $\dfrac{1}{2}\left(\sqrt{29}\right)\left(\dfrac{17}{\sqrt{29}}\right) = \dfrac{17}{2}$

67. $(y + 1)^2 = x + 1$;   parabola, vertex at $(-1, 1)$

69. $2(x - 2)^2 + 3(y + 1)^2 = 6$,   or $\dfrac{(x - 2)^2}{3} + \dfrac{(y + 1)^2}{2} = 1$;   ellipse, center at $(2, -1)$

71. $(y - 2)^2 - 4(x - 1)^2 = 4$,   or $\dfrac{(y - 2)^2}{4} - \dfrac{(x - 1)^2}{1} = 1$;   hyperbola, center at $(1, 2)$

73. $4(x-3)^2 - (y+2)^2 = 16$,   or $\dfrac{(x-3)^2}{4} - \dfrac{(y+2)^2}{16} = 1$;   hyperbola, center at $(3, -2)$

75. Substitute $y = m(x-5) + 12$ into $x^2 + y^2 = 169$ and you get a quadratic in x that involves m. That quadratic has a unique solution iff $m = -\frac{5}{12}$. (A quadratic $ax^2 + bx + c = 0$ has a unique solution iff $b^2 - 4ac = 0$. This is clear from the general quadratic formula.)

79. midpoint of line segment $\overline{PQ}$: $\left(\dfrac{5}{2}, \dfrac{5}{2}\right)$

slope of line segment $\overline{PQ}$: $\dfrac{13}{3}$

equation of the perpendicular bisector: $y - \dfrac{5}{2} = -\left(\dfrac{3}{13}\right)\left(x - \dfrac{5}{2}\right)$ or $3x + 13y - 40 = 0$

81. $d(P_0, P_1) = \sqrt{(-2-1)^2 + (5-3)^2} = \sqrt{13}$, $d(P_0, P_2) = \sqrt{[-2-(-1)]^2 + (5-0)^2} = \sqrt{26}$,

$d(P_1, P_2) = \sqrt{[1-(-1)]^2 + (3-0)^2} = \sqrt{13}$.

Since $d(P_0, P_1) = d(P_1, P_2)$,   the triangle is isosceles.

Since $[d(P_0, P_1)]^2 + [d(P_1, P_2)]^2 = [d(P_0, P_2)]^2$,   the triangle is a right triangle.

83. $d(P_0, P_1) = \sqrt{(3-1)^2 + (4-1)^2} = \sqrt{13}$, $d(P_0, P_2) = \sqrt{[3-(-2)]^2 + (4-3)^2} = \sqrt{26}$,

$d(P_1, P_2) = \sqrt{[1-(-2)]^2 + (1-3)^2} = \sqrt{13}$.

Since $d(P_0, P_1) = d(P_1, P_2)$,   the triangle is isosceles.

Since $[d(P_0, P_1)]^2 + [d(P_1, P_2)]^2 = [d(P_0, P_2)]^2$,   the triangle is a right triangle.

85. The coordinates of M are $\left(\dfrac{a}{2}, \dfrac{b}{2}\right)$;   and

$d(M, (0,b)) = d(M, (0,a)) = d(M, (0,0)) = \frac{1}{2}\sqrt{a^2 + b^2}$.

87. Denote the points $(0,1)$, $(3,4)$ and $(-1,6)$ by A, B and C, respectively. The midpoints of the line segments $\overline{AB}$, $\overline{AC}$, and $\overline{BC}$ are $P(2,2)$, $Q(0,3)$ and $R(1,5)$, respectively.
An equation for the line through A and R is: $x = 1$.
An equation for the line through B and Q is: $y = \frac{1}{3}x + 3$.
An equation for the line through C and P is: $y - 2 = -\frac{4}{3}(x - 2)$.
These three lines intersect at the point $(1, \frac{10}{3})$.

89. Let $A(0,0)$ and $B(a,0)$, $a > 0$, be adjacent vertices of a parallelogram. If $C(b,c)$ is the vertex opposite B, then the vertex D opposite A has coordinates $(a+b, c)$ [see the figure].

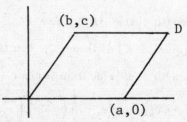

The line through A and D has equation: $y = \dfrac{c}{a+b}x$.

The line through B and C has equation: $y = -\dfrac{c}{a-b}(x - a)$.

These lines intersect at the point $\left(\dfrac{a+b}{2}, \dfrac{c}{2}\right)$ which is the midpoint of each of the line segments $\overline{AD}$ and $\overline{BC}$.

91. Since the relation between F and C is linear, $F = mC + b$ for some constants m and C. Setting $C = 0$ and $F = 32$ gives $b = 32$. Thus $F = mC + 32$. Now, letting $C = 100$ and $F = 212$ gives $m = (212 - 32)/100 = 9/5$. Therefore

$$F = \frac{9}{5}C + 32$$

The Fahrenheit and Centigrade temperatures are equal when

$$C = F = \frac{9}{5}C + 32$$

which implies $C = F = -40°$.

SECTION 1.5

1. (a) $f(0) = 2(0)^2 - 3(0) + 2 = 2$ (b) $f(1) = 2(1)^2 - 3(1) + 2 = 1$

(c) $f(-2) = 2(-2)^2 - 3(-2) + 2 = 16$ (d) $f(\frac{3}{2}) = 2(3/2)^2 - 3(3/2) + 2 = 2$

3. (a) $f(0) = \sqrt{0^2 + 2 \cdot 0} = 0$ (b) $f(1) = \sqrt{1^2 + 2 \cdot 1} = \sqrt{3}$

(c) $f(-2) = \sqrt{(-2)^2 + 2(-2)} = 0$ (d) $f(\frac{3}{2}) = \sqrt{(3/2)^2 + 2(3/2)} = \frac{1}{2}\sqrt{21}$

5. (a) $f(0) = \dfrac{2 \cdot 0}{|0 + 2| + 0^2} = 0$ (b) $f(1) = \dfrac{2 \cdot 1}{|1 + 2| + 1^2} = \dfrac{1}{2}$

(c) $f(-2) = \dfrac{2 \cdot (-2)}{|-2 + 2| + (-2)^2} = -1$ (d) $f(\frac{3}{2}) = \dfrac{2 \cdot (3/2)}{|(3/2) + 2| + (3/2)^2} = \dfrac{12}{23}$

7. (a) $f(-x) = (-x)^2 - 2(-x) = x^2 + 2x$ (b) $f(1/x) = (1/x)^2 - 2(1/x) = \dfrac{1 - 2x}{x^2}$

(c) $f(a + b) = (a + b)^2 - 2(a + b) = a^2 + 2ab + b^2 - 2a - 2b$

9. (a) $f(-x) = \sqrt{1 + (-x)^2} = \sqrt{1 + x^2}$ (b) $f(1/x) = \sqrt{1 + (1/x)^2} = |x|/\sqrt{1 + x^2}$

(c) $f(a + b) = \sqrt{1 + (a + b)^2} = \sqrt{a^2 + 2ab + b^2 + 1}$

11. (a) $f(a + h) = 2(a + h)^2 - 3(a + h) = 2a^2 + 4ah + 2h^2 - 3a - 3h$

(b) $\dfrac{f(a + h) - f(a)}{h} = \dfrac{[2(a + h)^2 - 3(a + h)] - [2a^2 - 3a]}{h} = \dfrac{4ah + 2h^2 - 3h}{h} = 4a + 2h - 3$

13. $x = 1, 3$ 15. $x = -2$

17. $x = -3, 3$ 19. $\text{dom}(f) = (-\infty, \infty); \quad \text{range}(f) = [0, \infty)$

21. $\text{dom}(f) = (-\infty, \infty); \quad \text{range}(f) = (-\infty, \infty)$

23. $\text{dom}(f) = (-\infty, 0) \cup (0, \infty); \quad \text{range}(f) = (0, \infty)$

25. $\text{dom}(f) = (-\infty, 1]; \quad \text{range}(f) = [0, \infty)$ 27. $\text{dom}(f) = (-\infty, 7]; \quad \text{range}(f) = [-1, \infty)$

29. $\text{dom}(f) = (-\infty, 2); \quad \text{range}(f) = (0, \infty)$

31. $\text{dom}\,(f) = (-\infty, \infty)$ **33.** $\text{dom}\,(f) = (-\infty, \infty)$ **35.** $\text{dom}\,(f) = (-\infty, \infty)$

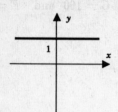 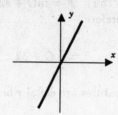 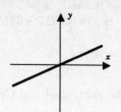

37. $\text{dom}\,(f) = (-\infty, \infty)$ **39.** $\text{dom}\,(f) = [-2, 2]$ **41.** $\text{dom}\,(f) = (-\infty, \infty)$

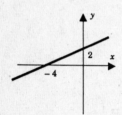 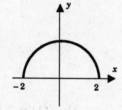 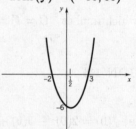

43. $\text{dom}\,(f) = (-\infty, \infty)$ **45.** $\text{dom}\,(f) = (-\infty, 0) \cup (0, \infty)$;

$\text{range}\,(f) = \{-1, 1\}$

 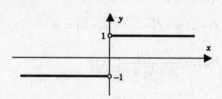

47. $\text{dom}\,(f) = [0, \infty)$; $\text{range}\,(f) = [1, \infty)$

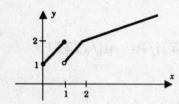

49. The curve is the graph of a function: domain $[-2, 2]$,   range $[-2, 2]$.

51. The curve is not the graph of a function; it fails the *vertical line test*.

53. odd: $f(-x) = (-x)^3 = -x^3 = -f(x)$

55. neither even nor odd: $g(-x) = -x(-x - 1) = x(x + 1)$; $g(-x) \neq g(x)$ and $g(-x) \neq -g(x)$

57. even: $f(-x) = \dfrac{(-x)^2}{1 - |-x|} = \dfrac{x^2}{1 - |x|}$

59. (a)

61. $-5 \le x \le 8, \quad 0 \le y \le 100$

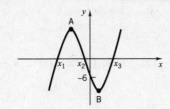

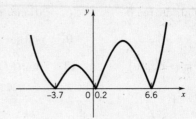

(b) $x_1 = -6.566, \quad x_2 = -0.493, \quad x_3 = 5.559$

(c) $A\,(-4, 28.667), \quad B\,(3, -28.500)$

63. $A = \dfrac{C^2}{4\pi}$, where C is the circumference;

    $\operatorname{dom}(A) = [0, \infty)$

65. $V = s^{3/2}$, where s is the area of a face;

    $\operatorname{dom} V = [0, \infty)$

67. $S = 3d^2$, where d is the diagonal of a face;

    $\operatorname{dom}(S) = [0, \infty)$

69. $A = \dfrac{\sqrt{3}}{4}\,x^2$, where x is the length of a side;

    $\operatorname{dom}(A) = [0, \infty)$

71. Let x be the edge length of the square end and let s be the length of the box. Then

$$4x + s = 108 \quad \text{and} \quad s = 108 - 4x, \qquad 0 < x \le 27.$$

The volume $V = x^2 s = x^2(108 - 4x) = 108x^2 - 4x^3, \quad 0 < x \le 27.$

73. Let y be the length of the rectangle. Then

$$x + 2y + \frac{\pi x}{2} = 15 \quad \text{and} \quad y = \frac{15}{2} - \frac{2 + \pi}{4}\,x, \qquad 0 < x < \frac{30}{2 + \pi}$$

The area $A = xy + \tfrac{1}{2}\,\pi\,(x/2)^2 = \left(\dfrac{15}{2} - \dfrac{2+\pi}{4}\,x\right)x + \dfrac{1}{8}\,\pi x^2 = \dfrac{15}{2}\,x - \dfrac{x^2}{2}\dfrac{\pi}{8}\,x^2 \quad 0 < x < \dfrac{30}{2+\pi}.$

75. Let y be the length of the beam. Then $y = \sqrt{d^2 - x^2}, \ 0 < x < d.$

The cross-sectional area $A = x\sqrt{d^2 - x^2}.$

77. The coordinates x and y are related by the equation $y = -\dfrac{b}{a}(x - a), \ 0 \le x \le a.$

The area A of the rectangle is given by $A = xy = x\left[-\dfrac{b}{a}(x - a)\right] = bx - \dfrac{b}{a}\,x^2, \ 0 \le x \le a.$

79. Let P be the perimeter of the square. Then the edge length of the square is $P/4$ and the area of the square is $A_s = (P/4)^2 = P^2/16$. Now, the circumference of the circle is $28 - P$ which implies that the radius is $\dfrac{1}{2\pi}(28 - \pi)$. Thus, the area of the circle is $A_c = \pi\left[\dfrac{1}{2\pi}(28 - P)\right]^2 = \dfrac{1}{4\pi}(28 - P)^2$. and the total area is $A_s + A_c = \dfrac{P^2}{16} + \dfrac{1}{4\pi}(28 - P)^2, \ 0 \le P \le 28.$

SECTION 1.6

1. polynomial, degree 0   3. rational function   5. neither

7. neither   9. neither

11. $\operatorname{dom}(f) = (-\infty, \infty)$ 13. $\operatorname{dom}(f) = (-\infty, \infty)$ 15. $\operatorname{dom}(f) = \{x : x \neq \pm 2\}$

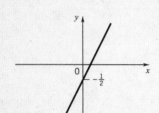 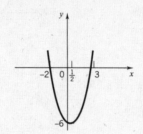 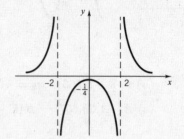

17. $225\left(\dfrac{\pi}{180}\right) = \dfrac{5\pi}{4}$ 19. $(-300)\left(\dfrac{\pi}{180}\right) = -\dfrac{5\pi}{3}$ 21. $15\left(\dfrac{\pi}{180}\right) = \dfrac{\pi}{12}$

23. $\left(-\dfrac{3\pi}{2}\right)\left(\dfrac{180}{\pi}\right) = -270°$ 25. $\left(\dfrac{5\pi}{3}\right)\left(\dfrac{180}{\pi}\right) = 300°$ 27. $2\left(\dfrac{180}{\pi}\right) \cong 114.59°$

29. $\sin x = \frac{1}{2};\ x = \pi/6,\ 5\pi/6$ 31. $\tan(x/2) = 1;\ x = \pi/2$

33. $\cos x = \sqrt{2}/2;\ x = \pi/4,\ 7\pi/4$ 35. $\cos 2x = 0;\ x = \pi/4,\ 3\pi/4,\ 5\pi/4,\ 7\pi/4$

37. $\sin 51° \cong 0.7772$ 39. $\sin(2.352) \cong 0.7101$

41. $\tan 72.4° \cong 3.1524$ 43. $\tan(11.249) \cong -3.8611$

45. $\sec(4.360) \cong -2.8974$ 47. $\sin x = 0.5231;\ x = 0.5505,\ \pi - 0.5505$

49. $\tan x = 6.7192;\ x = 1.4231,\ \pi + 1.4231$ 51. $\sec x = -4.4073;\ x = 1.7997,\ \pi + 1.7997$

53. $\operatorname{dom}(f) = (-\infty, \infty);\ \operatorname{range}(f) = [0,1]$ 55. $\operatorname{dom}(f) = (-\infty, \infty);\ \operatorname{range}(f) = [-2,2]$

57. $\operatorname{dom}(f) = \left(k\pi - \dfrac{\pi}{2}, k\pi + \dfrac{\pi}{2}\right),\ k = 0, \pm 1, \pm 2, \cdots;\ \operatorname{range}(f) = [1, \infty)$

59.   61.   63.

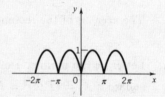

65. $f(-x) = \sin(-3x) = -\sin 3x = -f(x);\ f$ is odd

67. $f(-x) = 1 + \cos[2(-x)] = 1 + \cos[-2x] = 1 + \cos 2x = f(x);\ f$ is even

69. $f(-x) = (-x)^3 + \sin(-x) = -x^3 - \sin x = -f(x);$ f is odd

71.   Assume that $\theta_2 > \theta_1$. Let $m_1 = \tan\theta_1$, $m_2 = \tan\theta_2$.. The angle α between l_1 and l_2 is the smaller of $\theta_2 - \theta_1$ and $180° - [\theta_2 - \theta_1]$. In the first case

$$\tan\alpha = \tan[\theta_2 - \theta_1] = \frac{\tan\theta_2 - \tan\theta_1}{1 + \tan\theta_2 \tan\theta_1} = \frac{m_2 - m_1}{1 + m_2 m_1} > 0$$

In the second case

$$\tan\alpha = \tan[180° - (\theta_2 - \theta_1)] = -\tan(\theta_2 - \theta_1) = -\frac{m_2 - m_1}{1 + m_2 m_1} > 0$$

Thus

$$\tan\alpha = \left| \frac{m_2 - m_1}{1 + m_2 m_1} \right|$$

73. $h = b \sin A = a \sin B$ (see figure)

so $\dfrac{\sin A}{a} = \dfrac{\sin B}{b}$

Similarly, $\dfrac{\sin A}{a} = \dfrac{\sin C}{c}$

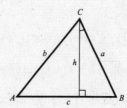

75. $A = \frac{1}{2} ah = \frac{1}{2} a^2 \sin\theta$

(see figure)

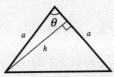

77.   (a)

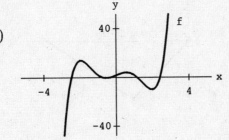

79.   (b)

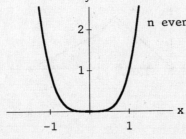

(c) $f_k(x) \geq f_{k+1}(x)$ on $[0,1]$; $f_{k+1}(x) \geq f_k(x)$ on $[1,\infty)$

SECTION 1.7

1. $(f+g)(2) = f(2) + g(2) = 3 + \dfrac{9}{2} = \dfrac{15}{2}$ 3. $(f \cdot g)(-2) = f(-2)g(-2) = 15 \cdot \dfrac{7}{2} = \dfrac{105}{2}$

5. $(2f - 3g)(\tfrac{1}{2}) = 2f(\tfrac{1}{2}) - 3g(\tfrac{1}{2}) = 2 \cdot 0 - 3 \cdot \dfrac{9}{4} = -\dfrac{27}{4}$

7. $(f \circ g)(1) = f[g(1)] = f(2) = 3$

9. $(f+g)(x) = f(x) + g(x) = x - 1;$ $\operatorname{dom}(f+g) = (-\infty, \infty)$

   $(f-g)(x) = f(x) - g(x) = 3x - 5;$ $\operatorname{dom}(f-g) = (-\infty, \infty)$

   $(f \cdot g)(x) = f(x)g(x) = -2x^2 + 7x - 6;$ $\operatorname{dom}(f \cdot g) = (-\infty, \infty)$

   $(f/g)(x) = \dfrac{2x - 3}{2 - x};$ $\operatorname{dom}(f/g) = \{x : x \neq 2\}$

11. $(f+g)(x) = x + \sqrt{x-1} - \sqrt{x+1};$ $\operatorname{dom}(f+g) = [1, \infty)$

    $(f-g)(x) = \sqrt{x-1} + \sqrt{x+1} - x;$ $\operatorname{dom}(f-g) = [1, \infty)$

    $(f \cdot g)(x) = \sqrt{x-1}\,(x - \sqrt{x+1}) = x\sqrt{x-1} - \sqrt{x^2 - 1};$ $\operatorname{dom}(f \cdot g) = [1, \infty)$

    $(f/g)(x) = \dfrac{\sqrt{x-1}}{x - \sqrt{x+1}};$ $\operatorname{dom}(f/g) = \{x : x \geq 1 \text{ and } x \neq \tfrac{1}{2}(1 + \sqrt{5})\}$

13. (a) $(6f + 3g)(x) = 6(x + 1/\sqrt{x}) + 3(\sqrt{x} - 2/\sqrt{x}) = 6x + 3\sqrt{x};$ $x > 0$

    (b) $(f - g)(x) = x + 1/\sqrt{x} - (\sqrt{x} - 2/\sqrt{x}) = x + 3/\sqrt{x} - \sqrt{x};$ $x > 0$

    (c) $(f/g)(x) = \dfrac{x\sqrt{x} + 1}{x - 2};$ $x > 0, x \neq 2$

15.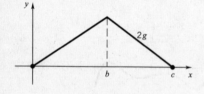

17.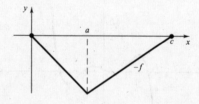

19.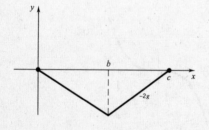

21.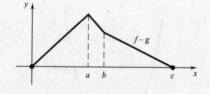

23. $(f \circ g)(x) = 2x^2 + 5;$ $\operatorname{dom}(f \circ g) = (-\infty, \infty)$ 25. $(f \circ g)(x) = \sqrt{x^2 + 5};$ $\operatorname{dom}(f \circ g) = (-\infty, \infty)$

27. $(f \circ g)(x) = \dfrac{x}{x-2}$; $\mathrm{dom}\,(f \circ g) = \{x : \ x \neq 0, 2\}$

29. $(f \circ g)(x) = \dfrac{1}{|x^2 - 1| - 3}$; $\mathrm{dom}\,(f \circ g) = \{x : \ x \neq \pm 2\}$

31. $(f \circ g)(x) = \sqrt{1 - \cos^2 2x} = |\sin 2x|$; $\mathrm{dom}\,(f \circ g) = (-\infty, \infty)$

33. $(f \circ g \circ h)(x) = 4\,[g(h(x))] = 4\,[h(x) - 1] = 4(x^2 - 1)$; $\mathrm{dom}\,(f \circ g \circ h) = (-\infty, \infty)$

35. $(f \circ g \circ h)(x) = \dfrac{1}{g(h(x))} = \dfrac{1}{1/[2h(x) + 1]} = 2h(x) + 1 = 2x^2 + 1$; $\mathrm{dom}\, f \circ g \circ h) = (-\infty, \infty)$

37. Take $f(x) = \dfrac{1}{x}$ since $\dfrac{1 + x^4}{1 + x^2} = F(x) = f(g(x)) = f\left(\dfrac{1 + x^2}{1 + x^4}\right)$.

39. Take $f(x) = 2\sin x$ since $2\sin 3x = F(x) = f(g(x)) = f(3x)$.

41. Take $g(x) = \left(1 - \dfrac{1}{x^4}\right)^{2/3}$ since $\left(1 - \dfrac{1}{x^4}\right)^2 = F(x) = f(g(x)) = [g(x)]^3$.

43. Take $g(x) = 2x^3 - 1$ (or $-(2x^3 - 1)$)   since $(2x^3 - 1)^2 + 1 = F(x) = f(g(x)) = [g(x)]^2 + 1$.

45. $(f \circ g)(x) = f(g(x)) = \sqrt{g(x)} = \sqrt{x^2} = |x|$;

   $(g \circ f)(x) = g(f(x)) = [f(x)]^2 = [\sqrt{x}]^2 = x$, $x \geq 0$

47. $(f \circ g)(x) = f(g(x)) = 1 - \sin^2 x = \cos x$; $(g \circ f)(x) = g(f(x)) = \sin f(x) = \sin(1 - x^2)$

49. $(f \circ g)(x) = f(g(x)) = (x - 1) + 1 = x$; $(g \circ f)(x) = g(f(x)) = \sqrt[3]{(x^3 + 1) - 1} = x$

51. fg is even since $(fg)(-x) = f(-x)g(-x) = f(x)g(x) = (fg)(x)$.

53. (a) If f is even, then
$$f(x) = \begin{cases} -x & -1 \leq x < 0 \\ 1 & x < -1 \end{cases}$$

   (b) If f is odd, then
$$f(x) = \begin{cases} x & -1 \leq x < 0 \\ -1 & x < -1 \end{cases}$$

55. $g(-x) = f(-x) + f[-(-x)] = f(-x) + f(x) = g(x)$

57.

	f_1	f_2	f_3	f_4	f_5	f_6
f_1	f_1	f_2	f_3	f_4	f_5	f_6
f_2	f_2	f_1	f_4	f_3	f_6	f_5
f_3	f_3	f_5	f_1	f_6	f_2	f_4
f_4	f_4	f_6	f_2	f_5	f_1	f_3
f_5	f_5	f_3	f_6	f_1	f_4	f_2
f_6	f_6	f_4	f_5	f_2	f_3	f_1

59. $(f \circ g)(x) = 3[\frac{1}{3}(x+5)] - 5 = x$ and $(g \circ f)(x) = \frac{1}{3}[(3x-5)+5] = x$

61. $(f \circ g)(x) = \dfrac{1}{\dfrac{1-x}{x}+1} = \dfrac{x}{(1-x)+x} = x$ and $(g \circ f)(x) = \dfrac{1 - \dfrac{1}{x+1}}{\dfrac{1}{x+1}} = (x+1) - 1 = x$

63. (a) For fixed b, varying a varies the x-coordinate of the vertex of the parabola.

  (b) For fixed a, varying b varies the y-coordinate of the parabola

65. (a) For $a > 0$, the graph of $f(x-a)$ is the graph of f shifted horizontally a units to the right; for $a < 0$, the graph of $f(x-a)$ is the graph of f shifted horizontally $|a|$ units to the left.

  (b) For $b > 1$, the graph of $f(bx)$ is the graph of f compressed horizontally;   for $0 < b < 1$, the graph of $f(bx)$ is the graph of f stretched horizontally;   for $-1 < b < 0$, the graph of $f(bx)$ is the graph of f stretched horizontally and reflected in the y-axis;   for $b < -1$, the graph of $f(bx)$ is the graph of f compressed horizontally and reflected in the y-axis.

  (c) The graph of $f(x)+c$ is the graph of f shifted c units up if $c > 0$ and shifted $|c|$ units down if $c < 0$.

67. (a) For $A > 0$, the graph of Af is the graph of f scaled vertically by the factor A;   for $A < 0$, the graph of Af is the graph of f scaled vertically by the factor $|A|$ and then reflected in the x-axis.

  (b) See Exercise 65(b).

SECTION 1.8

1. Let S be the set of integers for which the statement is true. Since $2(1) \leq 2^1$, S contains 1. Assume now that $k \in S$. This tells us that $2k \leq 2^k$, and thus

$$2(k+1) = 2k + 2 \leq 2^k + 2 \leq 2^k + 2^k = 2(2^k) = 2^{k+1}.$$

$$(k \geq 1)$$

This places $k+1$ in S.

We have shown that

$$1 \in S \quad \text{and that} \quad k \in S \quad \text{implies} \quad k+1 \in S.$$

It follows that S contains all the positive integers.

3. Let S be the set of integers for which the statement is true. Since $(1)(2) = 2$ is divisible by $2, 1 \in S$.

Assume now that $k \in S$. This tells us that $k(k+1)$ is divisible by 2 and therefore

$$(k+1)(k+2) = k(k+1) + 2(k+1)$$

is also divisible by 2. This places $k + 1 \in S$.

We have shown that

$1 \in S$ and that $k \in S$ implies $k + 1 \in S$.

It follows that S contains all the positive integers.

5.   Use
$$1^2 + 2^2 + \cdots + k^2 + (k+1)^2 = \tfrac{1}{6}k(k+1)(2k+1) + (k+1)^2$$
$$= \tfrac{1}{6}(k+1)[k(2k+1) + 6(k+1)]$$
$$= \tfrac{1}{6}(k+1)(2k^2 + 7k + 6)$$
$$= \tfrac{1}{6}(k+1)(k+2)(2k+3)$$
$$= \tfrac{1}{6}(k+1)[(k+1)+1][2(k+1)+1].$$

7.   By Exercise 6 and Example 1
$$1^3 + 2^3 + \cdots + (n-1)^3 = [\tfrac{1}{2}(n-1)n]^2 = \tfrac{1}{4}(n-1)^2 n^2 < \tfrac{1}{4}n^4$$
and
$$1^3 + 2^3 + \cdots + n^3 = [\tfrac{1}{2}n(n+1)]^2 = \tfrac{1}{4}n^2(n+1)^2 > \tfrac{1}{4}n^4.$$

9.   Use
$$\frac{1}{\sqrt{1}} + \frac{1}{\sqrt{2}} + \frac{1}{\sqrt{3}} + \cdots + \frac{1}{\sqrt{n}} + \frac{1}{\sqrt{n+1}}$$
$$> \sqrt{n} + \frac{1}{\sqrt{n+1}+\sqrt{n}}\left(\frac{\sqrt{n+1}-\sqrt{n}}{\sqrt{n+1}-\sqrt{n}}\right) = \sqrt{n+1}.$$

11.   Let S be the set of integers for which the statement is true. Since
$$3^{2(1)+1} + 2^{1+2} = 27 + 8 = 35$$
is divisible by 7, we see that $1 \in S$.

Assume now that $k \in S$. This tells us that

$3^{2k+1} + 2^{k+2}$ is divisible by 7.

It follows that

$$3^{2(k+1)+1} + 2^{(k+1)+2} = 3^2 \cdot 3^{2k+1} + 2 \cdot 2^{k+2}$$
$$= 9 \cdot 3^{2k+1} + 2 \cdot 2^{k+2}$$
$$= 7 \cdot 3^{2k+1} + 2(3^{2k+1} + 2^{k+2})$$

is also divisible by 7. This places $k + 1 \in S$.

We have shown that

$$1 \in S \qquad \text{and that} \qquad k \in S \quad \text{implies} \quad k+1 \in S.$$

It follows that S contains all the positive integers.

13. For all positive integers $n \geq 2$,

$$\left(1 - \frac{1}{2}\right)\left(1 - \frac{1}{3}\right) \cdots \left(1 - \frac{1}{n}\right) = \frac{1}{n}.$$

To see this, let S be the set of integers n for which the formula holds. Since $1 - \frac{1}{2} = \frac{1}{2}$, $2 \in S$. Suppose now that $k \in S$. This tells us that

$$\left(1 - \frac{1}{2}\right)\left(1 - \frac{1}{3}\right) \cdots \left(1 - \frac{1}{k}\right) = \frac{1}{k}$$

and therefore that

$$\left(1 - \frac{1}{2}\right)\left(1 - \frac{1}{3}\right) \cdots \left(1 - \frac{1}{k}\right)\left(1 - \frac{1}{k+1}\right) = \frac{1}{k}\left(1 - \frac{1}{k+1}\right) = \frac{1}{k}\left(\frac{k}{k+1}\right) = \frac{1}{k+1}.$$

This places $k + 1 \in S$ and verifies the formula for $n \geq 2$.

15. From the figure, observe that adding a vertex V_{N+1} to an N-sided polygon increases the number of diagonals by $(N - 2) + 1 = N - 1$. Then use the identity

$$\tfrac{1}{2}N(N - 3) + (N - 1) = \tfrac{1}{2}(N + 1)(N + 1 - 3).$$

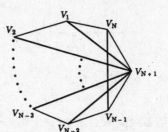

17. To go from k to $k+1$, take $A = \{a_1, \cdots, a_{k+1}\}$ and $B = \{a_1, \cdots, a_k\}$. Assume that B has 2^k subsets: $B_1, B_2, \cdots B_{2^k}$. The subsets of A are then $B_1, B_2, \cdots, B_{2^k}$ together with

$$B_1 \cup \{a_{k+1}\}, \ B_2 \cup \{a_{k+1}\}, \cdots, B_{2^k} \cup \{a_{k+1}\}.$$

This gives $2(2^k) = 2^{k+1}$ subsets for A.

PROJECTS AND EXPLORATIONS

1.1.   (a) Let $f(x) = x^4 - 3x^2 + 0.1x + 1$.

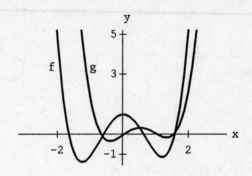

Let $f(x) = \dfrac{x^3 - 6}{x^4 + 3}$.

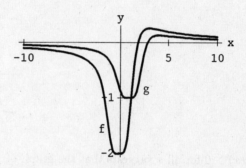

(b) In each of the given cases, $\text{dom}(f) = \text{dom}(g)$. In general, $\text{dom}(g)$ will be a subset of $\text{dom}(f)$, and, in particular, if $\text{dom}(f)$ is bounded below, then $\text{dom}(g) \subset \text{dom}(f)$.

(c) Setting $g(x) = f(x)$ gives

$$\frac{f(x-1) + f(x)}{2} = f(x)$$

which implies $f(x-1) = f(x)$. In the case of $f(x) = x^4 - 3x^2 = 0.1x + 1$, this occurs at the points where $x \cong -0.6280, 0.5200, 1.6081$.

For $f(x) = \dfrac{x^3 - 6}{x^4 + 3}$, f and g intersect at $x \cong 0.3533, 3.6009$.

(d) Suppose that f has one of the following two properties:

(1) $f(x_2) > f(x_1)$ whenever $x_2 > x_1$ $x_1, x_2 \in \text{dom}(f)$

(2) $f(x_2) < f(x_1)$ whenever $x_2 > x_1$ $x_1, x_2 \in \text{dom}(f)$

Then f is one-to-one. If f has either property (1) or (2) then g will have the same property and hence will be one-to-one. Now consider the function f defined on the set $S = \{1, 2, 3, 4\}$ by $f(1) = 0, f(2) = 1, f(3) = 2, f(4) = -1$. Then f is one-to-one, but $g(2) = g(4) = \frac{1}{2}$.

1.3.   (a) Let $f(x) = \sqrt{1+x}$

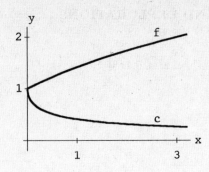

Let $f(x) = |x^2 - 7x + 2|$

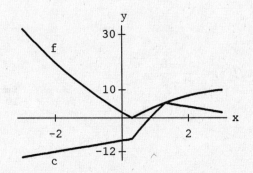

(b) $c(x) > 0$ for all x suggests that the graph of f is rising; $c(x) < 0$ for all x suggests that the graph of f is falling. If $c(x) = 0$ for all x, then $f(x) = f(x-1)$ for all x and f is a periodic function (with period 1 if 1 is the smallest such number with this property.

(c) Let $f(x) = x^2$. Then $c(x) = 2x - 1$. f is not one-to-one, but c is.

(d) Let f be an even function. Then

$$c(-x) = f(-x) - f(-x-1) = f(x) - f(x+1) = -c(x+1)$$

Thus, if c is periodic with period 1, then c will be odd. For a trivial example, let $f(x) = \cos(2\pi x)$.

CHAPTER 2

SECTION 2.1

1.   (a) 2   (b) -1 (c) does not exist   (d) -3

3.   (a) does not exist (b) -3 (c) does not exist   (d) -3

5.   (a) does not exist (b) does not exist   (c) does not exist   (d) 1

7.   (a) 2   (b) 2   (c) 2   (d) -1

9.   (a) -1 (b) -1 (c) -1 (d) undefined

11. (a) 0   (b) 0   (c) 0   (d) 0

13. $c = 0, 6$ 15. -1 17. 4   19. 1

21. $\frac{3}{2}$ 23. does not exist   25. $\lim\limits_{x\to 3}\dfrac{2x-6}{x-3}=\lim\limits_{x\to 3}2=2$

27. $\lim\limits_{x\to 3}\dfrac{x-3}{x^2-6x+9}=\lim\limits_{x\to 3}\dfrac{x-3}{(x-3)^2}=\lim\limits_{x\to 3}\dfrac{1}{x-3}$; does not exist

29. $\lim\limits_{x\to 2}\dfrac{x-2}{x^2-3x+2}=\lim\limits_{x\to 2}\dfrac{x-2}{(x-1)(x-2)}=\lim\limits_{x\to 2}\dfrac{1}{x-1}=1$

31. does not exist   33. $\lim\limits_{x\to 0}\dfrac{2x-5x^2}{x}=\lim\limits_{x\to 0}(2-5x)=2$

35. $\lim\limits_{x\to 1}\dfrac{x^2-1}{x-1}=\lim\limits_{x\to 1}\dfrac{(x-1)(x+1)}{x-1}=\lim\limits_{x\to 1}(x+1)=2$

37. 0   39. 1   41. 16

43. does not exist   45. does not exist   47. 4

49.

$$\lim_{x\to 1}\frac{\sqrt{x^2+1}-\sqrt{2}}{x-1}=\lim_{x\to 1}\frac{(\sqrt{x^2+1}-\sqrt{2})(\sqrt{x^2+1}+\sqrt{2})}{(x-1)(\sqrt{x^2+1}+\sqrt{2})}$$

$$=\lim_{x\to 1}\frac{x^2-1}{(x-1)(\sqrt{x^2+1}+\sqrt{2})}=\lim_{x\to 1}\frac{x+1}{\sqrt{x^2+1}+\sqrt{2}}=\frac{2}{2\sqrt{2}}=\frac{1}{\sqrt{2}}$$

51. $f(x)=x^2,\quad a=2,\quad f(2)=4$

$$\frac{f(2+h)-f(2)}{h}=\frac{(2+h)^2-4}{h}=\frac{4+4h+h^2-4}{h}=4+h$$

$$\lim_{h\to 0}\frac{f(2+h)-f(2)}{h}=\lim_{h\to 0}(4+h)=4$$

tangent line: $y-4=4(x-2)$ or $y=4x-4$

53. $f(x) = 1 - 2x + x^2, \quad a = -1, \quad f(-1) = 4$

$$\frac{f(-1+h) - f(-1)}{h} = \frac{1 - 2(-1+h) + (-1+h)^2 - 4}{h} = \frac{4 - 4h + h^2 - 4}{h} = -4 + h$$

$$\lim_{h \to 0} \frac{f(-1+h) - f(-1)}{h} = \lim_{h \to 0} (-4 + h) = -4$$

tangent line: $y - 4 = -4(x+1) \quad$ or $\quad y = -4x$

55. $f(x) = \sqrt{x}, \quad a = 1 \quad f(1) = 1$

$$\frac{f(1+h) - f(1)}{h} = \frac{\sqrt{1+h} - 1}{h} = \frac{\sqrt{1+h} - 1}{h} \cdot \frac{\sqrt{1+h} + 1}{\sqrt{1+h} + 1} = \frac{h}{h(1+h) + 1} = \frac{1}{\sqrt{1+h} + 1}$$

$$\lim_{h \to 0} \frac{f(1+h) - f(1)}{h} = \lim_{h \to 0} \frac{1}{\sqrt{1+h} + 1} = \frac{1}{2}$$

tangent line: $y - 1 = \frac{1}{2}(x-1) \quad$ or $\quad y = \frac{1}{2}x + \frac{1}{2}$

57. $f(x) = \sqrt[3]{x}, \quad a = 0, \quad f(0) = 0$

$$\frac{f(0+h) - f(0)}{h} = \frac{\sqrt[3]{h}}{h} = \frac{1}{h^{2/3}}$$

$$\lim_{h \to 0} \frac{f(0+h) - f(0)}{h} = \lim_{h \to 0} \frac{1}{h^{2/3}} \quad \text{does not exist}$$

59.   (a)   (b) neither limit exists

$$f(1/\pi) = f(2/\pi) = f(3/\pi) = f(4/\pi) = 0$$

$$f\left(\frac{1}{\pi/2}\right) = f\left(\frac{1}{5\pi/2}\right) = f\left(\frac{1}{9\pi/2}\right) = 1$$

$$f\left(\frac{1}{3\pi/2}\right) = f\left(\frac{1}{7\pi/2}\right) = f\left(\frac{1}{11\pi/2}\right) = -1$$

(c)

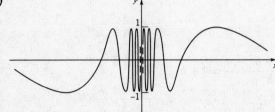

61. 2   **63.** $\frac{3}{2}$ **65.** 2.7182817

SECTION 2.2

1. $\frac{1}{2}$

3. $\lim\limits_{x\to 0}\dfrac{x(1+x)}{2x^2}=\lim\limits_{x\to 0}\dfrac{1+x}{2x}$; does not exist

5. $\dfrac{4}{\sqrt{5}}=\dfrac{4}{5}\sqrt{5}$

7. $\lim\limits_{x\to 1}\dfrac{x^4-1}{x-1}=\lim\limits_{x\to 1}(x^3+x^2+x+1)=4$

9. does not exist   **11.** -1 **13.** does not exist   **15.** 0

17. $\lim\limits_{x\to 2+}f(x)=\lim\limits_{x\to 2+}(x^2-x)=2$ **19.** 1

21. 1   **23.** δ_1 and δ_2 **25.** $\frac{1}{2}\epsilon$ **27.** 2ϵ

29. Since

$$|(2x-5)-3|=|2x-8|=2|x-4|,$$

we can take $\delta=\frac{1}{2}\epsilon$:

$$\text{if}\quad 0<|x-4|<\tfrac{1}{2}\epsilon\quad\text{then,}\quad |(2x-5)-3|=2|x-4|<\epsilon.$$

31. Since

$$|(6x-7)-11|=|6x-18|=6|x-3|,$$

we can take $\delta=\frac{1}{6}\epsilon$:

$$\text{if}\quad 0<|x-3|<\tfrac{1}{6}\epsilon\quad\text{then}\quad |(6x-7)-11|=6|x-3|<\epsilon.$$

33. Since

$$\big||1-3x|-5\big|=\big||3x-1|-5\big|\le|3x-6|=3|x-2|,$$

we can take $\delta=\frac{1}{3}\epsilon$:

$$\text{if}\quad 0<|x-2|<\tfrac{1}{3}\epsilon\quad\text{then}\quad \big||1-3x|-5\big|\le 3|x-2|<\epsilon.$$

35. Statements (b), (e), (g), and (i) are necessarily true.

37. (i) $\lim\limits_{x\to 3}\dfrac{1}{x-1}=\dfrac{1}{2}$ (ii) $\lim\limits_{h\to 0}\dfrac{1}{(3+h)-1}=\dfrac{1}{2}$

(iii) $\lim\limits_{x\to 3}\left(\dfrac{1}{x-1}-\dfrac{1}{2}\right)=0$ (iv) $\lim\limits_{x\to 3}\left|\dfrac{1}{x-1}-\dfrac{1}{2}\right|=0$

39. $\lim\limits_{h\to 0}\dfrac{f(x+h)-f(x)}{h}=\lim\limits_{h\to 0}\dfrac{[5(x+h)+2]-(5x+2)}{h}=\lim\limits_{h\to 0}\dfrac{5h}{h}=5$

41. $\displaystyle\lim_{h\to 0}\frac{f(x+h)-f(x)}{h}=\lim_{h\to 0}\frac{\left[4(x+h)+5(x+h)^2\right]-(4x+5x^2)}{h}$

$\displaystyle\qquad\qquad =\lim_{h\to 0}\frac{4x+4h+5x^2+10xh+5h^2-4x-5x^2}{h}$

$\displaystyle\qquad\qquad =\lim_{h\to 0}\frac{h(4+10x)+5h^2}{h}=\lim_{h\to 0}(4+10x+5h)=4+10x$

43. $\displaystyle\lim_{h\to 0}\frac{f(x+h)-f(x)}{h}=\lim_{h\to 0}\frac{\dfrac{1}{x+h+1}-\dfrac{1}{x+1}}{h}=\lim_{h\to 0}-\frac{h}{h(x+h+1)(x+1)}$

$\displaystyle\qquad\qquad =-\lim_{h\to 0}\frac{1}{(x+h+1)(x+1)}=-\frac{1}{(x+1)^2}$

45. By (2.2.5) parts (i) and (iv) with $L=0$

47. $\delta=0.001$ $\qquad\qquad\qquad\qquad$ **49.** $\delta=0.04$

51. Let $\epsilon>0$. If

$$\lim_{x\to c}f(x)=L,$$

then there must exist $\delta>0$ such that

(∗) $\qquad\qquad$ if $\quad 0<|x-c|<\delta\quad$ then $\quad |f(x)-L|<\epsilon.$

Suppose now that

$$0<|h|<\delta.$$

Then

$$0<|(c+h)-c|<\delta$$

and thus by (∗)

$$|f(c+h)-L|<\epsilon.$$

This proves that

$$\text{if}\quad \lim_{x\to c}f(x)=L\quad\text{then}\quad \lim_{h\to 0}f(c+h)=L.$$

If, on the other hand,

$$\lim_{h\to 0}f(c+h)=L,$$

then there must exist $\delta>0$ such that

(∗∗) $\qquad\qquad$ if $\quad 0<|h|<\delta\quad$ then $\quad |f(c+h)-L|<\epsilon.$

Suppose now that

$$0<|x-c|<\delta.$$

Then by (∗∗)

$$|f(c+(x-c))-L|<\epsilon.$$

More simply stated,

$$|f(x) - L| < \epsilon.$$

This proves that

$$\text{if} \quad \lim_{h \to 0} f(c + h) = L \quad \text{then} \quad \lim_{x \to c} f(x) = L.$$

53. (a) Set $\delta = \epsilon \sqrt{c}$. By the hint,

$$\text{if} \quad 0 < |x - c| < \epsilon \sqrt{c} \quad \text{then} \quad |\sqrt{x} - \sqrt{c}| < \frac{1}{\sqrt{c}}|x - c| < \epsilon.$$

(b) Set $\delta = \epsilon^2$. If $0 < x < \epsilon^2$, then $|\sqrt{x} - 0| = \sqrt{x} < \epsilon$.

55. Take $\delta = $ minimum of 1 and $\epsilon/7$. If $0 < |x - 1| < \delta$, then $0 < x < 2$

and $|x - 1| < \epsilon/7$. Therefore

$$|x^3 - 1| = |x^2 + x + 1||x - 1| < 7|x - 1| < 7(\epsilon/7) = \epsilon.$$

57. Set $\delta = \epsilon^2$. If $3 - \epsilon^2 < x < 3$, then $-\epsilon^2 < x - 3$, $0 < 3 - x < \epsilon^2$

and therefore $|\sqrt{3 - x} - 0| < \epsilon$.

59. Suppose, on the contrary, that $\lim_{x \to c} f(x) = L$ for some particular c. Taking $\epsilon = \frac{1}{2}$, there must exist $\delta > 0$ such that

$$\text{if} \quad 0 < |x - c| < \delta, \quad \text{then} \quad |f(x) - L| < \frac{1}{2}.$$

Let x_1 be a rational number satisfying $0 < |x_1 - c| < \delta$ and x_2 an irrational number satisfying $0 < |x_2 - c| < \delta$. (That such numbers exist follows from the fact that every interval contains both rational and irrational numbers.) Now $f(x_1) = L$ and $f(x_2) = 0$. Thus we must have both

$$|1 - L| < \frac{1}{2} \quad \text{and} \quad |0 - L| < \frac{1}{2}.$$

From the first inequality we conclude that $L > \frac{1}{2}$. From the second, we conclude that $L < \frac{1}{2}$. Clearly no such number L exists.

61. We begin by assuming that $\lim_{x \to c^+} f(x) = L$ and showing that

$$\lim_{h \to 0} f(c + |h|) = L.$$

Let $\epsilon > 0$. Since $\lim_{x \to c^+} f(x) = L$, there exists $\delta > 0$ such that

$$(*) \qquad\qquad \text{if} \quad c < x < c + \delta \quad \text{then} \quad |f(x) - L| < \epsilon.$$

Suppose now that $0 < |h| < \delta$. Then $c < c + |h| < c + \delta$ and, by $(*)$,

$$|f(c + |h|) - L| < \epsilon.$$

Thus $\lim_{h \to 0} f(c + |h|) = L$.

Conversely we now assume that $\lim_{h \to 0} f(c + |h|) = L$. Then for $\epsilon > 0$ there exists $\delta > 0$ such that

$$(**) \qquad\qquad \text{if} \quad 0 < |h| < \delta \quad \text{then} \quad |f(c + |h|) - L| < \epsilon.$$

Suppose now that $c < x < c + \delta$. Then $0 < x - c < \delta$ so that, by $(**)$,

$$|f(c + (x - c)) - L| = |f(x) - L| < \epsilon.$$

Thus $\lim_{x \to c^+} f(x) = L$.

63. (a) Let $\epsilon = L$. Since $\lim\limits_{x \to c} f(x) = L$, there exists $\delta > 0$ such that if $0 < |x - c| < \delta$ then

$$L - f(x) \le |L - f(x)| = |f(x) - L| < L$$

Therefore, $f(x) > L - L = 0$ for all $x \in (c - \delta, c + \delta)$; take $\gamma = \delta$.

(b) Let $\epsilon = -L$ and repeat the argument in part (a).

65. (a) Let $\lim\limits_{x \to c} f(x) = L$ and $\lim\limits_{x \to c} g(x) = M$, and let $\epsilon > 0$. There exist positive numbers δ_1 and δ_2 such that

$$|f(x) - L| < \epsilon/2 \quad \text{if} \quad 0 < |x - c| < \delta_1$$

and

$$|g(x) - M| < \epsilon/2 \quad \text{if} \quad 0 < |x - c| < \delta_2$$

Let $\delta = \min(\delta_1, \delta_2)$. Then

$$M - L = M - g(x) + g(x) - f(x) + f(x) - L \ge [M - g(x)] + [f(x) - L]$$

$$\ge -\epsilon/2 - \epsilon/2 = -\epsilon$$

for all x such that $0 < |x - c| < \delta$. Since ϵ is arbitrary, it follows that $M \ge L$.

(b) No. For example, if $f(x) = x^2$ and $g(x) = |x|$ on $(-1, 1)$, then $f(x) < g(x)$ on $(-1, 1)$ except at $x = 0$, but $\lim\limits_{x \to 0} x^2 = \lim\limits_{x \to 0} |x| = 0$.

67. $f(x) = 2x^2 - 3x, \quad a = 2, \quad f(2) = 2$

$$\lim_{x \to a} \frac{f(x) - f(a)}{x - a} = \lim_{x \to a} \frac{2x^2 - 3x - 2}{x - 2}$$

$$= \lim_{x \to 2} \frac{(x - 2)(2x + 1)}{x - 2} = \lim_{x \to 2} (2x + 1) = 5$$

tangent line: $y - 2 = 5(x - 2)$ or $y = 5x - 8$.

69. $f(x) = \sqrt{x}, \quad a = 4, \quad f(4) = 2$

$$\lim_{x \to a} \frac{f(x) - f(a)}{x - a} = \lim_{x \to 4} \frac{\sqrt{x} - 2}{x - 4} = \lim_{x \to 4} \frac{\sqrt{x} - 2}{x - 2} \cdot \frac{\sqrt{x} + 2}{\sqrt{x} + 2}$$

$$= \lim_{x \to 4} \frac{x - 4}{(x - 4)(\sqrt{x} + 2)} = \lim_{x \to 4} \frac{1}{\sqrt{x} + 2} = \frac{1}{4}$$

tangent line: $y - 2 = \frac{1}{4}(x - 4)$ or $y = \frac{1}{4}x + 1$.

71. $\lim\limits_{x \to \frac{1}{3}} \dfrac{\cos \pi x - \cos(\pi/3)}{x - \frac{1}{3}} \cong -2.72070$ **73.** $\lim\limits_{x \to 0} \dfrac{3^x - 3^0}{x} \cong 1.09861$

SECTION 2.3

1.   (a) 3   (b) 4   (c) −2   (d) 0   (e) does not exist   (f) $\frac{1}{3}$

3. $\lim\limits_{x\to 4}\left(\dfrac{1}{x}-\dfrac{1}{4}\right)\left(\dfrac{1}{x-4}\right)=\lim\limits_{x\to 4}\left(\dfrac{4-x}{4x}\right)\left(\dfrac{1}{x-4}\right)=\lim\limits_{x\to 4}\dfrac{-1}{4x}=-\dfrac{1}{16};$ Theorem 2.3.2 does not apply

since $\lim\limits_{x\to 4}\dfrac{1}{x-4}$ does not exist.

5. 3   7. −3   9. 5

11. does not exist   13. −1   15. does not exist

17. $\lim\limits_{h\to 0}h\left(1+\dfrac{1}{h}\right)=\lim\limits_{h\to 0}(h+1)=1$ 19. $\lim\limits_{x\to 2}\dfrac{x^2-4}{x-2}=\lim\limits_{x\to 2}\dfrac{x+2}{1}=4$

21. $\lim\limits_{x\to 4}\dfrac{\sqrt{x}-2}{x-4}=\lim\limits_{x\to 4}\dfrac{\sqrt{x}-2}{x-4}\cdot\dfrac{\sqrt{x}+2}{\sqrt{x}+2}=\lim\limits_{x\to 4}\dfrac{x-4}{(x-4)(\sqrt{x}+2)}=\dfrac{1}{4}$

23. $\lim\limits_{x\to 1}\dfrac{x^2-x-6}{(x+2)^2}=\lim\limits_{x\to 1}\dfrac{(x+2)(x-3)}{(x+2)^2}=\lim\limits_{x\to 1}\dfrac{x-3}{x+2}=-\dfrac{2}{3}$

25. $\lim\limits_{h\to 0}\dfrac{1-1/h^2}{1-1/h}=\lim\limits_{h\to 0}\dfrac{h^2-1}{h^2-h}=\lim\limits_{h\to 0}\dfrac{(h+1)(h-1)}{h(h-1)}=\lim\limits_{h\to 0}\dfrac{h+1}{h};$ does not exist

27. $\lim\limits_{h\to 0}\dfrac{1-1/h}{1+1/h}=\lim\limits_{h\to 0}\dfrac{h-1}{h+1}=-1$

29. $\lim\limits_{t\to -1}\dfrac{t^2+6t+5}{t^2+3t+2}=\lim\limits_{t\to -1}\dfrac{(t+1)(t+5)}{(t+1)(t+2)}=\lim\limits_{t\to -1}\dfrac{t+5}{t+2}=4$

31. $\lim\limits_{t\to 0}\dfrac{t+a/t}{t+b/t}=\lim\limits_{t\to 0}\dfrac{t^2+a}{t^2+b}=\dfrac{a}{b}$

33. $\lim\limits_{x\to 1}\dfrac{x^5-1}{x^4-1}=\lim\limits_{x\to 1}\dfrac{(x-1)(x^4+x^3+x^2+x+1)}{(x-1)(x^3+x^2+x+1)}=\lim\limits_{x\to 1}\dfrac{x^4+x^3+x^2+x+1}{x^3+x^2+x+1}=\dfrac{5}{4}$

35. $\lim\limits_{h\to 0}h\left(1+\dfrac{1}{h^2}\right)=\lim\limits_{h\to 0}\dfrac{h^2+1}{h};$ does not exist

37. $\lim\limits_{x\to -4}\left(\dfrac{2x}{x+4}+\dfrac{8}{x+4}\right)=\lim\limits_{x\to -4}\dfrac{2x+8}{x+4}=\lim\limits_{x\to -4}2=2$

39. (a) $\lim\limits_{x\to 4}\left(\dfrac{1}{x}-\dfrac{1}{4}\right)=\lim\limits_{x\to 4}\dfrac{4-x}{4x}=0$

(b) $\lim\limits_{x\to 4}\left[\left(\dfrac{1}{x}-\dfrac{1}{4}\right)\left(\dfrac{1}{x-4}\right)\right]=\lim\limits_{x\to 4}\left[\left(\dfrac{4-x}{4x}\right)\left(\dfrac{1}{x-4}\right)\right]=\lim\limits_{x\to 4}\left(-\dfrac{1}{4x}\right)=-\dfrac{1}{16}$

(c) $\lim\limits_{x\to 4}\left[\left(\dfrac{1}{x}-\dfrac{1}{4}\right)(x-2)\right]=\lim\limits_{x\to 4}\dfrac{(4-x)(x-2)}{4x}=0$

(d) $\lim\limits_{x \to 4} \left[\left(\dfrac{1}{x} - \dfrac{1}{4} \right) \left(\dfrac{1}{x-4} \right)^2 \right] = \lim\limits_{x \to 4} \dfrac{4-x}{4x(x-4)^2} = \lim\limits_{x \to 4} \dfrac{1}{4x(4-x)};$ does not exist

41. (a) $\lim\limits_{x \to 4} \dfrac{f(x) - f(4)}{x - 4} = \lim\limits_{x \to 4} \dfrac{(x^2 - 4x) - (0)}{x - 4} = \lim\limits_{x \to 4} x = 4$

(b) $\lim\limits_{x \to 1} \dfrac{f(x) - f(1)}{x - 1} = \lim\limits_{x \to 1} \dfrac{x^2 - 4x + 3}{x - 1} = \lim\limits_{x \to 1} \dfrac{(x-1)(x-3)}{x-1} = \lim\limits_{x \to 1} (x - 3) = -2$

(c) $\lim\limits_{x \to 3} \dfrac{f(x) - f(1)}{x - 3} = \lim\limits_{x \to 3} \dfrac{x^2 - 4x + 3}{x - 3} = \lim\limits_{x \to 3} \dfrac{(x-1)(x-3)}{x-3} = \lim\limits_{x \to 3} (x - 1) = 2$

(d) $\lim\limits_{x \to 3} \dfrac{f(x) - f(2)}{x - 3} = \lim\limits_{x \to 3} \dfrac{x^2 - 4x + 4}{x - 3};$ does not exist

43. $f(x) = 1/x, \quad g(x) = -1/x \quad$ with $c = 0$

45. True. Let $\lim\limits_{x \to c}[f(x) + g(x)] = L$. If $\lim\limits_{x \to c} g(x) = M$ exists, then $\lim\limits_{x \to c} f(x) = \lim\limits_{x \to c}[f(x) + g(x) - g(x)] =$ $L - M$ also exists. This contradicts the fact that $\lim\limits_{x \to c} f(x)$ does not exist.

47. True. If $\lim\limits_{x \to c} \sqrt{f(x)} = L$ exists, then $\lim\limits_{x \to c} \sqrt{f(x)} \sqrt{f(x)} = L^2$ also exists.

49. False; for example set $f(x) = x \quad$ and $\quad c = 0$

51. False; for example, set $f(x) = 1 - x^2$, $g(x) = 1 + x^2$, and $c = 0$.

53. If $\lim\limits_{x \to c} f(x) = L \quad$ and $\quad \lim\limits_{x \to c} g(x) = L, \quad$ then

$$\lim\limits_{x \to c} h(x) = \lim\limits_{x \to c} \tfrac{1}{2}\{[f(x) + g(x)] - |f(x) - g(x)|\}$$

$$= \lim\limits_{x \to c} \tfrac{1}{2}[f(x) + g(x)] - \lim\limits_{x \to c} \tfrac{1}{2}(x) - g(x)|$$

$$= \tfrac{1}{2}(L + L) - \tfrac{1}{2}(L - L) = L.$$

A similar argument works for H.

55. (a) Suppose on the contrary that $\lim\limits_{x \to c} g(x)$ does exist. Let $L = \lim\limits_{x \to c} g(x)$. Then

$$\lim\limits_{x \to c} f(x)g(x) = \lim\limits_{x \to c} f(x) \cdot \lim\limits_{x \to c} g(x) = 0 \cdot L = 0.$$

This contradicts the fact that $\lim\limits_{x \to c} f(x)g(x) = 1$

(b) $\lim\limits_{x \to c} g(x)$ exists since $\lim\limits_{x \to c} g(x) = \lim\limits_{x \to c} \dfrac{f(x)g(x)}{f(x)} = \dfrac{1}{L}.$

57. (a) $\lim\limits_{h \to 0} \dfrac{f(x+h) - f(x)}{h} = \lim\limits_{h \to 0} \dfrac{x + h - x}{h} = 1$

(b) $\lim\limits_{h \to 0} \dfrac{f(x+h) - f(x)}{h} = \lim\limits_{h \to 0} \dfrac{(x+h)^2 - x^2}{h} = \lim\limits_{h \to 0} \dfrac{x^2 + 2xh + h^2 - x^2}{h}$

$\qquad\qquad = \lim\limits_{h \to 0} (2x + h) = 2x$

(c) $\lim\limits_{h \to 0} \dfrac{f(x+h) - f(x)}{h} = \lim\limits_{h \to 0} \dfrac{(x+h)^3 - x^3}{h} = \lim\limits_{h \to 0} \dfrac{x^3 + 3x^2h + 3xh^2 + h^3 - x^3}{h}$

$\qquad\qquad = \lim\limits_{h \to 0} (3x^2 + 3xh + h^2) = 3x^2$

(d) $\lim\limits_{h \to 0} \dfrac{f(x+h) - f(x)}{h} = \lim\limits_{h \to 0} \dfrac{(x+h)^4 - x^4}{h} = \lim\limits_{h \to 0} \dfrac{x^4 + 4x^3h + 6x^2h^2 + 4xh^3 + h^4 - x^4}{h}$

$\qquad\qquad = \lim\limits_{h \to 0} (4x^3 + 6x^2h + 4xh^2 + h^3) = 4x^3$

(e) $\lim\limits_{h \to 0} \dfrac{f(x+h) - f(x)}{h} = \lim\limits_{h \to 0} \dfrac{(x+h)^n - x^n}{h} = nx^{n-1}$ for any positive integer n.

SECTION 2.4

1. (a) f is discontinuous at $x = -3,\ 0,\ 2,\ 6$

   (b) at -3, neither; f is continuous from the right at 0; at 2 and 6, neither

3. continuous   5. continuous   7. continuous

9. removable discontinuity   11. jump discontinuity

13. continuous   15. jump discontinuity

17.

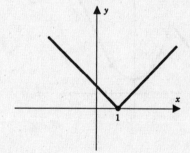

removable discontinuity at 2

19.

no discontinuities

21.

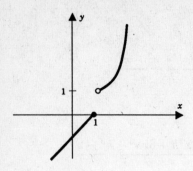

23.

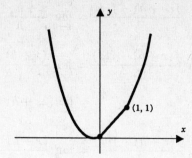

jump discontinuity at 1

no discontinuities

25.

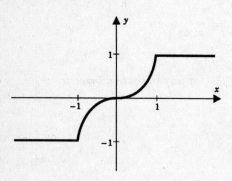

27.

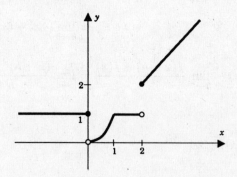

no discontinuities

jump discontinuities at 0 and 2

29.

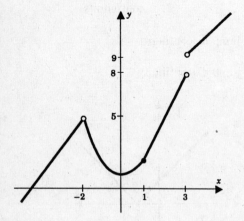

removable discontinuity at -2;   jump discontinuity at 3

31. $f(1) = 2$ **33.**   impossible; $\lim\limits_{x \to 1^-} f(x) = -1$ and $\lim\limits_{x \to 1^+} f(x) = 1$

35.   Since $\lim\limits_{x \to 1^-} f(x) = 1$ and $\lim\limits_{x \to 1^+} f(x) = A - 3 = f(1)$, take $A = 4$.

37. The function f is continuous at $x = 1$ iff

$$f(1) = \lim_{x \to 1^-} f(x) = A - B \quad \text{and} \quad \lim_{x \to 1^+} f(x) = 3$$

are equal;   that is, $A - B = 3$. The function f is discontinuous at $x = 2$ iff

$$\lim_{x \to 2^-} f(x) = 6 \quad \text{and} \quad \lim_{x \to 2^+} f(x) = f(2) = 4B - A$$

are unequal;   that is, iff $4B - A \neq 6$. More simply we have $A - B = 3$ with $B \neq 3$:

$$A - B = 3,\ 4B - A \neq 6 \implies A - B = 3,\ 3B - 3 \neq 6 \implies A - B = 3,\ B \neq 3.$$

39. $f(5) = \frac{1}{6}$ **41.** $f(5) = \frac{1}{3}$

43. nowhere;   see Figure 2.1.8

45. $x = 0,\quad x = 2,\quad$ and all nonintegral values of x

47. Refer to (2.2.5). Use the equivalence of (i) and (ii) setting $L = f(c)$.

49. Suppose that g does not have a non-removable discontinuity at c. Then either g is continuous at c or it has a removable discontinuity at c. In either case, $\lim g(x)$ as $x \to c$ exists. Since $g(x) = f(x)$ except at a finite set of points $x_1, x_2, \ldots, x_n$, $\lim f(x)$ exists as $x \to c$ by Exercise 54, Section 2.3.

51. By implication, f is defined on $(c - p, c + p)$. The given inequality implies that $B \geq 0$. If $B = 0$, then $f \equiv f(c)$ is a constant function and hence is continuous. Now assume that $B > 0$. Let $\epsilon > 0$ and let $\delta = \min\{\epsilon/B, p\}$. If $|x - c| < \delta$ then $x \in (c - p,\ c + p)$ and

$$|f(x) - f(c)| \leq B|x - c| < B \cdot \delta \leq B \cdot \frac{\epsilon}{B} = \epsilon$$

Thus, f is continuous at c.

53. $\displaystyle \lim_{h \to 0} [f(c + h) - f(c)] = \lim_{h \to 0} \left[\frac{f(c + h) - f(c)}{h} \cdot h \right] = \lim_{h \to 0} \left[\frac{f(c + h) - f(c)}{h} \right] \cdot \lim_{h \to 0} h = L \cdot 0 = 0.$
Therefore f is continuous at c by Exercise 47.

SECTION 2.5

1. $\displaystyle \lim_{x \to 0} \frac{\sin 3x}{x} = \lim_{x \to 0} 3\left(\frac{\sin 3x}{3x} \right) = 3(1) = 3$ **3.** $\displaystyle \lim_{x \to 0} \frac{3x}{\sin 5x} = \lim_{x \to 0} \frac{3}{5}\left(\frac{5x}{\sin 5x} \right) = \frac{3}{5}(1) = \frac{3}{5}$

5. $\displaystyle \lim_{x \to 0} \frac{\sin 4x}{\sin 2x} = \lim_{x \to 0} \frac{4x}{2x} \cdot \frac{\sin 4x}{4x} \cdot \frac{2x}{\sin 2x} = 2(1)(1) = 2$

7. $\displaystyle \lim_{x \to 0} \frac{\sin x^2}{x} = \lim_{x \to 0} x\left(\frac{\sin x^2}{x^2} \right) = \lim_{x \to 0} x \cdot \lim_{x \to 0} \frac{\sin x^2}{x^2} = 0(1) = 0$

9. $\displaystyle \lim_{x \to 0} \frac{\sin x}{x^2} = \lim_{x \to 0} \frac{(\sin x)/x}{x};\quad$ does not exist   **11.** $\displaystyle \lim_{x \to 0} \frac{\sin^2 3x}{5x^2} = \lim_{x \to 0} \frac{9}{5}\left(\frac{\sin 3x}{3x} \right)^2 = \frac{9}{5}(1) = \frac{9}{5}$

13. $\displaystyle \lim_{x \to 0} \frac{2x}{\tan 3x} = \lim_{x \to 0} \frac{2x \cos 3x}{\sin 3x} = \lim_{x \to 0} \frac{2}{3}\left(\frac{3x}{\sin 3x} \right) \cos 3x = \frac{2}{3}(1)(1) = \frac{2}{3}$

15. $\lim\limits_{x \to 0} x \csc x = \lim\limits_{x \to 0} \dfrac{x}{\sin x} = 1$

17. $\lim\limits_{x \to 0} \dfrac{x^2}{1 - \cos 2x} = \lim\limits_{x \to 0} \dfrac{x^2}{1 - \cos 2x} \cdot \left(\dfrac{1 + \cos 2x}{1 + \cos 2x} \right) = \lim\limits_{x \to 0} \dfrac{x^2(1 + \cos 2x)}{\sin^2 2x}$

$$= \lim\limits_{x \to 0} \dfrac{1}{4} \left(\dfrac{2x}{\sin 2x} \right)^2 (1 + \cos 2x) = \dfrac{1}{4}(1)(2) = \dfrac{1}{2}$$

19. $\lim\limits_{x \to 0} \dfrac{1 - \sec^2 2x}{x^2} = \lim\limits_{x \to 0} \dfrac{-\tan^2 2x}{x^2} = \lim\limits_{x \to 0} \dfrac{-\sin^2 2x}{x^2 \cos^2 2x} = \lim\limits_{x \to 0} \left[-4 \left(\dfrac{\sin 2x}{2x} \right)^2 \dfrac{1}{\cos^2 2x} \right] = -4$

21. $\lim\limits_{x \to 0} \dfrac{2x^2 + x}{\sin x} = \lim\limits_{x \to 0} (2x + 1) \dfrac{x}{\sin x} = 1$

23. $\lim\limits_{x \to 0} \dfrac{\tan 3x}{2x^2 + 5x} = \lim\limits_{x \to 0} \dfrac{1}{x(2x + 5)} \dfrac{\sin 3x}{\cos 3x} = \lim\limits_{x \to 0} \dfrac{3}{2x + 5} \left(\dfrac{\sin 3x}{3x} \right) \dfrac{1}{\cos 3x} = \dfrac{3}{5}(1)(1) = \dfrac{3}{5}$

25. $\lim\limits_{x \to 0} \dfrac{\sec x - 1}{x \sec x} = \lim\limits_{x \to 0} \dfrac{\dfrac{1}{\cos x} - 1}{x \left(\dfrac{1}{\cos x} \right)} = \lim\limits_{x \to 0} \dfrac{1 - \cos x}{x} = 0$ **27.** $\dfrac{2\sqrt{2}}{\pi}$

29. $\lim\limits_{x \to \pi/2} \dfrac{\cos x}{x - \pi/2} = \lim\limits_{h \to 0} \dfrac{\cos (h + \pi/2)}{h} = \lim\limits_{h \to 0} \dfrac{-\sin h}{h} = -1$

$\quad\quad\quad \underset{\displaystyle h = x - \pi/2}{\Big\uparrow} \quad\quad \underset{\displaystyle \cos (h + \pi/2) = \cos h \cos \pi/2 - \sin h \sin \pi/2}{\Big\uparrow}$

31. $\lim\limits_{x \to \pi/4} \dfrac{\sin (x + \pi/4) - 1}{x - \pi/4} = \lim\limits_{h \to 0} \dfrac{\sin (h + \pi/2) - 1}{h} = \lim\limits_{h \to 0} \dfrac{\cos h - 1}{h} = 0$

$\quad\quad\quad\quad\quad\quad \underset{\displaystyle h = x - \pi/4}{\Big\uparrow}$

33. Equivalently we will show that $\lim\limits_{h \to 0} \cos (c + h) = \cos c$. The identity

$$\cos (c + h) = \cos c \cos h - \sin c \sin h$$

gives

$$\lim\limits_{h \to 0} \cos (c + h) = \cos c \left(\lim\limits_{h \to 0} \cos h \right) - \sin c \left(\lim\limits_{h \to 0} \sin h \right)$$

$$= (\cos c)(1) - (\sin c)(0) = \cos c.$$

35. $f(x) = \sin x; \quad a = \pi/4$

$$\lim\limits_{h \to 0} \dfrac{f(a + h) - f(a)}{h} = \lim\limits_{h \to 0} \dfrac{\sin \left(\frac{\pi}{4} + h \right) - \sin \left(\frac{\pi}{4} \right)}{h} = \lim\limits_{h \to 0} \dfrac{\sin (\pi/4) \cos h + \cos (\pi/4) \sin h - \sin (\pi/4)}{h}$$

$$= \lim\limits_{h \to 0} \dfrac{-\sin (\pi/4)(1 - \cos h) + \cos (\pi/4) \sin h}{h}$$

$$= -\sin (\pi/4) \lim\limits_{h \to 0} \dfrac{1 - \cos h}{h} + \cos (\pi/4) \lim\limits_{h \to 0} \dfrac{\sin h}{h} = \cos (\pi/4) = \dfrac{\sqrt{2}}{2}$$

tangent line: $\quad y - \dfrac{\sqrt{2}}{2} = \dfrac{\sqrt{2}}{2} \left(x - \dfrac{\pi}{4} \right)$

37. $f(x) = \cos 2x; \quad a = \pi/6$

$$\lim_{h \to 0} \frac{f(a+h) - f(a)}{h} = \lim_{h \to 0} \frac{\cos 2\left(\frac{\pi}{6} + h\right) - \cos\left(2\frac{\pi}{6}\right)}{h} = \lim_{h \to 0} \frac{\cos(2h + \pi/3) - \cos(\pi/3)}{h}$$

$$= \lim_{h \to 0} \frac{\cos(\pi/3)\cos 2h - \sin(\pi/3)\sin 2h - \cos(\pi/3)}{h}$$

$$= -\cos(\pi/3) \lim_{h \to 0} 2\frac{1 - \cos 2h}{h} - \sin(\pi/3) \lim_{h \to 0} 2\frac{\sin 2h}{h}$$

$$= -\cos(\pi/3) \cdot 2 \cdot 0 - \sin(\pi/3) \cdot 2 \cdot 1 = -\sqrt{3}$$

tangent line: $y - \dfrac{1}{2} = -\sqrt{3}\left(x - \dfrac{\pi}{6}\right)$

39. For $x \neq 0$, $|x \sin(1/x)| = |x||\sin(1/x)| \leq |x|$. Thus,

$$-|x| \leq |x \sin(1/x)| \leq |x|$$

Since $\lim_{x \to 0}(-|x|) = \lim_{x \to 0}|x| = 0$, the result follows by the pinching theorem.

41. For x close to 1(radian), $0 < \sin x \leq 1$. Thus,

$$0 < |x - 1| \sin x \leq |x - 1|$$

and the result follows by the pinching theorem.

43. Suppose that there is a number B such that $|f(x)| \leq B$ for all $x \neq 0$. Then $|x f(x)| \leq B|x|$ and

$$-B|x| \leq x f(x) \leq B|x|$$

The result follows by the pinching theorem.

45. Suppose that there is a number B such that $\left|\dfrac{f(x) - L}{x - c}\right| \leq B$ for $x \neq c$. Then

$$0 \leq |f(x) - L| = \left|(x - c)\frac{f(x) - L}{x - c}\right| \leq B|x - c|$$

By the pinching theorem, $\lim_{x \to c} |f(x) - L| = 0$ which implies $\lim_{x \to c} f(x) = L$.

SECTION 2.6

1. 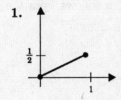 **3.** 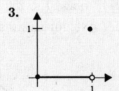 **5.** 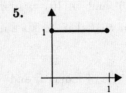 **7.**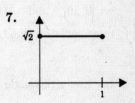

9. Impossible by the intermediate value theorem.

11. 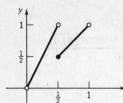   **13.** 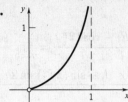   **15.**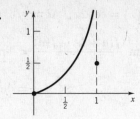

17. Let $f(x) = 2x^3 - 4x^2 + 5x - 4$. Then $f(1) = -1 < 0$ and $f(2) = 6 > 0$. Since f is a polynomial, f is continuous on $(-\infty, \infty)$ and so f is continuous on $[1, 2]$. By the intermediate-value theorem, there exists a number $c \in (1, 2)$ such that $f(c) = 0$.

19. The function $f(x) = \dfrac{1}{x-1} + \dfrac{1}{x-4}$ is continuous on $(1, 4)$. Since $f(2) = \dfrac{1}{2} > 0$ and $f(3) = -\dfrac{1}{2} < 0$, and since f is continuous on $[2, 3]$, it follows that there is a number $c \in (2, 3)$ such that $f(c) = 0$ by the intermediate-value theorem.

21. Set $g(x) = x - f(x)$. Since g is continuous on $[0, 1]$ and $g(0) \leq 0 \leq g(1)$, there exists c in $[0, 1]$ such that $g(c) = c - f(c) = 0$.

23. Since f is bounded on $(-p, p)$, it follows from Exercise 43, Section 2.5, that $\lim\limits_{x \to 0} x f(x) = 0$. Thus,
$$\lim_{x \to 0} g(x) = \lim_{x \to 0} x f(x) = 0 = g(0)$$
which implies that g is continuous at 0.

25. The cubic polynomial $P(x) = x^3 + ax^2 + bx + c$ is continuous on $(-\infty, \infty)$.. Writing P as
$$P(x) = x^3 \left(1 + \frac{a}{x} + \frac{b}{x^2} + \frac{c}{x^3} \right) \quad x \neq 0$$
it follows that $P(x) < 0$ for large negative values of x and $P(x) > 0$ for large positive values of x. Thus there exists a negative number N such that $P(x) < 0$ for $x < N$, and a positive number M such that $P(x) > 0$ for $x > M$. By the intermediate-value theorem, P has a zero in $[N, M]$.

27. Let $A(r)$ denote the area of a circle with radius r, $r \in [0, 10]$. Then $A(r) = \pi r^2$ is continuous on $[0, 10]$, and $A(0) = 0$ and $A(10) = 100\pi \cong 314$. Since $0 < 250 < 314$ it follows from the intermediate value theorem that there exists a number $c \in (0, 10)$ such that $A(c) = 250$.

29. Inscribe a rectangle in a circle of radius R and introduce a coordinate system as shown in the figure. Then the area of the rectangle is given by

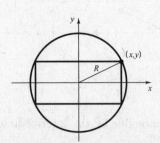

$$A(x) = 4x\sqrt{R^2 - x^2}, \quad x \in [0, R].$$

Since A is continuous on $[0, R]$, A has a maximum value.

31. $m_{10} = 1.7314$ **33.** $m_{10} = 0.7392$

35. $f(-3) = -9$, $f(-2) = 5$; $f(0) = 3$, $f(\ \ ) = -1$; $f(1) = -1$, $f(2) = 1$ Thus, f has a zero in $(-3, -2,)$ in $(0,1)$ and in $(1,2)$.

$r_1 = -2.4909$, $r_2 = 0.6566$,   and $r_3 = 1.8343$

37. $f(-2) = -5.6814$, $f(-1) = 1.1829$; $f(0) = 0.5$, $f(1) = -0.1829$; $f(1) = -0.1829$, $f(2) = 6.681$

Thus, f has a zero in $(-2, -1)$, in $(0,1)$ and in $(1,2)$.

$r_1 = -1.3482$, $r_2 = 0.2620$,   and $r_3 = 1.0816$

39. f is bounded.

$\max(f) = 1$ $[f(1) = 1]$

$\min(f) = -1$ $[f(-1) = -1]$

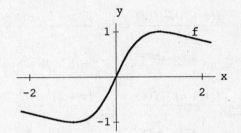

41. f is bounded.

$\max(f) = 0.5$

$\min(f) \cong 0.3540$

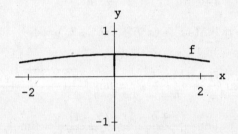

PROJECTS AND EXPLORATIONS

2.1.   (a) For $x \neq 0$, $x^7 \neq 0$ and so F is defined near 0. When x is close to 0 both $\sin x$ and $\tan x$ are close to 0 and so $\sin(\tan x)$ and $\tan(\sin x)$ are both close to 0.

(b) $\displaystyle\lim_{x \to 0} \frac{\sin(\tan x) - \tan(\sin x)}{x^7} = -\frac{1}{30}$

2.3.   (a) Let $F(x) = \cos x - x$. Then $F(0) = 1$ and $F(1) \cong -0.46$. Thus F must have a zero in $(0,1)$.

(b)   Sketch the graph of F and observe that it has exactly one zero.

$x \cong 0.74$ is the fixed point of $\cos x$ on $(0,1)$.

(d)   Let $G(x) = \dfrac{3}{x^2 + 1} - x$. Then $G(1) = 0.5$ and $G(2) = -1.4$. Thus G has a zero in $(1,2)$. $x \cong 1.21$ is the fixed point of g in $(1,2)$.

CHAPTER 3

SECTION 3.1

1. $f'(x) = \lim_{h \to 0} \dfrac{f(x+h) - f(x)}{h} = \lim_{h \to 0} \dfrac{4-4}{h} = \lim_{h \to 0} 0 = 0$

3. $f'(x) = \lim_{h \to 0} \dfrac{f(x+h) - f(x)}{h} = \lim_{h \to 0} \dfrac{[2-3(x+h)] - [2-3x]}{h}$

$\qquad = \lim_{h \to 0} \dfrac{-3h}{h} = \lim_{h \to 0} -3 = -3$

5. $f'(x) = \lim_{h \to 0} \dfrac{f(x+h) - f(x)}{h} = \lim_{h \to 0} \dfrac{[5(x+h) - (x+h)^2] - (5x - x^2)}{h}$

$\qquad = \lim_{h \to 0} \dfrac{5h - 2xh - h^2}{h} = \lim_{h \to 0}(5 - 2x - h) = 5 - 2x$

7. $f'(x) = \lim_{h \to 0} \dfrac{f(x+h) - f(x)}{h} = \lim_{h \to 0} \dfrac{(x+h)^4 - x^4}{h}$

$\qquad = \lim_{h \to 0} \dfrac{(x^4 + 4x^3 h + 6x^2 h^2 + 4xh^3 + h^4) - x^4}{h}$

$\qquad = \lim_{h \to 0}(4x^3 + 6x^2 h + 4xh^2 + h^3) = 4x^3$

9. $f'(x) = \lim_{h \to 0} \dfrac{f(x+h) - f(x)}{h} = \lim_{h \to 0} \dfrac{\sqrt{x+h-1} - \sqrt{x-1}}{h}$

$\qquad = \lim_{h \to 0} \dfrac{(x+h-1) - (x-1)}{h(\sqrt{x+h-1} + \sqrt{x-1})} = \lim_{h \to 0} \dfrac{1}{\sqrt{x+h-1} + \sqrt{x-1}} = \dfrac{1}{2\sqrt{x-1}}$

11. $f'(x) = \lim_{h \to 0} \dfrac{f(x+h) - f(x)}{h} = \lim_{h \to 0} \dfrac{\dfrac{1}{(x+h)^2} - \dfrac{1}{x^2}}{h}$

$\qquad = \lim_{h \to 0} \dfrac{x^2 - (x^2 + 2hx + h^2)}{hx^2(x+h)^2} = \lim_{h \to 0} \dfrac{-2x - h}{x^2(x+h)^2} = -\dfrac{2}{x^3}$

13. $f'(2) = \lim_{h \to 0} \dfrac{f(2+h) - f(2)}{h} = \lim_{h \to 0} \dfrac{(3h-1)^2 - 1}{h}$

$\qquad = \lim_{h \to 0} \dfrac{9h^2 - 6h}{h} = \lim_{h \to 0}(9h - 6) = -6$

15. $f'(2) = \lim_{h \to 0} \dfrac{f(2+h) - f(2)}{h} = \lim_{h \to 0} \dfrac{\dfrac{9}{6+h} - \dfrac{3}{2}}{h}$

$\qquad = \lim_{h \to 0} \dfrac{18 - 3(6+h)}{2h(6+h)} = \lim_{h \to 0} \dfrac{-3}{2(6+h)} = -\dfrac{1}{4}$

17. $f'(2) = \lim_{h \to 0} \dfrac{f(2+h) - f(2)}{h} = \lim_{h \to 0} \dfrac{(2 + h + \sqrt{4+2h}\,) - 4}{h}$

$= \lim_{h \to 0} \left(1 + \dfrac{\sqrt{4+2h} - 2}{h} \right) = \lim_{h \to 0} 1 + \dfrac{(4+2h) - 4}{h(\sqrt{4+2h} + 2)}$

$= \lim_{h \to 0} 1 + \dfrac{2}{\sqrt{4+2h} + 2} = \dfrac{3}{2}$

19. Slope of tangent at $(2, 4)$ is $f'(2) = 4$.   Tangent $y - 4 = 4(x - 2)$;

normal $y - 4 = -\frac{1}{4}(x - 2)$.

21. Slope of tangent at $(4, 4)$ is $f'(4) = -3$.   Tangent $y - 4 = -3(x - 4)$;

normal $y - 4 = \frac{1}{3}(x - 4)$.

23. Slope of tangent at $(-3, -1)$ is -1.   Tangent $y + 1 = -(x + 3)$;

normal $y + 1 = x + 3$.

25. (a) f is not continuous at $c = -1$ and $c = 1$. f has a removable discontinuity at $c = -1$

and a jump discontinuity at $c = 1$.

(b) f is continuous but not differentiable at $c = 0$ and $c = 3$.

27.   at $x = -1$ **29.**   at $x = 0$ **31.**   at $x = 1$

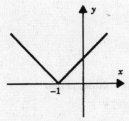 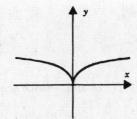 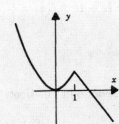

33. $f'(1) = 4$

$\lim_{h \to 0^-} \dfrac{f(1+h) - f(1)}{h} = \lim_{h \to 0^-} \dfrac{4(1+h) - 4}{h} = 4$

$\lim_{h \to 0^+} \dfrac{f(1+h) - f(1)}{h} = \lim_{h \to 0^+} \dfrac{2(1+h)^2 + 2 - 4}{h} = 4$

35. $f'(-1)$ does not exist

$\lim_{h \to 0^-} \dfrac{f(-1+h) - f(-1)}{h} = \lim_{h \to 0^-} \dfrac{h - 0}{h} = 1$

$\lim_{h \to 0^+} \dfrac{f(-1+h) - f(-1)}{h} = \lim_{h \to 0^+} \dfrac{h^2 - 0}{h} = 0$

37. 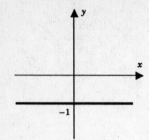   **39.** 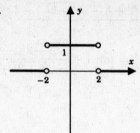   **41.**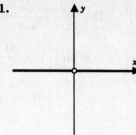

43. $f(x) = x^2; \quad c = 1$ **45.** $f(x) = \sqrt{x}; \quad c = 4$ **47.** $f(x) = \cos x; \quad c = \pi$

49. Since $f(1) = 1$ and $\lim\limits_{x \to 1^+} f(x) = 2$, f is not continuous at 1 and thus, by (3.1.4), is not differentiable at 1.

51. (a) $f'(x) = \begin{cases} 2(x+1), & x < 0 \\ 2(x-1), & x > 0 \end{cases}$

   (b) $\lim\limits_{h \to 0^-} \dfrac{f(0+h) - f(0)}{h} = \lim\limits_{h \to 0^-} \dfrac{(h+1)^2 - 1}{h} = \lim\limits_{h \to 0^-} (h+2) = 2,$

   $\lim\limits_{h \to 0^+} \dfrac{f(0+h) - f(0)}{h} = \lim\limits_{h \to 0^+} \dfrac{(h-1)^2 - 1}{h} = \lim\limits_{h \to 0^+} (h-2) = -2.$

53. $f(x) = c$, c any constant

55. $f(x) = |x+1|; \quad f(x) = \begin{cases} 0, & x \neq -1 \\ 1, & x = -1 \end{cases}$

57. $f(x) = 2x + 5$

59. Let f be an odd function: $f(-x) = -f(x)$, and assume that f is differentiable. Then

$$f'(-x) = \lim_{h \to 0} \frac{f(-x+h) - f(-x)}{h} = \lim_{h \to 0} \frac{f[-(x-h)] + f(x)}{h}$$

$$= \lim_{h \to 0} \frac{-[f(x-h) - f(x)]}{h} = \lim_{h \to 0} \frac{f(x-h) - f(x)}{-h}$$

$$= \lim_{k \to 0} \frac{f(x+k) - f(x)}{k} \quad \text{(set } k = -h\text{)}$$

$$= f'(x) \quad \text{and } f' \text{ is an even function.}$$

61. (a) $\lim\limits_{x \to 2^+} f(x) = \lim\limits_{x \to 2^-} f(x) = f(2) = 2$ Thus, f is continuous at $x = 2$.

   (b) $f'_-(2) = \lim\limits_{h \to 0^-} \dfrac{f(2+h) - f(2)}{h} = \lim\limits_{h \to 0^-} \dfrac{(2+h)^2 - (2+h) - 2}{h} = 3$

   $f'_+(2) = \lim\limits_{h \to 0^+} \dfrac{f(2+h) - f(2)}{h} = \lim\limits_{h \to 0^+} \dfrac{2(2+h) - 2 - 2}{h} = 2$

   (c) No, since $f'_-(2) \neq f'_+(2)$.

63. (a) $f'(x) = \lim\limits_{h \to 0} \dfrac{f(x+h) - f(x)}{h} = \lim\limits_{h \to 0} \dfrac{\sqrt{1-(x+h)} - \sqrt{1-x}}{h}$

$$= \lim_{h \to 0} \frac{-h}{h\left(\sqrt{1-(x+h)} + \sqrt{1-x}\right)} = \frac{-1}{2\sqrt{1-x}}$$

(b) $f'_+(0) = \lim\limits_{h \to 0^+} \dfrac{\sqrt{1-h} - 1}{h} = -\dfrac{1}{2}$

(c) $f'_-(1) = \lim\limits_{h \to 0^-} \dfrac{\sqrt{1-(1+h)}}{h} = \lim\limits_{h \to 0^-} \dfrac{\sqrt{-h}}{h} = \lim\limits_{h \to 0^-} \dfrac{-1}{\sqrt{-h}}$ does not exist.

65. Suppose $f'_-(c) = \lim\limits_{h \to 0^-} \dfrac{f(c+h) - f(c)}{h} = L = \lim\limits_{h \to 0^+} \dfrac{f(c+h) - f(c)}{h} = f'_+(c).$

Then $\lim\limits_{h \to 0} \dfrac{f(c+h) - f(c)}{h} = L$ exists and f is differentiable at c.

67. (a) Since $|\sin(1/x)| \le 1$ it follows that

$$-x \le f(x) \le x \quad \text{and} \quad -x^2 \le g(x) \le x^2$$

Thus $\lim\limits_{x \to 0} f(x) = f(0) = 0$ and $\lim\limits_{x \to 0} g(x) = g(0) = 0,$ which implies that f and g

are continuous at 0.

(b) $\lim\limits_{h \to 0} \dfrac{h\sin(1/h) - 0}{h} = \lim\limits_{h \to 0} \sin(1/h)$ does not exist.

(c) $\lim\limits_{h \to 0} \dfrac{h^2\sin(1/h) - 0}{h} = \lim\limits_{h \to 0} h\sin(1/h) = 0.$ Thus g is differentiable at 0 and $g'(0) = 0.$

69. $f'(-1) = \lim\limits_{h \to 0} \dfrac{f(1+h) - f(1)}{h} = \lim\limits_{h \to 0} \dfrac{[(1+h)^2 - 3(1+h)] - (-2)]}{h} = \lim\limits_{h \to 0} \dfrac{-h + h^2}{h} = -1$

$f'(1) = \lim\limits_{x \to 1} \dfrac{f(x) - f(1)}{x-1} = \lim\limits_{x \to 1} \dfrac{(x^2 - 3x) - (-2)}{x-1} = \lim\limits_{x \to 1} \dfrac{(x-2)(x-1)}{x-1} = \lim\limits_{x \to 1} (x-2) = -1$

71. $f'(-1) = \lim\limits_{h \to 0} \dfrac{f(-1+h) - f(-1)}{h} = \lim\limits_{h \to 0} \dfrac{(-1+h)^{1/3} + 1}{h}$

$$= \lim_{h \to 0} \frac{(-1+h)^{1/3} + 1}{h} \cdot \frac{(-1+h)^{2/3} - (-1+h)^{1/3} + 1}{(-1+h)^{2/3} - (-1+h)^{1/3} + 1}$$

$$= \lim_{h \to 0} \frac{h}{h\left((-1+h)^{2/3} - (-1+h)^{1/3} + 1\right)} = \frac{1}{3}$$

$$f'(-1) = \lim_{x \to -1} \frac{f(x) - f(-1)}{x - (-1)} = \lim_{x \to -1} \frac{x^{1/3} + 1}{x + 1} = \lim_{x \to -1} \frac{x^{1/3} + 1}{x + 1} \cdot \frac{x^{2/3} - x^{1/3} + 1}{x^{2/3} - x^{1/3} + 1}$$

$$= \lim_{x \to -1} \frac{x + 1}{(x + 1)(x^{2/3} - x^{1/3} + 1)} = \frac{1}{3}$$

73.   (a) $D = \dfrac{(2 + h)^{5/2} - 2^{5/2}}{h}$ $-1 \le h \le 1$

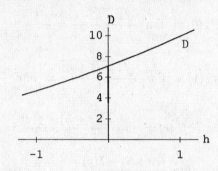

(b) $f'(2) \cong 7.071$ (c) $D(0.001) \cong 7.074$ and $D(-0.001) \cong 7.068$

75.   (a)   Let $f(x) = 4x - x^3$.   (b)

Then $f'(x) = 4 - 3x^2$; $f'(3/2) = \frac{11}{4}$

$T(x) = -\frac{11}{4}\left(x - \frac{3}{2}\right) + \frac{21}{8}$

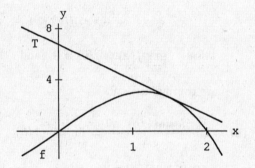

(c) $(1.453, 1.547)$

SECTION 3.2

1. $F'(x) = -1$

3. $F'(x) = 55x^4 - 18x^2$

5. $F'(x) = 2ax + b$

7. $F'(x) = 2x^{-3}$

9. $G'(x) = (x^2 - 1)(1) + (x - 3)(2x) = 3x^2 - 6x - 1$

11. $G'(x) = \dfrac{(1 - x)(3x^2) - x^3(-1)}{(1 - x)^2} = \dfrac{3x^2 - 2x^3}{(1 - x)^2}$

13. $G'(x) = \dfrac{(2x + 3)(2x) - (x^2 - 1)(2)}{(2x + 3)^2} = \dfrac{2(x^2 + 3x + 1)}{(2x + 3)^2}$

15. $G'(x) = (x - 1)(1) + (x - 2)(1) = 2x - 3$

17. $G'(x) = \dfrac{(x - 2)(1/x^2) - (6 - 1/x)(1)}{(x - 2)^2} = \dfrac{-2(3x^2 - x + 1)}{x^2(x - 2)^2}$

19. $G'(x) = (9x^8 - 8x^9)\left(1 - \dfrac{1}{x^2}\right) + \left(x + \dfrac{1}{x}\right)(72x^7 - 72x^8) = -80x^9 + 81x^8 - 64x^7 + 63x^6$

21. $f'(x) = -x(x-2)^{-2}$, $f'(0) = -\frac{1}{4}$, $f'(1) = -1$

23. $f'(x) = \dfrac{(1+x^2)(-2x) - (1-x^2)(2x)}{(1+x^2)^2} = \dfrac{-4x}{(1+x^2)^2}$, $f'(0) = 0$, $f'(1) = -1$

25. $f'(x) = \dfrac{(cx+d)a - (ax+b)c}{(cx+d)^2} = \dfrac{ad-bc}{(cx+d)^2}$, $f'(0) = \dfrac{ad-bc}{d^2}$, $f'(1) = \dfrac{ad-bc}{(c+d)^2}$

27. $f'(x) = xh'(x) + h(x)$, $f'(0) = 0h'(0) + h(0) = 0(2) + 3 = 3$

29. $f'(x) = h'(x) + \dfrac{h'(x)}{[h(x)]^2}$, $f'(0) = h'(0) + \dfrac{h'(0)}{[h(0)]^2} = 2 + \dfrac{2}{3^2} = \dfrac{20}{9}$

31. $f'(x) = \dfrac{(x+2)(1) - x(1)}{(x+2)^2} = \dfrac{2}{(x+2)^2}$,

slope of tangent at $(-4,2) : f'(-4) = 1/2$,

equation for tangent: $y - 2 = \frac{1}{2}(x+4)$

33. $f'(x) = (x^2 - 3)(5 - 3x^2) + (5x - x^3)(2x)$,

slope of tangent at $(1,-8) : f'(1) = (-2)(2) + (4)(2) = 4$,

equation for tangent: $y + 8 = 4(x-1)$

35. $f'(x) = -12x^{-3}$;   slope of tangent at $(3, \frac{2}{3}) : f'(3) = -\frac{4}{9}$,

equation for tangent: $y - \frac{2}{3} = -\frac{4}{9}(x-3)$

37. $f'(x) = (x-2)(2x-1) + (x^2 - x - 11)(1) = 3(x-3)(x+1)$,

$f'(x) = 0$ at $x = -1, 3$; $(-1, 27)$, $(3, -5)$

39. $f'(x) = \dfrac{(x^2+1)(5) - 5x(2x)}{(x^2+1)^2} = \dfrac{5(1-x^2)}{(x^2+1)^2}$, $f'(x) = 0$ at $x = \pm 1$; $(-1, -5/2)$, $(1, 5/2)$

41. $f'(x) = 1 - 8/x^3$, $f'(x) = 0$ at $x = 2$; $(2, 3)$

43. slope of line 4,

slope of tangent $-2x$,

$-2x = 4$ at $x = -2$; $(-2, -10)$

45. slope of line $-1/5$,

slope of tangent $3x^2 - 2x$,

$3x^2 - 2x = 5$ at $x = -1, 5/3$;

$(-1, -2)$, $\left(\frac{5}{3}, \frac{50}{27}\right)$

47.

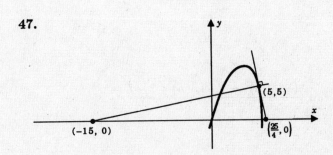

slope of tangent at $(5,5)$ is $f'(5) = -4$

tangent $y - 5 = -4(x-5)$ intersects

x-axis at $\left(\frac{25}{4}, 0\right)$

normal $y - 5 = \frac{1}{4}(x-5)$ intersects

x-axis at $(-15, 0)$

area of triangle is

$$\frac{1}{2}(5)\left(15 + \frac{25}{4}\right) = \frac{425}{8}$$

49.   If the point $(1,3)$ lies on the graph, we have $f(1) = 3$ and thus

(*) $$A + B + C = 3.$$

If the line $4x + y = 8$ (slope -4) is tangent to the graph at $(2,0)$, then

$f(2) = 0$ and $f'(2) = -4$. Thus,

(**) $$4A + 2B + C = 0 \quad \text{and} \quad 4A + B = -4.$$

Solving the equations in (*) and (**), we find that $A = -1, \quad B = 0, \quad C = 4$.

51.   Let $f(x) = ax^2 + bx + c$.   Then $f'(x) = 2ax + b$ and $f'(x) = 0$ at $x = -b/2a$.

53.   Let $f(x) = x^3 - x$. The secant line through $(-1, f(-1) = (-1,0)$ and $(2, f(2)) = (2,6)$ has slope $m = \dfrac{6 - 0}{2 - (-1)} = 2$.   Now, $f'(x) = 3x^2 - 1$ and $3c^2 - 1 = 2$ implies $c = -1, \ 1$.

55.   Let $f(x) = 1/x$, $x > 0$. Then $f'(x) = -1/x^2$. An equation for the tangent line to the graph of f at the point $(a, f(a))$, $a > 0$, is $y = (-1/a^2)x + 2/a$. The y-intercept is $2/a$ and the x-intercept is $2a$. The area of the triangle formed by this line and the coordinate axes is: $A = \frac{1}{2}(2/a)(2a) = 2$ square units.

57.   Since f and $f + g$ are differentiable, $g = (f + g) - f$ is differentiable. The functions $f(x) = |x|$ and $g(x) = -|x|$ are not differentiable at $x = 0$ yet their sum $f(x) + g(x) \equiv 0$ is differentiable for all x.

59.   Since

$$\left(\frac{f}{g}\right)(x) = \frac{f(x)}{g(x)} = f(x) \cdot \frac{1}{g(x)},$$

it follows from the product and reciprocal rules that

$$\left(\frac{f}{g}\right)'(x) = \left(f \cdot \frac{1}{g}\right)'(x) = f(x)\left(-\frac{g'(x)}{[g(x)]^2}\right) + f'(x) \cdot \frac{1}{g(x)} = \frac{g(x)f'(x) - f(x)g'(x)}{[g(x)]^2}.$$

61. $F'(x) = 2x\left(1 + \dfrac{1}{x}\right)(2x^3 - x + 1) + (x^2 + 1)\left(\dfrac{-1}{x^2}\right)(2x^3 - x + 1) + (x^2 + 1)\left(1 + \dfrac{1}{x}\right)(6x^2 - 1)$

63. $g(x) = [f(x)]^2 = f(x) \cdot f(x)$

$g'(x) = f(x)f'(x) + f(x)f'(x) = 2f(x)f'(x)$

65. $g'(x) = 3(x^3 - 2x^2 + x + 2)^2(3x^2 - 4x + 1)$

67.   (a) $f'_+(-1) = \lim\limits_{h \to 0^+} \dfrac{f(-1 + h) - f(-1)}{h} = \lim\limits_{h \to 0^+} \dfrac{(-1 + h)^2 - 4(-1 + h) + 2 - 7}{h}$

$= \lim\limits_{h \to 0^+} \dfrac{-6h + h^2}{h} = -6$

$f'_-(3) = \lim\limits_{h \to 0^-} \dfrac{f(3 + h) - f(3)}{h} = \lim\limits_{h \to 0^-} \dfrac{(3 + h)^2 - 4(3 + h) + 2 + 1}{h}$

$$= \lim_{h \to 0^-} \frac{2h + h^2}{h} = 2$$

(b) If $g(x) = x^2 - 4x + 2$ then $g'(x) = 2x - 4$, and $g'(-1) = -6$, $g'(3) = 2$.

69. We want f to be continuous at $x = 2$. That is, we want

$$\lim_{x \to 2^-} f(x) = f(2) = \lim_{x \to 2^+} f(x).$$

This gives

(1) $\qquad\qquad\qquad\qquad 8A + 2B + 2 = 4B - A.$

We also want

$$\lim_{x \to 2^-} f'(x) = \lim_{x \to 2^+} f'(x).$$

This gives

(2) $\qquad\qquad\qquad\qquad 12A + B = 4B.$

Equations (1) and (2) together imply that $A = -2$ and $B = -8$.

71. (a) $\dfrac{\sin(0 + 0.001) - \sin 0}{0.001} \cong 0.99999 \qquad \dfrac{\sin(0 - 0.001) - \sin 0}{-0.001} \cong 0.99999$

$\dfrac{\sin[(\pi/6) + 0.001] - \sin(\pi/6)}{0.001} \cong 0.86578 \qquad \dfrac{\sin[(\pi/6) - 0.001] - \sin(\pi/6)}{-0.001} \cong 0.86628$

$\dfrac{\sin[(\pi/4) + 0.001] - \sin(\pi/4)}{0.001} \cong 0.70675 \qquad \dfrac{\sin[(\pi/4) - 0.001] - \sin(\pi/4)}{-0.001} \cong 0.70746$

$\dfrac{\sin[(\pi/3) + 0.001] - \sin(\pi/3)}{0.001} \cong 0.49957 \qquad \dfrac{\sin[(\pi/3) - 0.001] - \sin(\pi/3)}{-0.001} \cong 0.50043$

$\dfrac{\sin[(\pi/2) + 0.001] - \sin(\pi/2)}{0.001} \cong -0.0005 \qquad \dfrac{\sin[(\pi/2) - 0.001] - \sin(\pi/2)}{-0.001} \cong 0.0005$

(b) $\cos 0 = 1$, $\cos(\pi/6) \cong 0.866025$, $\cos(\pi/4) \cong 0.707107$, $\cos(\pi/3) = 0.5$, $\cos(\pi/2) = 0$

(c) If $f(x) = \sin x$ then $f'(x) = \cos x$.

73. (a) $\dfrac{2^{0+0.001} - 2^0}{0.001} \cong 0.69339 \qquad \dfrac{2^{0-0.001} - 2^0}{-0.001} \cong 0.69291$

$\dfrac{2^{1+0.001} - 2^1}{0.001} \cong 1.38678 \qquad \dfrac{2^{1-0.001} - 2^1}{-0.001} \cong 1.38581$

$\dfrac{2^{2+0.001} - 2^2}{0.001} \cong 2.77355 \qquad \dfrac{2^{2-0.001} - 2^2}{-0.001} \cong 2.77163$

$\dfrac{2^{3+0.001} - 2^3}{0.001} \cong 5.54710 \qquad \dfrac{2^{3-0.001} - 2^0}{-0.001} \cong 5.54326$

(b) $\dfrac{f'(x)}{f(x)} \cong 0.693$ (c)   If $f(x) = 2^x$ then $f'(x) = 2^x K$, where $K \cong 0.693$.

SECTION 3.3

1. $\dfrac{dy}{dx} = 12x^3 - 2x$ **3.** $\dfrac{dy}{dx} = 1 + \dfrac{1}{x^2}$

5. $\dfrac{dy}{dx} = \dfrac{(1+x^2)(1) - x(2x)}{(1+x^2)^2} = \dfrac{1-x^2}{(1+x^2)^2}$ **7.** $\dfrac{dy}{dx} = \dfrac{(1-x)2x - x^2(-1)}{(1-x)^2} = \dfrac{x(2-x)}{(1-x)^2}$

9. $\dfrac{dy}{dx} = \dfrac{(x^3-1)3x^2 - (x^3+1)3x^2}{(x^3-1)^2} = \dfrac{-6x^2}{(x^3-1)^2}$ **11.** $\dfrac{d}{dx}(2x-5) = 2$

13. $\dfrac{d}{dx}[(3x^2 - x^{-1})(2x+5)] = (3x^2 - x^{-1})2 + (2x+5)(6x + x^{-2}) = 18x^2 + 30x + 5x^{-2}$

15. $\dfrac{d}{dt}\left(\dfrac{t^2+1}{t^2-1}\right) = \dfrac{(t^2-1)2t - (t^2+1)(2t)}{(t^2-1)^2} = \dfrac{-4t}{(t^2-1)^2}$

17. $\dfrac{d}{dt}\left(\dfrac{t^4}{2t^3-1}\right) = \dfrac{(2t^3-1)4t^3 - t^4(6t^2)}{(2t^3-1)^2} = \dfrac{2t^3(t^3-2)}{(2t^3-1)^2}$

19. $\dfrac{d}{du}\left(\dfrac{2u}{1-2u}\right) = \dfrac{(1-2u)2 - 2u(-2)}{(1-2u)^2} = \dfrac{2}{(1-2u)^2}$

21. $\dfrac{d}{du}\left(\dfrac{u}{u-1} - \dfrac{u}{u+1}\right) = \dfrac{(u-1)(1) - u}{(u-1)^2} - \dfrac{(u+1)(1) - u}{(u+1)^2}$

$$= -\dfrac{1}{(u-1)^2} - \dfrac{1}{(u+1)^2} = -\dfrac{2(1+u^2)}{(u^2-1)^2}$$

23. $\dfrac{d}{dx}\left(\dfrac{x^2}{1-x^2} - \dfrac{1-x^2}{x^2}\right) = \dfrac{(1-x^2)2x - x^2(-2x)}{(1-x^2)^2} - \dfrac{x^2(-2x) - (1-x^2)2x}{x^4}$

$$= \dfrac{2x}{(1-x^2)^2} + \dfrac{2}{x^3}$$

25. $\dfrac{d}{dx}\left(\dfrac{x^3+x^2+x+1}{x^3-x^2+x-1}\right) = \dfrac{(x^3-x^2+x-1)(3x^2+2x+1) - (x^3+x^2+x+1)(3x^2-2x+1)}{(x^3-x^2+x-1)^2}$

$$= \dfrac{-2(x^4+2x^2+1)}{(x^2+1)^2(x-1)^2} = \dfrac{-2}{(x-1)^2}$$

27. $\dfrac{dy}{dx} = (x+1)\dfrac{d}{dx}[(x+2)(x+3)] + (x+2)(x+3)\dfrac{d}{dx}(x+1)$

$$= (x+1)(2x+5) + (x+2)(x+3)$$

At $x = 2$, $\dfrac{dy}{dx} = (3)(9) + (4)(5) = 47$.

29. $\dfrac{dy}{dx} = \dfrac{(x+2)\dfrac{d}{dx}[(x-1)(x-2)] - (x-1)(x-2)(1)}{(x+2)^2}$

$\qquad = \dfrac{(x+2)(2x-3) - (x-1)(x-2)}{(x+2)^2}$

At $x = 2$, $\dfrac{dy}{dx} = \dfrac{4(1) - 1(0)}{16} = \dfrac{1}{4}$.

31. $f'(x) = 21x^2 - 30x^4$ **33.** $f'(x) = 1 + 3x^{-2}$ **35.** $f'(x) = 4x + 4x^{-3}$

$\quad f''(x) = 42x - 120x^3$ $\qquad f''(x) = -6x^{-3}$ $\qquad f''(x) = 4 - 12x^{-4}$

37. $\dfrac{dy}{dx} = x^2 + x + 1$ **39.** $\dfrac{dy}{dx} = 8x - 20$ **41.** $\dfrac{dy}{dx} = 3x^2 + 3x^{-4}$

$\quad \dfrac{d^2y}{dx^2} = 2x + 1$ $\qquad \dfrac{d^2y}{dx^2} = 8$ $\qquad \dfrac{d^2y}{dx^2} = 6x - 12x^{-5}$

$\quad \dfrac{d^3y}{dx^3} = 2$ $\qquad \dfrac{d^3y}{dx^3} = 0$ $\qquad \dfrac{d^3y}{dx^3} = 6 + 60x^{-6}$

43. $\dfrac{d}{dx}\left[x\dfrac{d}{dx}(x - x^2)\right] = \dfrac{d}{dx}[x(1 - 2x)] = \dfrac{d}{dx}[x - 2x^2] = 1 - 4x$

45. $\dfrac{d^4}{dx^4}[3x - x^4] = \dfrac{d^3}{dx^3}[3 - 4x^3] = \dfrac{d^2}{dx^2}[-12x^2] = \dfrac{d}{dx}[-24x] = -24$

47. $\dfrac{d^2}{dx^2}\left[(1 + 2x)\dfrac{d^2}{dx^2}(5 - x^3)\right] = \dfrac{d^2}{dx^2}[(1 + 2x)(-6x)] = \dfrac{d^2}{dx^2}[-6x - 12x^2] = -24$

49. Let $p(x) = ax^2 + bx + c$. Then $p'(x) = 2ax + b$ and $p''(x) = 2a$. Now

$$p''(1) = 2a = 4 \Longrightarrow a = 2$$

$$p'(1) = 2(2)(1) + b = -2 \Longrightarrow b = -6$$

$$p(1) = 2(1)^2 - 6(1) + c = 3 \Longrightarrow c = 7$$

Thus $p(x) = 2x^2 - 6x + 7$.

51. (a) If $k = n$, $f^{(n)}(x) = n!$ (b) If $k > n$, $f^n(x) = 0$.

(c) If $k < n$, $f^{(n)}(x) = n(n-1)(n-2)\cdots(n-k+1)x^{n-k}$.

53. Let $f(x) = \begin{cases} x^2 & x \ge 0 \\ 0 & x \le 0 \end{cases}$

(a) $f'_+(0) = \lim\limits_{h \to 0^+} \dfrac{f(0+h) - f(0)}{h} = \lim\limits_{h \to 0^+} \dfrac{h^2 - 0}{h} = 0$ and

$$f'_-(0) = \lim_{h \to 0-} \frac{f(0+h) - f(0)}{h} = \lim_{h \to 0-} \frac{0}{h} = 0$$

Therefore, f is differentiable at 0 and $f'(0) = 0$.

(b) $f'(x) = \begin{cases} 2x & x \geq 0 \\ 0 & x \leq 0 \end{cases}$

(c) $f''_+(0) = \lim_{h \to 0+} \frac{f'(0+h) - f,(0)}{h} = \lim_{h \to 0+} \frac{2h - 0}{h} = 2$ and

$$f''_-(0) = \lim_{h \to 0-} \frac{f'(0+h) - f'(0)}{h} = \lim_{h \to 0-} \frac{0}{h} = 0$$

Since $f''_+(0) \neq f''_-(0)$, $f''(0)$ does not exist.

(d)

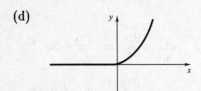

55. It suffices to give a single counterexample. For instance, if

$f(x) = g(x) = x$, then $(fg)(x) = x^2$ so that $(fg)''(x) = 2$ but

$f(x)g''(x) + f''(x)g(x) = x \cdot 0 + 0 \cdot x = 0.$

57. $f''(x) = 6x$; (a) $x = 0$ (b) $x > 0$ (c) $x < 0$

59. $f''(x) = 12x^2 + 12x - 24$; (a) $x = -2, 1$ (b) $x < -2$, $x > 1$ (c) $-2 < x < 1$

61. The result is true for $n = 1$:

$$\frac{d^1 y}{dx^1} = \frac{dy}{dx} = -x^{-2} = (-1)^1 1! \, x^{-1-1}.$$

If the result is true for $n = k$:

$$\frac{d^k y}{dx^k} = (-1)^k k! \, x^{-k-1}$$

then the result is true for $n = k + 1$:

$$\frac{d^{k+1} y}{dx^{k+1}} = \frac{d}{dx}\left[\frac{d^k y}{dx^k}\right] = \frac{d}{dx}\left[(-1)^k k! \, x^{-(k+1)}\right] = (-1)^{(k+1)}(k+1)! \, x^{-(k+1)-1}.$$

63. (a) $(f \cdot g)'' = [(f \cdot g)']' = [f'g + fg']' = f''g + f'g' + f'g' + fg'' = f''g + 2f'g' + fg''$

(b) $(f \cdot g)''' = [(f \cdot g)'']' = [f''g + 2f'g' + fg'']' = f'''g + f''g' + 2f''g' + 2f'g'' + f'g'' + fg'''$

$$= f'''g + 3f''g' + 3f'g'' + g'''$$

65. $\dfrac{d}{dx}(uvw) = uv\dfrac{dw}{dx} + uw\dfrac{dv}{dx} + vw\dfrac{du}{dx}$

67. (a) Let $f(x) = x^3 + x^2 - 4x + 1$. Then $f'(x) = 3x^2 + 2x - 4$.

(b)

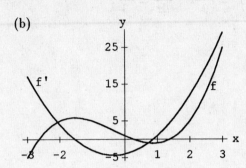

(c) The graph is "falling" when $f'(x) < 0$;

The graph is "rising" when $f'(x) > 0$.

69. (a) Let $f(x) = \frac{1}{2}x^3 - 3x^2 + 4x + 1$.

Then $f'(x) = \frac{3}{2}x^2 - 6x + 4$ and

$f'(0) = 4$.

Tangent line at $x = 0$: $y = 4x + 1$

(b)

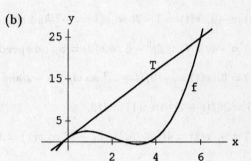

(c) Solving $\frac{1}{2}x^3 - 3x^2 + 4x + 1 = 4x + 1$ for x gives $x = 6$; the graph and the tangent line

intersect at $(6, 25)$.

SECTION 3.4

1. $A = \pi r^2$, $\dfrac{dA}{dr} = 2\pi r$. When $r = 2$, $\dfrac{dA}{dr} = 4\pi$.

3. $A = \dfrac{1}{2}z^2$, $\dfrac{dA}{dz} = z$. When $z = 4$, $\dfrac{dA}{dz} = 4$.

5. $y = \dfrac{1}{x(1+x)}$, $\dfrac{dy}{dx} = \dfrac{-(2x+1)}{x^2(1+x)^2}$. At $x = 2$, $\dfrac{dy}{dx} = -\dfrac{5}{36}$.

7. $V = \dfrac{4}{3}\pi r^3$, $\dfrac{dV}{dr} = 4\pi r^2 =$ the surface area of the ball.

9. $y = 2x^2 + x - 1$, $\dfrac{dy}{dx} = 4x + 1$. $\dfrac{dy}{dx} = 4$ at $x = \dfrac{3}{4}$. Therefore $x_0 = \dfrac{3}{4}$.

11. (a) $w = s\sqrt{2}$, $V = s^3 = \left(\dfrac{w}{\sqrt{2}}\right)^3 = \dfrac{\sqrt{2}}{4}w^3$, $\dfrac{dV}{dw} = \dfrac{3\sqrt{2}}{4}w^2$.

(b) $z^2 = s^2 + w^2 = 3s^2$, $z = s\sqrt{3}$. $V = s^3 = \left(\dfrac{z}{\sqrt{3}}\right)^3 = \dfrac{\sqrt{3}}{9}z^3$, $\dfrac{dV}{dz} = \dfrac{\sqrt{3}}{3}z^2$.

13. (a) $\dfrac{dA}{d\theta} = \dfrac{1}{2}r^2$ (b) $\dfrac{dA}{dr} = r\theta$

   (c) $\theta = \dfrac{2A}{r^2}$ so $\dfrac{d\theta}{dr} = \dfrac{-4A}{r^3} = \dfrac{-4}{r^3}\left(\dfrac{1}{2}r^2\theta\right) = \dfrac{-2\theta}{r}$

15. $y = ax^2 + bx + c$, $z = bx^2 + ax + c$.

   $\dfrac{dy}{dx} = 2ax + b$, $\dfrac{dz}{dx} = 2bx + a$.

   $\dfrac{dy}{dx} = \dfrac{dz}{dx}$ iff $2ax + b = 2bx + a$. With $a \neq b$, this occurs only at $x = \dfrac{1}{2}$.

17. $x(5) = -6$, $v(t) = 3 - 2t$ so $v(5) = -7$ and speed $= 7$, $a(t) = -2$ so $a(5) = -2$.

19. $x(2) = -4$, $v(t) = 3t^2 - 6$ so $v(2) = 6$ and speed $= 6$, $a(t) = 6t$ so $a(2) = 12$.

21. $x(1) = 6$, $v(t) = -18/(t+2)^2$ so $v(1) = -2$ and speed $= 2$,

   $a(t) = 36/(t+2)^3$ so $a(1) = 4/3$.

23. $x(1) = 0$, $v(t) = 4t^3 + 18t^2 + 6t - 10$ so $v(1) = 18$ and speed $= 18$,

   $a(t) = 12t^2 + 36t + 6$ so $a(1) = 54$.

25. $v(t) = 3t^2 - 6t + 3 = 3(t-1)^2 \geq 0$;   the object never changes direction.

27. $v(t) = 1 - \dfrac{5}{(t+2)^2}$;   the object changes direction (from left to right) at $t = -2 + \sqrt{5}$.

29. $v(t) = \dfrac{8 - t^2}{(t^2 + 8)^2}$;   the object changes direction (from right to left) at $t = 2\sqrt{2}$.

31. A 33. A 35. A and B 37. A 39. A and C

41. The object is moving right when $v(t) > 0$. Here,

   $v(t) = 4t^3 - 36t^2 + 56t = 4t(t-2)(t-7)$ and $v(t) > 0$ when $0 < t < 2$ and $7 < t$.

43. The object is speeding up when $v(t)$ and $a(t)$ have the same sign.

   $v(t) = 5t^3(4 - t)$ sign of $v(t)$: $+ + + + + + + + + + + + + + + + + + 0 - - - - -$

   $a(t) = 20t^2(3 - t)$ sign of $a(t)$: $+ + + + + + + + + + + + + 0 - - - - - - - - - - -$

   Thus, $0 < t < 3$ and $4 < t$.

45. The object is moving left and slowing down when $v(t) < 0$ and $a(t) > 0$.

$v(t) = 3(t-5)(t+1)$ sign of $v(t)$:

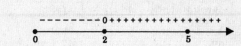

$a(t) = 6(t-2)$ sign of $a(t)$:

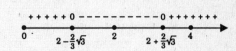

Thus, $2 < t < 5$.

47.   The object is moving right and speeding up when $v(t) > 0$ and $a(t) > 0$.

$v(t) = 4t(t-2)(t-4)$ sign of $v(t)$:

$a(t) = 4(3t^2 - 12t + 8)$ sign of $a(t)$:

Thus, $0 < t < 2 - \frac{2}{3}\sqrt{3}$ and $4 < t$.

49.   Since $v_0 = 0$ the equation of motion is

$$y(t) = -16t^2 + y_0.$$

We want to find y_0 so that $y(6) = 0$. From

$$0 = -16(6)^2 + y_0$$

we get $y_0 = 576$ feet.

51.   The object's height and velocity at time t are given by

$$y(t) = -\frac{1}{2}gt^2 + v_0 t \quad \text{and} \quad v(t) = -gt + v_0$$

Since the object's velocity at its maximum height is 0, it takes v_0/g seconds to reach

maximum height, and

$$y(v_0/g) = -\frac{1}{2}g(v_0/g)^2 + v_0(v_0/g) = v_0^2/2g \quad \text{or} \quad v_0^2/19.6 \quad \text{(meters)}$$

53.   At time t, the object's height is $y(t) = -\frac{1}{2}gt^2 + v_0 t + y_0$, and its velocity is $v(t) = -gt + v_0$. Suppose that $y(t_1) = y(t_2)$, $t_1 \neq t_2$. Then

$$-\frac{1}{2}gt_1^2 + v_0 t_1 + y_0 = -\frac{1}{2}gt_2^2 + v_0 t_2 + y_0$$

$$\frac{1}{2}g(t_2^2 - t_1^2) = v_0(t_2 - t_1)$$

$$gt_2 + gt_1) = 2v_0$$

From this equation, we get $-(-gt_1 + v_0) = -gt_2 + v_0$ and so $|v(t_1)| = |v(t_2)|$.

55.   In the equation

$$y(t) = -16t^2 + v_0 t + y_0$$

we take $v_0 = -80$ and $y_0 = 224$. The ball first strikes the ground when

$$-16t^2 - 80t + 224 = 0;$$

that is, at $t = 2$. Since

$$v(t) = y'(t) = -32t - 80,$$

we have $v(2) = -144$ so that the speed of the ball the first time it strikes the ground is 144 ft/sec. Thus, the speed of the ball the third time it strikes the ground is $\frac{1}{4}\left[\frac{1}{4}(144)\right] = 9$ ft/sec.

57. The equation is $y(t) = -16t^2 + 32t$.   (Here $y_0 = 0$ and $v_0 = 32$.)

 (a) We solve $y(t) = 0$ to find that the stone strikes the ground at $t = 2$ seconds.

 (b) The stone attains its maximum height when $v(t) = 0$. Solving

$$v(t) = -32t + 32 = 0, \quad \text{we get} \quad t = 1 \quad \text{and, thus, the maximum height is } y(1) = 16 \text{ feet.}$$

 (c) We want to choose v_0 in

$$y(t) = -16t^2 + v_0 t$$

so that $y(t_0) = 36$ when $v(t_0) = 0$ for some time t_0.

From $v(t) = -32t + v_0 = 0$ we get $t_0 = v_0/32$ so that

$$-16\left(\frac{v_0}{32}\right)^2 + v_0\left(\frac{v_0}{32}\right) = 36, \quad \text{or} \quad \frac{v_0{}^2}{64} = 36.$$

Thus, $v_0 = 48$ ft/sec.

59. For all three parts of the problem the basic equation is

$$y(t) = -16t^2 + v_0 t + y_0$$

with

(∗) $$y(t_0) = 100 \quad \text{and} \quad y(t_0 + 2) = 16$$

for some time $t_0 > 0$.

We are asked to find y_0 for a given value of v_0.

From (∗) we get

$$16 - 100 = y(t_0 + 2) - y(t_0)$$

$$= [-16(t_0 + 2)^2 + v_0(t_0 + 2) + y_0] - [-16t_0{}^2 + v_0 t_0 + y_0]$$

$$= -64t_0 - 64 + 2v_0$$

so that

$$t_0 = \tfrac{1}{32}(v_0 + 10).$$

Substituting this result in the basic equation and noting that $y(t_0) = 100$, we have

$$-16\left(\frac{v_0 + 10}{32}\right)^2 + v_0\left(\frac{v_0 + 10}{32}\right) + y_0 = 100$$

and therefore

(∗∗) $$y_0 = 100 - \frac{v_0{}^2}{64} + \frac{25}{16}.$$

We use (∗∗) to find the answer to each part of the problem.

 (a) $v_0 = 0$ so $y_0 = \frac{1625}{16}$ ft   (b) $v_0 = -5$ so $y_0 = \frac{6475}{64}$ ft   (c) $v_0 = 10$ so $y_0 = 100$ ft

61. $C(x) = 200 + 0.02x + 0.0001x^2,$ $C'(x) = 0.02 + 0.002x$

Marginal cost at $x = 100$ units: $C'(100) = 0.04$

Actual cost of 101st unit: $C(101) - C(100) = 0.0401$

63. $C(x) = 200 + 0.01x + \dfrac{100}{x},$ $C'(x) = 0.01 - \dfrac{100}{x^2}$

Marginal cost at $x = 100$ units: $C'(100) = 0$

Actual cost of producing the 101st unit: $C(101) - C(100) = 0$

65. $C(x) = 1000 + 25x - \dfrac{x^2}{10},$ $C'(x) = 25 - \dfrac{x}{5}$

Marginal cost of producing 10 motors: $C'(10) = 23\$$

Actual cost of producing the 10th motor: $C(11) - C(10) = 22.90\$$

67. (a) $\overline{C(x)} = \dfrac{C(x)}{x} = \dfrac{200}{x} + 0.02 + 0.0001x,$ $\overline{C'(x)} = -\dfrac{200}{x^2} + 0.0001$

(b) $\overline{C(x)} = \dfrac{C(x)}{x} = \dfrac{200}{x} + 0.01 + \dfrac{100}{x^2},$ $\overline{C'(x)} = -\dfrac{200}{x^2} - \dfrac{200}{x^3}$

69. (a) $v(t) = 3t^2 - 14t + 10,\ 0 \le t \le 5$

(b) The object is moving to the right when $0 < t < 0.88$ and when $3.79 < t < 5$.

The object is moving to the left when $0.88 < t < 3.79$

(c)

The object stops at times $t \cong 0.88$ and $t \cong 3.79$.

The maximum speed is $v \cong 6.33$ at $t \cong 2.33$.

(d) $a(t) = 6t - 14$

The object is speeding up when $v(t)$ and $a(t)$ have

the same sign: $0.88 < t < 2.33$ and $3.79 < t < 5$.

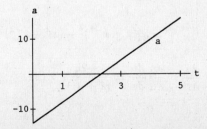

The object is slowing down when $v(t)$ and $a(t)$ have

opposite sign: $0 < t < 0.88$ and $2.33 < t < 3.79$.

SECTION 3.5

1. $f(x) = x^4 + 2x^2 + 1,$ $f'(x) = 4x^3 + 4x = 4x(x^2 + 1)$

   $f(x) = (x^2 + 1)^2,$ $f'(x) = 2(x^2 + 1)(2x) = 4x(x^2 + 1)$

3. $f(x) = 8x^3 + 12x^2 + 6x + 1,$ $f'(x) = 24x^2 + 24x + 6 = 6(2x + 1)^2$

   $f(x) = (2x + 1)^3,$ $f'(x) = 3(2x + 1)^2(2) = 6(2x + 1)^2$

5. $f(x) = x^2 + 2 + x^{-2},$ $f'(x) = 2x - 2x^{-3} = 2x(1 - x^{-4})$

   $f(x) = (x + x^{-1})^2,$ $f'(x) = 2(x + x^{-1})(1 - x^{-2}) = 2x(1 + x^{-2})(1 - x^{-2}) = 2x(1 - x^{-4})$

7. $f'(x)$ $= -1(1 - 2x)^{-2} \dfrac{d}{dx}(1 - 2x) = 2(1 - 2x)^{-2}$

9. $f'(x)$ $= 20(x^5 - x^{10})^{19} \dfrac{d}{dx}(x^5 - x^{10}) = 20(x^5 - x^{10})^{19}(5x^4 - 10x^9)$

11. $f'(x)$ $= 4\left(x - \dfrac{1}{x}\right)^3 \dfrac{d}{dx}\left(x - \dfrac{1}{x}\right) = 4\left(x - \dfrac{1}{x}\right)^3\left(1 + \dfrac{1}{x^2}\right)$

13. $f'(x)$ $= 4(x - x^3 - x^5)^3 \dfrac{d}{dx}(x - x^3 - x^5) = 4(x - x^3 - x^5)^3(1 - 3x^2 - 5x^4)$

15. $f'(t)$ $= 100(t^2 - 1)^{99} \dfrac{d}{dt}(t^2 - 1) = 200t(t^2 - 1)^{99}$

17. $f'(t)$ $= 4(t^{-1} + t^{-2})^3 \dfrac{d}{dt}(t^{-1} + t^{-2}) = 4(t^{-1} + t^{-2})^3(-t^{-2} - 2t^{-3})$

19. $f'(x)$ $= 4\left(\dfrac{3x}{x^2 + 1}\right)^3 \dfrac{d}{dx}\left(\dfrac{3x}{x^2 + 1}\right) = 4\left(\dfrac{3x}{x^2 + 1}\right)^3\left[\dfrac{(x^2 + 1)3 - 3x(2x)}{(x^2 + 1)^2}\right] = \dfrac{324x^3(1 - x^2)}{(x^2 + 1)^5}$

21. $f'(x)$ $= 2(x^4 + x^2 + x)^1 \dfrac{d}{dx}(x^4 + x^2 + x) = 2(x^4 + x^2 + x)(4x^3 + 2x + 1)$

23. $f'(x)$ $= -\left(\dfrac{x^3}{3} + \dfrac{x^2}{2} + \dfrac{x}{1}\right)^{-2} \dfrac{d}{dx}\left(\dfrac{x^3}{3} + \dfrac{x^2}{2} + \dfrac{x}{1}\right) = -\left(\dfrac{x^3}{3} + \dfrac{x^2}{2} + x\right)^{-2}(x^2 + x + 1)$

25. $f'(x)$ $= 3[(x + x^{-1})^2 - (x^2 + x^{-2})^{-1}]^2 \left\{\dfrac{d}{dx}(x + x^{-1})^2 - \dfrac{d}{dx}(x^2 + x^{-2})^{-1}\right\}$

   $= 3\left[(x + x^{-1})^2 - (x^2 + x^{-2})^{-1}\right]^2 [2(x + x^{-1})(1 - x^{-2}) + (x^2 + x^{-2})^{-2}(2x - 2x^{-3})]$

27. $\dfrac{dy}{dx} = \dfrac{dy}{du}\dfrac{du}{dx} = \dfrac{-2u}{(1 + u^2)^2} \cdot (2)$

At $x = 0$, we have $u = 1$ and thus $\dfrac{dy}{dx} = \dfrac{-4}{4} = -1$.

29. $\dfrac{dy}{dx} = \dfrac{dy}{du}\dfrac{du}{dx} = \dfrac{(1-4u)2 - 2u(-4)}{(1-4u)^2} \cdot 4(5x^2+1)^3(10x) = \dfrac{2}{(1-4u)^2} \cdot 40x(5x^2+1)^3$

At $x = 0$, we have $u = 1$ and thus $\dfrac{dy}{dx} = \dfrac{2}{9}(0) = 0$.

31. $\dfrac{dy}{dt} = \dfrac{dy}{du}\dfrac{du}{dx}\dfrac{dx}{dt} = \dfrac{(1+u^2)(-7) - (1-7u)(2u)}{(1+u^2)^2}(2x)(2)$

$\quad = \dfrac{7u^2 - 2u - 7}{(1+u^2)^2}(4x) = \dfrac{4x(7x^4 + 12x - 2)}{(x^4 + 2x^2 + 2)^2} = \dfrac{4(2t-5)[7(2t-5)^4 + 12(2t-5)^2 - 2]}{[(2t-5)^4 + 2(2t-5)^2 + 2]^2}$

33. $\dfrac{dy}{dx} = \dfrac{dy}{ds}\dfrac{ds}{dt}\dfrac{dt}{dx} = 2(s+3) \cdot \dfrac{1}{2\sqrt{t-3}} \cdot (2x)$

At $x = 2$, we have $t = 4$ so that $s = 1$ and thus $\dfrac{dy}{dx} = 2(4)\dfrac{1}{2 \cdot 1}(4) = 16$.

35. $(f \circ g)'(0) = f'(g(0))g'(0) = f'(2)g'(0) = (1)(1) = 1$

37. $(f \circ g)'(2) = f'(g(2))g'(2) = f'(2)g'(2) = (1)(1) = 1$

39. $(g \circ f)'(1) = g'(f(1))f'(1) = g'(0)f'(1) = (1)(1) = 1$

41. $(f \circ h)'(0) = f'(h(0))h'(0) = f'(1)h'(0) = (1)(2) = 2$

43. $(g \circ f \circ h)'(2) = g'(f(h(2))) \, f'(h(2))h'(2) = g'(1)f'(0)h'(2) = (0)(2)(2) = 0$

45. $f'(x) = 4(x^3 + x)^3(3x^2 + 1)$

$f''(x) = 3(4)(x^3+x)^2(3x^2+1)^2 + 4(x^3+x)^3(6x) = 12(x^3+x)^2[(3x^2+1)^2 + 2x(x^3+x)]$

47. $f'(x) = 3\left(\dfrac{x}{1-x}\right)^2 \cdot \dfrac{1}{(1-x)^2} = \dfrac{3x^2}{(1-x)^4}$

$f''(x) = \dfrac{6x(1-x)^4 - 3x^2(4)(1-x)^3(-1)}{(1-x)^8} = \dfrac{6x(1+x)}{(1-x)^5}$

49. $2xf'(x^2+1)$ **51.** $2f(x)f'(x)$

53. $f'(x) = -4x(1+x^2)^{-3}$; (a) $x = 0$ (b) $x < 0$ (c) $x > 0$

55. $f'(x) = \dfrac{1-x^2}{(1+x^2)^2}$; (a) $x = \pm 1$ (b) $-1 < x < 1$ (c) $x < -1, \ x > 1$

57. $v(t) = 5(t+1)(t-9)^2(t-3)$; the object changes direction (from left to right) at $t = 3$.

59. $v(t) = 12t^3(t^2 - 12)^3(t^2 - 4)$; the object changes direction (from right to left) at $t = 2$

and (from left to right) at $t = 2\sqrt{3}$.

61. $L'(x) = \dfrac{1}{x^2 + 1} \cdot 2x = \dfrac{2x}{x^2 + 1}$

63. $T'(x) = 2f(x) \cdot f'(x) + 2g(x) \cdot g'(x) = 2f(x) \cdot g(x) - 2g(x) \cdot f(x) = 0$

65. Suppose $p(x) = (x - a)^2 q(x),$ where $q(a) \neq 0.$ Then

$$p'(x) = 2(x - a)q(x) + (x - a)^2 q'(x) \quad \text{and} \quad p''(x) = 2q(x) + 4(x - a)q'(x) + (x - a)^2 q''(x),$$

and it follows that $p(a) = p'(a) = 0,$ and $p''(a) \neq 0.$

Now suppose that $p(a) = p'(a) = 0$ and $p''(a) \neq 0.$

$$p(a) = 0 \quad \Rightarrow \quad p(x) = (x - a)g(x) \quad \text{for some polynomial } g.$$

Then $p'(x) = g(x) + (x - a)g'(x)$ and

$$p'(a) = 0 \quad \Rightarrow \quad g(a) = 0 \text{ and so } g(x) = (x - a)q(x) \text{ for some polynomial } q.$$

Therefore, $p(x) = (x - a)^2 q(x).$ Finally, $p''(a) \neq 0$ implies $q(a) \neq 0.$

67. Let p be a polynomial function of degree n. The number a is a root of p of multiplicity $k,$ $(k < n)$ if and only if $p(a) = p'(a) = \cdots = p^{(k-1)}(a) = 0$ and $p^{(k)}(a) \neq 0.$

69. $\dfrac{dy}{dt} = \dfrac{dy}{dx} \cdot \dfrac{dx}{dt} = (3x^2 - 3)(4t - 1)$

At $t = 2,$ $x(2) = 8$ and $\dfrac{dy}{dt} = [3(8)^2 - 3][4(2) - 1] = 1323.$

71. $V = \tfrac{4}{3}\pi r^3$ and $\dfrac{dr}{dt} = 2$ cm/sec. By the chain rule, $\dfrac{dV}{dt} = \dfrac{dV}{dr}\dfrac{dr}{dt} = 4\pi r^2 \dfrac{dr}{dt} = 8\pi r^2.$

At the instant the radius is 10 centimeters, the volume is increasing at the rate

$$\dfrac{dV}{dt} = 8\pi(10)^2 = 800\pi \ \text{cm}^3/\text{sec}.$$

73. $KE = \tfrac{1}{2}mv^2;$ $\quad \dfrac{d(KE)}{dt} = \dfrac{d(KE)}{dv} \cdot \dfrac{dv}{dt} = mv\dfrac{dv}{dt}.$

SECTION 3.6

1. $\dfrac{dy}{dx} = -3\sin x - 4\sec x \tan x$

3. $\dfrac{dy}{dx} = 3x^2 \csc x - x^3 \csc x \cot x$

5. $\dfrac{dy}{dt} = -2\cos t \sin t$

7. $\dfrac{dy}{du} = 4\sin^3 \sqrt{u} \, \dfrac{d}{du}(\sin \sqrt{u}) = 4\sin^3 \sqrt{u} \, \cos \sqrt{u} \, \dfrac{d}{du}(\sqrt{u}) = 2u^{-1/2}\sin^3 \sqrt{u} \, \cos \sqrt{u}$

9. $\dfrac{dy}{dx} = \sec^2 x^2 \dfrac{d}{dx}(x^2) = 2x\sec^2 x^2$ 11. $\dfrac{dy}{dx} = 4[x + \cot \pi x]^3 [1 - \pi \csc^2 \pi x]$

13. $\dfrac{dy}{dx} = \cos x,\ \dfrac{d^2 y}{dx^2} = -\sin x$

15. $\dfrac{dy}{dx} = \dfrac{(1 + \sin x)(-\sin x) - \cos x\,(\cos x)}{(1 + \sin x)^2} = \dfrac{-\sin x - (\sin^2 x + \cos^2 x)}{(1 + \sin x)^2} = -(1 + \sin x)^{-1}$

$\dfrac{d^2 y}{dx^2} = (1 + \sin x)^{-2} \dfrac{d}{dx}(1 + \sin x) = \cos x\,(1 + \sin x)^{-2}$

17. $\dfrac{dy}{du} = 3\cos^2 2u\,\dfrac{d}{du}(\cos 2u) = -6\cos^2 2u\sin 2u$

$\dfrac{d^2 y}{du^2} = -6[\cos^2 2u\,\dfrac{d}{du}(\sin 2u) + \sin 2u\,\dfrac{d}{du}(\cos^2 2u)]$

$\qquad = -6[2\cos^3 2u + \sin 2u\,(-4\cos 2u\sin 2u)] = 12\cos 2u\,[2\sin^2 2u - \cos^2 2u]$

19. $\dfrac{dy}{dt} = 2\sec^2 2t,\quad \dfrac{d^2 y}{dt^2} = 4\sec 2t\,\dfrac{d}{dt}(\sec 2t) = 8\sec^2 2t\,\tan 2t$

21. $\dfrac{dy}{dx} = x^2(3\cos 3x) + 2x\sin 3x$

$\dfrac{d^2 y}{dx^2} = [x^2(-9\sin 3x) + 2x(3\cos 3x)] + [2x(3\cos 3x) + 2(\sin 3x)]$

$\qquad = (2 - 9x^2)\sin 3x + 12x\cos 3x$

23. $y = \sin^2 x + \cos^2 x = 1\ $ so $\ \dfrac{dy}{dx} = \dfrac{d^2 y}{dx^2} = 0$

25. $\dfrac{d^4}{dx^4}(\sin x) = \dfrac{d^3}{dx^3}(\cos x) = \dfrac{d^2}{dx^2}(-\sin x) = \dfrac{d}{dx}(-\cos x) = \sin x$

27. $\dfrac{d}{dt}\left[t^2\dfrac{d^2}{dt^2}(t\cos 3t)\right] = \dfrac{d}{dt}\left[t^2\dfrac{d}{dt}(\cos 3t - 3t\sin 3t)\right]$

$\qquad\qquad = \dfrac{d}{dt}[t^2(-3\sin 3t - 3\sin 3t - 9t\cos 3t)]$

$\qquad\qquad = \dfrac{d}{dt}[-6t^2\sin 3t - 9t^3\cos 3t]$

$\qquad\qquad = (-18t^2\cos 3t - 12t\sin 3t) + (27t^3\sin 3t - 27t^2\cos 3t)$

$\qquad\qquad = (27t^3 - 12t)\sin 3t - 45t^2\cos 3t$

29. $\dfrac{d}{dx}[f(\sin 3x)] = f'(\sin 3x)\dfrac{d}{dx}(\sin 3x) = 3\cos 3x\,f'(\sin 3x)$

31. $\dfrac{dy}{dx} = \cos x;\ $ slope of tangent at $(0,0)$ is 1, an equation for tangent is $y = x$.

33. $\dfrac{dy}{dx} = -\csc^2 x;$ slope of tangent at $\left(\dfrac{\pi}{6}, \sqrt{3}\right)$ is -4, an equation for

tangent is $y - \sqrt{3} = -4\left(x - \dfrac{\pi}{6}\right)$.

35. $\dfrac{dy}{dx} = \sec x \tan x,$ slope of tangent at $\left(\dfrac{\pi}{4}, \sqrt{2}\right)$ is $\sqrt{2}$, an equation for

tangent is $y - \sqrt{2} = \sqrt{2}\left(x - \dfrac{\pi}{4}\right)$.

37. $\dfrac{dy}{dx} = -\sin x;$ $x = \pi$

39. $\dfrac{dy}{dx} = \cos x - \sqrt{3}\sin x;$ $\dfrac{dy}{dx} = 0$ gives $\tan x = \dfrac{1}{\sqrt{3}};$ $x = \dfrac{\pi}{6}, \dfrac{7\pi}{6}$

41. $\dfrac{dy}{dx} = 2\sin x \cos x = \sin 2x;$ $x = \dfrac{\pi}{2}, \pi, \dfrac{3\pi}{2}$

43. $\dfrac{dy}{dx} = \sec^2 x - 2;$ $\dfrac{dy}{dx} = 0$ gives $\sec x = \pm\sqrt{2};$ $x = \dfrac{\pi}{4}, \dfrac{3\pi}{4}, \dfrac{5\pi}{4}, \dfrac{7\pi}{4}$

45. $\dfrac{dy}{dx} = 2\sec x \tan x + \sec^2 x;$ since $\sec x$ is never zero, $\dfrac{dy}{dx} = 0$ gives

$2\tan x + \sec x = 0$ so that $\sin x = -1/2;$ $x = \dfrac{7\pi}{6}, \dfrac{11\pi}{6}$

47. We want $v(t) > 0$ and $a(t) > 0$.

$v(t) = 3\cos 3t$ sign of $v(t)$:

$a(t) = -9\sin 3t$ sign of $a(t)$:

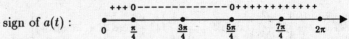

Thus, $\pi < t < \dfrac{2\pi}{3}, \quad \dfrac{7\pi}{6} < t < \dfrac{4\pi}{3}, \quad \dfrac{11\pi}{6} < t < 2\pi.$

49. We want $v(t) > 0$ and $a(t) > 0$.

$v(t) = \cos t + \sin t$ sign of $v(t)$:

$a(t) = -\sin t + \cos t$ sign of $a(t)$:

Thus, $0 < t < \dfrac{\pi}{4}$ and $\dfrac{7\pi}{4} < t < 2\pi.$

51. We want $v(t) > 0$ and $a(t) > 0$.

$v(t) = 1 - 2\sin t$ sign of $v(t)$:

$a(t) = -2\cos t$ sign of $a(t)$:

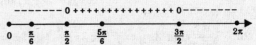

Thus, $\dfrac{5\pi}{6} < t < \dfrac{3\pi}{2}.$

53. (a) $\dfrac{dy}{dt} = \dfrac{dy}{du}\dfrac{du}{dx}\dfrac{dx}{dt} = (2u)(\sec x \tan x)\pi = 2\pi \sec^2 \pi t \tan \pi t$

(b) $y = \sec^2 \pi t - 1,\quad \dfrac{dy}{dt} = 2\sec \pi t\,(\sec \pi t \tan \pi t)\pi = 2\pi \sec^2 \pi t \tan \pi t$

55. (a) $\dfrac{dy}{dt} = \dfrac{dy}{du}\dfrac{du}{dx}\dfrac{dx}{dt} = 4\left[\dfrac{1}{2}(1-u)\right]^3\left(-\dfrac{1}{2}\right)(-\sin x)(2) = 4\left[\dfrac{1}{2}(1-\cos 2t)\right]^3 \sin 2t$

$$= 4\sin^6 t\,(2\sin t \cos t) = 8\sin^7 t \cos t$$

(b) $y = \left[\dfrac{1}{2}(1-\cos 2t)\right]^4 = \sin^8 t,\quad \dfrac{dy}{dt} = 8\sin^7 t \cos t$

57. $\dfrac{d^n}{dx^n}(\cos x) = \left\{\begin{array}{l}(-1)^{(n+1)/2}\sin x,\ n\text{ odd}\\(-1)^{n/2}\cos x,\ n\text{ even}\end{array}\right]$

59. (a) $f'(x) = \sin(1/x) + x\,\cos(1/x)(-1/x^2)$

$$= \sin(1/x) - (1/x)\cos(1/x)$$

$g'(x) = 2x\,\sin(1/x) + x^2\cos(1/x)(-1/x^2)$

$$= 2x\,\sin(1/x) - \cos(1/x)$$

(b) $\displaystyle\lim_{x\to 0} g'(x) = \lim_{x\to 0}[2x\,\sin(1/x) - \cos(1/x)] = -\lim_{x\to 0}\cos(1/x)$ does not exist

61. Let $y(t) = A\sin\omega t + B\cos\omega t$. Then

$$y'(t) = \omega A\cos\omega t - \omega B\sin\omega t\quad\text{and}\quad y''(t) = -\omega^2 A\sin\omega t - \omega^2 B\cos\omega t$$

Thus,

$$\frac{d^2 y}{dt^2} + \omega^2 y = 0.$$

63. $A = \dfrac{1}{2}c^2\sin x;\quad \dfrac{dA}{dx} = \dfrac{1}{2}c^2\cos x$

65.

$\cos\theta = \dfrac{x}{13}$ and $\dfrac{dx}{dt} = 2$ (ft/sec)

$-\sin\theta\,\dfrac{d\theta}{dt} = \dfrac{1}{13}\dfrac{dx}{dt}$

$\dfrac{d\theta}{dt} = \dfrac{-2}{13}\sin\theta$

When $x = 5$, $\sin\theta = 12/13$ and $\dfrac{d\theta}{dt} = -\dfrac{1}{6}$ (rad/sec).

67.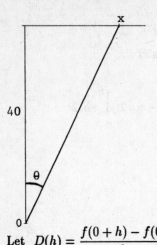

$\tan \theta = \dfrac{x}{40}$ (see figure), and $\dfrac{dx}{dt} = 4$

$\sec^2 \theta \, \dfrac{d\theta}{dt} = \dfrac{1}{40} \dfrac{dx}{dt} = \dfrac{1}{10}$

$\dfrac{d\theta}{dt} = \dfrac{1}{10} \cos^2 \theta$

When $t = 6$, $x = 24$ and $\cos^2 \theta = \dfrac{25}{34}$

Thus, $\dfrac{d\theta}{dt} = \dfrac{25}{340} \cong 0.074$ (rad/sec).

69. Let $D(h) = \dfrac{f(0+h) - f(0)}{h} = \dfrac{\cos^2 h - 1}{h}$. Then

$D(0.1) \cong -0.0005$ $\qquad$ $D(0.01) \cong 0$ $\qquad$ $D(0.001) \cong 0$

$D(-0.1) \cong 0.0005$ $\qquad$ $D(-0.01) \cong 0$ $\qquad$ $D(0.001) \cong 0$

By the chain rule, $f'(x) = -2x \sin x^2$, and $f'(0) = 0$.

71. (a)

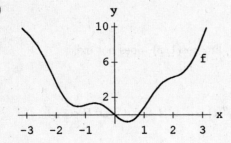

(b) $f(x) = 0$ at $x = 0$ and $x \cong 0.81$

(c) $f'(x) = 0$ at $x \cong -1.25$, $x \cong -0.68$, and $x \cong 0.43$

73.

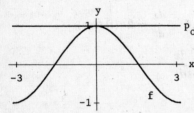 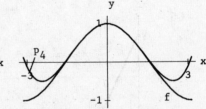

1. $$x^2 + y^2 = 4$$ **3.** $$4x^2 + 9y^2 = 36$$

$$2x + 2y\frac{dy}{dx} = 0$$ $$8x + 18y\frac{dy}{dx} = 0$$

$$\frac{dy}{dx} = \frac{-x}{y}$$ $$\frac{dy}{dx} = \frac{-4x}{9y}$$

5. $$x^4 + 4x^3y + y^4 = 1$$

$$4x^3 + 12x^2y + 4x^3\frac{dy}{dx} + 4y^3\frac{dy}{dx} = 0$$

$$\frac{dy}{dx} = -\frac{x^3 + 3x^2y}{x^3 + y^3}$$

7. $$(x - y)^2 - y = 0$$

$$2(x - y)\left(1 - \frac{dy}{dx}\right) - \frac{dy}{dx} = 0$$

$$\frac{dy}{dx} = \frac{2(x - y)}{2(x - y) + 1}$$

9. $$\sin(x + y) = xy$$

$$\cos(x + y)\left(1 + \frac{dy}{dx}\right) = x\frac{dy}{dx} + y$$

$$\frac{dy}{dx} = \frac{y - \cos(x + y)}{\cos(x + y) - x}$$

11. $$y^2 + 2xy = 16$$

$$2y\frac{dy}{dx} + 2x\frac{dy}{dx} + 2y = 0$$

$$(x + y)\frac{dy}{dx} + y = 0.$$

Differentiating a second time, we have

$$(x + y)\frac{d^2y}{dx^2} + \frac{dy}{dx}\left(2 + \frac{dy}{dx}\right) = 0.$$

Substituting $\dfrac{dy}{dx} = \dfrac{-y}{x + y}$,   we have

$$(x + y)\frac{d^2y}{dx^2} - \frac{y}{(x + y)}\left(\frac{2x + y}{x + y}\right) = 0, \quad \frac{d^2y}{dx^2} = \frac{2xy + y^2}{(x + y)^3} = \frac{16}{(x + y)^3}.$$

13. $$y^2 + xy - x^2 = 9$$

$$2y\frac{dy}{dx} + x\frac{dy}{dx} + y - 2x = 0.$$

Differentiating a second time, we have

$$\left[2\left(\frac{dy}{dx}\right)^2 + 2y\frac{d^2y}{dx^2}\right] + \left[x\frac{d^2y}{dx^2} + \frac{dy}{dx}\right] + \frac{dy}{dx} - 2 = 0$$

$$(2y + x)\frac{d^2y}{dx^2} + 2\left[\left(\frac{dy}{dx}\right)^2 + \frac{dy}{dx} - 1\right] = 0.$$

Substituting $\quad \dfrac{dy}{dx} = \dfrac{2x - y}{2y + x}, \quad$ we have

$$(2y + x)\frac{d^2y}{dx^2} + 2\left[\frac{(2x - y)^2 + (2x - y)(2y + x) - (2y + x)^2}{(2y + x)^2}\right] = 0$$

$$\frac{d^2y}{dx^2} = \frac{10(y^2 + xy - x^2)}{(2y + x)^3} = \frac{90}{(2y + x)^3}.$$

15.
$$4\tan y = x^3$$

$$4\sec^2 y\,\frac{dy}{dx} = 3x^2$$

$$\frac{dy}{dx} = \frac{3}{4}x^2\cos^2 y$$

$$\frac{d^2y}{dx^2} = \frac{3}{2}x\cos^2 y + \frac{3}{4}x^2\left(2\cos y(-\sin y)\frac{dy}{dx}\right)$$

$$= \frac{3}{2}x\cos^2 y - \frac{9}{8}x^4\sin y\cos^3 y$$

17. $x^2 - 4y^2 = 9,\qquad 2x - 8y\dfrac{dy}{dx} = 0.$

At $(5, 2)$, we get $\dfrac{dy}{dx} = \dfrac{5}{8}.$ Then,

$$2 - 8\left[y\frac{d^2y}{dx^2} + \left(\frac{dy}{dx}\right)^2\right] = 0.$$

At $(5, 2)$ we get

$$2 - 8\left[2\frac{d^2y}{dx^2} + \frac{25}{64}\right] = 0 \quad \text{so that} \quad \frac{d^2y}{dx^2} = -\frac{9}{128}.$$

19. $\cos(x + 2y) = 0 \qquad -\sin(x + 2y)\left(1 + 2\dfrac{dy}{dx}\right) = 0.$

At $(\pi/6, \pi/6)$,   we get $\dfrac{dy}{dx} = -1/2.$ Then,

$$-\cos(x + 2y)\left(1 + 2\frac{dy}{dx}\right)^2 - \sin(x + 2y)\left(2\frac{d^2y}{dx^2}\right) = 0.$$

At $(\pi/6, \pi/6)$,   we get

$$-\cos\frac{\pi}{2}(0)^2 - \sin\frac{\pi}{2}\left(2\frac{d^2y}{dx^2}\right) = 0 \quad \text{so that} \quad \frac{d^2y}{dx^2} = 0.$$

21.
$$2x + 3y = 5$$
$$2 + 3\frac{dy}{dx} = 0$$
slope of tangent at $(-2, 3)$: $-2/3$

tangent: $y - 3 = -\frac{2}{3}(x + 2)$

normal: $y - 3 = \frac{3}{2}(x + 2)$

23.
$$x^2 + xy + 2y^2 = 28$$
$$2x + x\frac{dy}{dx} + y + 4y\frac{dy}{dx} = 0$$
slope of tangent at $(-2, -3)$: $-\frac{1}{2}$

tangent: $y + 3 = -\frac{1}{2}(x + 2)$

normal: $y + 3 = 2(x + 2)$

25.
$$x = \cos y$$
$$1 = -\sin y \frac{dy}{dx}$$
$$\frac{dy}{dx} = -\frac{1}{\sin y}$$

slope of tangent at $\left(\frac{1}{2}, \frac{\pi}{3}\right)$: $\frac{-2}{\sqrt{3}}$

tangent: $y - \frac{\pi}{3} = -\frac{2}{\sqrt{3}}\left(x - \frac{1}{2}\right)$

normal: $y - \frac{\pi}{3} = \frac{\sqrt{3}}{2}\left(x - \frac{1}{2}\right)$

27. $\frac{dy}{dx} = \frac{1}{2}(x^3 + 1)^{-1/2}\frac{d}{dx}(x^3 + 1) = \frac{3}{2}x^2(x^3 + 1)^{-1/2}$

29. $\frac{dy}{dx} = x\left(\frac{1}{2}(x^2 + 1)^{-1/2}(2x)\right) + (x^2 + 1)^{1/2} = (1 + 2x^2)(x^2 + 1)^{-1/2}$

31. $\frac{dy}{dx} = \frac{1}{4}(2x^2 + 1)^{-3/4}\frac{d}{dx}(2x^2 + 1) = x(2x^2 + 1)^{-3/4}$

33. $\frac{dy}{dx} = \sqrt{2 - x^2}\left[\frac{-x}{\sqrt{3 - x^2}}\right] + \sqrt{3 - x^2}\left[\frac{-x}{\sqrt{2 - x^2}}\right] = \frac{x(2x^2 - 5)}{\sqrt{2 - x^2}\sqrt{3 - x^2}}$

35. $\frac{d}{dx}\left(\sqrt{x} + \frac{1}{\sqrt{x}}\right) = \frac{d}{dx}(x^{1/2} + x^{-1/2}) = \frac{1}{2}x^{-1/2} - \frac{1}{2}x^{-3/2} = \frac{1}{2}x^{-3/2}(x - 1)$

37. $\frac{d}{dx}\left(\frac{x}{\sqrt{x^2 + 1}}\right) = \frac{d}{dx}\left(x(x^2 + 1)^{-1/2}\right)$
$$= x\left(-\frac{1}{2}(x^2 + 1)^{-3/2}(2x)\right) + (x^2 + 1)^{-1/2} = (x^2 + 1)^{-3/2}$$

39. $\frac{d}{dx}(x^{1/3} + x^{-1/3}) = \frac{1}{3}x^{-2/3} - \frac{1}{3}x^{-4/3} = \frac{1}{3}x^{-4/3}(x^{2/3} - 1)$

41. (a)   (b)   (c)

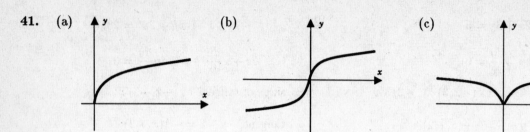

43. $y = (a + bx)^{1/3}$; $\dfrac{dy}{dx} = \dfrac{b}{3}(a + bx)^{-2/3}$; $\dfrac{d^2y}{dx^2} = \dfrac{-2b^2}{9}(a + bx)^{-5/3}$

45. $y = \sqrt{x}\,\tan\sqrt{x}$

$$\frac{dy}{dx} = \frac{1}{2\sqrt{x}}\tan\sqrt{x} + \sqrt{x}\,\sec^2\sqrt{x}\left(\frac{1}{2\sqrt{x}}\right) = \frac{1}{2\sqrt{x}}\tan\sqrt{x} + \frac{1}{2}\sec^2\sqrt{x}$$

$$\frac{d^2y}{dx^2} = \frac{2\sqrt{x}\,\sec^2\sqrt{x}\,(1/2\sqrt{x}) - \tan\sqrt{x}(1/\sqrt{x})}{4x} + \sec\sqrt{x}\,\sec\sqrt{x}\,\tan\sqrt{x}(1/2\sqrt{x})$$

$$= \frac{\sqrt{x}\,\sec^2\sqrt{x} - \tan\sqrt{x} + 2x\,\sec^2\sqrt{x}\,\tan\sqrt{x}}{4x\sqrt{x}}$$

47. Differentiation of $x^2 + y^2 = r^2$ gives $2x + 2y\dfrac{dy}{dx} = 0$ so that the slope of the normal line is

$$\frac{-1}{dy/dx} = \frac{y}{x} \quad (x \neq 0).$$

Let (x_0, y_0) be a point on the circle. Clearly, if $x_0 = 0$, the normal line, $x = 0$, passes through the origin. If $x_0 \neq 0$, the normal line is

$$y - y_0 = \frac{y_0}{x_0}(x - x_0), \quad \text{which simplifies to} \quad y = \frac{y_0}{x_0}x,$$

a line through the origin.

49. For the parabola $y^2 = 2px + p^2$, we have $2y\dfrac{dy}{dx} = 2p$ and the slope of a tangent is given by $m_1 = p/y$.

For the parabola $y^2 = p^2 - 2px$, we obtain $m_2 = -p/y$ as the slope of a tangent. The parabolas intersect at the points $(0, \pm p)$. At each of these points $m_1 m_2 = -1$; the parabolas intersect at right angles.

51. For $y = x^2$ we have $m_1\dfrac{dy}{dx} = 2x$; for $x = y^3$ we have $3y^2\dfrac{dy}{dx} = 1$ or $m_2 = \dfrac{dy}{dx} = 1/3y^2$.

At $(1, 1)$, $m_1 = 2$, $m_2 = 1/3$ and

$$\tan\alpha = \left|\frac{m_1 - m_2}{1 - m_1 m_2}\right| = \left|\frac{2 - (1/3)}{1 + 2(1/3)}\right| = 1 \;\Rightarrow\; \alpha = \frac{\pi}{4}$$

At $(0, 0)$, $m_1 = 0$ and m_2 is undefined. Thus $\alpha = \pi/2$.

53. The hyperbola and the ellipse intersect at the points $(\pm 3, \pm 2)$. For the hyperbola, $\dfrac{dy}{dx} = \dfrac{x}{y}$ and for

the ellipse $\dfrac{dy}{dx} = -\dfrac{4x}{9y}$. The product of these slopes is $-\dfrac{4x^2}{9y^2}$. This product is -1 at each of the

points of intersection. Therefore the hyperbola and ellipse are orthogonal.

55. For the circles, $\dfrac{dy}{dx} = -\dfrac{x}{y}$, $y \neq 0$, and for the straight lines, $\dfrac{dy}{dx} = m = \dfrac{y}{x}$, $x \neq 0$. Since the product

of the slopes is -1, it follows that the two families are orthogonal trajectories.

57. The line $x + 2y + 3 = 0$ has slope $m = -1/2$. Thus, a line perpendicular to this line will have slope

2. A tangent line to the ellipse $4x^2 + y^2 = 72$ has slope $m = \dfrac{dy}{dx} = -\dfrac{4x}{y}$. Setting $-\dfrac{4x}{y} = 2$ gives

$y = -2x$. Substituting into the equation for the ellipse, we have

$$4x^2 + 4x^2 = 72 \quad \Rightarrow \quad 8x^2 = 72 \quad \Rightarrow \quad x = \pm 3$$

It now follows that $y = \mp 6$ and the equations of the tangents are:

at $(3, -6)$: $y + 6 = 2(x - 3)$ or $y = 2x - 12$;

at $(-3, 6)$: $y - 6 = 2(x + 3)$ or $y = 2x + 12$.

59. Differentiate the equation $(x^2 + y^2)^2 = x^2 - y^2$ implicitly with respect to x :
$$2(x^2 + y^2)\left(2x + 2y\,\dfrac{dy}{dx}\right) = 2x - 2y\,\dfrac{dy}{dx}$$
Now set $dy/dx = 0$. This gives
$$2x(x^2 + y^2) = x$$
$$x^2 + y^2 = \dfrac{1}{2} \quad (x \neq 0)$$

Substituting this result into the original equation, we get
$$x^2 - y^2 = \dfrac{1}{4}$$

Now
$$\begin{array}{c} x^2 + y^2 = 1/2 \\ x^2 - y^2 = 1/4 \end{array} \Rightarrow \quad x = \pm\dfrac{\sqrt{6}}{4}, \quad y = \pm\dfrac{\sqrt{2}}{4}$$

Thus, the points on the curve at which the tangent line is horizontal are:

$(\sqrt{6}/4, \sqrt{2}/4)$, $(\sqrt{6}/4, -\sqrt{2}/4)$, $(-\sqrt{6}/4, \sqrt{2}/4)$, $(-\sqrt{6}/4, -\sqrt{2}/4)$.

61. Differentiate the equation $x^{1/2} + y^{1/2} = c^{1/2}$ implicitly with respect to x :
$$\dfrac{1}{2}\,x^{-1/2} + \dfrac{1}{2}\,y^{-1/2}\,\dfrac{dy}{dx} = 0 \quad \text{which implies} \quad \dfrac{dy}{dx} = -\left(\dfrac{y}{x}\right)^{1/2}$$
An equation for the tangent line to the graph at the point (x_0, y_0) is
$$y - y_0 = -\left(\dfrac{y_0}{x_0}\right)^{1/2}(x - x_0)$$

The x- and y-intercepts of this line are

$$a = (x_0 y_0)^{1/2} + x_0 \quad \text{and} \quad b = (x_0 y_0)^{1/2} + y_0 \quad \text{respectively.}$$

Now

$$a + b = 2(x_0 y_0)^{1/2} + x_0 + y_0 = \left(x_0^{1/2} + y_0^{1/2}\right)^2 = c.$$

63.   (a)

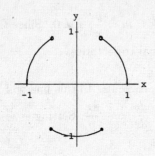

(b) $2x + 2y\dfrac{dy}{dx} = 0 \quad \text{and} \quad \dfrac{dy}{dx} = -\dfrac{x}{y}$

At $(-\sqrt{3}/2, 1/2),\quad \dfrac{dy}{dx} = \sqrt{3}.$

At $(\sqrt{3}/2, 1/2),\quad \dfrac{dy}{dx} = -\sqrt{3}.$

At $(0, -1),\quad \dfrac{dy}{dx} = 0.$

(c) $y = -\sqrt{1 - x^2} \quad \text{for} \quad -\tfrac{1}{2} \le x \le \tfrac{1}{2}$

$$y'_+(-1/2) = \lim_{h \to 0^+} \frac{y(-\tfrac{1}{2} + h) - y(-\tfrac{1}{2})}{h} = \lim_{h \to 0^+} \frac{-\sqrt{1 - (-\tfrac{1}{2} + h)^2} + \frac{\sqrt{3}}{2}}{h}$$

$$= \lim_{h \to 0^+} \frac{-\sqrt{3 + 4h - 4h^2} + \sqrt{3}}{2h} = -\frac{1}{\sqrt{3}}$$

$$y'_-(1/2) = \lim_{h \to 0^-} \frac{y(\tfrac{1}{2} + h) - y(\tfrac{1}{2})}{h} = \lim_{h \to 0^-} \frac{-\sqrt{1 - (\tfrac{1}{2} + h)^2} - \frac{\sqrt{3}}{2}}{h}$$

$$= \lim_{h \to 0^-} \frac{-\sqrt{3 - 4h - 4h^2} + \sqrt{3}}{2h} = \frac{1}{\sqrt{3}}$$

65.   By numerical work, $f'(16) \cong 0.375;$ from (3.7.1)

$$f'(x) = \frac{3}{4} x^{-1/4}, \quad \text{and} \quad f'(16) = \tfrac{3}{8} = 0.375.$$

67. $x = t, \quad y = \sqrt{4 - t^2}$ 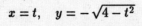 $x = t, \quad y = -\sqrt{4 - t^2}$

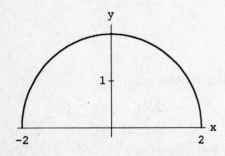   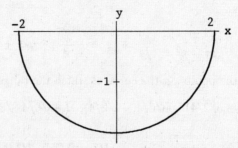

69.   (a) The graph of $x^4 = x^2 - y^2$ is:

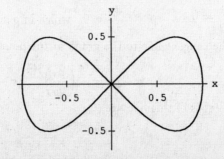

(b) Differentiate the equation $x^4 = x^2 - y^2$ implicitly with respect to x :

$$4x^3 = 2x - 2y\frac{dy}{dx}$$

Now set $dy/dx = 0$. This gives

$$4x^3 = 2x \quad \text{which implies} \quad x = \pm\frac{\sqrt{2}}{2}$$

SECTION 3.8

1. $x + 2y = 2, \quad \dfrac{dx}{dt} + 2\dfrac{dy}{dt} = 0$

    (a) If $\dfrac{dx}{dt} = 4$, then $\dfrac{dy}{dt} = -2$ units/sec.   (b)   If $\dfrac{dy}{dt} = -2$, then $\dfrac{dx}{dt} = 4$ units/sec.

3. $y^2 = 4(x+2), \quad 2y\dfrac{dy}{dt} = 4\dfrac{dx}{dt} \quad$ and $\quad \dfrac{dx}{dt} = \frac{1}{2}\, y\dfrac{dy}{dt}$

    At the point $(7,6)$, $\quad \dfrac{dy}{dt} = 3$. Therefore $\quad \dfrac{dx}{dt} = \frac{1}{2} \cdot 6 \cdot 3 = 9$ units/sec.

5. Let $s = \sqrt{x^2 + y^2}$ denote the distance to the origin at time t. Since $x = 4\cos t$ and $y = 2\sin t$, we have

    $$s(t) = \sqrt{16\cos^2 t + 4\sin^2 t} = \sqrt{12\cos^2 t + 4}$$

    $$\frac{ds}{dt} = \frac{1}{2}\left(12\cos^2 t + 4\right)^{-1/2}(-24\cos t \sin t)$$

    $$= \frac{-12\cos t \sin t}{\sqrt{12\cos^2 t + 4}}$$

    At $t = \pi/4$, $\quad \dfrac{ds}{dt} = \dfrac{-12\cos(\pi/4)\sin(\pi/4)}{\sqrt{12\cos^2(\pi/4) + 4}} = -\frac{3}{5}\sqrt{10}.$

7. 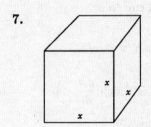 Find $\quad \dfrac{dx}{dt}$ and $\dfrac{dS}{dt} \quad$ when $\quad V = 27\text{m}^3$

    given that $\quad \dfrac{dV}{dt} = -2\text{m}^3/\text{min}.$

    (*) $V = x^3, \quad S = 6x^2$

Differentiation of equations (*) gives

$$\frac{dV}{dt} = 3x^2\frac{dx}{dt} \quad \text{and} \quad \frac{dS}{dt} = 12x\frac{dx}{dt}.$$

When $V = 27$, $x = 3$. Substituting $x = 3$ and $dV/dt = -2$, we get

$$-2 = 27\frac{dx}{dt} \quad \text{so that} \quad \frac{dx}{dt} = -2/27 \quad \text{and} \quad \frac{dS}{dt} = 12(3)\left(\frac{-2}{27}\right) = -8/3.$$

The rate of change of an edge is $-2/27$ m/min;   the rate of change of the surface area is $-8/3$ m^2/min.

9.

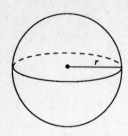

Find $\dfrac{dr}{dt}$ and $\dfrac{dS}{dt}$ when $r = 10$ ft

given that $\dfrac{dV}{dt} = 8$ ft^3/min.

(∗) $V = \frac{4}{3}\pi r^3$, $S = 4\pi r^2$

Differentiation of equations (∗) with respect to t gives

$$\frac{dV}{dt} = 4\pi r^2 \frac{dr}{dt} \quad \text{and} \quad \frac{dS}{dt} = 8\pi r \frac{dr}{dt}.$$

Substituting $r = 10$ and $dV/dt = 8$,   we get

$$8 = 4\pi(10)^2 \frac{dr}{dt} \quad \text{so that} \quad \frac{dr}{dt} = \frac{1}{50\pi} \quad \text{and} \quad \frac{dS}{dt} = 8\pi(10)\frac{1}{50\pi} = \frac{8}{5}.$$

The radius is increasing $\dfrac{1}{50\pi}$ ft/min; the surface area is increasing $\dfrac{8}{5}$ ft^2/min.

11.

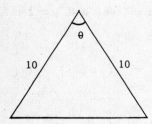

(a) $A = \frac{1}{2} \cdot 10 \cdot 10 \cdot \sin\theta = 50\sin\theta$ (see the figure)

(b) $\dfrac{d\theta}{dt} = 10° = \frac{10}{360}(2\pi) = \dfrac{\pi}{18}$ radians

$\dfrac{dA}{dt} = 50\cos\theta\,\dfrac{d\theta}{dt}$

At the instant $\theta = 60° = \pi/3$ radians, $\dfrac{dA}{dt} = 50\cos(\pi/3)\dfrac{\pi}{18} \cong 4.36$ cm^2/min

(c) $\dfrac{dA}{d\theta} = 50\cos\theta = 0 \;\Rightarrow\; \theta = \pi/2;$ the triangle has maximum area when $\theta = \pi/2$.

13.

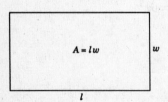

We will find the values of l for which $\dfrac{dA}{dt} < 0$

given that $\dfrac{dl}{dt} = 1$ cm/sec and

$P = 2(l + w) = 24.$

We combine $A = lw$ and $l + w = 12$ to write $A = 12l - l^2$. Differentiating with respect to t, we have

$$\frac{dA}{dt} = 12\frac{dl}{dt} - 2l\frac{dl}{dt} = 2(6 - l)\frac{dl}{dt}.$$

Since $dl/dt = 1$, $\dfrac{dA}{dt} < 0$ for $l > 6$. The area of the rectangle starts to decrease when the length is 6 cm.

15.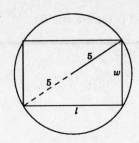

Find $\dfrac{dA}{dt}$ when $l = 6$ in.

given that $\dfrac{dl}{dt} = -2$ in./sec.

By the Pythagorean theorem

$$l^2 + w^2 = 100.$$

Also, $A = lw$. Thus, $A = l\sqrt{100 - l^2}$.   Differentiation with respect to t gives

$$\frac{dA}{dt} = l\left(\frac{-l}{\sqrt{100 - l^2}}\right)\frac{dl}{dt} + \sqrt{100 - l^2}\,\frac{dl}{dt}.$$

Substituting $l = 6$ and $dl/dt = -2$, we get

$$\frac{dA}{dt} = 6\left(\frac{-6}{8}\right)(-2) + (8)(-2) = -7.$$

The area is decreasing at the rate of 7 in.^{2}/sec.

17.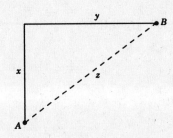

Compare $\dfrac{dy}{dt}$ to $\dfrac{dx}{dt} = -13$ mph

given that $z = 16$ and $\dfrac{dz}{dt} = -17$

when $x = y$.

By the Pythagorean theorem $x^2 + y^2 = z^2$. Thus,

$$2x\frac{dx}{dt} + 2y\frac{dy}{dt} = 2z\frac{dz}{dt}.$$

Since $x = y$ when $z = 16$, we have $x = y = 8\sqrt{2}$ and

$$2(8\sqrt{2})(-13) + 2(8\sqrt{2})\frac{dy}{dt} = 2(16)(-17).$$

Solving for dy/dt, we get

$$-13\sqrt{2} + \sqrt{2}\,\frac{dy}{dt} = -34 \quad \text{or} \quad \frac{dy}{dt} = \frac{1}{\sqrt{2}}(13\sqrt{2} - 34) \cong -11.$$

Thus, boat A wins the race.

19. We want to find dV/dt when $V = 1000$ ft^3 and $P = 5$ lb/in.^{2} given that $dP/dt = -0.05$ lb/in.^{2}/hr.

Differentiating $PV = C$ with respect to t, we get

$$P\frac{dV}{dt} + V\frac{dP}{dt} = 0 \quad \text{so that} \quad 5\frac{dV}{dt} + 1000(-0.05) = 0. \quad \text{Thus,}\ \frac{dV}{dt} = 10.$$

The volume increases at the rate of 10 ft^3/hr.

21.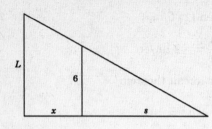

Find $\dfrac{ds}{dt}$ when $x = 3$ ft (and $s = 4$ ft)

given that $\dfrac{dx}{dt} = 400$ ft/min.

By similar triangles

$$\frac{L}{x+s} = \frac{6}{s}.$$

Substitution of $x = 3$ and $s = 4$ gives us $\dfrac{L}{7} = \dfrac{6}{4}$ so that the lamp post is

$L = 10.5$ ft tall. Rewriting

$$\frac{10.5}{x+s} = \frac{6}{s} \quad \text{as} \quad s = \frac{4}{3}x$$

and differentiating with respect to t, we find that

$$\frac{ds}{dt} = \frac{4}{3}\frac{dx}{dt} = \frac{1600}{3}.$$

The shadow lengthens at the rate of 1600/3 ft/min.

23. Let $W(t) = 150 \left(1 + \frac{1}{4000} r\right)^{-2}$. We want to find dW/dt when $r = 400$ given that

$dr/dt = 10$ mi/sec. Differentiating with respect to t, we get

$$\frac{dW}{dt} = -300 \left(1 + \frac{1}{4000} r\right)^{-3} \left(\frac{1}{4000}\right) \frac{dr}{dt}$$

Now set $r = 400$ and $dr/dt = 10$. Then

$$\frac{dW}{dt} = -300 \left(1 + \frac{400}{4000}\right)^{-3} \frac{10}{4000} \cong 0.5634 \text{lbs/sec}$$

25.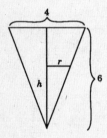

Find $\dfrac{dh}{dt}$ when $h = 3$ in.

given that $\dfrac{dV}{dt} = -\dfrac{1}{2}$ cu in./min.

By similar triangles

$$r = \tfrac{1}{3}h.$$

Thus $V = \frac{1}{3}\pi r^2 h = \frac{1}{27}\pi h^3$. Differentiating with respect to t, we get

$$\frac{dV}{dt} = \frac{1}{9}\pi h^2 \frac{dh}{dt}.$$

When $h = 3$,

$$-\frac{1}{2} = \frac{1}{9}\pi(9)\frac{dh}{dt} \quad \text{and} \quad \frac{dh}{dt} = -\frac{1}{2\pi}.$$

The water level is dropping at the rate of $1/2\pi$ inches per minute.

27.

Find $\dfrac{d\theta}{dt}$ when $x = 4$ ft

given that $\dfrac{dx}{dt} = 2\text{in./min.}$

(*) $\tan \dfrac{\theta}{2} = \dfrac{3}{x}$

Differentiation of (*) with respect to t gives

$$\frac{1}{2}\sec^2\frac{\theta}{2}\frac{d\theta}{dt} = -\frac{3}{x^2}\frac{dx}{dt} \quad \text{or} \quad \frac{d\theta}{dt} = -\frac{6}{x^2}\cos^2\frac{\theta}{2}\frac{dx}{dt}.$$

Note that $dx/dt = 2$ in./min=1/6 ft/min. When $x = 4$, we have $\cos\theta/2 = 4/5$ and thus

$$\frac{d\theta}{dt} = -\frac{6}{16}\left(\frac{4}{5}\right)^2\left(\frac{1}{6}\right) = -\frac{1}{25}.$$

The vertex angle decreases at the rate of 0.04 rad/min.

29.

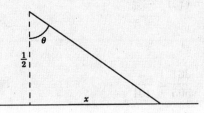

Find $\dfrac{dx}{dt}$ when $x = 1$ mi

given that $\dfrac{d\theta}{dt} = 2\pi$ rad/min.

(*) $\tan\theta = \dfrac{x}{1/2} = 2x$

Differentiation of (*) with respect to t gives

$$\sec^2\theta\,\frac{d\theta}{dt} = 2\frac{dx}{dt}.$$

When $x = 1$, we get $\sec\theta = \sqrt{5}$ and thus $\dfrac{dx}{dt} = 5\pi$. The light is traveling at 5π mi/min.

31.

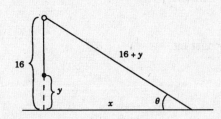

Find $\dfrac{d\theta}{dt}$ when $y = 4$ ft

given that $\dfrac{dx}{dt} = 3$ ft/sec.

$\tan\theta = \dfrac{16}{x}, \quad x^2 + (16)^2 = (16 + y)^2$

Differentiating $\tan\theta = 16/x$ with respect to t, we obtain

$$\sec^2\theta\frac{d\theta}{dt} = \frac{-16}{x^2}\frac{dx}{dt} \quad \text{and thus} \quad \frac{d\theta}{dt} = \frac{-16}{x^2}\cos^2\theta\frac{dx}{dt}.$$

From $x^2 + (16)^2 = (16 + y)^2$ we conclude that $x = 12$, when $y = 4$. Thus

$$\cos\theta = \frac{x}{16+y} = \frac{12}{20} = \frac{3}{5} \quad \text{and} \quad \frac{d\theta}{dt} = \frac{-16}{(12)^2}\left(\frac{3}{5}\right)^2(3) = \frac{-3}{25}.$$

The angle decreases at the rate of 0.12 rad/sec.

33.

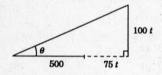

Find $\dfrac{d\theta}{dt}$ when $t = 6$ min.

$$\tan\theta = \frac{100t}{500+75t} = \frac{4t}{20+3t}$$

Differentiation with respect to t gives

$$\sec^2\theta\,\frac{d\theta}{dt} = \frac{(20+3t)4 - 4t(3)}{(20+3t)^2} = \frac{80}{(20+3t)^2}.$$

When $t = 6$

$$\tan\theta = \tfrac{24}{38} = \tfrac{12}{19} \quad \text{and} \quad \sec^2\theta = 1 + \left(\tfrac{12}{19}\right)^2 = \tfrac{505}{361}$$

so that

$$\frac{d\theta}{dt} = \frac{80}{(20+3t)^2}\cdot\frac{1}{\sec^2\theta} = \frac{80}{(38)^2}\cdot\frac{361}{505} = \frac{4}{101}.$$

The angle increases at the rate of 4/101 rad/min.

35.   length of arc $= r\theta$, speed $= \dfrac{d}{dt}[r\theta] = r\dfrac{d\theta}{dt} = r\omega$

37.   We know that $d\theta/dt = \omega$ and, at time t, $\theta = \theta_0$. Therefore $\theta = \omega t + \theta_0$. It follows that

$$x(t) = r\cos(\omega t + \theta_0) \quad \text{and} \quad y(t) = r\sin(\omega t + \theta_0).$$

39.   For the sector $\quad A = \dfrac{1}{2}r^2\theta, \quad \dfrac{dA}{dt} = \dfrac{1}{2}r^2\dfrac{d\theta}{dt} = \dfrac{1}{2}r^2\omega \quad$ is constant.

For triangle T

$$A = \tfrac{1}{2}(2r\sin\tfrac{1}{2}\theta)(r\cos\tfrac{1}{2}\theta)$$

$$= \tfrac{1}{2}r^2(2\sin\tfrac{1}{2}\theta\cos\tfrac{1}{2}\theta) = \tfrac{1}{2}r^2\sin\theta,$$

$$\frac{dA}{dt} = \frac{1}{2}r^2\cos\theta\,\frac{d\theta}{dt} = \frac{1}{2}r^2\omega\cos\theta \quad \text{varies with } \theta.$$

For segment S

$$A = \tfrac{1}{2}r^2\theta - \tfrac{1}{2}r^2\sin\theta = \tfrac{1}{2}r^2(\theta - \sin\theta),$$

$$\frac{dA}{dt} = \frac{1}{2}r^2\left(\frac{d\theta}{dt} - \cos\frac{d\theta}{dt}\right) = \frac{1}{2}r^2\omega(1 - \cos\theta) \quad \text{varies with } \theta.$$

SECTION 3.9

1.
$$\Delta V = (x + h)^3 - x^3$$

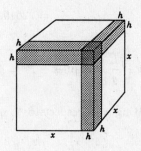

$$= (x^3 + 3x^2h + 3xh^2 + h^3) - x^3$$

$$= 3x^2h + 3xh^2 + h^3,$$

$$dV = 3x^2h,$$

$$\Delta V - dV = 3xh^2 + h^3 \quad \text{(see figure)}$$

3. $f(x) = x^{1/3}, \quad x = 1000, \quad h = 10, \quad f'(x) = \tfrac{1}{3}x^{-2/3}$

$$\sqrt[3]{1010} = f(x + h) \cong f(x) + hf'(x) = \sqrt[3]{1000} + 10\left(\tfrac{1}{3}(1000)^{-2/3}\right) = 10\tfrac{1}{30}$$

5. $f(x) = x^{1/4}, \quad x = 16, \quad h = -1, \quad f'(x) = \tfrac{1}{4}x^{-3/4}$

$$(15)^{1/4} = f(x + h) \cong f(x) + hf'(x) = (16)^{1/4} + (-1)\left(\tfrac{1}{4}(16)^{-3/4}\right) = 1\tfrac{31}{32}$$

7. $f(x) = x^{1/5}, \quad x = 32, \quad h = -2, \quad f'(x) = \tfrac{1}{5}x^{-4/5}$

$$(30)^{1/5} = f(x + h) \cong f(x) + hf'(x) \doteq (32)^{1/5} + (-2)\left(\tfrac{1}{5}(32)^{-4/5}\right) = 1.975$$

9. $f(x) = x^{3/5}, \quad x = 32, \quad h = 1, \quad f'(x) = \tfrac{3}{5}x^{-2/5}$

$$(33)^{3/5} = f(x + h) \cong f(x) + hf'(x) = (32)^{3/5} + (1)\left(\tfrac{3}{5}(32)^{-2/5}\right) = 8.15$$

11. $f(x) = \sin x, \quad x = \dfrac{\pi}{4}, \quad h = \dfrac{\pi}{180}, \quad f'(x) = \cos x$

$$\sin 46° = f(x + h) \cong f(x) + hf'(x) = \sin\frac{\pi}{4} + \frac{\pi}{180}\cos\frac{\pi}{4} = \frac{\sqrt{2}}{2}\left(1 + \frac{\pi}{180}\right) \cong 0.719$$

13. $f(x) = \tan x, \quad x = \dfrac{\pi}{6}, \quad h = \dfrac{-\pi}{90}, \quad f'(x) = \sec^2 x$

$$\tan 28° = f(x + h) \cong f(x) + hf'(x) = \tan\frac{\pi}{6} + \left(\frac{-\pi}{90}\right)\left(\frac{4}{3}\right) = \frac{\sqrt{3}}{3} - \frac{2\pi}{135} \cong 0.531$$

15. $f(2.8) \cong f(3) + (-0.2)f'(3) = 2 + (-0.2)(2) = 1.6$

17. $V(x) = \pi x^2 h; \quad \text{volume} = V(r + t) - V(r) \cong tV'(r) = 2\pi rht$

19. $V(x) = x^3, \quad V'(x) = 3x^2, \quad \Delta V \cong dV = V'(10)h = 300h$

$$|dV| \leq 3 \quad \implies \quad |300h| \leq 3 \quad \implies \quad |h| \leq 0.01, \quad \text{error} \leq 0.01 \text{ feet}$$

21. $V(r) = \frac{2}{3}\pi r^3$ and $dr = 0.01$.

$$V(r + 0.01) - V(r) \cong V'(r)(0.01) = 2\pi r^2 (0.01)$$

$$= 2\pi(600)^2(0.01) \quad (50 \text{ ft} = 600 \text{ in})$$

$$= 22619.5 \text{ in}^3 \quad \text{or} \quad 98 \text{ gallons (approx.)}$$

23. $P = 2\pi\sqrt{\dfrac{L}{g}}$ implies $P^2 = 4\pi^2 \dfrac{L}{g}$

Differentiating with respect to t, we have

$$2P\frac{dP}{dt} = \frac{4\pi^2}{g} \cdot \frac{dL}{dt} = \frac{P^2}{L} \cdot \frac{dL}{dt} \quad \text{since} \quad \frac{P^2}{L} = \frac{4\pi^2}{g}.$$

Thus $\dfrac{dP}{P} = \dfrac{1}{2} \cdot \dfrac{dL}{L}$

25. $L = 3.26$ ft, $P = 2$ sec, and $dL = 0.01$ ft

$$\frac{dP}{P} = \frac{1}{2} \cdot \frac{dL}{L}$$

$$dP = \frac{1}{2} \cdot \frac{dL}{L} \cdot P = \frac{1}{2} \cdot \frac{0.01}{3.26} \cdot 2 \quad dP \cong 0.00307 \text{ sec}$$

27. $A(x) = \dfrac{1}{4}\pi x^2$, $dA = \dfrac{1}{2}\pi x h$, $\dfrac{dA}{A} = 2\dfrac{h}{x}$

$$\frac{dA}{A} \leq 0.01 \quad \Longleftrightarrow \quad 2\frac{h}{x} \leq 0.01 \quad \Longleftrightarrow \quad \frac{h}{x} \leq 0.005 \quad \text{within } \frac{1}{2}\%$$

29.   (a) $x_{n+1} = \dfrac{1}{2}x_n + 12\left(\dfrac{1}{x_n}\right)$ (b) $x_4 \cong 4.89898$

31.   (a) $x_{n+1} = \dfrac{2}{3}x_n + \dfrac{25}{3}\left(\dfrac{1}{x_n}\right)^2$ (b) $x_4 \cong 2.92402$

33.   (a) $x_{n+1} = \dfrac{x_n \sin x_n + \cos x_n}{\sin x_n + 1}$ (b) $x_4 \cong 0.73909$

35.   (a) $x_{n+1} = \dfrac{2x_n \cos x_n - 2\sin x_n}{2\cos x_n - 1}$ (b) $x_4 \cong 1.89549$

37.   Let $f(x) = x^{1/3}$. Then $f'(x) = \frac{1}{3}x^{-2/3}$. The Newton-Raphson method applied to this function gives:

$$x_{n+1} = x_n - \frac{f(x_n)}{f'(x_n)} = x_n - \frac{x_n^{1/3}}{\frac{1}{3}x_n^{-2/3}} = -2x_n.$$

Choose any $x_1 \neq 0$. Then $x_2 = -2x_1$, $x_3 = -2x_2 = 4x_1$, $\cdots$,

$$x_n = -2x_{n-1} = (-1)^{n-1}2^n\, x_1, \cdots.$$

39. (a) Let $f(x) = x^4 - 2x^2 - \frac{17}{16}$. Then $f'(x) = 4x^3 - 4x$. The Newton-Raphson method

applied to this function gives:

$$x_{n+1} = x_n - \frac{x_n^4 - 2x_n^2 - \frac{17}{16}}{4x_n^3 - 4x_n}$$

If $x_1 = \frac{1}{2}$, then $x_2 = -\frac{1}{2}$, $x_3 = \frac{1}{2}$, $\cdots x_n = (-1)^{n-1}\frac{1}{2}$, $\cdots$.

(b) $x_1 = 2$, $x_2 = 1.71094$, $x_3 = 1.58569$, $x_4 = 1.56165$; $f(x_4) = 0.00748$

41. (a) Let $f(x) = x^k - a$. Then $f'(x) = kx^{k-1}$. The Newton-Raphson method applied to

this function gives:

$$x_{n+1} = x_n - \frac{x_n^k - a}{kx_n^{k-1}} = x_n - \frac{1}{k}x_n + \frac{1}{k}\frac{a}{x_n^{k-1}}$$

$$= \frac{1}{k}\left[(k-1)x_n + \frac{a}{x_n^{k-1}}\right]$$

(b) Let $a = 23$, $k = 3$ and $x_1 = 3$. Then

$$x_1 = 3, \quad x_2 = 2.85185, \quad x_3 = 2.84389, \quad x_4 = 2.84382; \quad f(x_4) = -0.00114$$

43. (a) Let $F(x) = \frac{1}{2}\cos x - x$. Then $F(0) = \frac{1}{2} > 0$ and $F(\pi/2) = -\pi/2 < 0$. Thus F has

a zero in $(0, \pi/2)$ which implies that $f(x) = \frac{1}{2}\cos x$ has a fixed point on $[0, \pi/2]$.

(b) $F'(x) = -\frac{1}{2}\sin x - 1$ and

$$x_{n+1} = x_n - \frac{\frac{1}{2}\cos x_n - x_n}{-\frac{1}{2}\sin x_n - 1} = x_n + \frac{\cos x_n - 2x_n}{\sin x_n + 2}$$

$$x_1 = 0, \quad x_2 = 0.5, \quad x_3 = 0.4506, \quad x_4 = 0.4502; \quad f(x_4) = 0.45018$$

45. (a) and (b)

47. $\displaystyle \lim_{h\to 0}\frac{g_1(h)+g_2(h)}{h} = \lim_{h\to 0}\frac{g_1(h)}{h} + \lim_{h\to 0}\frac{g_2(h)}{h} = 0 + 0 = 0$

$$\lim_{h \to 0} \frac{g_1(h)g_2(h)}{h} = \lim_{h \to 0} h \frac{g_1(h)g_2(h)}{h^2} = \left(\lim_{h \to 0} h \right) \left(\lim_{h \to 0} \frac{g_1(h)}{h} \right) \left(\lim_{h \to 0} \frac{g_2(h)}{h} \right) = (0)(0)(0) = 0$$

PROJECTS AND EXPLORATIONS

3.1. (a) The system of equations generated by the specified conditions is:

$$a + b + c + d = 3$$

$$27a + 9b + 3c + d = 7$$

$$6a + 2b = 0$$

$$27\alpha + 9\beta + 3\gamma + \delta = 7$$

$$729\alpha + 81\beta + 9\gamma + \delta = -2$$

$$54\alpha + 2\beta = 0$$

$$27a + 6b + c = 27\alpha + 6\beta + \gamma$$

$$18a + 2b = 18\alpha + 2\beta$$

$a \cong -0.1094$ $\alpha \cong 0.0365$

$b \cong 0.3281$ $\beta \cong -0.9844$

$c \cong 2.1094$ $\gamma \cong 6.0469$

$d \cong 0.6719$ $\delta \cong -3.2656$

(b)

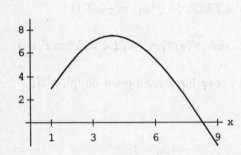

(c) Clearly p and q are continuous on their respective intervals. The conditions $p(3) = q(3)$, $p'(3) = q'(3)$ and $p''(3) = q''(3)$ imply that F, F', and F''' are continuous on $[1, 9]$.

3.3. (a)

$$\left(x^2 + y^2 + 3x \right)^2 = 2(x^2 + y^2)$$

$$2(x^2 + y^2 + 3x) \left(2x + 2y \frac{dy}{dx} + 3 \right) = 4x + 4y \frac{dy}{dx}$$

$$\frac{dy}{dx} = -\frac{2x^3 + 9x^2 + 7x + 2xy^2 + 3y^2}{2y(x^2 + y^2 + 3x - 1)}$$

(d) Your values should indicate that the graph of the equation lies inside the rectangle

    $R:\ -5 \le x \le 1,\ \ -3 \le y \le 3.$

(e) Vertical lines intersect the graph in 0, 2, or 4 points (1 or 3 points at the extremes);

    similarly for horizontal lines.

(h)

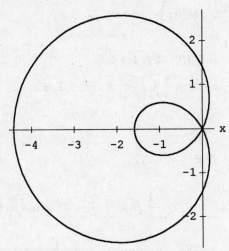

CHAPTER 4

SECTION 4.1

1. f is differentiable on $(0,1)$, continuous on $[0,1]$; and $f(0) = f(1) = 0$.

$$f'(c) = 3c^2 - 1; \quad 3c^2 - 1 = 0 \Longrightarrow c = \frac{\sqrt{3}}{3} \quad \left(-\frac{\sqrt{3}}{3} \notin (0,1)\right)$$

3. f is differentiable on $(0, 2\pi)$, continuous on $[0, 2\pi]$; and $f(0) = f(2\pi) = 0$.

$$f'(c) = 2\cos 2c; \quad 2\cos 2c = 0 \Longrightarrow 2c = \frac{\pi}{2} + n\pi, \quad \text{and} \quad c = \frac{\pi}{4} + \frac{n\pi}{2}, \quad n = 0, \pm 1, \pm 2 \dots$$

Thus, $c = \dfrac{\pi}{4}, \dfrac{3\pi}{4}, \dfrac{5\pi}{4}, \dfrac{7\pi}{4}$

5. $f'(c) = 2c, \quad \dfrac{f(b) - f(a)}{b - a} = \dfrac{4 - 1}{2 - 1} = 3; \quad 2c = 3 \implies c = 3/2$

7. $f'(c) = 3c^2, \quad \dfrac{f(b) - f(a)}{b - a} = \dfrac{27 - 1}{3 - 1} = 13; \quad 3c^2 = 13 \implies c = \dfrac{1}{3}\sqrt{39} \quad \left(-\dfrac{1}{3}\sqrt{39} \text{ is not in } [a, b]\right)$

9. $f'(c) = \dfrac{-c}{\sqrt{1 - c^2}}, \quad \dfrac{f(b) - f(a)}{b - a} = \dfrac{0 - 1}{1 - 0} = -1; \quad \dfrac{-c}{\sqrt{1 - c^2}} = -1 \implies c = \dfrac{1}{2}\sqrt{2}$

$\left(-\dfrac{1}{2}\sqrt{2} \text{ is not in } [a, b]\right)$

11. f is continuous on $[-1, 1]$, differentiable on $(-1, 1)$ and $f(-1) = f(1) = 0$.

$$f'(x) = \frac{-x(5 - x^2)}{(3 + x^2)^2\sqrt{1 - x^2}}, \quad f'(c) = 0 \text{ for } c \text{ in } (-1, 1) \text{ implies } c = 0.$$

13.   No. By the mean-value theorem there exists at least one number $c \in (0, 2)$ such that

$$f'(c) = \frac{f(2) - f(0)}{2 - 0} = \frac{3}{2} > 1.$$

15. f is everywhere continuous and everywhere differentiable except possibly at $x = -1$.

f is continuous at $x = -1$: as you can check,

$$\lim_{x \to -1^-} f(x) = 0, \quad \lim_{x \to -1^+} f(x) = 0, \quad \text{and} \quad f(-1) = 0.$$

f is differentiable at $x = -1$ and $f'(-1) = 2$: as you can check,

$$\lim_{h \to 0^-} \frac{f(-1 + h) - f(-1)}{h} = 2 \quad \text{and} \quad \lim_{h \to 0^+} \frac{f(-1 + h) - f(-1)}{h} = 2.$$

Thus f satisfies the conditions of the mean-value theorem on every closed interval $[a, b]$.

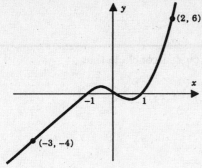

$$f'(x) = \left\{ \begin{array}{cc} 2, & x \le -1 \\ 3x^2 - 1, & x > -1 \end{array} \right];$$

$$\frac{f(2) - f(-3)}{2 - (-3)} = \frac{6 - (-4)}{2 - (-3)} = 2.$$

$$f'(c) = 2 \quad \text{with} \quad c \in (-3, 2) \quad \text{iff} \quad c = 1 \quad \text{or} \quad -3 < c \le -1.$$

17. Let $f(x) = Ax^2 + Bx + C$. Then $f'(x) = 2Ax + B$. By the mean-value theorem

$$f'(c) = \frac{f(b) - f(a)}{b - a} = \frac{(Ab^2 + Bb + C) - (Aa^2 + Ba + C)}{b - a}$$

$$= \frac{A(b^2 - a^2) + B(b - a)}{b - a} = A(b + a) + B$$

Therefore, we have

$$2Ac + B = A(b + a) + B \Longrightarrow c = \frac{a + b}{2}$$

19. $\dfrac{f(1) - f(-1)}{1 - (-1)} = 0$ and $f'(x)$ is never zero. This result does not violate the mean-value theorem

since f is not differentiable at 0; the theorem does not apply.

21. Set $P(x) = 6x^4 - 7x + 1$. If there existed three numbers $a < b < c$ at which $P(x) = 0$, then by Rolle's theorem $P'(x)$ would have to be zero for some x in (a, b) and also for some x in (b, c). This is not the case: $P'(x) = 24x^3 - 7$ is zero only at $x = (7/24)^{1/3}$.

23. Set $P(x) = x^3 + 9x^2 + 33x - 8$. Note that $P(0) < 0$ and $P(1) > 0$. Thus, by the intermediate-value theorem, there exists some number c between 0 and 1 at which $P(x) = 0$. If the equation $P(x) = 0$ had an additional real root, then by Rolle's theorem there would have to be some real number at which $P'(x) = 0$. This is not the case: $P'(x) = 3x^2 + 18x + 33$ is never zero since the discriminant $b^2 - 4ac = (18)^2 - 12(33) < 0$.

25. Let c and d be two consecutive roots of the equation $P'(x) = 0$. The equation $P(x) = 0$ cannot have two or more roots between c and d for then, by Rolle's theorem, $P'(x)$ would have to be zero somewhere between these two roots and thus between c and d. In this case c and d would no longer be consecutive roots of $P'(x) = 0$.

27. Suppose that f has two fixed points a, $b \in I$, with $a < b$. Let $g(x) = f(x) - x$. Then $g(a) = f(a) - a = 0$ and $g(b) = f(b) - b = 0$. Since f is differentiable on I, we can conclude that g is differentiable on (a, b) and continuous on $[a, b]$. By Rolle's theorem, there exists a number $c \in (a, b)$ such that $g'(c) = f'(c) - 1 = 0$ or $f'(c) = 1$. This contradicts the assumption that $f'(x) < 1$ on I.

29. If $x_1 = x_2$, then $|f(x_1) - f(x_2)|$ and $|x_1 - x_2|$ are both 0 and the inequality holds. If $x_1 \ne x_2$, then you know by the mean-value theorem that

$$\frac{f(x_1) - f(x_2)}{x_1 - x_2} = f'(c)$$

for some number c between x_1 and x_2. Since $|f'(c)| \le 1$, you can conclude that

$$\left| \frac{f(x_1) - f(x_2)}{x_1 - x_2} \right| \le 1 \quad \text{and thus that} \quad |f(x_1) - f(x_2)| \le |x_1 - x_2|.$$

31.   Set, for instance, $f(x) = \begin{cases} 1, & a < x < b \\ 0, & x = a, b \end{cases}$.

33.   (a)   By the mean-value theorem, there exists a number $c \in (a, b)$ such that $f(b) - f(a) = f'(c)(b-a)$. If $f'(x) \le M$ for all $x \in (a, b)$, then it follows that

$$f(b) \le f(a) + M(b - a)$$

(b)   If $f'(x) \ge m$ for all $x \in (a, b)$, then it follows that

$$f(b) \ge f(a) + m(b - a)$$

(c)   If $|f'(x)| \le L$ on (a, b), then $-L \le f'(x) \le L$ on (a, b) and the result follows from parts (a) and (b).

35.   Let $f(x) = \cos x$ and $g(x) = \sin x$ on $I = (-\infty, \infty)$. Then

$$f(x)g'(x) - g(x)f'(x) = \cos^2 x + \sin^2 x = 1 \ \text{ for all } x \in I$$

The result follows from Exercise 34.

37.
$$f'(x_0) = \lim_{y \to 0} \frac{f(x_0 + y) - f(x_0)}{y} = \lim_{y \to 0} \frac{f'(x_0 + \theta y)y}{y} = \lim_{y \to 0} f'(x_0 + \theta y)$$

(by the hint)

$$= \lim_{x \to x_0} f'(x) = L$$

(by 2.2.5)

39.   (a)   Between any two times that the object is at the origin there is at least one instant when the velocity is zero.

(b)   During any time interval there is at least one instant when the instantaneous velocity equals the average velocity over that interval.

41.   Let $s(t)$ denote the distance that the car has traveled in t seconds since applying the brakes, $0 \le t \le 6$. Then $s(0) = 0$ and $s(6) = 280$. Assume that s is differentiable on $(0, 6)$ and continuous on $[0, 6]$. Then, by the mean-value theorem, there exists a time $c \in (0, 6)$ such that

$$s'(c) = v(c) = \frac{s(6) - s(0)}{6 - 0} = \frac{280}{6} \cong 46.67 \text{ ft/sec}$$

Now $v(0) \ge v(c) = 46.7$ ft/sec. Thus, the driver must have been exceeding the speed limit (44 ft/sec) at the instant he applied his brakes.

43. Let $f(x) = \sqrt{x}$. Then $f'(x) = \dfrac{1}{2\sqrt{x}}$. Using Exercise 42, we have

$$\sqrt{65} = \sqrt{64+1} = f(64+1) \cong f(64) + f'(64)(1) = \sqrt{64} + \frac{1}{2\sqrt{64}} = 8.0625$$

45. **(a)** Let $f(x) = 1 + 4x - 2\cos x$, $x \in I = (-\infty, \infty)$. If f had two (or more) zeros on I, then, by Rolle's theorem, f' would have to have a zero on I. But, $f'(x) = 4 + 2\sin x > 0$ on I. Thus f has at most one zero on I.

(b) $f(0) = -1$ and $f(1) \cong 3.92$. Thus f has a zero in $(0,1)$.

(b) $x_{n+1} = x_n - \dfrac{1 + 4x_n - 2\cos x_n}{4 + 2\sin x_n}$; $x_1 = 0$, $x_2 = 0.25$, $x_3 \cong 0.2361$

47. $f(x) = 1 - x^3 - \cos(\pi x/2)$ is differentiable on $(0,1)$, continuous on $[0,1]$, and $f(0) = f(1) = 0$.

$f'(x) = -3x^2 + \dfrac{\pi}{2}\sin(\pi x/2)$

$f'(c) = 0$ at $c \cong 0.676$

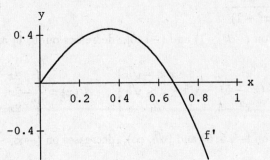

49. Let $f(x) = x^4 - 7x^2 + 2$. Then $f'(x) = 4x^3 - 14x$ and

$$g(x) = 4x^3 - 14x - \frac{f(3) - f(1)}{3 - 1} = 4x^3 - 14x - 12$$

$g(c) = 0$ at $c \cong 2.205$

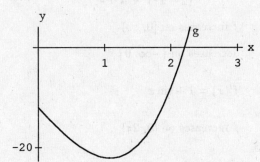 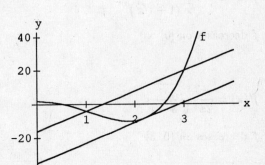

SECTION 4.2

1. $f'(x) = 3x^2 - 3 = 3\left(x^2 - 1\right) = 3(x+1)(x-1)$

f increases on $(-\infty, -1]$ and $[1, \infty)$, decreases on $[-1, 1]$

3. $f'(x) = 1 - \dfrac{1}{x^2} = \dfrac{x^2 - 1}{x^2} = \dfrac{(x+1)(x-1)}{x^2}$

f increases on $(-\infty, -1]$ and $[1, \infty)$, decreases on $[-1, 0)$ and $(0, 1]$ (f is not defined at 0)

5. $f'(x) = 3x^2 + 4x^3 = x^2(3 + 4x)$

f increases on $\left[-\frac{3}{4}, \infty\right)$,   decreases on $\left(-\infty, -\frac{3}{4}\right]$

7. $f'(x) = 4(x + 1)^3$

f increases on $[-1, \infty)$,   decreases on $(-\infty, -1]$

9. $f(x) = \begin{cases} \dfrac{1}{2-x}, & x < 2 \\ \dfrac{1}{x-2}, & x > 2 \end{cases}$ $f'(x) = \begin{cases} \dfrac{1}{(2-x)^2}, & x < 2 \\ \dfrac{-1}{(x-2)^2}, & x > 2 \end{cases}$

f increases on $(-\infty, 2)$,   decreases on $(2, \infty)$ (f is not defined at 2)

11. $f'(x) = -\dfrac{4x}{(x^2 - 1)^2}$

f increases on $(-\infty, -1)$ and $(-1, 0]$, decreases on $[0, 1)$ and $(1, \infty)$ (f is not defined at ± 1)

13. $f(x) = \begin{cases} x^2 - 5, & x < -\sqrt{5} \\ -(x^2 - 5), & -\sqrt{5} \le x \le \sqrt{5} \\ x^2 - 5, & \sqrt{5} < x \end{cases}$, $f'(x) = \begin{cases} 2x, & x < -\sqrt{5} \\ -2x, & -\sqrt{5} < x < \sqrt{5} \\ 2x, & \sqrt{5} < x \end{cases}$

f increases on $[-\sqrt{5}, 0]$ and $[\sqrt{5}, \infty)$, decreases on $(-\infty, -\sqrt{5}]$ and $[0, \sqrt{5}]$

15. $f'(x) = \dfrac{2}{(x + 1)^2}$

f increases on $(-\infty, -1)$ and $(-1, \infty)$ (f is not defined at -1)

17. $f'(x) = -\dfrac{7(1 - \sqrt{x})^6}{\sqrt{x}\,(1 + \sqrt{x})^8}$ **19.** $f'(x) = \dfrac{x}{(2 + x^2)^2}\sqrt{\dfrac{2 + x^2}{1 + x^2}}$

f decreases on $[0, \infty)$ f increases on $[0, \infty)$

                                            decreases on $(-\infty, 0]$

21. $f'(x) = \dfrac{-3}{2x^2}\sqrt{\dfrac{x}{3 - x}}$ **23.** $f'(x) = 1 + \sin x$

f decreases on $(0, 3]$ f increases on $[0, 2\pi]$

25. $f'(x) = -2\sin 2x - 2\sin x = -2\sin x\,(2\cos x + 1)$

f increases on $\left[\frac{2}{3}\pi, \pi\right]$,   decreases on $\left[0, \frac{2}{3}\pi\right]$

27. $f'(x) = \sqrt{3} + 2\sin 2x$

f increases on $\left[0, \frac{2}{3}\pi\right]$ and $\left[\frac{5}{6}\pi, \pi\right]$, decreases on $\left[\frac{2}{3}\pi, \frac{5}{6}\pi\right]$

29. $\dfrac{d}{dx}\left(\dfrac{x^3}{3} - x\right) = f'(x) \implies f(x) = \dfrac{x^3}{3} - x + C$

$f(1) = 2 \implies 2 = \frac{1}{3} - 1 + C$, so $C = \frac{8}{3}$. Thus, $f(x) = \frac{1}{3}x^3 - x + \frac{8}{3}$.

31. $\dfrac{d}{dx}\left(x^5 + x^4 + x^3 + x^2 + x\right) = f'(x) \implies f(x) = x^5 + x^4 + x^3 + x + C$

$f(0) = 5 \implies 5 = 0 + C$, so $C = 5$. Thus, $f(x) = x^5 + x^4 + x^3 + x^2 + x + 5$.

33. $\dfrac{d}{dx}\left(\dfrac{3}{4}x^{4/3} - \dfrac{2}{3}x^{3/2}\right) = f'(x) \implies f(x) = \dfrac{3}{4}x^{4/3} - \dfrac{2}{3}x^{3/2} + C$

$f(0) = 1 \implies 1 = 0 + C$, so $C = 1$. Thus, $f(x) = \frac{3}{4}x^{4/3} - \frac{2}{3}x^{3/2} + 1$, $x \geq 0$.

35. $\dfrac{d}{dx}\left(2x - \cos x\right) = f'(x) \implies f(x) = 2x - \cos x + C$

$f(0) = 3 \implies 3 = 0 - 1 + C$, so $C = 4$. Thus, $f(x) = 2x - \cos x + 4$.

37. $f'(x) = \begin{cases} 1, & x < -3 \\ -1, & -3 < x < -1 \\ 1, & -1 < x < 1 \\ -2, & 1 < x \end{cases}$

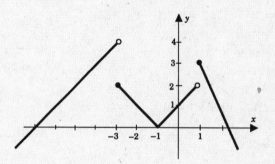

f increases on $(-\infty, -3)$ and $[-1, 1]$,

decreases on $[-3, -1]$ and $[1, \infty)$

39. $f'(x) = \begin{cases} -2x, & x < 1 \\ -2, & 1 < x < 3 \\ 3, & 3 < x \end{cases}$

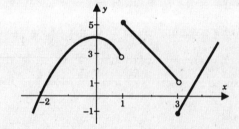

f increases on $(-\infty, 0]$ and $[3, \infty)$,

decreases on $[0, 1)$ and $[1, 3]$

41.

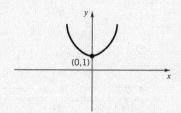

43.

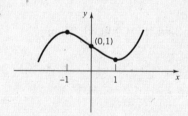

45.

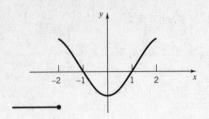

47.

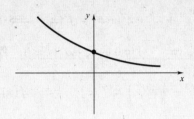

49. Not possible; f is increasing, so $f(2)$ must be greater than $f(-1)$.

51. Let $x(t) = t^3 - 6t^2 + 9t + 2$. Then

$v(t) = 3t^2 - 12t + 9 = 3(t-1)(t-3)$

sign of v :

$a(t) = 6t - 12 = 6(t-2)$

sign of a :

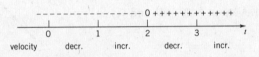

sign of v : $+ + + 0 - - - - - - 0 + + +$

sign of a : $- - - - - - 0 + + + + + +$

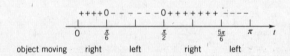

Let $x(t) = 2\sin 3t, \quad t \in [0, \pi]$. Then

$v(t) = 6\cos 3t$

sign of v :

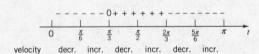

$a(t) = -18\sin 3t$

sign of a :

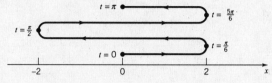

sign of v: $+++0----0++++0--$

sign of a: $-----0++++0----$

speed   decr   incr   decr   incr   decr   incr

55. (a) $M \leq L \leq N$ (b) none   (c) $M = L = N$

57. Set, for instance, $f(x) = \begin{cases} 1, & x \text{ rational} \\ 0, & x \text{ irrational} \end{cases}$.

59. (a) $f'(x) = 2\sec x(\sec x \tan x) = 2\sec^2 x \tan x$ and $g'(x) = 2\tan x \sec^2 x$.

    Therefore, $f'(x) = g'(x)$ for all $x \in I$.

    (b) Evaluating $\sec^2 x - \tan^2 x = C$ at $x = 0$ gives $C = 1$.

61. Let f and g be functions such that $f'(x) = -g(x)$ and $g'(x) = f(x)$. Then:

    (a) Differentiating $f^2(x) + g^2(x)$ with repsect to x, we have

    $$2f(x)f'(x) + 2g(x)g'(x) = -2f(x)g(x) + 2g(x)f(x) = 0.$$

    Thus, $f^2(x) + g^2(x) = C$ (constant).

    (b) $f(a) = 1$ and $g(a) = 0$ implies $C = 1$. The functions $f(x) = \cos(x - a)$, $g(x) = \sin(x - a)$

    have these properties.

63. Let $f(x) = x - \sin x$. Then $f'(x) = 1 - \cos x$.

    (a) $f'(x) \geq 0$ for all $x \in (-\infty, \infty)$ and $f'(x) = 0$ only at $x = \dfrac{\pi}{2} + n\pi$, $n = 0, \pm 1, \pm 2, \dots$

    It follows from Theorem 4.2.3 that f is increasing on $(-\infty, \infty)$.

    (b) Since f is increasing on $(-\infty, \infty)$ and $f(0) = 0 - \sin 0 = 0$, we have:

    $$f(x) > 0 \text{ for all } x > 0 \Rightarrow x > \sin x \text{ on } (0, \infty);$$
    $$f(x) < 0 \text{ for all } x < 0 \Rightarrow x < \sin x \text{ on } (-\infty, 0).$$

65. Let $f(x) = \tan x$ and $g(x) = x$ for $x \in [0, \pi/2)$. Then $f(0) = g(0) = 0$ and $f'(x) = \sec^2 x > g'(x) = 1$
    for $x \in (0, \pi/2)$. Thus, $\tan x > x$ for $x \in (0, \pi/2)$ by Exercise 64(a).

67. Choose an integer $n > 1$. Let $f(x) = (1 + x)^n$ and $g(x) = 1 + nx$, $x > 0$. Then, $f(0) = g(0) = 1$ and
    $f'(x) = n(1 + x)^{n-1} > g'(x) = n$ since $(1 + x)^{n-1} > 1$ for $x > 0$. The result follows from Exercise
    64(a).

69. $4° \cong 0.06981$ radians. By Exercises 63 and 68,
$$0.6981 - \frac{(0.6981)^3}{6} = 0.06975 < \sin 4° < 0.6981$$

71.   Let $f(x) = 3x^4 - 10x^3 - 4x^2 + 10x + 9$, $x \in [-2, 5]$. Then $f'(x) = 12x^3 - 30x^2 - 8x + 10$.

$f'(x) = 0$ at $x \cong -0.633$, 0.5, 2.633

f is decreasing on $[-2, -0.633]$

and $[0.5, 2.633]$

f is increasing on $[-0.633, 0.5]$

and $[2.633, 5]$

73.   Let $f(x) = x \cos x - 3 \sin 2x$, $x \in [0, 6]$. Then $f'(x) = \cos x - x \sin x - 6 \cos 2x$.

$f'(x) = 0$ at $x \cong 0.770$, 2.155, 3.798, 5.812

f is decreasing on $[0, 0.770]$, $[2.155, 3.798]$

and $[5.812, 6]$

f is increasing on $[0.770, 2.155]$

and $[3.798, 5.812]$

75.   By Exercise 74, $mgy + \frac{1}{2}mv^2 = C$ (constant). Since $v = 0$ at height $y = y_0$, we have $C = mgy_0$. Thus,
$$mgy_0 = mgy + \frac{1}{2}mv^2 \quad \text{and} \quad |v| = \sqrt{2g(y_0 - y)}$$

77.   Set $y_0 = 150$ $y = 0$ and $g = 9.8$ in the equation $|v| = \sqrt{2g(y_0 - y)}$. Then
$$|v| = \sqrt{2(9.8)(150)} \cong 54.22 \text{ m/sec}$$

SECTION 4.3

1. $f'(x) = 3x^2 + 3 > 0$;   no critical nos, no local extreme values

3. $f'(x) = 1 - \dfrac{1}{x^2}$

critical nos $-1, 1$

$f''(x) = \dfrac{2}{x^3}$, $f''(-1) = -2$, $f''(1) = 2$

$f(-1) = -2$ local max, $f(1) = 2$ local min

5. $f'(x) = 2x - 3x^2 = x(2 - 3x)$

critical nos $0, \frac{2}{3}$

$f''(x) = 2 - 6x$, $f''(0) = 2$,

$f''\left(\frac{2}{3}\right) = -2$

$f(0) = 0$ local min,

$f\left(\frac{2}{3}\right) = \frac{4}{27}$ local max

7. $f'(x) = \dfrac{2}{(1-x)^2}$;   no critical nos, no local extreme values

9. $f'(x) = -\dfrac{2(2x+1)}{x^2(x+1)^2}$;   critical no $-\dfrac{1}{2}$

$f\left(-\dfrac{1}{2}\right) = -8$ local max

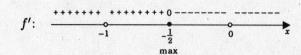

11. $f'(x) = x^2(5x-3)(x-1)$;   critical nos $0, \dfrac{3}{5}, 1$

$f\left(\dfrac{3}{5}\right) = \dfrac{2^2 3^3}{5^5}$ local max

$f(1) = 0$ local min

no local extreme at 0

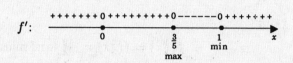

13. $f'(x) = (5-8x)(x-1)^2$;   critical nos $\dfrac{5}{8}, 1$

$f\left(\dfrac{5}{8}\right) = \dfrac{27}{2048}$ local max

no local extreme at 1

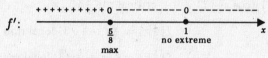

15. $f'(x) = \dfrac{x(2+x)}{(1+x)^2}$;   critical nos $-2, 0$

$f(-2) = -4$ local max

$f(0) = 0$ local min

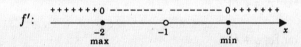

17. $f'(x) = \begin{cases} 2x+1, & x < -2, x > 1 \\ -(2x+1), & -2 < x < 1 \end{cases}$;   critical nos $-2, -\dfrac{1}{2}, 1$

$f(-2) = 0$ local min

$f\left(-\dfrac{1}{2}\right) = \dfrac{9}{4}$ local max

$f(1) = 0$ local min

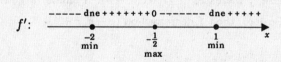

19. $f'(x) = \dfrac{1}{3}x(7x+12)(x+2)^{-2/3}$;   critical nos $-2, -\dfrac{12}{7}, 0$

$f\left(-\dfrac{12}{7}\right) = \dfrac{144}{49}\left(\dfrac{2}{7}\right)^{1/3}$ local max

$f(0) = 0$ local min

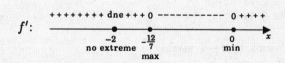

21. $f(x) = \begin{cases} 2-3x, & x \le -\dfrac{1}{2} \\ x+4, & -\dfrac{1}{2} < x < 3 \\ 3x-2, & 3 \le x \end{cases}$, $f'(x) = \begin{cases} -3, & x < -\dfrac{1}{2} \\ 1, & -\dfrac{1}{2} < x < 3 \\ 3, & 3 < x \end{cases}$;

critical nos $-\dfrac{1}{2}, 3$

$f\left(-\dfrac{1}{2}\right) = \dfrac{7}{2}$ local min

no local extreme at 3

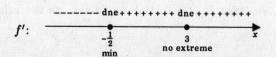

23. $f'(x) = \frac{2}{3}x^{-4/3}(x-1);$ critical nos $0, -1$

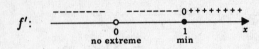

$f(1) = 3$ local min

no local extreme at 0

25. $f'(x) = \cos x - \sin x;$ critical nos $\frac{1}{4}\pi, \frac{5}{4}\pi$

$f''(x) = -\sin x - \cos x,$ $f''\left(\frac{1}{4}\pi\right) = -\sqrt{2},$ $f''\left(\frac{5}{4}\pi\right) = \sqrt{2}$

$f\left(\frac{1}{4}\pi\right) = \sqrt{2}$ local max, $f\left(\frac{5}{4}\pi\right) = -\sqrt{2}$ local min

27. $f'(x) = \cos x\,(2\sin x - \sqrt{3}\,);$ critical nos $\frac{1}{2}\pi, \frac{1}{3}\pi, \frac{2}{3}\pi$

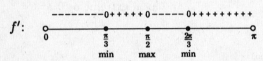

$f\left(\frac{1}{3}\pi\right) = f\left(\frac{2}{3}\pi\right) = -\frac{3}{4}$ local mins

$f\left(\frac{1}{2}\pi\right) = 1 - \sqrt{3}$ local max

29. $f'(x) = \cos^2 x - \sin^2 x - 3\cos x + 2 = (2\cos x - 1)(\cos x - 1)$ critical pts $\frac{1}{3}\pi, \frac{5}{3}\pi$

$f\left(\frac{1}{3}\pi\right) = \frac{2}{3}\pi - \frac{5}{4}\sqrt{3}$ local min

$f\left(\frac{5}{3}\pi\right) = \frac{10}{3}\pi + \frac{5}{4}\sqrt{3}$ local max

$f':$ ```---- 0+ + + + + + + + + + + + + + + + + + + +0 -----```
    0    $\frac{\pi}{3}$ $\frac{5\pi}{3}$ 2π
       min   min

31.   (i) f increases on $(c - \delta, c]$ and decreases on $[c, c + \delta)$.

   (ii) f decreases on $(c - \delta, c]$ and increases on $[c, c + \delta)$.

   (iii)   If $f'(x) > 0$ on $(c - \delta, c) \cup (c, c + \delta)$, then, since f is continuous at c, f increases on $(c - \delta, c]$ and also on $[c, c + \delta)$. Therefore, in this case, f increases on $(c - \delta, c + \delta)$. A similar argument shows that, if $f'(x) < 0$ on $(c - \delta, c) \cup (c, c + \delta)$, then f decreases on $(c - \delta, c + \delta)$.

33.

$$P(x) = x^4 - 8x^3 + 22x^2 - 24x + 4$$

$$P'(x) = 4x^3 - 24x^2 + 44x - 24$$

$$P''(x) = 12x^2 - 48x + 44$$

Since $P'(1) = 0$, $P'(x)$ is divisible by $x - 1$. Division by $x - 1$ gives

$$P'(x) = (x - 1)\left(4x^2 - 20x + 24\right) = 4(x - 1)(x - 2)(x - 3).$$

The critical pts are $1, 2, 3$. Since

$$P''(1) > 0, \quad P''(2) < 0, \quad P''(3) > 0,$$

$P(1) = -5$ is a local min, $P(2) = -4$ is a local max, and $P(3) = -5$ is a local min.

Since $P'(x) < 0$ for $x < 0$, P decreases on $(-\infty, 0]$. Since $P(0) > 0$, P does not take on the value 0 on $(-\infty, 0]$.

Since $P(0) > 0$ and $P(1) < 0$, P takes on the value 0 at least once on $(0, 1)$. Since $P'(x) < 0$ on $(0, 1)$, P decreases on $[0, 1]$. It follows that P takes on the value zero only once on $[0, 1]$.

Since $P'(x) > 0$ on $(1, 2)$ and $P'(x) < 0$ on $(2, 3)$, P increases on $[1, 2]$ and decreases on $[2, 3]$. Since $P(1)$, $P(2)$, $P(3)$ are all negative, P cannot take on the value 0 between 1 and 3.

Since $P(3) < 0$ and $P(100) > 0$, P takes on the value 0 at least once on $(3, 100)$. Since $P'(x) > 0$ on $(3, 100)$, P increases on $[3, 100]$. It follows that P takes on the value zero only once on $[3, 100]$.

Since $P'(x) > 0$ on $(100, \infty)$, P increases on $[100, \infty)$. Since $P(100) > 0$, P does not take on the value 0 on $[100, \infty)$.

35.   (a)

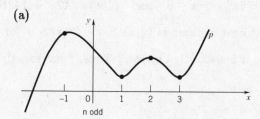

   (b)

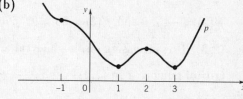

37.   Let $f(x) = \dfrac{ax}{x^2 + b^2}$. Then $f'(x) = \dfrac{a\left(b^2 - x^2\right)}{\left(b^2 + x^2\right)^2}$. Now

$$f'(0) = \frac{a}{b^2} = 1 \Rightarrow a = b^2 \quad \text{and} \quad f'(x) = \frac{b^2\left(b^2 - x^2\right)}{\left(b^2 + x^2\right)^2}$$

$$f'(-2) = \frac{b^2\left(b^2 - 4\right)}{\left(b^2 + 4\right)^2} = 0 \Rightarrow b = \pm 2$$

Thus, $a = 4$ and $b = \pm 2$.

39.   Let δ be any positive number and consider f on the interval $(-\delta, \delta)$. Let n be a positive integer such that $0 < \dfrac{1}{\frac{\pi}{2} + 2n\pi} < \delta$ and $0 < \dfrac{1}{\frac{-\pi}{2} + 2n\pi} < \delta$. Then $f\left(\dfrac{1}{\frac{\pi}{2} + 2n\pi}\right) > 0$ and $f\left(\dfrac{1}{\frac{-\pi}{2} + 2n\pi}\right) < 0$. Thus f takes on both positive and negative values in every interval centered at 0 and it follows that f cannot have a local maximum or minimum at 0.

41.   The function $D(x) = \sqrt{x^2 + [f(x)]^2}$ gives the distance from the origin to the point $(x, f(x))$ on the graph of f. Since the graph of f does not pass through the origin,

$$D'(x) = \frac{x + f(x)f'(x)}{\sqrt{x^2 + [f(x)]^2}}$$

is defined for all $x \in \operatorname{dom}(f)$. Suppose that D has a local extreme value at c. Then

$$D'(c) = \frac{c + f(c)f'(c)}{\sqrt{c^2 + [f(c)]^2}} = 0 \Rightarrow c + f(c)f'(c) = 0 \quad \text{and} \quad f'(c) = -\frac{c}{f(c)}$$

Suppose that $c \neq 0$. The slope of the line through $(0,0)$ and $(c, f(c))$ is given by $m_1 = \dfrac{f(c)}{c}$ and the slope of the tangent line to the graph of f at $x = c$ is given by $m_2 = f'(c) = -\dfrac{c}{f(c)}$. Since

$m_1 m_2 = -1$, these two lines are perpendicular. If $c = 0$, then the tangent line to the graph of f is horizontal and the line through $(0,0)$ and $(0, f(0))$ is vertical.

43. (a) Let $f(x) = x^4 - 2x^2 - 3x + 2$. Then $f'(x) = 4x^3 - 4x - 3$ and $f''(x) = 12x^2 - 4$. Since $f'(1) = -3 < 0$ and $f'(2) = 21 > 0$, f' has at least one zero in $(1,2)$. Since $f''(x) > 0$ for $x \in (1,2)$, f' is increasing on this interval and so it has exactly one zero. Thus, f has exactly one critical number c in $(1,2)$.

(b) $c \cong 1.3125$; f has a local minimum at c.

45. (a) Let $f(x) = x^4 - 7x^2 - 8x - 3$. Then $f'(x) = 4x^3 - 14x - 8$ and $f''(x) = 12x^2 - 14$. Since $f'(2) = -4 < 0$ and $f'(3) = 58 > 0$, f' has at least one zero in $(2,3)$. Since $f''(x) > 0$ for $x \in (2,3)$, f' is increasing on this interval and so it has exactly one zero. Thus, f has exactly one critical number c in $(2,3)$.

(b) $c \cong 2.1091$; f has a local minimum at c.

47. (a) Let $f(x) = \sin x + \dfrac{x^2}{2} - 2x$. Then $f'(x) = \cos x + x - 2$ and $f''(x) = -\sin x + 1$. Since $f'(2) = -0.4161 < 0$ and $f'(3) = 0.01 > 0$, f' has at least one zero in $(2,3)$. Since $f''(x) > 0$ for $x \in (2,3)$, f' is increasing on this interval and so it has exactly one zero. Thus, f has exactly one critical number c in $(2,3)$.

(b) $x_{n+1} = x_n - \dfrac{\cos x_n + x_n - 2}{-\sin x_n + 1}$; $x_1 = 3$, $x_2 = 2.9883$, $x_3 = 2.9883$. Thus $c \cong 2.9883$; f has a local minimum at c.

49. (a)

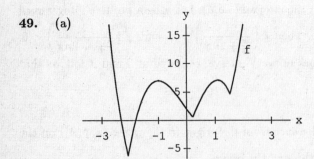

critical numbers: $x_1 \cong -2.085$, $x_2 \cong -1$, $x_3 \cong 0.207$, $x_4 \cong 1.096$, $x_5 = 1.544$
local extreme values: $f(-2.085) \cong -6.255$, $f(-1) = 7$, $f(0.207) \cong 0.621$, $f(1.096) \cong 7.097$, $f(1.544) \cong 4.635$

(b) f is increasing on $[-2.085, -1]$, $[0.207, 1.096]$, and $[1.544, 4]$
f is decreasing on $[-4, -2.085]$, $[-1, 0.207]$, and $[1.096, 1.544]$

51.   (a)

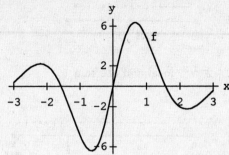

critical numbers: $x_1 \cong -2.204$, $x_2 \cong -0.654$, $x_3 \cong 0.654$, $x_4 \cong 2.204$

local extreme values: $f(-2.204) \cong 2.226$, $f(-0.654) \cong -6.634$, $f(0.654) \cong 6.634$,

$f(2.204) \cong -2.226$

(b) f is increasing on $[-3. -2.204]$, $[-0.654, 0.654]$, and $[2.204, 3]$

f is decreasing on $[-2.204, -0.654]$, and $[0.654, 2.204]$

53.

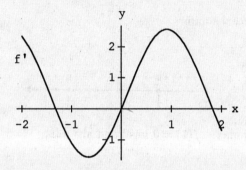

critical numbers of f : $x_1 \cong -1.326$, $x_2 = 0$, $x_3 \cong 1.816$

$f''(-1.326) \cong -4 < 0$ $\Rightarrow$ f has a local maximum at $x = -1.326$

$f''(0) = 4 > 0$ $\Rightarrow$ f has a local minimum at $x = 0$

$f''(1.816) \cong -4$ $\Rightarrow$ f has a local maximum at $x = 1.816$

SECTION 4.4

1. $f'(x) = \frac{1}{2}(x+2)^{-1/2}$, $x > -2$; f' :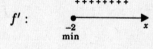

critical no -2;

$f(-2) = 0$ endpt and abs min;   as $x \to \infty$, $f(x) \to \infty$;   so no abs max

3. $f'(x) = 2x - 4$, $x \in (0,3)$; f' :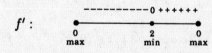

critical nos 0, 2, 3;

$f(0) = 1$ endpt and abs max, $f(2) = -3$ local and abs min, $f(3) = -2$ endpt max

5. $f'(x) = 2x - \dfrac{1}{x^2} = \dfrac{2x^3 - 1}{x^2}$, $x \neq 0$; $f'(x) = 0$ at $x = 2^{-1/3}$

critical no $2^{-1/3}$; $f''(x) = 2 + \dfrac{2}{x^3}$, $f''(2^{-1/3}) = 6$

$f(2^{-1/3}) = 2^{-2/3} + 2^{1/3} = 2^{-2/3} + 2 \cdot 2^{-2/3} = 3 \cdot 2^{-2/3}$ local min

7. $f'(x) = \dfrac{2x^3 - 1}{x^2}$, $x \in \left(\dfrac{1}{10}, 2\right)$; $\quad f':$

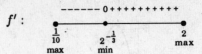

critical nos $\frac{1}{10}$, $2^{-1/3}$, 2;

$f\left(\frac{1}{10}\right) = 10\frac{1}{100}$ endpt and abs max, $f\left(2^{-1/3}\right) = 3 \cdot 2^{-2/3}$ local and abs min,

$f(2) = 4\frac{1}{2}$ endpt max

9. $f'(x) = 2x - 3$, $x \in (0, 2)$; $\quad f':$

critical nos 0, $\frac{3}{2}$, 2;

$f(0) = 2$ endpt and abs max, $f\left(\frac{3}{2}\right) = -\frac{1}{4}$ local and abs min,

$f(2) = 0$ endpt max

11. $f'(x) = \dfrac{(2-x)(2+x)}{(4+x^2)^2}$, $x \in (-3, 1)$; $\quad f':$

critical nos -3, -2, 1;

$f(-3) = -\frac{3}{13}$ endpt max, $f(-2) = -\frac{1}{4}$ local and abs min,

$f(1) = \frac{1}{5}$ endpt and abs max

13. $f'(x) = 2\left(x - \sqrt{x}\right)\left(1 - \dfrac{1}{2\sqrt{x}}\right)$, $x > 0$; $\quad f':$

critical nos 0, $\frac{1}{4}$, 1;

$f(0) = 0$ endpt and abs min, $f\left(\frac{1}{4}\right) = \frac{1}{16}$ local max, $f(1) = 0$ local and abs min;

as $x \to \infty$, $f(x) \to \infty$; so no abs max

15. $f'(x) = \dfrac{3(2-x)}{2\sqrt{3-x}}$, $x < 3$ $\quad f':$

critical nos 2, 3;

$f(2) = 2$ local and abs max, $f(3) = 0$ endpt min;

as $x \to -\infty$, $f(x) \to -\infty$; so no abs min

17. $f'(x) = -\frac{1}{3}(x-1)^{-2/3}$, $x \ne 1$; $\quad f':$

critical no 1;

no local extremes; $\left.\begin{array}{l} \text{as } x \to \infty, \ f(x) \to -\infty, \\ \text{as } x \to -\infty, \ f(x) \to \infty \end{array}\right\}$ no abs extremes

19. $f'(x) = \sin x \left(2\cos x + \sqrt{3}\right)$, $x \in (0, \pi)$; $\quad f':$

critical nos 0, $\frac{5}{6}\pi$, π;

$f(0) = -\sqrt{3}$ endpt and abs min, $f\left(\frac{5}{6}\pi\right) = \frac{7}{4}$ local and abs max,

$f(\pi) = \sqrt{3}$ endpt min

21. $f'(x) = -3 \sin x \left(2 \cos^2 x + 1\right) < 0, \quad x \in (0, \pi); \quad \text{critical nos } 0, \pi;$

$f(0) = 5$ endpt and abs max, $f(\pi) = -5$ endpt and abs min

23. $f'(x) = \sec^2 x - 1 \geq 0, \quad x \in \left(-\frac{1}{3}\pi, \frac{1}{2}\pi\right); \quad \text{critical nos } -\frac{1}{3}\pi, 0;$

$f\left(-\frac{1}{3}\pi\right) = \frac{1}{3}\pi - \sqrt{3}$ endpt and abs min, no abs max

25.

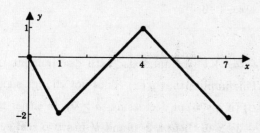

$$f'(x) = \begin{cases} -2, & 0 < x < 1 \\ 1, & 1 < x < 4 \\ -1, & 4 < x < 7 \end{cases}$$

critical nos 0, 1, 4, 7

$f(0) = 0$ endpt max, $f(1) = -2$ local and abs min,

$f(4) = 1$ local and absolute max, $f(7) = -2$ endpt and abs min

27.

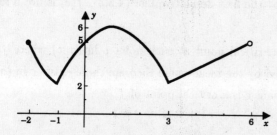

$$f'(x) = \begin{cases} 2x, & -2 < x < -1 \\ 2 - 2x, & -1 < x < 3 \\ 1, & 3 < x < 6 \end{cases}$$

critical nos $-2, -1, 1, 3$

$f(-2) = 5$ endpt max, $f(-1) = 2$ local and abs min,

$f(1) = 6$ local and abs max, $f(3) = 2$ local and abs min

29.

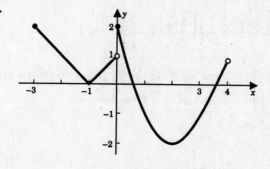

$$f'(x) = \begin{cases} -1, & -3 < x < -1 \\ 1, & -1 < x < 0 \\ 2x - 4, & 0 < x < 3 \\ 2, & 3 \leq x < 4 \end{cases}$$

critical nos $-3, -1, 0, 2$

$f(-3) = 2$ endpt and abs max, $f(-1) = 0$ local min,

$f(0) = 2$ local and abs max, $f(2) = -2$ local and abs min

31.

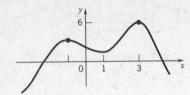

33. $f(-3) = 0$ and $f'(x) > 0$ on $(-3, -1)$

$\Rightarrow f(-1) > 0.$

$f(3) = 0$ and $f'(x) > 0$ on $(1, 3) \Rightarrow f(1) < 0.$

It now follows that f has a zero on $(-1, 1)$,

contradicting the fact that $f(x) \neq 0$ for

$x \in (-3, 3).$

35. Let $p(x) = x^3 + ax^2 + bx + c$. Then $p'(x) = 3x^2 + 2ax + b$ is a quadratic with discriminant $\Delta = 4a^2 - 12b = 4(a^2 - 3b)$. If $a^2 \leq 3b$, then $\Delta \leq 0$. This implies that $p'(x)$ does not change sign on $(-\infty, \infty)$ and hence p is either increasing on $(-\infty, \infty)$ (if $a \leq 0$) or decreasing ($a \geq 0$). In either case, p has no extreme values. On the other hand, if $a^2 - 3b > 0$, then $\Delta > 0$ and p' has two real zeros, c_1 and c_2, from which it follows that p has extreme values at c_1 and c_2. Thus, if p has no extreme values, then we must have $a^2 - 3b \leq 0$.

37. By contradiction. If f is continuous at c, then, by the first-derivative test (4.3.3), $f(c)$ is not a local maximum.

39. If f is not differentiable on (a, b), then f has a critical point at each point c in (a, b) where $f'(c)$ does not exist. If f is differentiable on (a, b), then by the mean-value theorem there exists c in (a, b) where $f'(c) = [f(b) - f(a)]/(b - a) = 0$. This means c is a critical point of f.

41. Let M be a positive number. Then

$$P(x) - M \geq a_n x^n - \left(|a_{n-1}|x^{n-1} + \cdots + |a_1|x + |a_0| + M\right) \quad \text{for} \quad x > 0$$

$$\geq a_n x^n - \left(|a_{n-1}| + \cdots + |a_1| + |a_0| + M\right) \quad \text{for} \quad x > 1$$

It now follows that

$$P(x) - M \geq 0 \quad \text{for} \quad x \geq K = \left(\frac{|a_{n-1}| + \cdots + |a_1| + |a_0| + M}{a_n}\right)^{1/n} + 1.$$

43. Let R be a rectangle with its diagonals having length c, and let x be the length of one of its sides. Then the length of the other side is $y = \sqrt{c^2 - x^2}$ and the area of R is given by

$$A(x) = x\sqrt{c^2 - x^2}$$

Now

$$A'(x) = \sqrt{c^2 - x^2} - \frac{x^2}{\sqrt{c^2 - x^2}}$$

$$= \frac{c^2 - 2x^2}{\sqrt{c^2 - x^2}},$$

and

$$A'(x) = 0 \Longrightarrow x = \frac{\sqrt{2}}{2}c$$

It is easy to verify that A has a maximum at $x = \dfrac{\sqrt{2}}{2} c$. Since $y = \dfrac{\sqrt{2}}{2} c$ when $x = \dfrac{\sqrt{2}}{2} c$, it follows that the rectangle of maximum area is a square

45. Cut the wire into two pieces, one of length x and the other of length $L - x$. Suppose that the wire of length x is used to form the equilateral triangle, and the other piece is used to form the square. Then the area of the triangle is $\sqrt{3}\,x^2/36$, and the area of the square is $(L - x^2)/16$. Now, let

$$S(x) = \frac{\sqrt{3}}{36}\, x^2 + \frac{1}{16}\,(L - x)^2$$

Then

$$S'(x) = \frac{\sqrt{3}}{18}\, x - \frac{1}{8}\,(L - x)$$

$$= \frac{4\sqrt{3} + 9}{72}\, x - \frac{1}{8}\, L$$

Setting $S'(x) = 0$ we find that

$$x = \frac{9}{4\sqrt{3} + 9}\, L. \cong 0.5650\, L$$

Now,

$$S(0) = \frac{1}{16}\, L^2 = 0.0625 L^2 \quad \text{(absolute maximum)}$$

$$S\left(\frac{9}{4\sqrt{3} + 9}\, L\right) \cong 0.0390 L^2 \quad \text{(absolute minimum)}$$

$$S(L) = \frac{\sqrt{3}}{36}\, L^2 = 0.0481 L^2$$

To maximize the sum of the areas, use the wire to form a square; to minimize the sum, use $x \cong 0.5650\, L$ to form the triangle and the remainder to form the square.

47.

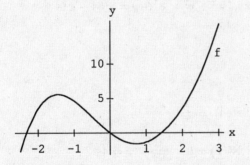

critical numbers: $x_1 = -1.452$, $x_2 = 0.760$

$f(-1.452)$ local maximum

$f(0.727)$ local minimum

$f(3)$ absolute maximum

$f(-2.5)$ absolute minimum

49.

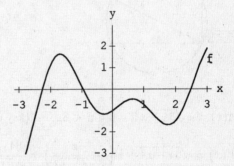

critical numbers: $x_1 = -1.683$, $x_2 = -0.284$,

$x_3 = 0.645$, $x_4 = 1.760$

$f(-1.683), f(0.645)$ local maxima

$f(-0.284), f(1.760)$ local minima

$f(\pi)$ absolute maximum

$f(-\pi)$ absolute minimum

SECTION 4.5

1. Set $P = xy$ and $y = 40 - x$. We want to maximize

    $$P(x) = x(40 - x), \quad 0 < x < 40. \quad \text{(key step completed)}$$
    $$P'(x) = 40 - 2x, \quad P'(x) = 0 \implies x = 20.$$

    Since P increases on $(0, 20]$ and decreases on $[20, 40)$, the abs max of P occurs when $x = 20$. Then, $y = 20$ and $xy = 400$.

    The maximal value of xy is 400.

3.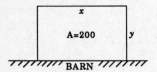

    <u>Minimize P</u>

    $$P = x + 2y, \quad 200 = xy, \quad y = 200/x$$

    $$P(x) = x + \frac{400}{x}, \quad 0 < x. \quad \text{(key step completed)}$$

    $$P'(x) = 1 - \frac{400}{x^2}, \quad P'(x) = 0 \implies x = 20.$$

    Since P decreases on $(0, 20]$ and increases on $[20, \infty)$, the abs min of P occurs when $x = 20$.

    To minimize the fencing, make the garden 20 ft (parallel to barn) by 10 ft.

5.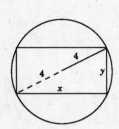

    <u>Maximize A</u>

    $$A = xy, \quad x^2 + y^2 = 8^2, \quad y = \sqrt{64 - x^2}$$

    $$A(x) = x\sqrt{64 - x^2}, \quad 0 < x < 8. \quad \text{(key step completed)}$$

    $$A'(x) = \sqrt{64 - x^2} + x\left(\frac{-x}{\sqrt{64 - x^2}}\right) = \frac{64 - 2x^2}{\sqrt{64 - x^2}}, \quad A'(x) = 0 \implies x = 4\sqrt{2}.$$

    Since A increases on $(0, 4\sqrt{2}]$ and decreases on $[4\sqrt{2}, 8)$, the abs max of A occurs when $x = 4\sqrt{2}$. Then, $y = 4\sqrt{2}$ and $xy = 32$.

    The maximal area is 32.

7.

<u>Maximize A</u>

$$A = xy, \quad 2y + 3x = 600, \quad y = \frac{600 - 3x}{2}$$

$A(x) = x\left(300 - \frac{3}{2}x\right), \quad 0 < x < 200.$ (key step completed)

$A'(x) = 300 - 3x, \quad A'(x) = 0 \implies x = 100.$

Since A increases on $(0, 100]$ and decreases on $[100, 200)$, the abs max of A occurs when $x = 100$. Then, $y = 150$.

The playground of greatest area measures 100 ft by 150 ft. (The fence divider is 100 ft long.)

9.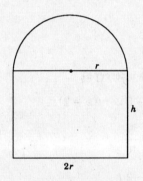

<u>Maximize L</u>

To account for the semi-circular portion admitting less light per square foot, we multiply its area by 1/3.

$$L = 2rh + \frac{1}{3}\left(\frac{\pi r^2}{2}\right),$$

$$2r + 2h + \pi r = 30, \quad h = \tfrac{1}{2}(30 - 2r - \pi r)$$

$$L = 2r\left(\frac{30 - 2r - \pi r}{2}\right) + \frac{1}{6}\pi r^2$$

$L(r) = 30r - \left(2 + \frac{5}{6}\pi\right)r^2, \quad 0 < r < \frac{30}{2 + \pi}.$ (key step completed)

$L'(r) = 30 - \left(4 + \frac{5}{3}\pi\right)r, \quad L'(r) = 0 \implies r = \frac{90}{12 + 5\pi}.$

Since $L''(r) < 0$ for all r in the domain of L, the local max at $r = 90/(12 + 5\pi)$ is the abs max.

For the window that admits the most light, take the radius of the semicircle as $\dfrac{90}{12 + 5\pi} \cong 3.25$ ft and the height of the rectangular portion as $\dfrac{90 + 30\pi}{12 + 5\pi} \cong 6.65$ ft.

11.

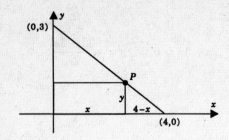

<u>Maximize A</u>

$$A = xy, \quad \frac{3}{4} = \frac{y}{4-x} \quad \text{(similar triangles)}$$

$$y = \frac{3}{4}(4-x)$$

$A(x) = \dfrac{3x}{4}(4-x), \quad 0 < x < 4.$ (key step completed)

$A'(x) = 3 - \dfrac{3x}{2}, \quad A'(x) = 0 \implies x = 2.$

Since A increases on $(0,2]$ and decreases on $[2,4)$, the abs max of A occurs when $x = 2$.

To maximize the area of the rectangle, take P as the point $\left(2, \frac{3}{2}\right)$.

13.

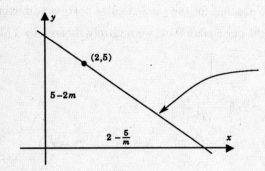

<u>Minimize A</u>

$A = \frac{1}{2}(x\text{-intercept})\,(y\text{-intercept})$

Equation of line: $y - 5 = m(x - 2)$

x-intercept: $2 - \dfrac{5}{m}$

y-intercept: $5 - 2m$

$$A = \frac{1}{2}\left(2 - \frac{5}{m}\right)(5 - 2m) = 10 - 2m - \frac{25}{2m}$$

$A(m) = 10 - 2m - \dfrac{25}{2m}, \quad m < 0.$ (key step completed)

$A'(m) = -2 + \dfrac{25}{2m^2}, \quad A'(m) = 0 \implies m = -\dfrac{5}{2}.$

Since $A''(m) = -25/m^3 > 0$ for $m < 0$, the local min at $m = -5/2$ is the abs min.

The triangle of minimal area is formed by the line of slope $-5/2$.

15.

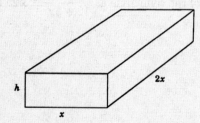

<u>Maximize V</u>

$V = 2x^2 h, \quad 2\left(2x^2 + xh + 2xh\right) = 100, \quad h = \dfrac{50 - 2x^2}{3x}$

$V = 2x^2 \left(\dfrac{50 - 2x^2}{3x}\right)$

$V(x) = \frac{100}{3}x - \frac{4}{3}x^3, \quad 0 < x < 5.$ (key step completed)

$V'(x) = \frac{100}{3} - 4x^2$, $V'(x) = 0 \implies x = \frac{5}{3}\sqrt{3}$.

Since $V''(x) = -8x < 0$ on $(0,5)$, the local max at $x = \frac{5}{3}\sqrt{3}$ is the abs max.

The base of the box of greatest volume measures $\frac{5}{3}\sqrt{3}$ in. by $\frac{10}{3}\sqrt{3}$ in.

17.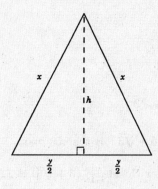

Maximize A

$A = \frac{1}{2}hy$

$2x + y = 12 \implies y = 12 - 2x$

Pythagorean Theorem:

$$h^2 + \left(\frac{y}{2}\right)^2 = x^2 \implies h = \sqrt{x^2 - \left(\frac{y}{2}\right)^2}$$

Thus, $h = \sqrt{x^2 - (6-x)^2} = \sqrt{12x - 36}$.

$A(x) = (6-x)\sqrt{12x-36}$, $3 < x < 6$.   (key step completed)

$A'(x) = -\sqrt{12x-36} + (6-x)\left(\frac{6}{\sqrt{12x-36}}\right) = \frac{72-18x}{\sqrt{12x-36}}$,

$A'(x) = 0 \implies x = 4$.

Since A increases on $(3,4]$ and decreases on $[4,6)$, the abs max of A occurs at $x = 4$.

The triangle of maximal area is equilateral with side of length 4.

19.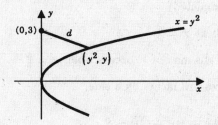

Minimize d

$$d = \sqrt{(y^2 - 0)^2 + (y-3)^2}$$

The square-root function is increasing;

d is minimal when $D = d^2$ is minimal.

$D(y) = y^4 + (y-3)^2$, y real.   (key step completed)

$D'(y) = 4y^3 + 2(y-3) = (y-1)\left(4y^2 + 4y + 6\right)$, $D'(y) = 0$ at $y = 1$.

Since $D''(y) = 12y^2 + 2 > 0$, the local min at $y = 1$ is the abs min.

The point $(1,1)$ is the point on the parabola closest to $(0,3)$.

21.

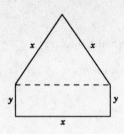

<u>Maximize A</u>

$$A = xy + \frac{\sqrt{3}}{4}x^2, \quad 30 = 3x + 2y, \quad y = \frac{30 - 3x}{2}$$

$$A(x) = 15x - \frac{3}{2}x^2 + \frac{\sqrt{3}}{4}x^2, \quad 0 < x < 10. \qquad \text{(key step completed)}$$

$$A'(x) = 15 - 3x + \frac{\sqrt{3}}{2}x, \quad A'(x) = 0 \implies x = \frac{30}{6 - \sqrt{3}} = \frac{10}{11}\left(6 + \sqrt{3}\right).$$

Since $A''(x) = -3 + \frac{\sqrt{3}}{2} < 0$ on $(0, 10)$, the local max at $x = \frac{10}{11}\left(6 + \sqrt{3}\right)$ is the abs max.

The pentagon of greatest area is composed of an equilateral triangle with side $\frac{10}{11}\left(6 + \sqrt{3}\right) \cong 7.03$ in. and rectangle with height $\frac{15}{11}\left(5 - \sqrt{3}\right) \cong 4.46$ in.

23.

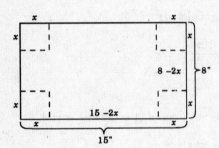

<u>Maximize V</u>

$$V = x(8 - 2x)(15 - 2x)$$

$$\begin{bmatrix} x > 0 \\ 8 - 2x > 0 \\ 15 - 2x > 0 \end{bmatrix} \implies 0 < x < 4$$

$$V(x) = 120x - 46x^2 + 4x^3, \quad 0 < x < 4. \qquad \text{(key step completed)}$$

$$V'(x) = 120 - 92x + 12x^2 = 4(3x - 5)(x - 6), \quad V'(x) = 0 \text{ at } x = \tfrac{5}{3}.$$

Since V increases on $\left(0, \frac{5}{3}\right]$ and decreases on $\left[\frac{5}{3}, 4\right)$, the abs max of V occurs when $x = \frac{5}{3}$.

The box of maximal volume is made by cutting out squares 5/3 inches on a side.

25.

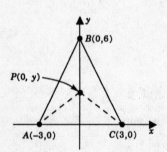

<u>Minimize $\overline{AP} + \overline{BP} + \overline{CP} = S$</u>

length $AP = \sqrt{9 + y^2}$

length $BP = 6 - y$

length $CP = \sqrt{9 + y^2}$

$$S(y) = 6 - y + 2\sqrt{9 + y^2}, \quad 0 \le y \le 6. \qquad \text{(key step completed)}$$

$$S'(y) = -1 + \frac{2y}{\sqrt{9 + y^2}}, \quad S'(y) = 0 \implies y = \sqrt{3}.$$

Since
$$S(0) = 12, \quad S\left(\sqrt{3}\right) = 6 + 3\sqrt{3} \cong 11.2, \quad \text{and} \quad S(6) = 6\sqrt{5} \cong 13.4,$$
the abs min of S occurs when $y = \sqrt{3}$.

To minimize the sum of the distances, take P as the point $\left(0, \sqrt{3}\right)$.

27.

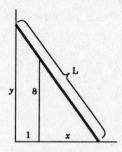

Minimize L

$$L^2 = y^2 + (x+1)^2.$$

By similar triangles $\dfrac{y}{x+1} = \dfrac{8}{x}, \quad y = \dfrac{8}{x}(x+1).$

$$L^2 = \left[\left(\frac{8}{x}\right)(x+1)\right]^2 + (x+1)^2 = (x+1)^2\left(\frac{64}{x^2} + 1\right)$$

Since L is minimal when L^2 is minimal, we consider the function

$$f(x) = (x+1)^2\left(\frac{64}{x^2} + 1\right), \quad x > 0. \qquad \text{(key step completed)}$$

$$f'(x) = 2(x+1)\left(\frac{64}{x^2} + 1\right) + (x+1)^2\left(\frac{-128}{x^3}\right)$$

$$= \frac{2(x+1)}{x^3}\left[x^3 - 64\right], \quad f'(x) = 0 \implies x = 4.$$

Since f decreases on $(0,4]$ and increases on $[4,\infty)$, the abs min of f occurs when $x = 4$.

The shortest ladder is $5\sqrt{5}$ ft long.

29.

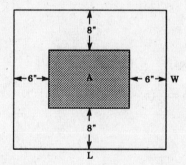

Maximize A

(We use feet rather than inches to reduce arithmetic.)

$$A = (L-1)\left(W - \frac{4}{3}\right)$$

$$LW = 27 \implies W = \frac{27}{L}$$

$$A = (L-1)\left(\frac{27}{L} - \frac{4}{3}\right) = \frac{85}{3} - \frac{27}{L} - \frac{4}{3}L$$

$$A(L) = \frac{85}{3} - \frac{27}{L} - \frac{4}{3}L, \quad 1 < L < \frac{81}{4}. \qquad \text{(key step completed)}$$

$$A'(L) = \frac{27}{L^2} - \frac{4}{3}, \quad A'(L) = 0 \implies L = \frac{9}{2}.$$

Since $A'(L) = -54/L^3 < 0$ for $1 < L < \frac{81}{4}$, the max at $L = \frac{9}{2}$ is the abs max.

The banner has length $9/2$ ft $= 54$ in. and height 6 ft $= 72$ in.

31.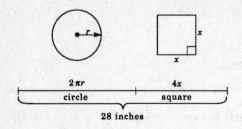

Find the extreme values of A

$A = \pi r^2 + x^2$

$2\pi r + 4x = 28 \implies x = 7 - \frac{1}{2}\pi r.$

$$A(r) = \pi r^2 + \left(7 - \frac{1}{2}\pi r\right)^2, \quad 0 \le r \le \frac{14}{\pi}. \qquad \text{(key step completed)}$$

Note: the endpoints of the domain correspond to the instances when the string is not cut: $r = 0$ when no circle is formed, $r = 14/\pi$ when no square is formed.

$$A'(r) = 2\pi r - \pi\left(7 - \frac{1}{2}\pi r\right), \quad A'(r) = 0 \implies r = \frac{14}{4 + \pi}.$$

Since $A''(r) = 2\pi + \pi^2/2 > 0$ on $(0, 14/\pi)$, the abs min of A occurs when $r = 14/(4 + \pi)$ and the abs max of A occurs at one of the endpts: $A(0) = 49$, $A(14/\pi) = 196/\pi > 49$.

(a) To maximize the sum of the two areas, use all of the string to form the circle.

(b) To minimize the sum of the two areas, use $2\pi r = 28\pi/(4 + \pi) \cong 12.32$ inches of string for the circle.

33.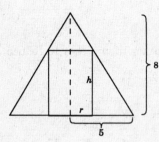

Maximize V

$V = \pi r^2 h$

By similar triangles

$\dfrac{8}{5} = \dfrac{h}{5 - r}$ or $h = \dfrac{8}{5}(5 - r).$

$$V(r) = \frac{8\pi}{5} r^2 (5 - r), \quad 0 < r < 5. \qquad \text{(key step completed)}$$

$$V'(r) = \frac{8\pi}{5}\left(10r - 3r^2\right), \quad V'(r) = 0 \implies r = 10/3.$$

Since V increases on $(0, 10/3]$ and decreases on $[10/3, 5)$, the abs max of V occurs when $r = 10/3$.

The cylinder with maximal volume has radius $10/3$ and height $8/3$.

35.

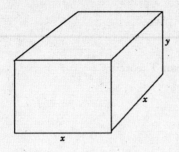

<u>Minimize C</u>

In dollars,

$$C = \text{cost base} + \text{cost top} + \text{cost sides}$$

$$= .35\left(x^2\right) + .15\left(x^2\right) + .20(4xy)$$

$$= \tfrac{1}{2}x^2 + \tfrac{4}{5}xy$$

$$\text{Volume} = x^2 y = 1250 \quad y = \frac{1250}{x^2}$$

$$C(x) = \frac{1}{2}x^2 + \frac{1000}{x}, \quad x > 0. \quad \text{(key step completed)}$$

$$C'(x) = x - \frac{1000}{x^2}, \quad C'(x) = 0 \implies x = 10.$$

Since $C''(x) = 1 + 2000/x^3 > 0$ for $x > 0$, the local min of C at $x = 10$ is the abs min.

The least expensive box is 12.5 ft tall with a square base 10 ft on a side.

37.

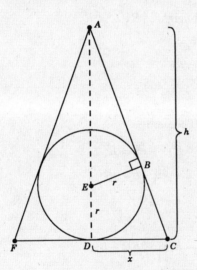

<u>Minimize A</u>

$$A = \tfrac{1}{2}(h)(2x) = hx$$

Triangles ADC and ABE are similar:

$$\frac{AD}{DC} = \frac{AB}{BE} \quad \text{or} \quad \frac{h}{x} = \frac{AB}{r}.$$

Pythagorean Theorem:

$$r^2 + (AB)^2 = (h - r)^2.$$

Thus

$$r^2 + \left(\frac{hr}{x}\right)^2 = (h - r)^2.$$

Solving this equation for h we find that

$$h = \frac{2x^2 r}{x^2 - r^2}.$$

$$A(x) = \frac{2x^3 r}{x^2 - r^2}, \quad x > r. \quad \text{(key step completed)}$$

$$A'(x) = \frac{\left(x^2 - r^2\right)\left(6x^2 r\right) - 2x^3 r(2x)}{\left(x^2 - r^2\right)^2} = \frac{2x^2 r\left(x^2 - 3r^2\right)}{\left(x^2 - r^2\right)^2},$$

$$A'(x) = 0 \implies x = r\sqrt{3}.$$

Since A decreases on $\left(r, r\sqrt{3}\,\right]$ and increases on $\left[r\sqrt{3}, \infty\right)$, the local min at $x = r\sqrt{3}$ is the abs min of A. When $x = r\sqrt{3}$, we get $h = 3r$ so that $FC = 2r\sqrt{3}$ and $AF = FC = \sqrt{h^2 + x^2} = 2r\sqrt{3}$.

The triangle of least area is equilateral with side of length $2r\sqrt{3}$.

39.

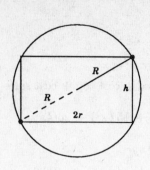

<u>Maximize V</u>

$$V = \pi r^2 h$$

By the Pythagorean Theorem,

$$(2r)^2 + h^2 = (2R)^2$$

so

$$h = 2\sqrt{R^2 - r^2}.$$

$V(r) = 2\pi r^2 \sqrt{R^2 - r^2}, \quad 0 < r < R.$ (key step completed)

$V'(r) = 2\pi \left[2r\sqrt{R^2 - r^2} - \dfrac{r^3}{\sqrt{R^2 - r^2}} \right] = \dfrac{2\pi r \left(2R^2 - 3r^2 \right)}{\sqrt{R^2 - r^2}}$

$V'(r) = 0 \implies r = \frac{1}{3}R\sqrt{6}.$

Since V increases on $\left(0, \frac{1}{3}R\sqrt{6}\right]$ and decreases on $\left[\frac{1}{3}R\sqrt{6}, R\right)$, the local max at $r = \frac{1}{3}R\sqrt{6}$ is the abs max.

The cylinder of maximal volume has base radius $\frac{1}{3}R\sqrt{6}$ and height $\frac{2}{3}R\sqrt{3}$.

41.

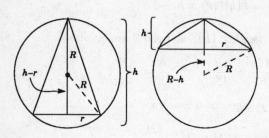

<u>Maximize V</u>

$$V = \frac{1}{3}\pi r^2 h$$

Pythagorean Theorem

Case 1 : $(h - R)^2 + r^2 = R^2$

Case 2 : $(R - h)^2 + r^2 = R^2$

Case 1 : $h \geq R$ Case 2 : $h \leq R$

In both cases

$$r^2 = R^2 - (R - h)^2 = 2hR - h^2.$$

$V(h) = \frac{1}{3}\pi \left(2h^2 R - h^3 \right), \quad 0 < h < 2R.$ (key step completed)

$V'(h) = \dfrac{1}{3}\pi \left(4hR - 3h^2 \right), \quad V'(h) = 0 \text{ at } h = \dfrac{4R}{3}.$

Since V increases on $\left(0, \frac{4}{3}R\right]$ and decreases on $\left[\frac{4}{3}R, 2R\right)$, the local max at $h = \frac{4}{3}R$ is the abs max.

The cone of maximal volume has height $\frac{4}{3}R$ and radius $\frac{2}{3}R\sqrt{2}$.

43.

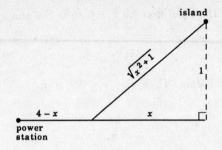

<u>Minimize C</u>

In units of $\$10,000$,

$$C = \frac{\text{cost of cable}}{\text{underground}} + \frac{\text{cost of cable}}{\text{under water}}$$

$$= \quad 3(4-x) \quad + \quad 5\sqrt{x^2+1}.$$

Clearly, the cost is unnecessarily high if

$$x > 4 \quad \text{or} \quad x < 0.$$

$$C(x) = 12 - 3x + 5\sqrt{x^2+1}, \quad 0 \le x \le 4. \qquad \text{(key step completed)}$$

$$C'(x) = -3 + \frac{5x}{\sqrt{x^2+1}}, \quad C'(x) = 0 \implies x = 3/4.$$

Since the domain of C is closed, the abs min can be identified by evaluating C at each critical point:

$$C(0) = 17, \quad C\left(\tfrac{3}{4}\right) = 16, \quad C(4) = 5\sqrt{17} \cong 20.6.$$

The minimum cost is $\$160,000$.

45. $\quad P'(\theta) = \dfrac{-mW(m\cos\theta - \sin\theta)}{(m\sin\theta + \cos\theta)^2}; \quad P$ is minimized when $\tan\theta = m$.

47.

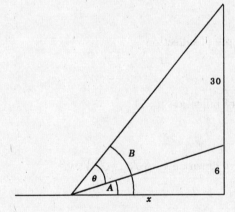

<u>Maximize θ</u>

Since the tangent function increases on

$[0, \pi/2)$, we can maximize θ by maximizing $\tan\theta$.

$$\tan\theta = \tan(B - A)$$

$$= \frac{\tan B - \tan A}{1 + \tan B \tan A}$$

$$= \frac{36/x - 6/x}{1 + (36/x)(6/x)} = \frac{30x}{x^2 + 216}.$$

Thus, we consider

$$f(x) = \frac{30x}{x^2 + 216}, \quad x \ge 0. \qquad \text{(key step completed)}$$

$$f'(x) = \frac{(x^2 + 216)\,30 - 30x(2x)}{(x^2 + 216)^2} = \frac{30\left(216 - x^2\right)}{(x^2 + 216)^2},$$

$$f'(x) = 0 \implies x = 6\sqrt{6}.$$

Since f increases on $[0, 6\sqrt{6}]$ and decreases on $[6\sqrt{6}, \infty)$, the local max at $x = 6\sqrt{6}$ is the abs max. The observer should sit $6\sqrt{6}$ ft from the screen.

49. Let x be the number of customers and P the net profit in dollars. Then $0 \le x \le 250$ and

$$P(x) = \begin{cases} 12x, & 0 \le x \le 50 \\ [12 - 0.06(x - 50)]x, & 50 < x \le 250 \end{cases};$$

$$P'(x) = \begin{cases} 12, & 0 \le x \le 50 \\ 62x - 0.06x^2, & 50 < x \le 250 \end{cases}.$$

The critical numbers are $x = 0$, $x = 50$, $x = 125$, and $x = 250$. From $P(0) = 0$, $P(50) = 600$, $P(125) = 937.50$, and $P(250) = 0$, we conclude that the net profit is maximized by servicing 125 customers.

SECTION 4.6

1. (a) f is increasing on $[a, b]$, $[d, n]$; f is decreasing on $[b, d]$, $[n, p]$.

(b) The graph of f is concave up on (c, k), (l, m);

The graph of f is concave down on (a, c), (k, l), (m, p).

The x-coordinates of the points of inflection are: $x = c$, $x = k$, $x = l$, $x = m$.

3. $f'(x) = -x^{-2}$, $f''(x) = 2x^{-3}$;

concave down on $(-\infty, 0)$, concave up on $(0, \infty)$; no pts of inflection

5. $f'(x) = 3x^2 - 3$, $f''(x) = 6x$;

concave down on $(-\infty, 0)$, concave up on $(0, \infty)$; pt of inflection $(0, 2)$

7. $f'(x) = x^3 - x$, $f''(x) = 3x^2 - 1$;

concave up on $\left(-\infty, -\frac{1}{3}\sqrt{3}\right)$ and on $\left(\frac{1}{3}\sqrt{3}, \infty\right)$, concave down on $\left(-\frac{1}{3}\sqrt{3}, \frac{1}{3}\sqrt{3}\right)$;

pts of inflection $\left(-\frac{1}{3}\sqrt{3}, -\frac{5}{36}\right)$ and $\left(\frac{1}{3}\sqrt{3}, -\frac{5}{36}\right)$

9. $f'(x) = -\dfrac{x^2 + 1}{(x^2 - 1)^2}$, $f''(x) = \dfrac{2x(x^2 + 3)}{(x^2 - 1)^3}$;

concave down on $(-\infty, -1)$ and on $(0, 1)$, concave up on $(-1, 0)$ and on $(1, \infty)$;

pt of inflection $(0, 0)$

11. $f'(x) = 4x^3 - 4x$, $f''(x) = 12x^2 - 4$;

concave up on $\left(-\infty, -\frac{1}{3}\sqrt{3}\right)$ and on $\left(\frac{1}{3}\sqrt{3}, \infty\right)$, concave down on $\left(-\frac{1}{3}\sqrt{3}, \frac{1}{3}\sqrt{3}\right)$;

pts of inflection $\left(-\frac{1}{3}\sqrt{3}, \frac{4}{9}\right)$ and $\left(\frac{1}{3}\sqrt{3}, \frac{4}{9}\right)$

13. $f'(x) = \dfrac{-1}{\sqrt{x}\,(1+\sqrt{x}\,)^2}, \quad f''(x) = \dfrac{1+3\sqrt{x}}{2x\sqrt{x}\,(1+\sqrt{x}\,)^3};$

concave up on $(0, \infty)$;   no pts of inflection

15. $f'(x) = \frac{5}{3}(x+2)^{2/3}, \quad f''(x) = \frac{10}{9}(x+2)^{-1/3};$

concave down on $(-\infty, -2)$, concave up on $(-2, \infty)$;   pt of inflection $(-2, 0)$

17. $f'(x) = 2\sin x \cos x = \sin 2x, \quad f''(x) = 2\cos 2x;$

concave up on $\left(0, \frac{1}{4}\pi\right)$ and $\left(\frac{3}{4}\pi, \pi\right)$, concave down on $\left(\frac{1}{4}\pi, \frac{3}{4}\pi\right)$;

pts of inflection $\left(\frac{1}{4}\pi, \frac{1}{2}\right)$ and $\left(\frac{3}{4}\pi, \frac{1}{2}\right)$

19. $f'(x) = 2x + 2\cos 2x, \quad f''(x) = 2 - 4\sin 2x;$

concave up on $\left(0, \frac{1}{12}\pi\right)$ and on $\left(\frac{5}{12}\pi, \pi\right)$, concave down on $\left(\frac{1}{12}\pi, \frac{5}{12}\pi\right)$;

pts of inflection $\left(\dfrac{1}{12}\pi, \dfrac{72+\pi^2}{144}\right)$ and $\left(\dfrac{5}{12}\pi, \dfrac{72+25\pi^2}{144}\right)$

21. $f(x) = x^3 - 9x$

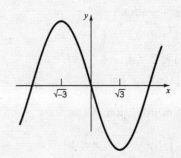

(a) $f'(x) = 3x^2 - 9 = 3(x^2 - 3)$

$f'(x) \geq 0 \Rightarrow x \leq -\sqrt{3} \text{ or } x \geq \sqrt{3};$

$f'(x) \leq 0 \Rightarrow -\sqrt{3} \leq x \leq \sqrt{3}.$

Thus, f is increasing on $(-\infty, -\sqrt{3}] \cup [\sqrt{3}, \infty)$

and decreasing on $[-\sqrt{3}, \sqrt{3}]$.

(b) $f(-\sqrt{3}) \cong 10.39$ is a local maximum;

$f(\sqrt{3}) \cong -10.39$ is a local minimum.

(c) $f''(x) = 6x;$

The graph of f is concave up on $(0, \infty)$ and concave down on $(-\infty, 0)$.

(d) point of inflection: $(0, 0)$

23. $f(x) = \dfrac{2x}{x^2 + 1}$

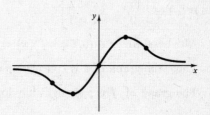

(a) $f'(x) = -\dfrac{2(x+1)(x-1)}{(x^2+1)^2}$

$f'(x) \geq 0 \Rightarrow -1 \leq x \leq 1;$

$f'(x) \leq 0 \Rightarrow x \leq -1 \text{ or } x \geq 1.$

Thus, f is increasing on $[-1, 1];$

and decreasing on $(-\infty, -1] \cup [1, \infty).$

(b) $f(-1) = -1$ is a local minimum;

$f(1) = 1$ is a local maximum.

(c) $f''(x) = \dfrac{4x(x + \sqrt{3})(x - \sqrt{3})}{(x^2 + 1)^3}$

$f''(x) > 0 \Rightarrow x \leq -\sqrt{3}$ or $x \geq \sqrt{3}$;

$f''(x) < 0 \Rightarrow -\sqrt{3} < x < \sqrt{3}.$

The graph of f is concave up on $(-\sqrt{3}, 0) \cup (\sqrt{3}, \infty)$ and concave down

on $(-\infty, -\sqrt{3}) \cup (0, \sqrt{3}).$

(d) points of inflection: $(-\sqrt{3}, -\sqrt{3}/2)$, $(0, 0)$, $(\sqrt{3}, \sqrt{3}/2)$

25. $f(x) = x + \sin x, \quad x \in [-\pi, \pi]$

(a) $f'(x) = 1 + \cos x$

$f'(x) > 0$ on $(-\pi, \pi)$

Thus, f is increasing on $[-\pi, \pi].$

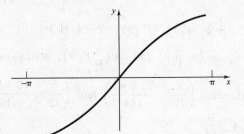

(b) No local extrema

(c) $f''(x) = -\sin x$

$f''(x) > 0$ for $x \in (-\pi, 0)$;

$f''(x) < 0$ for $x \in (0, \pi).$

The graph of f is concave up on $(\pi, 0)$ and concave down on $(0, \pi).$

(d) point of inflection: $(0, 0)$

27. $f(x) = \begin{cases} x^3, & x < 1 \\ 3x - 2, & x \geq 1 \end{cases}$

(a) $f'(x) = \begin{cases} 3x^2, & x < 1 \\ 3, & x \geq 1 \end{cases}$

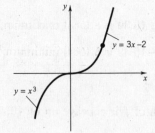

$f'(x) > 0$ on $(-\infty, 0) \cup (0, \infty)$

Thus, f is increasing on $(-\infty, \infty).$

(b) No local extrema

(c) $f''(x) = \begin{cases} 6x, & x < 1 \\ 0, & x \geq 1 \end{cases}$

$f''(x) > 0$ for $x \in (0, 1)$; $f''(x) < 0$ for $x \in (-\infty, 0).$

Thus, the graph of f is concave up on $(0, 1)$ and concave down on $(-\infty, 0).$

The graph of f is a straight line for $x \geq 1.$

(d) point of inflection: $(0, 0)$

29.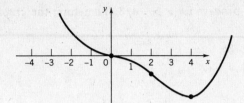

31.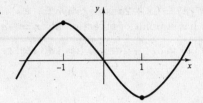

33. Since $f''(x) = 6x - 2(a + b + c)$, set $d = \frac{1}{3}(a + b + c)$. Note that $f''(d) = 0$ and that f is concave down on $(-\infty, d)$ and concave up on (d, ∞); $(d, f(d))$ is a point of inflection.

35. Since $(-1, 1)$ lies on the graph, $1 = -a + b$.

Since $f''(x)$ exists for all x and there is a pt of inflection at $x = \frac{1}{3}$, we must have $f''\left(\frac{1}{3}\right) = 0$.

Therefore

$$0 = 2a + 2b.$$

Solving these two equations, we find $a = -\frac{1}{2}$ and $b = \frac{1}{2}$.

Verification: the function

$$f(x) = -\frac{1}{2}x^3 + \frac{1}{2}x^2$$

has second derivative $f''(x) = -3x + 1$. This does change sign at $x = \frac{1}{3}$.

37. First, we require that $\left(\frac{1}{6}\pi, 5\right)$ lie on the curve:

$$5 = \frac{1}{2}A + B.$$

Next we require that $\dfrac{d^2y}{dx^2} = -4A\cos 2x - 9B\sin 3x$ be zero (and change sign) at $x = \dfrac{1}{6}\pi$:

$$0 = -2A - 9B.$$

Solving these two equations, we find $A = 18$, $B = -4$.

Verification: the function

$$f(x) = 18\cos 2x - 4\sin 3x$$

has second derivative $f''(x) = -72\cos 2x + 36\sin 3x$. This does change sign at $x = \frac{1}{6}\pi$.

39. Let $f'(x) = 3x^2 - 6x + 3$. Then we must have $f(x) = x^3 - 3x^2 + 3x + c$ for some constant c. Note that $f''(x) = 6x - 6$ and $f''(1) = 0$. Since $(1, -2)$ is a point of inflection of the graph of f, $(1, -2)$ must lie on the graph. Therefore,

$$1^3 - 3(1)^2 + 3(1) + c = -2 \quad \text{which implies} \quad c = -3$$

and so $f(x) = x^3 - 3x^2 + 3x - 3$.

41.   (a) $p''(x) = 6x + 2a$ is negative for $x < -a/3$, and positive for $x > -a/3$. Therefore, the graph of p has a point of inflection at $x = -a/3$.

(b) $p'(x) = 3x^2 + 2ax + b$ has two real zeros iff $a^2 > 3b$.

43.

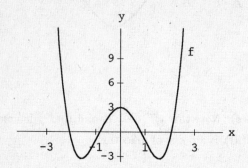

(a) concave up on $(-4, -0.913) \cup (0.913, 4)$

concave down on $(-0.913, 0.913)$

(b) points of inflection at $x = -0.913, \ 0.913$

45.

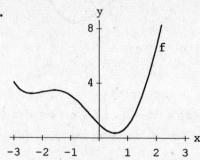

(a) concave up on $(-\pi, -1.996) \cup (-.0345, 2.550)$

concave down on $(-1.996, -0.345) \cup (2.550, \pi)$

(b) points of inflection at $x = -1.996, \ -0.345, \ 2,550$

47.

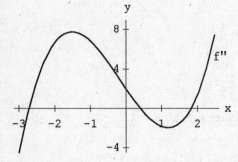

(a) concave up on $(-2.726, 0.402) \cup (1.823, 2.5)$;   concave down on $(-3, -2.726) \cup (0.402, 1.823)$.

(b) points of inflection at $x = -2.726, \ 0.402, \ 1.823$.

SECTION 4.7

1.   (a) ∞ (b) $-\infty$ (c) ∞ (d) 1

(e) 0   (f) $x = -1, \ x = 1$ (g) $y = 0, \ y = 1$

3.   vertical: $x = \frac{1}{3}$;   horizontal: $y = \frac{1}{3}$

5.   vertical: $x = 2$;   horizontal: none

7.   vertical: $x = \pm 3$;   horizontal: $y = 0$

9.   vertical: $x = -\frac{4}{3}$;   horizontal: $y = \frac{4}{9}$

11.   vertical: $x = \frac{5}{2}$;   horizontal: $y = 0$

13.   vertical: none;   horizontal: $y = \pm\frac{3}{2}$

15.   vertical: $x = 1$;   horizontal: $y = 0$

17.   vertical: none;   horizontal: $y = 0$

19.   vertical: $x = \left(2n + \frac{1}{2}\right)\pi$;   horizontal: none

21. $f'(x) = \frac{4}{3}(x + 3)^{1/3}$; neither

23. $f'(x) = -\frac{4}{5}(2 - x)^{-1/5}$; cusp

25. $f'(x) = \frac{6}{5}x^{-2/5}\left(1 - x^{3/5}\right)$; tangent

27. $f(-2)$ undefined; neither

29. $f'(x) = \begin{cases} \frac{1}{2}(x-1)^{-1/2}, & x > 1 \\ -\frac{1}{2}(1-x)^{-1/2}, & x < 1 \end{cases}$; cusp

31. $f'(x) = \begin{cases} \frac{1}{3}(x+8)^{-23}, & x > -8 \\ -\frac{1}{3}(x+8)^{-2/3}, & x < -8 \end{cases}$; cusp

33. f not continuous at 0; neither

35. 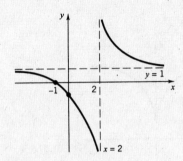   **37.**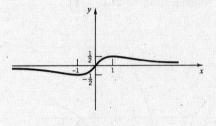

39. $f(x) = x - 3x^{1/3}$

(a) $f'(x) = 1 - \dfrac{1}{x^{2/3}}$

f is increasing on $(-\infty, -1] \cup [1, \infty)$

f is decreasing on $[-1, 1]$

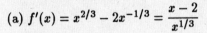

(b) $f''(x) = \frac{2}{3} x^{-5/3}$

concave up on $(0, \infty)$; concave down on $(-\infty, 0)$

vertical tangent at $(0, 0)$

41. $f(x) = \frac{3}{5} x^{5/3} - 3x^{2/3}$

(a) $f'(x) = x^{2/3} - 2x^{-1/3} = \dfrac{x-2}{x^{1/3}}$

f is increasing on $(-\infty, 0] \cup [2, \infty)$

f is decreasing on $[0, 2]$

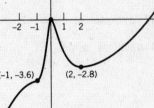

(b) $f''(x) = \frac{2}{3} x^{-1/3} + \frac{2}{3} x^{-4/3} = \dfrac{2x+2}{3x^{4/3}}$

concave up on $(-1, \infty)$; concave down on $(-\infty, -1)$

vertical cusp at $(0, 0)$

43. $f(x) = \dfrac{x^{2/3} - 1}{|x^{1/3} - 1|}$

$= \begin{cases} x^{1/3} + 1, & x > 1 \\ -(x^{1/3} + 1), & x < 1 \end{cases}$

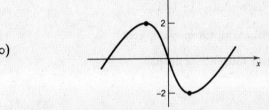
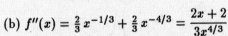
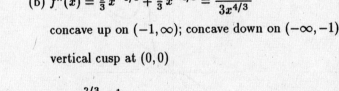

(a) $f'(x) = \left\{ \begin{array}{ll} \frac{1}{3} x^{-2/3}, & x > 1 \\ -\frac{1}{3} x^{-2/3}, & x < 1 \end{array} \right]$

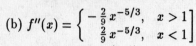

f is increasing on $[1, \infty)$

f is decreasing on $(-\infty, 1]$

(b) $f''(x) = \left\{ \begin{array}{ll} -\frac{2}{9} x^{-5/3}, & x > 1 \\ \frac{2}{9} x^{-5/3}, & x < 1 \end{array} \right]$

concave up on $(0, 1)$; concave down on $(-\infty, 0) \cup (1, \infty)$

vertical tangent at $(0, -1)$

45.

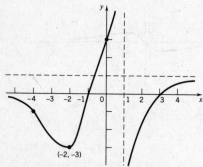

vertical asymptote: x=1

horizontal asymptotes: y=0, y=2

no vertical tangents or cusps

47.

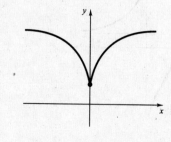

vertical cusp at $x = 0$

49.

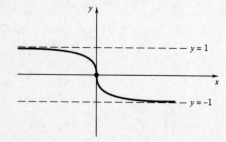

horizontal asymptotes: $y = -1$, $y = 1$

vertical tangent at $x = 0$

51.   (a) p odd; (b) p even.

53.

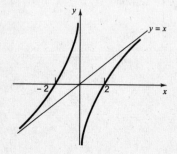

55.

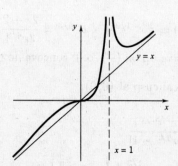

SECTION 4.8

[Rough sketches; not scale drawings]

1. $f(x) = (x - 2)^2$

$f'(x) = 2(x - 2)$

$f''(x) = 2$

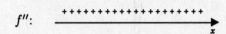

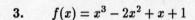

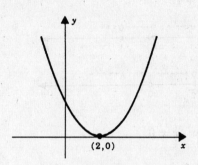

3. $f(x) = x^3 - 2x^2 + x + 1$

$f'(x) = (3x - 1)(x - 1)$

$f''(x) = 6x - 4$

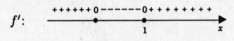

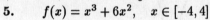

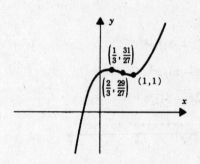

5. $f(x) = x^3 + 6x^2, \quad x \in [-4, 4]$

$f'(x) = 3x(x + 4)$

$f''(x) = 6x + 12$

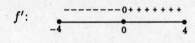

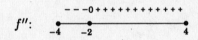

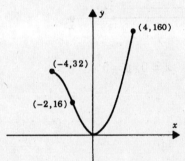

7. $f(x) = \frac{2}{3}x^3 - \frac{1}{2}x^2 - 10x - 1$

$f'(x) = (2x - 5)(x + 2)$

$f''(x) = 4x - 1$

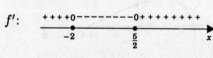

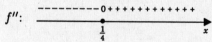

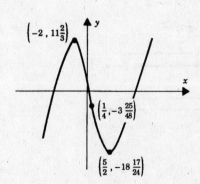

9. $f(x) = x^2 + 2x^{-1}$

$f'(x) = 2x - 2x^{-2} = 2\left(x^3 - 1\right)/x^2$

$f''(x) = 2 + 4x^{-3}$

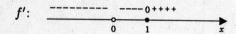

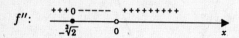

asymptote: $x = 0$

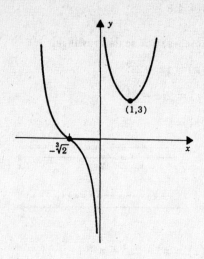

11. $f(x) = (x - 4)/x^2$

$f'(x) = (8 - x)/x^3$

$f''(x) = (2x - 24)/x^4$

$f':$

$f'':$

asymptotes: $x = 0, \ y = 0$

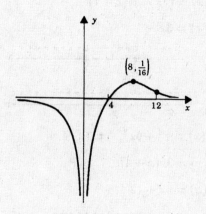

13. $f(x) = 2x^{1/2} - x, \quad x \in [0, 4]$

$f'(x) = x^{-1/2}\left(1 - x^{1/2}\right)$

$f''(x) = -\tfrac{1}{2}x^{-3/2}$

$f':$

$f'':$

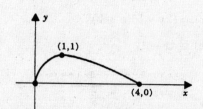

15. $f(x) = 2 + (x+1)^{6/5}$

$f'(x) = \frac{6}{5}(x+1)^{1/5}$

$f''(x) = \frac{6}{25}(x+1)^{-4/5}$

f': `--------- 0+ + + + + + + + + + +`
      -1

f'': `+ + + + + + + dne+ + + + + + + + + +`
      -1

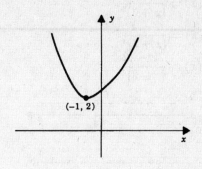

17. $f(x) = 3x^5 + 5x^3$

$f'(x) = 15x^2\left(x^2+1\right)$

$f''(x) = 30x\left(2x^2+1\right)$

f': `+ + + + + + + + 0 + + + + + + + + + + +`
      0

f'': `--------- 0+ + + + + + + + + + +`
      0

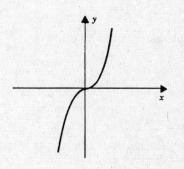

19. $f(x) = 1 + (x-2)^{5/3}$

$f'(x) = \frac{5}{3}(x-2)^{2/3}$

$f''(x) = \frac{10}{9}(x-2)^{-1/3}$

f': `+ + + + + + + + 0 + + + + + + + + + + + +`
      2

f'': `--------- dne+ + + + + + + + + +`
      2

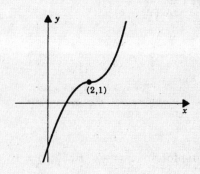

21. $f(x) = \dfrac{2x}{4x-3}$

$f'(x) = -6(4x-3)^{-2}$

$f''(x) = 48(4x-3)^{-3}$

f': `--------   --------`
      $\frac{3}{4}$

f'': `--------   + + + + + + +`
      $\frac{3}{4}$

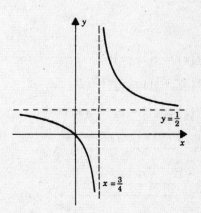

asymptotes: $x = 3/4,\; y = 1/2$

23. $f(x) = \dfrac{x}{(x+3)^2}$

$f'(x) = \dfrac{3-x}{(x+3)^3}$

$f''(x) = \dfrac{2x-12}{(x+3)^4}$

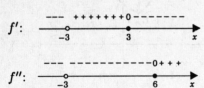

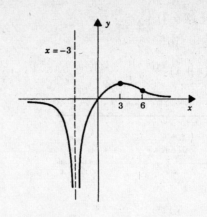

$f':$

$f'':$

asymptotes: $x = -3,\ y = 0$

25. $f(x) = \dfrac{x^2}{x^2-4}$

$f'(x) = \dfrac{-8x}{\left(x^2-4\right)^2}$

$f''(x) = \dfrac{8\left(3x^2+4\right)}{\left(x^2-4\right)^3}$

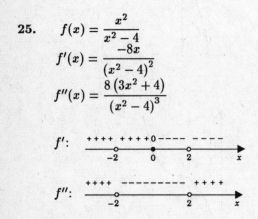

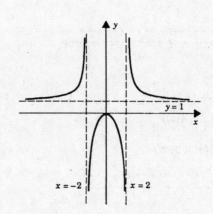

asymptotes: $x = -2,\ x = 2,\ y = 1$

27. $f(x) = x(1-x)^{1/2}$

$f'(x) = \tfrac{1}{2}(1-x)^{-1/2}(2-3x)$

$f''(x) = \tfrac{1}{4}(1-x)^{-3/2}(3x-4)$

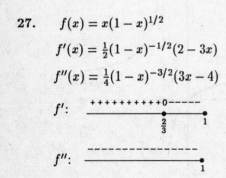

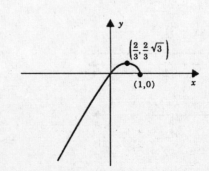

29. $f(x) = x + \sin 2x, \quad x \in [0, \pi]$

$f'(x) = 1 + 2\cos 2x$

$f''(x) = -4\sin 2x$

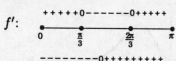

$f':$

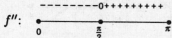

$f'':$

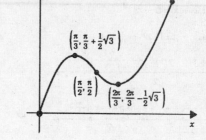

31. $f(x) = \cos^4 x, \quad x \in [0, \pi]$

$f'(x) = -4\cos^3 x \sin x$

$f''(x) = 4\cos^2 x \left(3\sin^2 x - \cos^2 x\right)$

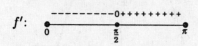

$f':$

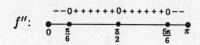

$f'':$

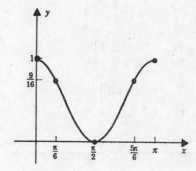

33. $f(x) = 2\sin^3 x + 3\sin x, \quad x \in [0, \pi]$

$f'(x) = 3\cos x \left(2\sin^2 x + 1\right)$

$f''(x) = 9\sin x \left(1 - 2\sin^2 x\right)$

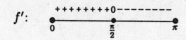

$f':$

$f'':$

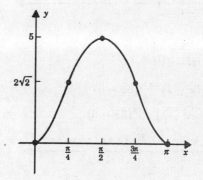

35. $f(x) = [(x+1) - 1]^3 + 1$

$f'(x) = 3x^2$

$f''(x) = 6x$

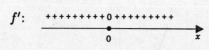

$f':$

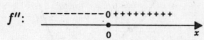

$f'':$

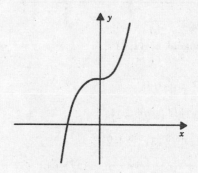

37. $f(x) = x^2(5-x)^3$

$f'(x) = 5x(2-x)(5-x)^2$

$f''(x) = 10(5-x)\left(2x^2 - 8x + 5\right)$

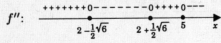

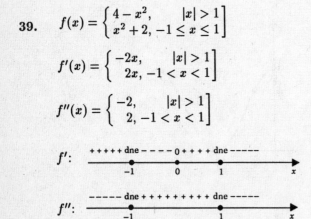

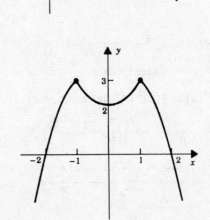

39. $f(x) = \begin{cases} 4-x^2, & |x| > 1 \\ x^2+2, & -1 \le x \le 1 \end{cases}$

$f'(x) = \begin{cases} -2x, & |x| > 1 \\ 2x, & -1 < x < 1 \end{cases}$

$f''(x) = \begin{cases} -2, & |x| > 1 \\ 2, & -1 < x < 1 \end{cases}$

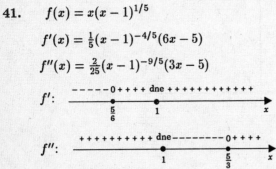

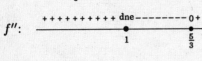

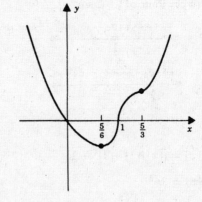

41. $f(x) = x(x-1)^{1/5}$

$f'(x) = \frac{1}{5}(x-1)^{-4/5}(6x-5)$

$f''(x) = \frac{2}{25}(x-1)^{-9/5}(3x-5)$

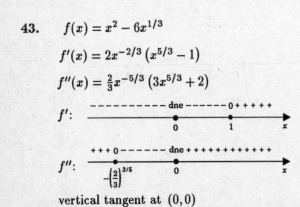

vertical tangent at $(1,0)$

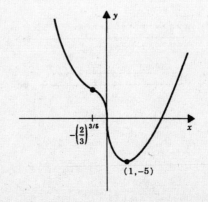

43. $f(x) = x^2 - 6x^{1/3}$

$f'(x) = 2x^{-2/3}\left(x^{5/3} - 1\right)$

$f''(x) = \frac{2}{3}x^{-5/3}\left(3x^{5/3} + 2\right)$

vertical tangent at $(0,0)$

45. $f(x) = \left(\dfrac{x}{x-2}\right)^{1/2}$; $x \le 0,\ x > 2$

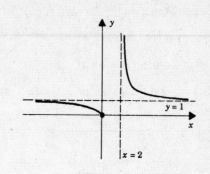

$f'(x) = -\left(\dfrac{x}{x-2}\right)^{-1/2}(x-2)^{-2}$

$f''(x) = (2x-1)\left(\dfrac{x}{x-2}\right)^{-3/2}(x-2)^{-4}$

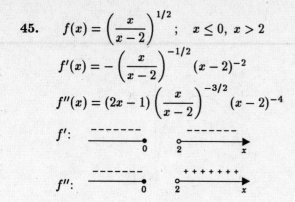

asymptotes: $x = 2,\ y = 1$

47. $f(x) = x^2\left(x^2 - 2\right)^{-1/2}$, $|x| > \sqrt{2}$

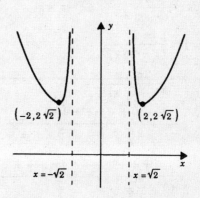

$f'(x) = x\left(x^2 - 4\right)\left(x^2 - 2\right)^{-3/2}$

$f''(x) = 2\left(x^2 + 4\right)\left(x^2 - 2\right)^{-5/2}$

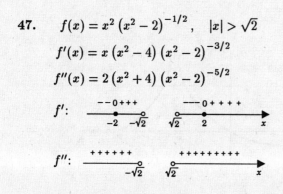

asymptotes: $x = -\sqrt{2},\ x = \sqrt{2}$

49. $f(x) = 2\sin 3x$, $x \in [0, \pi]$

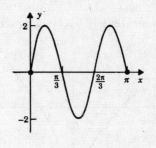

$f'(x) = 6\cos 3x$

$f''(x) = -18\sin 3x$

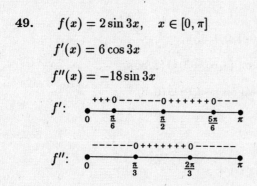

51. $f(x) = 2 \tan x - \sec^2 x, \quad x \in (0, \pi/2)$

$\qquad = -(1 - \tan x)^2$

$\quad f'(x) = 2 \sec^2 x \, (1 - \tan x)$

$\quad f''(x) = -2 \sec^2 x \, (3 \tan^2 x - 2 \tan x + 1)$

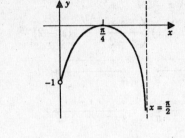

f':

```
+++++++0--------
0        π/4       π/2
```

f'':

```
----------------
0                π/2
```

asymptote: $x = \frac{1}{2}\pi$

53. $f(x) = \dfrac{\sin x}{1 - \sin x}, \quad x \in (-\pi, \pi)$

$\quad f'(x) = \dfrac{1}{(1 - \sin x)^2}$

$\quad f''(x) = \dfrac{2 \cos x}{(1 - \sin x)^3}$

f':

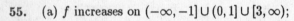

```
---- 0 + + + + dne ----
-π    -π/2   0    π/2    π
```

f'':

```
+ + + + + + + + + dne + + +
-π     -π/2    0     π/2     π
```

asymptote: $x = \frac{1}{2}\pi$

55. (a) f increases on $(-\infty, -1] \cup (0, 1] \cup [3, \infty)$;

$\qquad f$ decreases on $[-1, 0) \cup [1, 3]$;   critical numbers: $x = -1, 0, 1, 3.$

(b)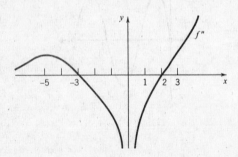

concave up on $(-\infty, -3) \cup (2, \infty)$

concave down on $(-3, 0) \cup (0, 2).$

(c)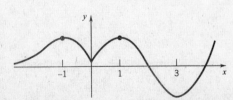

The graph does not necessarily have

a horizontal asymptote.

57. Solve the equation

$$\frac{x^2}{a^2} - \frac{y^2}{b^2} = 1$$

for y :

$$y^2 = \frac{b^2(x^2 - a^2)}{a^2} \quad \text{and}$$

$$y = \pm \frac{b}{a} \sqrt{x^2 - a^2} = \pm \frac{b}{a} x \sqrt{1 - \frac{a^2}{x^2}}$$

Now, for $|x|$ large, $y \cong \pm \dfrac{b}{a} x$.

PROJECTS AND EXPLORATIONS

4.1. (a) $A = wd$ where $d = \sqrt{4r^2 - w^2}$, $\quad 0 \le w \le 2r$.

Thus,

$$A = w\sqrt{4r^2 - w^2} \quad \text{and} \quad A' = \frac{4r^2 - 2w^2}{\sqrt{4r^2 - w^2}}$$

The absolute maximum of A on $[0, 2r]$ occurs at $w = \sqrt{2}\,r$. At this value of w,

$d = \sqrt{2}\,r$, and so the beam that has maximum cross-sectional area is square.

(b) The four planks are identical. Let w be the width and d the depth of one of the planks. Then

$A = wd$ where $d = \frac{1}{2}\left[\sqrt{4r^2 - w^2} - \sqrt{2}\,r\right]$, $\quad 0 \le w \le \sqrt{2}\,r$.

Thus,

$$A = \frac{1}{2}\,w\left[\sqrt{4r^2 - w^2} - \sqrt{2}\,r\right]$$

and

$$A' = \frac{4r^2 - 2w^2 - \sqrt{2}\,r\sqrt{4r^2 - w^2}}{2\sqrt{4r^2 - w^2}}.$$

The absolute maximum of A on $[0, \sqrt{2}\,r]$ occurs at $w = \dfrac{r}{2}\sqrt{7 - \sqrt{17}} \cong 3.3923\,r$.

At this value of w, $d = \dfrac{r}{4}\left[\sqrt{9 + \sqrt{17}} - 2\sqrt{2}\right] \cong 0.7942\,r$ and $w/d \cong 4.2713$.

(c) $S = wd^2$, where $d^2 = 4r^2 - w^2$, $\quad 0 \le w \le 2r$.

Thus,

$$S = w(4r^2 - w^2) \quad \text{and} \quad S' = 4r^2 - 3w^2.$$

The absolute maximum of S on $[0, 2r]$ occurs at $w = \dfrac{2}{\sqrt{3}}\,r$.

At this value of w, $d = \dfrac{2\sqrt{2}}{\sqrt{3}}\,r$, and $w/d = \sqrt{2}/2$.

(d) Since the beam of maximum strength is not square, there will be two types of planks (see figure).

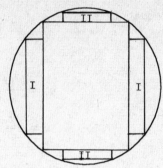

For plank I, $S = wd^2$, where $d^2 = 4 \left[r^2 - \left(\frac{1}{\sqrt{3}} r + w \right)^2 \right]$, $\quad 0 \leq w \leq \left(1 - \frac{1}{\sqrt{3}} \right) r$.

Thus,

$$S = 4w \left[r^2 - \left(\frac{1}{\sqrt{3}} r + w \right)^2 \right] = 4 \left[\frac{2}{3} r^2 w - \frac{2}{\sqrt{3}} rw^2 - w^3 \right]$$

and

$$S' = 4 \left[\frac{2}{3} r^2 - \frac{4}{\sqrt{3}} rw - 3w^2 \right]$$

The absolute maximum of S on $\left[0, \left(1 - \frac{1}{\sqrt{3}} \right) r \right]$ occurs at $w = \frac{\sqrt{10} - 2}{3\sqrt{3}} r \cong 0.2237\, r$.

At this value of w, $d = \frac{2\sqrt{2}}{3\sqrt{3}} \left(\sqrt{8 - \sqrt{10}} \right) r \cong 1.1972\, r$, and $w/d \cong 0.1869$.

For plank II, $S = wd^2$, where $d^2 = 4 \left[r^2 - \left(\frac{\sqrt{2}}{\sqrt{3}} r + w \right)^2 \right]$, $\quad 0 \leq w \leq \left(1 - \frac{\sqrt{2}}{\sqrt{3}} \right) r$.

Thus,
$$S = 4w \left[r^2 - \left(\frac{\sqrt{2}}{\sqrt{3}} r + w \right)^2 \right] = 4 \left[\frac{1}{3} r^2 w - \frac{2\sqrt{2}}{\sqrt{3}} rw^2 - w^3 \right]$$

and

$$S' = 4 \left[\frac{1}{3} r^2 - \frac{4\sqrt{2}}{\sqrt{3}} rw - 3w^2 \right]$$

The absolute maximum of S on $\left[0, \left(1 - \frac{\sqrt{2}}{\sqrt{3}} \right) r \right]$ occurs at $w = \frac{\sqrt{11} - 2\sqrt{2}}{3\sqrt{3}} r \cong 0.0940\, r$.

At this value of w, $d = \frac{2\sqrt{2}}{3\sqrt{3}} \left(\sqrt{7 - \sqrt{22}} \right) r \cong 0.8272\, r$, and $w/d \cong 0.1136$.

(e) Proceed as in parts (a)-(d). The ratio w/d for the beam of maximum cross-sectional area is $\sqrt{2}$; the beam is not square! There are two types of planks. The ratios w/d for the two types of planks with maximum cross-sectional area are (approximately) 6.0406 and 0.2376.

The ratio w/d for the beam of maximum strength is $1/2$. The ratios w/d for the two types of planks are (approximately) 0.1225 and 0.1495.

4.3.   (a)

h	$D_L(2)$	$D_R(2)$	$D_C(2)$
0.1	11.41	12.61	12.01
0.01	11.9401	12.0601	12.0001
0.001	11.994001	12.006001	12.000001
0.0005	11.997.00025	12.00300025	12.00000025

For a given h, $D_C(x)$ seems to give the best result.

(c) The graph of $f(x) = x^3$ is concave up on $(0, \infty)$. From the figure below, it is easy to see that

$$D_L(2) < f'(2) < D_R(2)$$

We need more than concavity to handle $D_C(2)$.

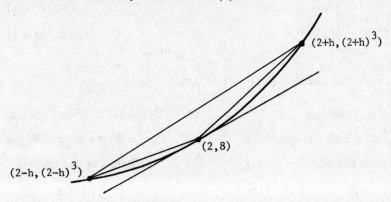

(d) $E(h) = -6h + h^2$. The graph of $E(h)$, $-0.1 \le h \le 0.1$ is shown below. It is difficult to see the concavity unless "larger" values of h are included.

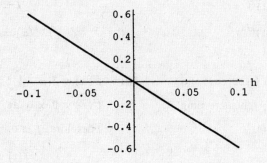

CHAPTER 5

SECTION 5.1

1. Suppose $f(x_1) = f(x_2)$ $x_1 \neq x_2$. Then

$5x_1 + 3 = 5x_2 + 3 \Rightarrow x_1 = x_2$;

f is one-to-one

$$f(t) = x$$
$$5t + 3 = x$$
$$5t = x - 3$$
$$t = \tfrac{1}{5}(x - 3)$$
$$f^{-1}(x) = \tfrac{1}{5}(x - 3)$$

3. Suppose $f(x_1) = f(x_2)$ $x_1 \neq x_2$. Then

$4x_1 - 7 = 4x_2 - 7 \Rightarrow x_1 = x_2$;

f is one-to-one

$$f(t) = x$$
$$4t - 7 = x$$
$$4t = x + 7$$
$$t = \tfrac{1}{4}(x + 7)$$
$$f^{-1}(x) = \tfrac{1}{4}(x + 7)$$

5. f is not one-to-one;   for instance, $f(1) = f(-1) = 0$

7. $f'(x) = 5x^4 \geq 0$ on $(-\infty, \infty)$ and

$f'(x) = 0$ only at $x = 0$; f is increasing.

Therefore, f is one-to-one.

$$f(t) = x$$
$$t^5 + 1 = x$$
$$t^5 = x - 1$$
$$t = (x - 1)^{1/5}$$
$$f^{-1}(x) = (x - 1)^{1/5}$$

9. $f'(x) = 9x^2 \geq 0$ on $(-\infty, \infty)$ and

$f'(x) = 0$ only at $x = 0$; f is increasing.

Therefore, f is one-to-one.

$$f(t) = x$$
$$1 + 3t^3 = x$$
$$t^3 = \tfrac{1}{3}(x - 1)$$
$$t = \left[\tfrac{1}{3}(x - 1)\right]^{1/3}$$
$$f^{-1}(x) = \left[\tfrac{1}{3}(x - 1)\right]^{1/3}$$

11. $f'(x) = 3(1 - x)^2 \geq 0$ on $(-\infty, \infty)$ and

$f'(x) = 0$ only at $x = 1$; f is increasing.

Therefore, f is one-to-one.

$$f(t) = x$$
$$(1 - t)^3 = x$$
$$1 - t = x^{1/3}$$
$$t = 1 - x^{1/3}$$
$$f^{-1}(x) = 1 - x^{1/3}$$

13. $f'(x) = 3(x + 1)^2 \geq 0$ on $(-\infty, \infty)$ and

$f'(x) = 0$ only at $x = -1$; f is increasing.

Therefore, f is one-to-one.

$$f(t) = x$$
$$(t + 1)^3 + 2 = x$$
$$(t + 1)^3 = x - 2$$
$$t + 1 = (x - 2)^{1/3}$$
$$t = (x - 2)^{1/3} - 1$$
$$f^{-1}(x) = (x - 2)^{1/3} - 1$$

15. $f'(x) = \dfrac{3}{5x^{2/5}} > 0$ for all $x \neq 0$;

    f is increasing on $(-\infty, \infty)$

$$f(t) = x$$
$$t^{3/5} = x$$
$$t = x^{5/3}$$
$$f^{-1}(x) = x^{5/3}$$

17. $f'(x) = 3(2 - 3x)^2 \geq 0$ for all x and

    $f'(x) = 0$ only at $x = 2/3$; f is increasing

$$f(t) = x$$
$$(2 - 3t)^3 = x$$
$$2 - 3t = x^{1/3}$$
$$3t = 2 - x^{1/3}$$
$$t = \tfrac{1}{3}(2 - x^{1/3})$$
$$f^{-1}(x) = \tfrac{1}{3}(2 - x^{1/3})$$

19. $f'(x) = -\dfrac{1}{x^2} < 0$ for all $x \neq 0$;

    f is decreasing on $(-\infty, 0) \cup (0, \infty)$

$$f(t) = x$$
$$\frac{1}{t} = x$$
$$t = \frac{1}{x}$$
$$f^{-1}(x) = \frac{1}{x}$$

21. f is not one-to-one; for instance

    $f\left(\tfrac{1}{2}\right) = f(2) = \tfrac{5}{2}$

23. $f'(x) = -\dfrac{3x^2}{(x^3 + 1)^2} \leq 0$ for all $x \neq -1$;

    f is decreasing on $(-\infty, -1) \cup (-1, \infty)$

$$f(t) = x$$
$$\frac{1}{t^3 + 1} = x$$
$$t^3 + 1 = \frac{1}{x}$$
$$t^3 = \frac{1}{x} - 1$$
$$t = \left(\frac{1}{x} - 1\right)^{1/3}$$
$$f^{-1}(x) = \left(\frac{1}{x} - 1\right)^{1/3}$$

25. $f'(x) = \dfrac{1}{(x + 2)^2} > 0$ for all $x \neq -2$;

    f is increasing on $(-\infty, -2) \cup (-2, \infty)$

$$f(t) = x$$
$$\frac{t + 2}{t + 1} = x$$
$$t + 2 = xt + x$$
$$t(1 - x) = x - 2$$
$$t = \frac{x - 2}{1 - x}$$
$$f^{-1}(x) = \frac{x - 2}{1 - x}$$

27. They are equal.

29.   **31.** 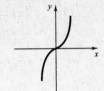   **33.**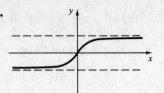

35. $f'(x) = 3x^2 \geq 0$ on $I = (-\infty, \infty)$ and $f'(x) = 0$ only at $x = 0$; f is increasing on I and so it has an inverse.

$$f(2) = 9 \quad \text{and} \quad f'(2) = 12; \quad (f^{-1})'(9) = \frac{1}{f'(2)} = \frac{1}{12}$$

37. $f'(x) = 1 + \dfrac{1}{\sqrt{x}} > 0$ on $I = (0, \infty)$; f is increasing on I and so it has an inverse.

$$f(4) = 8 \quad \text{and} \quad f'(4) = 1 + \frac{1}{2} = \frac{3}{2}; \quad (f^{-1})'(8) = \frac{1}{f'(4)} = \frac{1}{3/2} = \frac{2}{3}$$

39. $f'(x) = 2 - \sin x > 0$ on $I = (-\infty, \infty)$; f is increasing on I and so it has an inverse.

$$f(\pi/2) = \pi \quad \text{and} \quad f'(\pi/2) = 1; \quad (f^{-1})'(\pi) = \frac{1}{f'(\pi/2)} = 1$$

41. $f'(x) = \sec^2 x > 0$ on $I = (-\pi/2, \pi/2)$; f is increasing on I and so it has an inverse.

$$f(\pi/3) = \sqrt{3} \quad \text{and} \quad f'(\pi/3) = 4; \quad (f^{-1})'(\sqrt{3}) = \frac{1}{f'(\pi/3)} = \frac{1}{4}$$

43. Let $x \in \text{dom}(f^{-1})$ and let $f(z) = x$. Then

$$(f^{-1})'(x) = \frac{1}{f'(z)} = \frac{1}{f(z)} = \frac{1}{x}$$

45. Let $x \in \text{dom}(f^{-1})$ and let $f(z) = x$. Then

$$(f^{-1})'(x) = \frac{1}{f'(z)} = \frac{1}{\sqrt{1 - [f(z)]^2}} = \frac{1}{\sqrt{1 - x^2}}$$

47. (a)

$$f'(x) = \frac{(cx + d)a - (ax + b)c}{(cx + d)^2} = \frac{ad - bc}{(cx + d)^2}, \quad x \neq -d/c$$

Thus, $f'(x) \neq 0$ iff $ad - bc \neq 0$.

   (b)

$$\frac{at + b}{ct + d} = x$$

$$at + b = ctx + dx$$

$$(a - cx)t = dx - b$$

$$t = \frac{dx - b}{a - cx}; \quad f^{-1}(x) = \frac{dx - b}{a - cx}$$

49.   Differentiating implicitly, we have

$$y' \sin x + y \cos x - (1 - y') \sin(x - y) = 2x + 2yy'$$

Solving for $y' =$ and evaluating at $(\pi/2, 0)$ we get

$$y' = \frac{2x - y \cos x + \sin(x - y)}{\sin x + \sin(x - y) - 2y} \quad \text{and} \quad y' = \frac{\pi + 1}{2} \quad \text{at} \quad (\pi/2, 0)$$

Now, $(f^{-1})'(0) = \dfrac{1}{f'(\pi/2)} = \dfrac{2}{\pi + 1}$.

51.   (a) $g'(x) = \dfrac{1}{f'[g(x)]};\quad g''(x) = -\dfrac{1}{(f'[g(x)])^2} f''[g(x)]g'(x) = -\dfrac{f''[g(x)]}{(f'[g(x)])^3}$

(b) If f is increasing and its graph is concave up (down), then the graph of g is concave down (up). On the other hand, if f is decreasing then the graphs of f and g have the the same concavity.

53.   Let $f(x) = \sin x$ and let $y = f^{-1}(x)$. Then

$$\sin y = x$$
$$\cos y \, \frac{dy}{dx} = 1$$
$$\frac{dy}{dx} = \frac{1}{\cos y} \quad (y \neq \pm \pi/2)$$
$$= \frac{1}{\sqrt{1 - \sin^2 y}} = \frac{1}{\sqrt{1 - x^2}} \quad (x \neq \pm 1)$$

55. $f'(x) = 3x^2 + 3 > 0$ for all x;   **57.** $f'(x) = 8 \cos 2x > 0$ for $x \in (-\pi/4, \pi/4)$;

f is increasing on $(-\infty, \infty)$ f is increasing on $[-\pi/4, \pi/4]$

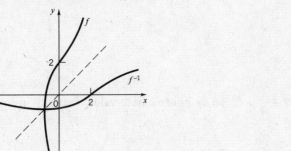   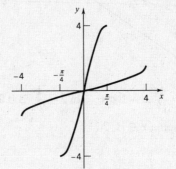

SECTION 5.2

1. $\ln 20 = \ln 2 + \ln 10 \cong 2.99$

3. $\ln 1.6 = \ln \frac{16}{10} = 2\ln 4 - \ln 10 \cong 0.48$

5. $\ln 0.1 = \ln \frac{1}{10} = \ln 1 - \ln 10 \cong -2.30$

7. $\ln 7.2 = \ln \frac{72}{10} = \ln 8 + \ln 9 - \ln 10 \cong 1.98$

9. $\ln \sqrt{2} = \frac{1}{2} \ln 2 \cong 0.35$

11. $\ln 2^5 = 5 \ln 2 \cong 3.45$

13. (a) $\ln 5.2 \cong \ln 5 + \frac{1}{5}(0.2) \cong 1.61 + 0.04 = 1.65$

    (b) $\ln 4.8 \cong \ln 5 - \frac{1}{5}(0.2) \cong 1.61 - 0.04 = 1.57$

    (c) $\ln 5.5 \cong \ln 5 + \frac{1}{5}(0.5) \cong 1.61 + 0.1 = 1.71$

15. $x = e^2$ **17.** $2 - \ln x = 0$ or $\ln x = 0$.   Thus $x = e^2$ or $x = 1$.

19.
$$\ln[(2x+1)(x+2)] = 2\ln(x+2)$$
$$\ln[(2x+1)(x+2)] = \ln\left[(x+2)^2\right]$$
$$(2x+1)(x+2) = (x+2)^2$$
$$x^2 + x - 2 = 0$$
$$(x+2)(x-1) = 0$$
$$x = -2, 1$$

We disregard the solution $x = -2$ since it does not satisfy the initial equation.

Thus, the only solution is $x = 1$.

21. See Exercises 3.1, Definition (3.1.5).

$$\lim_{x \to 1} \frac{\ln x}{x-1} = \frac{d}{dx}(\ln x)\bigg|_{x=1} = \frac{1}{x}\bigg|_{x=1} = 1$$

23. (a) $\ln 1 - g(1) = 1 > 0$, $\ln 2 - g(2) = \ln 2 - 2 < 0$,   so by intermediate-value theorem

        $\ln r - g(r) = 0$ for some $r \in [1, 2]$.

    (b) $r \cong 1.7915$

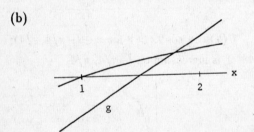

25. (a) $\ln 1 - \frac{1}{1^2} = -1 < 0$, $\ln 2 - \frac{1}{2^2} \cong 0.69 - \frac{1}{4} > 0$,   so by intermediate-value theorem $\ln r - \frac{1}{r^2} = 0$

        for some $r \in [1, 2]$.

    (b) $r \cong 1.5316$

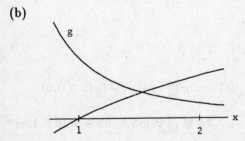

27. $\lim\limits_{x \to 0} x \ln|x| = 0$

29. (a) On $\left[1, 1+\dfrac{1}{n}\right]$: $y(x) = y'(x_0)(x-1) + y(1) = 1(x-1) = x - 1$; $y\left(1+\dfrac{1}{n}\right) = \dfrac{1}{n}$

(b) On $\left[1+\dfrac{1}{n}, 1+\dfrac{2}{n}\right]$:

$$y(x) = y'(x_1)(x - [1+1/n]) + y(1+1/n) = \frac{1}{1+1/n}\left(x - \frac{n+1}{n}\right) + \frac{1}{n}$$

$$= \frac{n}{n+1}\left(x - \frac{n+1}{n}\right) + \frac{1}{n} = \frac{n}{n+1}x - 1 + \frac{1}{n}$$

$$y\left(1+\frac{2}{n}\right) = \frac{n}{n+1}\left(1+\frac{2}{n}\right) - 1 + \frac{1}{n} = \frac{n}{n+1}\cdot\frac{n+2}{n} - 1 + \frac{1}{n} = \frac{1}{n} + \frac{1}{n+1}$$

(c) On $\left[1+\dfrac{2}{n}, 1+\dfrac{3}{n}\right]$:

$$y(x) = y'(x_2)(x - [1+2/n]) + y(1+2/n) = \frac{1}{1+2/n}\left(x - \frac{n+2}{n}\right) + \frac{1}{n} + \frac{1}{n+1}$$

$$= \frac{n}{n+2}\left(x - \frac{n+2}{n}\right) + \frac{1}{n} + \frac{1}{n+1}$$

$$= \frac{n}{n+2}x - 1 + \frac{1}{n} + \frac{1}{n+1}$$

$$y\left(1+\frac{3}{n}\right) = \frac{n}{n+2}\left(1+\frac{3}{n}\right) - 1 + \frac{1}{n} + \frac{1}{n+1} = \frac{n}{n+2}\cdot\frac{n+3}{n} - 1 + \frac{1}{n} + \frac{1}{n+2} = \frac{1}{n} + \frac{1}{n+1} + \frac{1}{n+2}$$

Continuing, it is easy to verify that

$$y(2) = y\left(1+\frac{n}{n}\right) = \frac{1}{n} + \frac{1}{n+1} + \cdots + \frac{1}{2n-1}$$

(d) $n = 20$: $y(2) \cong 0.7508$; $n = 30$: $y(2) = 0.7015$; $n = 40$: $y(2) = 0.6994$

SECTION 5.3

1. $\operatorname{dom}(f) = (0,\infty)$, $f'(x) = \dfrac{1}{4x}(4) = \dfrac{1}{x}$

3. $\operatorname{dom}(f) = (-1,\infty)$, $f'(x) = \dfrac{1}{x^3+1}\dfrac{d}{dx}(x^3+1) = \dfrac{3x^2}{x^3+1}$

5. $\operatorname{dom}(f) = (-\infty,\infty)$, $f(x) = \dfrac{1}{2}\ln(1+x^2)$ so $f'(x) = \dfrac{1}{2}\left[\dfrac{1}{1+x^2}(2x)\right] = \dfrac{x}{1+x^2}$

7. $\operatorname{dom}(f) = \{x \mid x \neq \pm 1\}$, $f'(x) = \dfrac{1}{x^4-1}\dfrac{d}{dx}(x^4-1) = \dfrac{4x^3}{x^4-1}$

9. $\operatorname{dom}(f) = (0,\infty)$, $f'(x) = x^2\dfrac{d}{dx}(\ln x) + 2x(\ln x) = x + 2x\ln x$

11. $\operatorname{dom}(f) = (0,1) \cup (1, \infty)$, $f(x) = (\ln x)^{-1}$ so $f'(x) = -(\ln x)^{-2} \dfrac{d}{dx} (\ln x) = -\dfrac{1}{x (\ln x)^2}$

13. $\operatorname{dom}(f) = (-1, \infty)$, $f'(x) = \dfrac{(x+1) \dfrac{d}{dx} [\ln (x+1)] - 1 [\ln (x+1)]}{(x+1)^2} = \dfrac{1 - \ln (x+1)}{(x+1)^2}$

15. $\operatorname{dom}(f) = (0, \infty)$, $f'(x) = \cos (\ln x) \left(\dfrac{1}{x} \right) = \dfrac{\cos (\ln x)}{x}$

17. $\operatorname{dom}(f) = \{x : |\cos x| > 0\}$; that is, the union of intervals of the form $\left(\dfrac{(2k-1)\pi}{2}, \dfrac{(2k+1)\pi}{2} \right)$,

$k = 0, \pm 1, \pm 2, \ldots$; $f'(x) = \dfrac{1}{\cos x} (-\sin x) = -\tan x$

19. $\operatorname{dom}(f) = \{x : |\sec x + \tan x|\} > 0$; that is, the union of intervals of the form

$\left(\dfrac{(2k-1)\pi}{2}, \dfrac{(2k+1)\pi}{2} \right)$, $k = 0, \pm 1, \pm 2, \ldots$;

$f'(x) = \dfrac{1}{\sec x + \tan x} (\sec x \tan x + \sec^2 x) = \sec x$

21.
$$\ln |g(x)| = 2 \ln (x^2 + 1) + 5 \ln |x - 1| + 3 \ln |x|$$

$$\frac{g'(x)}{g(x)} = 2 \left(\frac{2x}{x^2 + 1} \right) + \frac{5}{x - 1} + \frac{3}{x}$$

$$g'(x) = (x^2 + 1)^2 (x - 1)^5 x^3 \left(\frac{4x}{x^2 + 1} + \frac{5}{x - 1} + \frac{3}{x} \right)$$

23.
$$\ln |g(x)| = 4 \ln |x| + \ln |x - 1| - \ln |x + 2| - \ln (x^2 + 1)$$

$$\frac{g'(x)}{g(x)} = \frac{4}{x} + \frac{1}{x - 1} - \frac{1}{x + 2} - \frac{2x}{x^2 + 1}$$

$$g'(x) = \frac{x^4 (x - 1)}{(x + 2)(x^2 + 1)} \left(\frac{4}{x} + \frac{1}{x - 1} - \frac{1}{x + 2} - \frac{2x}{x^2 + 1} \right)$$

25.
$$\ln |g(x)| = \tfrac{1}{2} (\ln |x - 1| + \ln |x - 2| - \ln |x - 3| - \ln |x - 4|)$$

$$\frac{g'(x)}{g(x)} = \frac{1}{2} \left(\frac{1}{x - 1} + \frac{1}{x - 2} - \frac{1}{x - 3} - \frac{1}{x - 4} \right)$$

$$g'(x) = \frac{1}{2} \sqrt{\frac{(x - 1)(x - 2)}{(x - 3)(x - 4)}} \left(\frac{1}{x - 1} + \frac{1}{x - 2} - \frac{1}{x - 3} - \frac{1}{x - 4} \right)$$

27. $\dfrac{d}{dx}(\ln x) = \dfrac{1}{x}$

$\dfrac{d^2}{dx^2}(\ln x) = -\dfrac{1}{x^2}$

$\dfrac{d^3}{dx^3}(\ln x) = \dfrac{2}{x^3}$

$\dfrac{d^4}{dx^4}(\ln x) = -\dfrac{2 \cdot 3}{x^4}$

$\dfrac{d^5}{dx^5}(\ln x) = \dfrac{2 \cdot 3 \cdot 4}{x^5}$

$\vdots$

$\dfrac{d^n}{dx^n}(\ln x) = (-1)^{n-1}\dfrac{(n-1)!}{x^n}$

29. $\dfrac{d^n}{dx^n}(\ln 2x) = \dfrac{d^n}{dx^n}[\ln 2 + \ln x]$

$= 0 + \dfrac{d^n}{dx^n}(\ln x)$

$= (-1)^{n-1}\dfrac{(n-1)!}{x^n}$

See Exercise 27

31.   (a) If $g(x) = g_1(x)g_2(x)$,   5.3.3   gives $g'(x) = g(x)\left[\dfrac{g_1'(x)}{g_1(x)} + \dfrac{g_2'(x)}{g_2(x)}\right]$

$\implies g'(x) = g_1(x)g_2(x)\left[\dfrac{g_1'(x)}{g_1(x)} + \dfrac{g_2'(x)}{g_2(x)}\right] = g_1'(x)\,g_2(x) + g_1(x)\,g_2'(x).$

(b) If $g(x) = g_1(x)\left(\dfrac{1}{g_2(x)}\right)$

$\implies g'(x) = g_1(x)\left(\dfrac{1}{g_2(x)}\right)\left(\dfrac{g_1'(x)}{g_1(x)} + \dfrac{\dfrac{d}{dx}\left(\dfrac{1}{g_2(x)}\right)}{\dfrac{1}{g_2(x)}}\right) = \dfrac{g_1'(x)g_2(x) - g_1(x)g_2'(x)}{[g_2(x)]^2}$

33. $f'(x) = 1 - \dfrac{1}{x}$

$f''(x) = \dfrac{1}{x^2}$

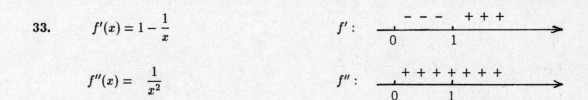

(i)   domain $(0, \infty)$

(ii)   decreases on $(0, 1]$,   increases on $[1, \infty)$

(iii) $f(1) = 1$ local and absolute min

(iv)   concave up on $(0, \infty)$;

no pts of inflection

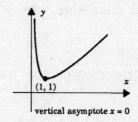

vertical asymptote $x = 0$

35. $f'(x) = -\dfrac{2x}{4 - x^2}$

$f''(x) = -\dfrac{4(2 + x^2)}{(4 - x^2)^2}$

(i)   domain $(-2, 2)$

(ii)   increases on $(-2, 0]$,   decreases on $[0, 2)$

(iii) $f(0) = \ln 4$ local and absolute max

(iv)   concave down on $(-2, 2)$;
        no pts of inflection

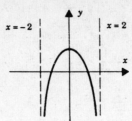

37. $f'(x) = \dfrac{8 - 2x}{8x - x^2}$

f' :

$f''(x) = -\dfrac{2(x^2 - 8x + 32)}{(8x - x^2)^2}$

f'' :

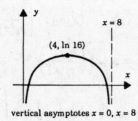

(i)   domain $(0, 8)$

(ii)   increases on $(0, 4]$,   decreases on $[4, 8)$

(iii) $f(4) = \ln 16$ local and absolute max

(iv)   concave down on $(0, 8)$;
        no pts of inflection

vertical asymptotes $x = 0$, $x = 8$

39.   (a) $f'(x) = \dfrac{1}{2x}(2) = \dfrac{1}{x}$; $g'(x) = \dfrac{1}{3x}(3) = \dfrac{1}{x}$

(b) $F'(x) = \dfrac{1}{kx}(k) = \dfrac{1}{x}$

(c) $F(x) = \ln kx = \ln k + \ln x$,   so $F'(x) = 0 + \dfrac{d}{dx}(\ln x) = \dfrac{1}{x}$.

41.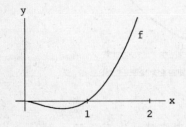

43.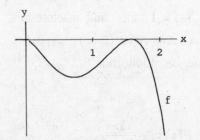

x-intercept at $x = 1$; abs min at $x \cong 0.6065$;   x-intercept at $x = \pi/2$; abs max at $x = \pi/2$;

abs max at $x = 2$ local min at $x \cong 0.7269$; abs min at $x = 2$;

SECTION 5.4

1. $\dfrac{dy}{dx} = e^{-2x} \dfrac{d}{dx}(-2x) = -2e^{-2x}$

3. $\dfrac{dy}{dx} = e^{x^2-1} \dfrac{d}{dx}(x^2-1) = 2xe^{x^2-1}$

5. $\dfrac{dy}{dx} = e^x \dfrac{d}{dx}(\ln x) + \ln x \dfrac{d}{dx}(e^x) = e^x \left(\dfrac{1}{x} + \ln x \right)$

7. $\dfrac{dy}{dx} = x^{-1} \dfrac{d}{dx}(e^{-x}) + e^{-x} \dfrac{d}{dx}(x^{-1}) = -x^{-1}e^{-x} - e^{-x}x^{-2} = -\left(x^{-1} + x^{-2} \right)e^{-x}$

9. $\dfrac{dy}{dx} = \dfrac{1}{2}\left(e^x - e^{-x} \right)$

11. $\dfrac{dy}{dx} = e^{\sqrt{x}} \dfrac{d}{dx}\left(\ln \sqrt{x} \right) + \ln \sqrt{x} \dfrac{d}{dx}\left(e^{\sqrt{x}} \right)$

$$= e^{\sqrt{x}} \left(\dfrac{1}{\sqrt{x}} \cdot \dfrac{1}{2\sqrt{x}} \right) + \ln \sqrt{x} \dfrac{e^{\sqrt{x}}}{2\sqrt{x}} = \dfrac{1}{2}e^{\sqrt{x}} \left(\dfrac{1}{x} + \dfrac{\ln \sqrt{x}}{\sqrt{x}} \right)$$

13. $\dfrac{dy}{dx} = 2\left(e^x + e^{-x} \right) \dfrac{d}{dx}\left(e^x + e^{-x} \right) = 2\left(e^x + e^{-x} \right)\left(e^x - e^{-x} \right) = 2\left(e^{2x} - e^{-2x} \right)$

15. $\dfrac{dy}{dx} = 2\left(e^{x^2} + 1 \right) \dfrac{d}{dx}\left(e^{x^2} + 1 \right) = 2\left(e^{x^2} + 1 \right)e^{x^2} \dfrac{d}{dx}\left(x^2 \right) = 4xe^{x^2}\left(e^{x^2} + 1 \right)$

17. $\dfrac{dy}{dx} = (x^2 - 2x + 2) \dfrac{d}{dx}(e^x) + e^x(2x - 2) = x^2 e^x$

19. $\dfrac{dy}{dx} = \dfrac{(e^x + 1)e^x - (e^x - 1)e^x}{(e^x + 1)^2} = \dfrac{2e^x}{(e^x + 1)^2}$

21. $\dfrac{dy}{dx} = \dfrac{\left(e^{ax} + e^{bx} \right)\left(ae^{ax} - be^{bx} \right) - \left(e^{ax} - e^{bx} \right)\left(ae^{ax} + be^{bx} \right)}{\left(e^{ax} + e^{bx} \right)^2} = 2(a-b)\dfrac{e^{(a+b)x}}{\left(e^{ax} + e^{bx} \right)^2}$

23. $y = e^{4\ln x} = \left(e^{\ln x} \right)^4 = x^4$ so $\dfrac{dy}{dx} = 4x^3.$

25. $f'(x) = \cos\left(e^{2x} \right)\left(e^{2x} \right)2 = 2e^{2x}\cos\left(e^{2x} \right)$

27. $f'(x) = e^{-2x}(-\sin x) + e^{-2x}(-2)\cos x = -e^{-2x}(2\cos x + \sin x)$

29. $f(x) = \tan\left(\sqrt{e^{-3x}} \right) = \tan\left(e^{-3x/2} \right)$

$f'(x) = \sec^2\left(e^{-3x/2} \right)e^{-3x/2}\left(-\dfrac{3}{2} \right) = -\dfrac{3}{2}e^{-3x/2}\sec^2\left(e^{-3x/2} \right)$

31. $e^{-0.4} = \dfrac{1}{e^{0.4}} \cong \dfrac{1}{1.49} \cong 0.67$

33. $e^{2.8} = \left(e^2 \right)\left(e^{0.8} \right) \cong (7.39)(2.23) \cong 16.48$

35. $x(t) = Ae^{ct} + Be^{-ct};$ $v(t) = x'(t) = Ace^{ct} - Bce^{-ct}$

$a(t) = v'(t) = x''(t) = Ac^2 e^{ct} + Bc^2 e^{-ct} = c^2\left(Ae^{ct} + Be^{-ct} \right) = c^2 x(t)$

37. $f(x) = \frac{1}{2}(e^x + e^{-x})$

$f'(x) = \frac{1}{2}(e^x - e^{-x})$

$f''(x) = \frac{1}{2}(e^x + e^{-x})$

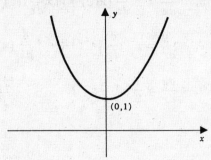

(i)   domain $(-\infty, \infty)$

(ii)   decreases on $(-\infty, 0]$,   increases on $[0, \infty)$

(iii) $f(0) = 1$ local and absolute min

(iv)   concave up everywhere

39. $f(x) = e^{(1/x)^2}$

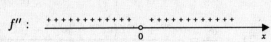

$f'(x) = \frac{-2}{x^3}e^{(1/x)^2}$

$f''(x) = \frac{6x^2 + 4}{x^6}e^{(1/x)^2}$

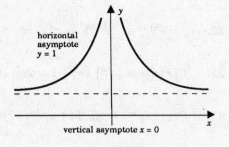

(i)   domain $(-\infty, 0) \cup (0, \infty)$

(ii)   increases on $(-\infty, 0)$,   decreases on $(0, \infty)$

(iii)   no extreme values

(iv)   concave up on $(-\infty, 0)$ and on $(0, \infty)$

41.   Numerically, $L \cong 10$;

$$\lim_{x \to 0} \frac{e^{10x} - 1}{x} = \lim_{x \to 0} \frac{e^{10x} - e^{10(0)}}{x - 0}$$

is the derivative of $f(x) = e^{10x}$ at $x = 0$. Note that

$$f'(x) = 10e^{10x} \quad \text{and} \quad f'(x) = 10$$

43.   Numerically, $L \cong 2.72$;

$$\lim_{x \to 1} \frac{e^x - e}{\ln x} = \left(\lim_{x \to 1} \frac{e^x - e^1}{x - 1}\right)\left(\lim_{x \to 1} \frac{x - 1}{\ln x - \ln 1}\right) = (e)(1) = e \cong 2.72$$

(The first limit is the derivative of $f(x) = e^x$ at $x = 1$. This is e. The second limit is the reciprocal of the derivative of $g(x) = \ln x$ at $x = 1$. This is 1.)

45. (a)

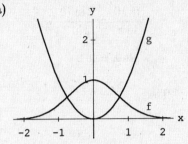

(b) Intersect at $x \cong \pm 0.7531$

(c) Area $\cong 0.98$

47.

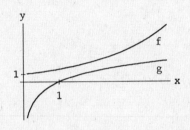

49.

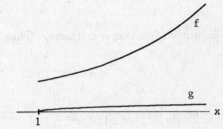

$$f\left(g(x)\right) = e^{\frac{\ln x^2}{2}} = e^{\frac{2\ln x}{2}} = e^{\ln x} = x$$

$$f\left(g(x)\right) = e^{2+\ln x - 2} = e^{\ln x} = x$$

SECTION 5.5

1. $\log_2 64 = \log_2 \left(2^6\right) = 6$

3. $\log_{64} (1/2) = \dfrac{\ln (1/2)}{\ln 64} = \dfrac{-\ln 2}{6\ln 2} = -\dfrac{1}{6}$

5. $\log_5 1 = \log_5 \left(5^0\right) = 0$

7. $\log_5 (125) = \log_5 \left(5^3\right) = 3$

9. $\log_{32} 8 = \dfrac{\ln 8}{\ln 32} = \dfrac{3\ln 2}{5\ln 2} = \dfrac{3}{5}$

11. $\log_{10} 100^{-4/5} = \log_{10} 10^{-8/5} = -\dfrac{8}{5}$

13. $\log_p xy = \dfrac{\ln xy}{\ln p} = \dfrac{\ln x + \ln y}{\ln p} = \dfrac{\ln x}{\ln p} + \dfrac{\ln y}{\ln p} = \log_p x + \log_p y$

15. $\log_p x^y = \dfrac{\ln x^y}{\ln p} = y\dfrac{\ln x}{\ln p} = y\log_p x$

17.
$$10^x = e^x$$
$$\left(e^{\ln 10}\right)^x = e^x$$
$$e^{x \ln 10} = e^x$$
$$x \ln 10 = x$$
$$x(\ln 10 - 1) = 0$$
Thus, $x = 0$.

19.
$$\log_x 10 = \log_4 100$$
$$\frac{\ln 10}{\ln x} = \frac{\ln 100}{\ln 4}$$
$$\frac{\ln 10}{\ln x} = \frac{2 \ln 10}{2 \ln 2}$$
$$\ln x = \ln 2$$
Thus, $x = 2$.

21.   Let $y = \log_2 x$. Then
$$2^y = x \implies (2^y)^2 = x^2 \text{ and } 2^{2y} = 4^y = x^2 \implies y = \log_4 x^2$$
Thus, $\log_2 x = \log_4 x^2$ for all $x > 0$.

23.   The logarithm function is increasing. Thus,
$$e^{t_1} < a < e^{t_2} \implies t_1 = \ln e^{t_1} < \ln a < \ln e^{t_2} = t_2.$$

25. $f'(x) = 3^{2x}(\ln 3)(2) = 2(\ln 3)3^{2x}$

27. $f'(x) = 2^{5x}(\ln 2)(5)3^{\ln x} + 2^{5x}3^{\ln x}(\ln 3)\dfrac{1}{x} = 2^{5x}3^{\ln x}\left(5 \ln 2 + \dfrac{\ln 3}{x}\right)$

29. $g'(x) = \frac{1}{2}\left(\log_3 x\right)^{-1/2}\left(\dfrac{1}{\ln 3}\right)\dfrac{1}{x} = \dfrac{1}{2(\ln 3)x\sqrt{\log_3 x}}$

31. $f'(x) = \sec^2\left(\log_5 x\right)(1 \ln 5)\dfrac{1}{x} = \dfrac{\sec^2\left(\log_5 x\right)}{x \ln 5}$

33. $F'(x) = -\sin\left(2^x + 2^{-x}\right)\left[2^x \ln 2 - 2^{-x}\ln 2\right] = \ln 2\left(2^{-x} - 2^x\right)\sin\left(2^x + 2^{-x}\right)$

35. $F'(x) = \dfrac{1}{\ln 2} \cdot \dfrac{1}{\log_4(2x+1)} \cdot \dfrac{1}{\ln 4} \cdot \dfrac{1}{2x+1} \cdot 2 = \dfrac{2}{(\ln 2)(\ln 4)(2x+1)\log_4(2x+1)}$

37.   Write $c = b^{\log_b c}$ Then $\log_a c = \log_a\left(b^{\log_b c}\right) = (\log_b c)(\log_a b)$ by log property.

39. $f(x) = x \log_3 x;$ $f'(x) = \log_3 x + x \cdot \dfrac{1}{x \ln 3} = \dfrac{\ln x + 1}{\ln 3};$ $f'(e) = \dfrac{2}{\ln 3}$

41. $f(x) = \log_3\left(\log_2 x\right) = \dfrac{\ln\left(\frac{\ln x}{\ln 2}\right)}{\ln 3} = \dfrac{\ln(\ln x) - \ln(\ln 2)}{\ln 3};$ $f'(x) = \dfrac{1}{\ln 3} \cdot \dfrac{1}{\ln x} \cdot \dfrac{1}{x} \implies f'(e) = \dfrac{1}{e \ln 3}$

43.
$$f(x) = p^{g(x)}$$

$$\ln f(x) = g(x) \ln p$$

$$\frac{f'(x)}{f(x)} = g'(x) \ln p$$

$$f'(x) = f(x)g'(x) \ln p = p^{g(x)} g'(x) \ln p$$

45.
$$y = (\ln x)^x$$

$$\ln y = x \ln(\ln x)$$

$$\frac{1}{y}\frac{dy}{dx} = \ln(\ln x) + x \frac{1}{\ln x} \cdot \frac{1}{x}$$

$$\frac{dy}{dx} = (\ln x)^x \left[\ln(\ln x) + \frac{1}{\ln x} \right]$$

47.
$$y = \left(\frac{1}{x}\right)^x$$

$$\ln y = x \ln \frac{1}{x} = -x \ln x$$

$$\frac{1}{y}\frac{dy}{dx} = -\ln x - x \cdot \frac{1}{x} = -\ln x - 1$$

$$\frac{dy}{dx} = -\left(\frac{1}{x}\right)^x [1 + \ln x]$$

49.
$$y = (\ln x)^{x^2+2}$$

$$\ln y = (x^2 + 2)\ln(\ln x)$$

$$\frac{1}{y}\frac{dy}{dx} = 2x \ln(\ln x) + \frac{x^2 + 2}{x \ln x}$$

$$\frac{dy}{dx} = (\ln x)^{x^2+2} \left[2x \ln(\ln x) + \frac{x^2 + 2}{x \ln x} \right]$$

51.
$$y = (\cos x)^{x^2+1}$$

$$\ln y = (x^2 + 1)\ln(\cos x)$$

$$\frac{1}{y}\frac{dy}{dx} = 2x \ln(\cos x) + (x^2 + 1)\left(-\frac{\sin x}{\cos x}\right)$$

$$\frac{dy}{dx} = (\cos x)^{x^2+1} \left[2x \ln(\cos x) - (x^2 + 1)\tan x \right]$$

53.
$$y = x^{x^2}$$

$$\ln y = x^2 \ln x$$

$$\frac{1}{y}\frac{dy}{dx} = 2x \ln x + x^2 \cdot \frac{1}{x}$$

$$\frac{dy}{dx} = x^{x^2+1}(2 \ln x + 1)$$

55.
$$y = (\tan x)^{\sec x}$$

$$\ln y = \sec x \, \ln(\tan x)$$

$$\frac{1}{y}\frac{dy}{dx} = \sec x \tan x \, \ln(\tan x) + \sec x \cdot \frac{\sec^2 x}{\tan x}$$

$$\frac{dy}{dx} = (\tan x)^{\sec x} \left[\sec x \tan x \, \ln(\tan x) + \sec^3 x \cot x \right]$$

57. 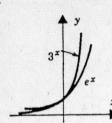   **59.** 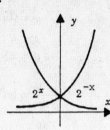   **61.** 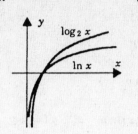   **63.**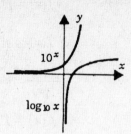

65. $f'(x) = 10^{1/(1-x^2)}(\ln 10) \cdot \dfrac{2x}{(1-x^2)^2}$

$f':$ $\underset{\substack{\\ -1 \qquad\quad 0 \qquad\quad 1}}{- \;\; - \;\; - \;\; - \;\; - \;\; + \;\; + \;\; + \;\; + \;\; +}$

domain all $x \neq \pm 1$;   decreases on $(-\infty, -1)$ and on $(-1, 0]$, increases on $[0, 1)$ and on $(1, \infty)$; $f(0) = 10$ local min

67. $f'(x) = \dfrac{1}{\ln 10} \cdot \dfrac{1}{\sqrt{1-x^2}} \cdot \dfrac{-x}{\sqrt{1-x^2}} = \dfrac{-x}{\ln 10(1-x^2)}$

$f':$ $\underset{\substack{\\ -1 \qquad\quad 0 \qquad\quad 1}}{+ \;\; + \;\; + \qquad\quad - \;\; - \;\; - \;\; -}$

domain $(-1, 1)$;   increases on $(-1, 0]$, decreases on $[0, 1)$; $f(0) = 0$ local and absolute max.

69. $\log_{10} 7 = \dfrac{\ln 7}{\ln 10} \cong \dfrac{1.95}{2.30} \cong 0.85$ **71.** $\log_{10} 45 = \dfrac{\ln 45}{\ln 10} = \dfrac{\ln 9 + \ln 5}{\ln 10} \cong \dfrac{2.20 + 1.61}{2.30} \cong 1.66$

73. approx 16.99999; $5^{\ln 17/\ln 5} = \left(e^{\ln 5}\right)^{\ln 17/\ln 5} = e^{\ln 17} = 17$

75. approx 54.59815; $16^{1/\ln 2} = \left(e^{\ln 16}\right)^{1/\ln 2} = e^{\ln 16/\ln 2} = e^{4\ln 2/\ln 2} = e^4 \cong 54.59815$

SECTION 5.6

1. We begin with

$$A(t) = A_0 e^{rt}$$

and take $A_0 = \$500$ and $t = 10$. The interest earned is given by

$$A(10) - A_0 = 500\left(e^{10r} - 1\right).$$

Thus, (a) $500\left(e^{0.6} - 1\right) \cong \$411.06$ (b) $500\left(e^{0.8} - 1\right) \cong \$612.77$

     (c) $500\left(e - 1\right) \cong \$859.14.$

3. In general, $A(t) = A_0 e^{rt}$. We set

$$3A_0 = A_0 e^{20r}$$

and solve for r:

$$3 = e^{20r}, \quad \ln 3 = 20r, \quad r = \frac{\ln 3}{20} \cong 5\tfrac{1}{2}\%.$$

5.   Let $P = P(t)$ denote the bacteria population at time t. Then $P'(t) = kP(t)$ (k constant) $P(0) = 1000$, and $P(t) = 1000\,e^{kt}$. Since $P(1/2) = 2P(0) = P(0)e^{k/2}$, it follows that $e^{k/2} = 2$ or $k = 2\ln 2$. Now

$$P(t) = 1000e^{(2\ln 2)t} \quad \text{and} \quad P(2) = 1000\,e^{(2\ln 2)2} = 16{,}000$$

7.   Let $s(t)$ be the number of pounds of salt present after t minutes. Since

$$s'(t) = \text{ rate in } - \text{ rate out } = 3\,(0.2) - 3\left(\frac{s(t)}{100}\right),$$

we have

$$s'(t) + 0.03s(t) = 0.6.$$

Using the approach in the proof of Theorem 7.6.1, multiply by $e^{\int 0.03dt} = e^{0.03t}$ to obtain

$$e^{0.03t}s'(t) + 0.03e^{0.03t}s(t) = 0.6e^{0.03t}$$

$$\frac{d}{dt}\left[e^{0.03t}s(t)\right] = 0.6e^{0.03t}$$

$$e^{0.03t}s(t) = 20e^{0.03t} + C$$

$$s(t) = 20 + Ce^{-0.03t}.$$

We use the initial condition $s(0) = 100(0.25) = 25$ to determine C:

$$25 = 20 + Ce^0 \quad \text{so} \quad C = 5.$$

Thus, $s(t) = 20 + 5e^{-0.03t}$ lb.

9.

$$V'(t) = ktV(t)$$

$$V'(t) - ktV(t) = 0$$

$$e^{-kt^2/2}V'(t) - kte^{-kt^2/2}V(t) = 0$$

$$\frac{d}{dt}\left[e^{-kt^2/2}V(t)\right] = 0$$

$$e^{-kt^2/2}V(t) = C$$

$$V(t) = Ce^{kt^2/2}.$$

Since $V(0) = C = 200$,

$$V(t) = 200e^{kt^2/2}.$$

Since $V(5) = 160$,

$$200e^{k(25/2)} = 160, \quad e^{k(25/2)} = \tfrac{4}{5}, \quad e^k = \left(\tfrac{4}{5}\right)^{2/25}$$

and therefore

$$V(t) = 200\left(\tfrac{4}{5}\right)^{t^2/25} \text{ liters.}$$

11.   Let $A(t)$ be the amount of radioactive substance present after t years.

In general

$$A(t) = A_0 e^{kt}.$$

Since one quarter of the substance decays in 4 years,

$$A(4) = \tfrac{3}{4} A_0$$

and thus

$$\tfrac{3}{4} A_0 = A_0 e^{4k} \quad \text{so that} \quad e^k = \left(\tfrac{3}{4}\right)^{1/4}.$$

To determine the half-life of the substance we must find the value of t for which

$$\tfrac{1}{2} A_0 = A_0 e^{kt} = A_0 \left(\tfrac{3}{4}\right)^{t/4}.$$

Solving for t, we have

$$\frac{1}{2} = \left(\frac{3}{4}\right)^{t/4}, \quad \ln \frac{1}{2} = \frac{t}{4} \ln \frac{3}{4}, \quad t = \frac{4 \ln (1/2)}{\ln (3/4)} \cong 9.64.$$

The half-life is a little more than $9\tfrac{1}{2}$ years.

13.   Take two years ago as time $t = 0$. In general

$$(*) \qquad\qquad A(t) = A_0 e^{kt}.$$

We are given that

$$A_0 = 5 \quad \text{and} \quad A(2) = 4.$$

Thus,

$$4 = 5e^{2k} \quad \text{so that} \quad \tfrac{4}{5} = e^{2k} \quad \text{or} \quad e^k = \left(\tfrac{4}{5}\right)^{1/2}.$$

We can write

$$A(t) = 5 \left(\tfrac{4}{5}\right)^{t/2}$$

and compute $A(5)$ as follows:

$$A(5) = 5 \left(\tfrac{4}{5}\right)^{5/2} = 5 e^{\frac{5}{2} \ln(4/5)} \cong 5 e^{-0.56} \cong 2.86.$$

About 2.86 gm will remain 3 years from now.

15.   A fundamental property of radioactive decay is that the percentage of substance that decays during any year is constant:

$$100 \left[\frac{A(t) - A(t+1)}{A(t)} \right] = 100 \left[\frac{A_0 e^{kt} - A_0 e^{k(t+1)}}{A_0 e^{kt}} \right] = 100(1 - e^k)$$

If the half-life is n years, then

$$\tfrac{1}{2} A_0 = A_0 e^{kn} \quad \text{so that} \quad e^k = \left(\tfrac{1}{2}\right)^{1/n}.$$

Thus, $100 \left[1 - \left(\tfrac{1}{2}\right)^{1/n} \right]$ % of the material decays during any one year.

17. (a) $x_1(t) = 10^6 t, \quad x_2(t) = e^t - 1$

(b) $\dfrac{d}{dt}[x_1(t) - x_2(t)] = \dfrac{d}{dt}[10^6 t - (e^t - 1)] = 10^6 - e^t$

This derivative is zero at $t = 6 \ln 10 \cong 13.8$. After that the derivative is negative.

(c) $x_2(15) < e^{15} = (e^3)^5 \cong 20^5 = 2^5 (10^5) = 3.2(10^6) < 15(10^6) = x_1(15)$

$x_2(18) = e^{18} - 1 = (e^3)^6 - 1 \cong 20^6 - 1 = 64(10^6) - 1 > 18(10^6) = x_1(18)$

$x_2(18) - x_1(18) \cong 64(10^6) - 1 - 18(10^6) \cong 46(10^6)$

(d) If by time t_1 EXP has passed LIN, then $t_1 > 6 \ln 10$. For all $t \geq t_1$ the speed of EXP is greater than the speed of LIN:

$$\text{for} \quad t \geq t_1 > 6 \ln 10, \quad v_2(t) = e^t > 10^6 = v_1(t).$$

19. Let $p(h)$ denote the pressure at altitude h. The equation $\dfrac{dp}{dh} = kp$ gives

(*) $$p(h) = p_0 e^{kh}$$

where p_0 is the pressure at altitude zero (sea level).

Since $p_0 = 15$ and $p(10000) = 10$,

$$10 = 15 e^{10000k}, \quad \tfrac{2}{3} = e^{10000k}, \quad \tfrac{1}{10000} \ln \tfrac{2}{3} = k.$$

Thus, (*) can be written

$$p(h) = 15 \left(\tfrac{2}{3}\right)^{h/10000}.$$

(a) $p(5000) = 15 \left(\tfrac{2}{3}\right)^{1/2} \cong 12.25 \text{ lb/in.}^2$.

(b) $p(15000) = 15 \left(\tfrac{2}{3}\right)^{3/2} \cong 8.16 \text{ lb/in.}^2$.

21. From Exercise 20, we have $6000 = 10,000 e^{-8r}$. Thus

$$e^{-8r} = \frac{6000}{10,000} = \frac{3}{5} \Rightarrow -8r = \ln(3/5) \quad \text{and} \quad r \cong 0.064 \text{ or } r = 6.4\%$$

23. The future value of $16,000 at an interest rate r, t years from now is given by $Q(t) = 16,000 e^{rt}$. Thus

(a) For $r = 0.05$: $P(3) = 16,000 e^{(0.05)3} \cong 18,589.35$ or $18,589.35.

(b) For $r = 0.08$: $P(3) = 16,000 e^{(0.08)3} \cong 20,339.99$ or $20,339.99.

(c) For $r = 0.12$: $P(3) = 16,000 e^{(0.12)3} \cong 22,933.27$ or $22,933.27.

25. By Exercise 24

(*) $$v(t) = Ce^{-kt}, \quad t \text{ in seconds.}$$

We use the initial conditions

$$v\,(0) = C = 4 \text{ mph } = \tfrac{1}{900} \text{ mi/sec} \quad \text{and} \quad v\,(60) = 2 = \tfrac{1}{1800} \text{ mi/sec}$$

to determine e^{-k}:

$$\tfrac{1}{1800} = \tfrac{1}{900}e^{-60k}, \qquad e^{60k} = 2, \qquad e^k = 2^{1/60}.$$

Thus, (∗) can be written

$$v\,(t) = \tfrac{1}{900}\, 2^{-t/60}.$$

Now let $s = s(t)$ denote the distance traveled by the boat. Then

$$\frac{ds}{dt} = v(t) = \frac{1}{900}\, 2^{-t/60} \implies s(t) = \frac{1}{900}\left[\frac{-60}{\ln 2}\, 2^{-t/60}\right] + C.$$

Since $s(0) = 0$, $C = \dfrac{1}{15\ln 2}$ and $s(t) = \dfrac{1}{15\ln 2}\left[1 - 2^{-t/60}\right].$

Finally, $s(60) = \dfrac{1}{15\ln 2}\left[1 - 2^{-1}\right] = \dfrac{1}{30\ln 2}\text{mi} = \dfrac{176}{\ln 2}$ ft   (about 254 ft).

27.   about 284 million; $203\,(227/203)^3 \cong 283.85$

29.   Let $A(t)$ denote the amount of ^{14}C remaining t years after the organism dies. Then $A(t) = A(0)e^{kt}$ for some constant k. Since the half-life of ^{14}C is 5700 years, we have

$$\frac{1}{2} = e^{5700k} \Rightarrow k = -\frac{\ln 2}{5700} \cong 0.000122 \text{ and } A(t) = A(0)e^{-0.000122t}$$

If 25% of the original amount of ^{14}C remains after t years, then

$$0.25A(0) = A(0)e^{-0.000122t} \Rightarrow t = \frac{\ln 0.25}{-0.000122} \cong 11,400 \text{ (years)}$$

31.   Future dollars are discounted by a factor of $e^{-0.05t}$. We want to maximize the function

$$f(t) = V(t)e^{-0.05t}$$
$$= V_0\left(\tfrac{3}{2}\right)^{\sqrt{t}} e^{-0.05t}$$
$$= V_0 e^{\sqrt{t}\left(\ln\frac{3}{2}\right)-0.05t}, \qquad t \geq 0.$$

Differentiation gives

$$f'(t) = V_0 e^{\sqrt{t}\left(\ln\frac{3}{2}\right)-0.05t}\left[\frac{1}{2\sqrt{t}}\ln\left(\frac{3}{2}\right) - 0.05\right].$$

Setting $f'(t) = 0$, we find that

$$\frac{\ln(3/2)}{2\sqrt{t}} = \frac{1}{20}, \quad \sqrt{t} = 10\ln\frac{3}{2}, \quad t = 100\left(\ln\frac{3}{2}\right)^2 \cong 16.44.$$

Clearly this value of t maximizes f. The company should wait about $16\frac{1}{2}$ years before cutting the timber.

33.   Since the amount $A(t)$ of raw sugar present after t hours decreases at a rate proportional to A, we have

$$A(t) = A_0 e^{kt}.$$

We are given $A_0 = 1000$ and $A(10) = 800$.   Thus,

$$800 = 1000e^{10k}, \quad \tfrac{4}{5} = e^{10k}, \quad e^k = \left(\tfrac{4}{5}\right)^{1/10}$$

so that

$$A(t) = 1000 \left(\tfrac{4}{5}\right)^{t/10}.$$

Now,

$$A(20) = 1000 \left(\frac{4}{5}\right)^{20/10} = 640;$$

after 10 more hours of inversion there will remain 640 pounds.

35.   By adding a constant to a function, we don't change its derivative. Therefore, we can write

$$f'(t) = k_1 f(t) + k_2 \quad \text{as} \quad \frac{d}{dt}[f(t) + k_2/k_1] = k_1[f(t) + k_2/k_1].$$

Now, we know from Theorem 5.6.1 that

$$f(t) + k_2/k_1 = Ae^{k_1(t+k_2/k_1)} = Ae^{k^1 t + k_2} = \left(Ae^{k_2}\right)e^{k_1 t} = Ce^{k_1 t} \quad (C = Ae^{k_2}).$$

Therefore $f(t) = Ce^{k_1 t} - k_2/k_1$, C an arbitrary constant.

37.   (a) From Exercise 35 you can determine that

$$v(t) = \frac{32}{K}\left(1 - e^{-Kt}\right).$$

(b) At each time t, $1 - e^{-Kt} < 1$.   With $K > 0$,

$$v(t) = \frac{32}{K}\left(1 - e^{-Kt}\right) < \frac{32}{K} \quad \text{and} \quad \lim_{t \to \infty} v(t) = \frac{32}{K}$$

(c) Graph of v with $K = 2$.

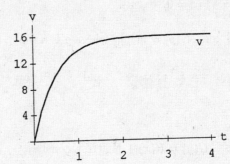

39.   From (5.6.5),

$$T(t) = \tau + (T(0) - \tau)e^{-kt}$$

Since $\tau = 65$ and $T(0) = 185$, we have

$$T(t) = 65 + (185 - 85)e^{-kt} = 65 + 120e^{-kt}.$$

Since $T(2) = 155$,

$$155 = 65 + 120e^{-2k}, \quad \tfrac{90}{120} = e^{-2k}, \quad e^k = \left(\tfrac{4}{3}\right)^{1/2}.$$

Thus,

$$T(t) = 65 - 120 \left(\tfrac{4}{3}\right)^{-t/2}.$$

We want to find t so that $T = 105$:

$$105 = 65 + 120 \left(\frac{4}{3}\right)^{-t/2}, \quad \frac{1}{3} = \left(\frac{4}{3}\right)^{-t/2}, \quad \ln\frac{1}{3} = -\frac{t}{2}\ln\frac{4}{3}.$$

Thus, $t = \dfrac{2\ln 3}{\ln(4/3)}$. You would expect to wait $\dfrac{2\ln 3}{\ln(4/3)} - 2 \cong 5.64$ minutes more for the coffee to cool to $105° F$.

41. From (5.6.5),

$$T(t) = \tau + (T(0) - \tau)e^{-kt}$$

Since $\tau = 100$ and $T(0) = 20$ it follows that

$$T(t) = 100 - 80e^{-kt}$$

Now, $T(1/3) = 30$ implies that

$$30 = 100 - 80e^{-k/3} \quad \text{and} \quad k = 3\ln(8/7)$$

Thus $T(t) = 100 - 80e^{-3\ln(8/7)t}$.

(a) $T(1) = 100 - 80e^{-3\ln(8/7)} \cong 46.41°C$

(b) $T(t) = 98 = 100 - 80e^{-3\ln(8/7)t} \quad \Rightarrow \quad t = 9.2$ min.

43. $\dfrac{dy}{dx} = \dfrac{y}{x} \implies \dfrac{1}{y}\dfrac{dy}{dx} = \dfrac{1}{x} \quad (x \neq 0, \ y \neq 0)$

Now, if $F(x) = \ln|x|$ and $G(y) = \ln|y|$, where y is a function of x, then

$$F'(x) = \frac{1}{x} \quad \text{and} \quad G'(y) = \frac{1}{y}\frac{dy}{dx}$$

Thus, we have

$$G'(y) = F'(x) \implies G(y) = F(x) + K \ \ (K \text{ a constant}) \quad \text{that is} \quad \ln|y| = \ln|x| + K,$$

which is equivalent to $y = Cx$, where C is any constant.

45. $\dfrac{dy}{dx} = e^y\cos x \implies e^{-y}\dfrac{dy}{dx} = \cos x$

Now, if $F(x) = \sin x$ and $G(y) = -e^{-y}$, where y is a function of x, then

$$F'(x) = \cos x \quad \text{and} \quad G'(y) = e^{-y}\frac{dy}{dx}$$

Thus, we have

$$G'(y) = F'(x) \implies G(y) = F(x) + C \ \ (C \text{ a constant}) \quad \text{that is} \quad -e^{-y} = \sin x + C,$$

which is equivalent to $y = -\ln|C - \sin x|$.

47. From Example 11, $\ln|y - a| = rx + C$

Therefore,

$$y - a = e^{rx+C} = Be^{rx} \quad (B = e^C) \quad \text{and} \quad y = a + Be^{rx}$$

SECTION 5.7

1. 0   3. $\pi/3$ 5. $2\pi/3$ 7. $-\pi/4$ 9. $-2/\sqrt{3}$

11. $1/2$ 13. 1.1630   15. -0.4580 17. 1.2002   19. $\dfrac{1}{\sqrt{1 + x^2}}$

21. x 23. $\sqrt{1 + x^2}$ 25. $\sqrt{1 - x^2}$ 27. $\dfrac{\sqrt{1 + x^2}}{x}$ 29. $\dfrac{x}{\sqrt{1 - x^2}}$

31. $\dfrac{dy}{dx} = \dfrac{1}{1 + (x + 1)^2} = \dfrac{1}{x^2 + 2x + 2}$ 33. $f'(x) = \dfrac{1}{|2x^2|\sqrt{(2x^2)^2 - 1}} \dfrac{d}{dx}(2x^2) = \dfrac{2x}{x\sqrt{4x^4 - 1}}$

35. $f'(x) = \sin^{-1} 2x + x\dfrac{1}{\sqrt{1 - (2x)^2}}\dfrac{d}{dx}(2x) = \sin^{-1} 2x + \dfrac{2x}{\sqrt{1 - 4x^2}}$

37. $\dfrac{du}{dx} = 2\left(\sin^{-1} x\right)\dfrac{d}{dx}\left(\sin^{-1} x\right) = \dfrac{2\sin^{-1} x}{\sqrt{1 - x^2}}$

39. $\dfrac{dy}{dx} = \dfrac{x\left(\dfrac{1}{1 + x^2}\right) - (1)\tan^{-1} x}{x^2} = \dfrac{x - \left(1 + x^2\right)\tan^{-1} x}{x^2\left(1 + x^2\right)}$

41. $f'(x) = \dfrac{1}{2}\left(\tan^{-1} 2x\right)^{-1/2}\dfrac{d}{dx}\left(\tan^{-1} 2x\right) = \dfrac{1}{2}\left(\tan^{-1} 2x\right)^{-1/2}\dfrac{2}{1 + (2x)^2} = \dfrac{1}{\left(1 + 4x^2\right)\sqrt{\tan^{-1} 2x}}$

43. $\dfrac{dy}{dx} = \dfrac{1}{1 + (\ln x)^2}\dfrac{d}{dx}(\ln x) = \dfrac{1}{x\left[1 + (\ln x)^2\right]}$

45. $\dfrac{d\theta}{dr} = \dfrac{1}{\sqrt{1 - \left(\sqrt{1 - r^2}\right)^2}}\dfrac{d}{dr}\left(\sqrt{1 - r^2}\right) = \dfrac{1}{\sqrt{r^2}} \cdot \dfrac{-r}{\sqrt{1 - r^2}} = -\dfrac{r}{|r|\sqrt{1 - r^2}}$

47. $g'(x) = 2x\sec^{-1}\left(\dfrac{1}{x}\right) + x^2 \cdot \dfrac{1}{\left|\dfrac{1}{x}\right|\sqrt{\dfrac{1}{x^2} - 1}} \cdot \left(-\dfrac{1}{x^2}\right) = 2x\sec^{-1}\left(\dfrac{1}{x}\right) - \dfrac{x^2}{\sqrt{1 - x^2}}$

49. $\dfrac{dy}{dx} = \cos\left[\sec^{-1}(\ln x)\right] \cdot \dfrac{1}{|\ln x|\sqrt{(\ln x)^2 - 1}} \cdot \dfrac{1}{x} = \dfrac{\cos\left[\sec^{-1}(\ln x)\right]}{x|\ln x|\sqrt{(\ln x)^2 - 1}}$

51. $f'(x) = \dfrac{-x}{\sqrt{c^2 - x^2}} + \dfrac{c}{\sqrt{1 - (x/c)^2}} \cdot \left(\dfrac{1}{c}\right) = \dfrac{c - x}{\sqrt{c^2 - x^2}} = \sqrt{\dfrac{c - x}{c + x}}$

53. $\dfrac{dy}{dx} = \dfrac{\sqrt{c^2 - x^2}\,(1) - x\left(\dfrac{-x}{\sqrt{c^2 - x^2}}\right)}{\left(\sqrt{c^2 - x^2}\,\right)^2} - \dfrac{1}{\sqrt{1 - (x/c)^2}}\left(\dfrac{1}{c}\right)$

$= \dfrac{c^2}{(c^2 - x^2)^{3/2}} - \dfrac{1}{(c^2 - x^2)^{1/2}} = \dfrac{x^2}{(c^2 - x^2)^{3/2}}$

55. $\mathrm{dom}(f) = [-1, 1],\quad \mathrm{range}(f) = [0, \pi].\quad y = \cos^{-1} x \implies \cos y = x \implies -\sin y\,\dfrac{dy}{dx} = 1$

$\implies \dfrac{dy}{dx} = -\dfrac{1}{\sin y} = -\dfrac{1}{\sqrt{1 - \cos^2 y}} = -\dfrac{1}{\sqrt{1 - x^2}}.$

57. $\mathrm{dom}(f) = (-\infty, -1] \cup [1, \infty),\quad \mathrm{range}(f) = \left[-\dfrac{\pi}{2}, 0\right) \cup \left(0, \dfrac{\pi}{2}\right]$

$y = \csc^{-1} x \implies \csc y = x \implies -\csc y \cot y\,\dfrac{dy}{dx} = 1 \implies \dfrac{dy}{dx} = \dfrac{-1}{\csc y \cot y} = \dfrac{-1}{|x|\sqrt{x^2 - 1}}$

59. $y = \tan^{-1} \dfrac{x}{30}\qquad \dfrac{dy}{dt} = \dfrac{1}{1 + \dfrac{x^2}{900}} \cdot \dfrac{1}{30} \cdot \dfrac{dx}{dt}$

If $\dfrac{dy}{dt} = 6$ and $x = 50$, then $\dfrac{dy}{dt} = \dfrac{30}{900 + (50)^2} \cdot 6 = \dfrac{9}{170}$ rad/sec

61. (a) $f'(x) = \dfrac{1}{1 + \left(\dfrac{a + x}{1 - ax}\right)^2} \cdot \dfrac{1 + a^2}{(1 - ax)^2} = \dfrac{1 + a^2}{(1 + a^2)(1 + x^2)} = \dfrac{1}{1 + x^2}$

(b) and (c) From part (a), $f(x)$ and $\tan^{-1}$ have the same derivative for $x \neq \dfrac{1}{a}$.

Therefore, it follows that

$$f(x) = \tan^{-1} x + C_1 \quad \text{for} \quad x < \dfrac{1}{a} \quad \text{and} \quad f(x) = \tan^{-1} x + C_2 \quad \text{for} \quad x > \dfrac{1}{a}.$$

Suppose that $a > 0$. Let $x = 0 \ \left(< \tfrac{1}{a}\right)$. Then

$$f(0) = \tan^{-1}(a) = \tan^{-1}(0) + C_1 = C_1 \quad \Rightarrow \quad C_1 = \tan^{-1}(a).$$

Now let $x \to \infty$. Then

$$\lim_{x \to \infty} f(x) = \lim_{x \to \infty} \tan^{-1}\left(\dfrac{a + x}{1 - ax}\right) = \tan^{-1}\left(-\dfrac{1}{a}\right)$$

$$= \lim_{x \to \infty} \tan^{-1} x + C_2 = \dfrac{\pi}{2} + C_2$$

$$\Rightarrow \quad C_2 = \tan^{-1}\left(-\dfrac{1}{a}\right) - \dfrac{\pi}{2}.$$

Thus, $C_1 = \tan^{-1} a$ for $x \in (-\infty, 1/a)$ and $C_2 = \tan^{-1}\left(-\frac{1}{a}\right) - \frac{\pi}{2}$ for $x \in (1/a, \infty)$.

A similar argument holds for the case $a < 0$.

63.   Numerical work suggests limit $\cong 1$. One way to see this is to note that the limit is the derivative

of $f(x) = \sin^{-1} x$ at $x = 0$ and this derivative is 1:

$$f'(x) = \frac{1}{\sqrt{1 - x^2}}, \quad f'(0) = 1$$

SECTION 5.8

1. $\dfrac{dy}{dx} = \cosh x^2 \dfrac{d}{dx}\left(x^2\right) = 2x \cosh x^2$

3. $\dfrac{dy}{dx} = \dfrac{1}{2}\left(\cosh ax\right)^{-1/2}\left(a \sinh ax\right) = \dfrac{a \sinh ax}{2\sqrt{\cosh ax}}$

5. $\dfrac{dy}{dx} = \dfrac{\left(\cosh x - 1\right)\left(\cosh x\right) - \sinh x\left(\sinh x\right)}{\left(\cosh x - 1\right)^2} = \dfrac{1}{1 - \cosh x}$

7. $\dfrac{dy}{dx} = ab \cosh bx - ab \sinh ax = ab\left(\cosh bx - \sinh ax\right)$

9. $\dfrac{dy}{dx} = \dfrac{1}{\sinh ax}\left(a \cosh ax\right) = \dfrac{a \cosh ax}{\sinh ax}$

11. $\dfrac{dy}{dx} = \cosh\left(e^{2x}\right) e^{2x}(2) = 2e^{2x} \cosh\left(e^{2x}\right)$

13. $\dfrac{dy}{dx} = -e^{-x} \cosh 2x + 2e^{-x} \sinh 2x$

15. $\dfrac{dy}{dx} = \dfrac{1}{\cosh x}\left(\sinh x\right) = \tanh x$

17. $\ln y = x \ln \sinh x; \quad \dfrac{1}{y}\dfrac{dy}{dx} = \ln \sinh x + x\dfrac{\cosh x}{\sinh x}$ and $\dfrac{dy}{dx} = \left(\sinh x\right)^x\left[\ln \sinh x + x \coth x\right]$

19. $\cosh^2 t - \sinh^2 t = \left(\dfrac{e^t + e^{-t}}{2}\right)^2 - \left(\dfrac{e^t + e^{-t}}{2}\right)^2 = \dfrac{1}{4}\left[\left(e^{2t} + 2 + e^{-2t}\right) - \left(e^{2t} - 2 + e^{-2t}\right)\right] = \dfrac{4}{4} = 1$

21. $\cosh t \cosh s + \sinh t \sinh s$

$$= \left(\frac{e^t + e^{-t}}{2}\right)\left(\frac{e^s + e^{-s}}{2}\right) + \left(\frac{e^t - e^{-t}}{2}\right)\left(\frac{e^s - e^{-s}}{2}\right)$$

$$= \frac{1}{4}\left[\left(e^{t+s} + e^{t-s} + e^{s-t} + e^{-t-s}\right) + \left(e^{t+s} - e^{t-s} - e^{s-t} + e^{-s-t}\right)\right]$$

$$= \frac{1}{4}\left[2e^{t+s} + 2e^{-(t+s)}\right] = \frac{e^{t+s} + e^{-(t+s)}}{2} = \cosh(t+s)$$

23. Set $s = t$ in $\cosh(t+s) = \cosh t \cosh s + \sinh t \sinh s$ to get $\cosh(2t) = \cosh^2 t + \sinh^2 t$.

Then use Exercise 19 to obtain the other two identities.

25. $\sinh(-t) = \dfrac{e^{(-t)} - e^{-(-t)}}{2} = -\dfrac{e^t - e^{-t}}{2} = -\sinh t$

27.
$$y = -5\cosh x + 4\sinh x = -\tfrac{5}{2}\left(e^x + e^{-x}\right) + \tfrac{4}{2}\left(e^x - e^{-x}\right) = -\tfrac{1}{2}e^x - \tfrac{9}{2}e^{-x}$$

$$\frac{dy}{dx} = -\frac{1}{2}e^x + \frac{9}{2}e^{-x} = \frac{e^{-x}}{2}\left(9 - e^{2x}\right); \quad \frac{dy}{dx} = 0 \implies e^x = 3 \text{ or } x = \ln 3$$

$$\frac{d^2 y}{dx^2} = -\frac{1}{2}e^x - \frac{9}{2}e^{-x} < 0 \quad \text{all} \quad x \quad \text{so abs max occurs when} \quad x = \ln 3.$$

The abs max is $y = -\tfrac{1}{2}e^{\ln 3} - \tfrac{9}{2}e^{-\ln 3} = -\tfrac{1}{2}(3) - \tfrac{9}{2}\left(\tfrac{1}{3}\right) = -3.$

29.
$$\left[\cosh x + \sinh x\right]^n = \left[\frac{e^x + e^{-x}}{2} + \frac{e^x - e^{-x}}{2}\right]^n$$

$$= [e^x]^n = e^{nx} = \frac{e^{nx} + e^{-nx}}{2} + \frac{e^{nx} - e^{-nx}}{2} = \cosh nx + \sinh nx$$

31.
$$y = A\cosh cx + B\sinh cx; \qquad\qquad y(0) = 2 \implies 2 = A.$$

$$y' = Ac\sinh cx + Bc\cosh cx; \qquad\qquad y'(0) = 1 \implies 1 = Bc.$$

$$y'' = Ac^2\cosh cx + Bc^2\sinh cx = c^2 y; \qquad y'' - 9y = 0 \implies (c^2 - 9)y = 0.$$

Thus, $c = 3$, $B = \tfrac{1}{3}$, and $A = 2$.

33. (a) $\displaystyle \lim_{x\to\infty} \frac{\sinh x}{e^x} = \lim_{x\to\infty} \frac{e^x - e^{-x}}{2e^x} = \lim_{x\to\infty}\left(\frac{1}{2} - \frac{e^{-2x}}{2}\right) = \frac{1}{2}$

(b) $\displaystyle \lim_{x\to\infty} \frac{\cosh x}{e^{ax}} = \lim_{x\to\infty} \frac{e^x + e^{-x}}{2e^{ax}} = \lim_{x\to\infty} \frac{1}{2}\left(e^{x-ax} + e^{-x-ax}\right)$

For $0 < a < 1$, limit $= \infty$. For $a > 1$, limit $= 0$.

SECTION * 5.9

1. $\dfrac{dy}{dx} = 2\tanh x \operatorname{sech}^2 x$

3. $\dfrac{dy}{dx} = \dfrac{1}{\tanh x}\operatorname{sech}^2 x = \operatorname{sech} x \operatorname{csch} x$

5. $\dfrac{dy}{dx} = \cosh\left(\tan^{-1} e^{2x}\right)\dfrac{d}{dx}\left(\tan^{-1} e^{2x}\right) = \dfrac{2e^{2x}\cosh\left(\tan^{-1} e^{2x}\right)}{1 + e^{4x}}$

7. $\dfrac{dy}{dx} = -\operatorname{csch}^2\left(\sqrt{x^2 + 1}\right)\dfrac{d}{dx}\left(\sqrt{x^2 + 1}\right) = -\dfrac{x}{\sqrt{x^2 + 1}}\operatorname{csch}^2\left(\sqrt{x^2 + 1}\right)$

9. $\dfrac{dy}{dx} = \dfrac{(1 + \cosh x)(-\operatorname{sech} x \tanh x) - \operatorname{sech} x(\sinh x)}{(1 + \cosh x)^2}$

$$= \frac{-\operatorname{sech} x \, (\tanh x + \cosh x \tanh x + \sinh x)}{(1 + \cosh x)^2} = \frac{-\operatorname{sech} x \, (\tanh x + 2 \sinh x)}{(1 + \cosh x)^2}$$

11.
$$\frac{d}{dx} (\coth x) = \frac{d}{dx} \left[\frac{\cosh x}{\sinh x} \right] = \frac{\sinh x \, (\sinh x) - \cosh x \, (\cosh x)}{\sinh^2 x}$$

$$= -\frac{\cosh^2 x - \sinh^2 x}{\sinh^2 x} = \frac{-1}{\sinh^2 x} = -\operatorname{csch}^2 x$$

13. $\quad \dfrac{d}{dx} (\operatorname{csch} x) = \dfrac{d}{dx} \left[\dfrac{1}{\sinh x} \right] = -\dfrac{\cosh x}{\sinh^2 x} = -\operatorname{csch} x \coth x$

15. (a)   By the hint $\operatorname{sech}^2 x_0 = \dfrac{9}{25}$. Take $\operatorname{sech} x_0 = \dfrac{3}{5}$ since $\operatorname{sech} x = \dfrac{1}{\cosh x} > 0$ for all x.

(b) $\cosh x_0 = \dfrac{1}{\operatorname{sech} x_0} = \dfrac{5}{3}$
$\qquad$ (c) $\sinh x_0 = \cosh x_0 \tanh x_0 = \left(\dfrac{5}{3} \right) \left(\dfrac{4}{5} \right) = \dfrac{4}{3}$

(d) $\coth x_0 = \dfrac{\cosh x_0}{\sinh x_0} = \dfrac{5/3}{4/3} = \dfrac{5}{4}$
$\qquad$ (e) $\operatorname{csch} x_0 = \dfrac{1}{\sinh x_0} = \dfrac{3}{4}$

17.   If $x \le 0$, the result is obvious. Suppose then that $x > 0$. Since $x^2 \ge 1$, we have $x \ge 1$.
Consequently

$$x - 1 = \sqrt{x-1} \, \sqrt{x-1} \le \sqrt{x-1} \, \sqrt{x+1} = \sqrt{x^2 - 1}$$

and therefore $\qquad x - \sqrt{x^2 - 1} \le 1.$

19.   By Theorem 5.9.2,

$$\frac{d}{dx} (\sinh^{-1} x) = \frac{d}{dx} \left[\ln \left(x + \sqrt{x^2 + 1} \,\right) \right] = \frac{1}{x + \sqrt{x^2 + 1}} \left(1 + \frac{x}{\sqrt{x^2 + 1}} \right) = \frac{1}{\sqrt{x^2 + 1}}.$$

21.   By Theorem 5.9.2

$$\frac{d}{dx} (\tan^{-1} x) = \frac{d}{dx} \left[\frac{1}{2} \ln \left(\frac{1+x}{1-x} \right) \right] = \frac{1}{2} \frac{1}{\left(\dfrac{1+x}{1-x} \right)} \left(\frac{(1-x)(1) - (1+x)(-1)}{(1-x)^2} \right)$$

$$= \frac{1}{\left(\dfrac{1+x}{1-x} \right) (1-x)^2} = \frac{1}{1-x^2}.$$

23.   Let $y = \operatorname{csch}^{-1} x$. Then $\operatorname{csch} y = x$ and $\sinh y = \dfrac{1}{x}$.

$$\sinh y = \frac{1}{x}$$

$$\cosh y \, \frac{dy}{dx} = -\frac{1}{x^2}$$

$$\frac{dy}{dx} = -\frac{1}{x^2 \cosh y} = -\frac{1}{x^2 \sqrt{1 + (1/x)^2}} = -\frac{1}{|x| \sqrt{1 + x^2}}$$

25. (a) $\dfrac{dy}{dx} = -\operatorname{sech} x \tanh x = -\dfrac{\sinh x}{\cosh^2 x}$

$\dfrac{dy}{dx} = 0$ at $x = 0$; $\dfrac{dy}{dx} > 0$ if $x < 0$; $\dfrac{dy}{dx} < 0$ if $x > 0$

f is increasing on $(-\infty, 0]$ and decreasing on $[0, \infty)$; $f(0) = 1$ is the absolute maximum of f.

(b) $\dfrac{d^2 y}{dx^2} = -\dfrac{\cosh^2 x - 2\sinh^2 x}{\cosh^3 x} = \dfrac{\sinh^2 x - 1}{\cosh^3 x}$

$\dfrac{d^2 y}{dx^2} = 0 \Rightarrow \sinh x = \pm 1$

$\sinh x = 1 \Rightarrow \dfrac{e^x - e^{-x}}{2} = 1 \Rightarrow e^{2x} - 2e^x - 1 = 0 \Rightarrow x = \ln(1 + \sqrt{2}) \cong 0.881;$

$\sinh x = -1 \Rightarrow \dfrac{e^x - e^{-x}}{2} = -1 \Rightarrow e^{2x} + 2e^x - 1 = 0 \Rightarrow x = -\ln(1 + \sqrt{2}) = -0.881$

(c) The graph of f is concave up on $(-\infty, -0.881) \cup (0.881, \infty)$ and concave down on $(-0.881, 0.881)$;

points of inflection at $x = \pm 0.881$

(d)

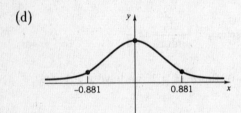

27. $y = \sinh x$; $\dfrac{dy}{dx} = \cosh x$; $\dfrac{d^2 y}{dx^2} = \sinh x$.

$\dfrac{d^2 y}{dx^2} = 0 \Rightarrow \sinh x = 0 \Rightarrow x = 0.$

$y = \sinh^{-1} x = \ln\left(x + \sqrt{x^2 + 1}\right);$ $\dfrac{dy}{dx} = \dfrac{1}{\sqrt{x^2 + 1}};$ $\dfrac{d^2 y}{dx^2} = -\dfrac{x}{(x^2 + 1)^{3/2}}.$

$\dfrac{d^2 y}{dx^2} = 0 \Rightarrow -\dfrac{x}{(x^2 + 1)^{3/2}} = 0 \Rightarrow x = 0.$

It is easy to verify that $(0, 0)$ is a point of inflection for both graphs.

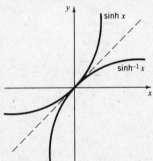

29.   (a) $\tan\phi = \sinh x$ (b) $\sinh x = \tan\phi$

$$\phi = \tan^{-1}(\sinh x)$$ $$x = \sinh^{-1}(\tan\phi)$$

$$\frac{d\phi}{dx} = \frac{\cosh x}{1 + \sinh^2 x}$$ $$= \ln\left(\tan\phi + \sqrt{\tan^2\phi + 1}\right)$$

$$= \frac{\cosh x}{\cosh^2 x} = \frac{1}{\cosh x} = \operatorname{sech} x$$ $$= \ln(\tan\phi + \sec\phi)$$

$$= \ln(\sec\phi + \tan\phi)$$

(c) $x = \ln(\sec\phi + \tan\phi)$

$$\frac{dx}{d\phi} = \frac{\sec\phi\tan\phi + \sec^2\phi}{\tan\phi + \sec\phi} = \sec\phi$$

PROJECTS AND EXPLORATIONS

5.1.   Assume throughout that the constant a is nonzero.

(a) To be tangent to the line $y = x$, we must have

$$f'(x) = \frac{a}{x} = 1 \implies x = a.$$

Now the graph of $f(x) = a\ln x$ passes through the point (a, a) iff $a\ln a = a$,

which implies $a = e$. Thus, $f(x) = e\ln x$ is tangent to the line $y = x$.

(b) Graph $F(x) = a\ln x - x$ for several values of a. Based on your graphs, you can verify that

$f(x) = x$ has:

(1) Two solutions, $x_1 < e$ and $x_2 > e$ if $a > e$; $f'(x_1) > 1$ and $f'(x_2) < 1$.

(2) One solution, $x_1 = e$ if $a = e$; $f'(e) = 1$.

(3) No solutions if $0 < a < e$.

(4) One solution, x_1 if $a < 0$; $f'(x_1) < 0$.

(c) Let $G(x) = f(f(x)) = a\ln[a\ln x]$. If $a > 0$, then G is defined only for $x > 1$.

If $a < 0$, then G is defined only for $0 < x < 1$.

A solution of $f(x) = x$ must also be a solution of $f(f(x)) = x$, but the converse need not hold.

(d) If $a = \dfrac{1}{\ln b}$ then $f(x) = \log_b x$.

5.3.   (a) Fix x. Then

$$\lim_{h\to\infty} D_x(h) = \lim_{h\to\infty} \frac{f(x + e^{-h}) - f(x)}{e^{-h}} = \lim_{k\to 0} \frac{f(x + k) - f(x)}{k} = f'(x)$$

$$k = e^{-h}$$

(c) For the functions given in part (b), it appears that E decreases to 0. However, this may not always be true. For example, try $f(x) = \sin(1/x)$.

(d) Some trial and error with different values of n suggests that $n = 1$ is the smallest value of n for which the limit is nonzero.

CHAPTER 6

SECTION 6.2

1. $L_f(P) = 0(\frac{1}{4}) + \frac{1}{2}(\frac{1}{4}) + 1(\frac{1}{2}) = \frac{5}{8}$, $U_f(P) = \frac{1}{2}(\frac{1}{4}) + 1(\frac{1}{4}) + 2(\frac{1}{2}) = \frac{11}{8}$

3. $L_f(P) = \frac{1}{4}(\frac{1}{2}) + \frac{1}{16}(\frac{1}{4}) + 0(\frac{1}{4}) = \frac{9}{64}$, $U_f(P) = 1(\frac{1}{2}) + \frac{1}{4}(\frac{1}{4}) + \frac{1}{16}(\frac{1}{4}) = \frac{37}{64}$

5. $L_f(P) = 1(\frac{1}{2}) + \frac{9}{8}(\frac{1}{2}) = \frac{17}{16}$, $U_f(P) = \frac{9}{8}(\frac{1}{2}) + 2(\frac{1}{2}) = \frac{25}{16}$

7. $L_f(P) = \frac{1}{2}(\frac{1}{2}) + 0(\frac{1}{2}) + 0(\frac{1}{4}) + \frac{1}{4}(\frac{3}{4}) = \frac{7}{16}$, $U_f(P) = 1(\frac{1}{2}) + \frac{1}{2}(\frac{1}{2}) + \frac{1}{4}(\frac{1}{4}) + 1(\frac{3}{4}) = \frac{25}{16}$

9. $L_f(P) = \frac{1}{16}(\frac{3}{4}) + 0(\frac{1}{2}) + \frac{1}{16}(\frac{1}{4}) + \frac{1}{4}(\frac{1}{2}) = \frac{3}{16}$, $U_f(P) = 1(\frac{3}{4}) + \frac{1}{16}(\frac{1}{2}) + \frac{1}{4}(\frac{1}{4}) + 1(\frac{1}{2}) = \frac{43}{32}$

11. $L_f(P) = 0\left(\frac{\pi}{6}\right) + \frac{1}{2}\left(\frac{\pi}{3}\right) + 0\left(\frac{\pi}{2}\right) = \frac{\pi}{6}$, $U_f(P) = \frac{1}{2}\left(\frac{\pi}{6}\right) + 1\left(\frac{\pi}{3}\right) + 1\left(\frac{\pi}{2}\right) = \frac{11\pi}{12}$

13. (a) $L_f(P) \le U_f(P)$ but $3 \not\le 2$.

    (b) $L_f(P) \le \int_{-1}^{1} f(x)\,dx \le U_f(P)$ but $3 \not\le 2 \le 6$.

    (c) $L_f(P) \le \int_{-1}^{1} f(x)\,dx \le U_f(P)$ but $3 \le 10 \not\le 6$.

15. (a) $L_f(P) = -3x_1(x_1 - x_0) - 3x_2(x_2 - x_1) - \cdots - 3x_n(x_n - x_{n-1})$,

        $U_f(P) = -3x_0(x_1 - x_0) - 3x_1(x_2 - x_1) - \cdots - 3x_{n-1}(x_n - x_{n-1})$

    (b)   For each index i

$$-3x_i \le -\tfrac{3}{2}(x_i + x_{i-1}) \le -3x_{i-1}.$$

Multiplication by $\Delta x_i = x_i - x_{i-1}$ gives

$$-3x_i\,\Delta x_i \le -\tfrac{3}{2}(x_i^2 - x_{i-1}^2) \le -3x_{i-1}\,\Delta x_i.$$

Summing from $i = 1$ to $i = n$, we find that

$$L_f(P) \le -\tfrac{3}{2}(x_1^2 - x_0^2) - \cdots - \tfrac{3}{2}(x_n^2 - x_{n-1}^2) \le U_f(P).$$

The middle sum collapses to

$$-\tfrac{3}{2}(x_n^2 - x_0^2) = -\tfrac{3}{2}(b^2 - a^2).$$

Thus

$$L_f(P) \le -\frac{3}{2}(b^2 - a^2) \le U_f(P) \quad \text{so that} \quad \int_a^b -3x\,dx = -\frac{3}{2}(b^2 - a^2).$$

17. $L_f(P) = x_0^3(x_1 - x_0) + x_1^3(x_2 - x_1) + \cdots + x_{n-1}^3(x_n - x_{n-1})$

      $U_f(P) = x_1^3(x_1 - x_0) + x_2^3(x_2 - x_1) + \cdots + x_n^3(x_n - x_{n-1})$

For each index i

$$x_{i-1}^3 \le \tfrac{1}{4}\left(x_i^3 + x_i^2 x_{i-1} + x_i x_{i-1}^2 + x_{i-1}^3\right) \le x_i^3$$

and thus by the hint

$$x_{i-1}^3(x_i - x_{i-1}) \le \tfrac{1}{4}\left(x_i{}^4 - x_{i-1}^4\right) \le x_i{}^3(x_i - x_{i-1}).$$

Adding up these inequalities, we find that

$$L_f(P) \le \tfrac{1}{4}\left(x_n{}^4 - x_0{}^4\right) \le U_f(P).$$

Since $x_n = 1$ and $x_0 = 0$, the middle term is $\dfrac{1}{4}$: $\displaystyle\int_0^1 x^3\,dx = \dfrac{1}{4}$.

19. Let $P = \{x_0, x_1, x_2, \ldots, x_n\}$ be a regular partition of $[a, b]$ and let $\Delta x = (b-a)/n$.

Since f is increasing on $[a, b]$,

$$L_f(P) = f(x_0)\Delta x + f(x_1)\Delta x + \cdots + f(x_{n-1})\Delta x$$

and

$$U_f(P) = f(x_1)\Delta x + f(x_2)\Delta x + \cdots + f(x_n)\Delta x.$$

Now,

$$U_f(P) - L_f(P) = [f(x_n) - f(x_0)]\Delta x = [f(b) - f(a)]\Delta x.$$

21. (a) By Definition 5.2.3, $L_f(P) \le I \le U_f(P)$. Subtracting $L_f(P)$ from these inequalities gives

$$0 \le I - L_f(P) \le U_f(P) - L_f(P)$$

(b) From Definition 5.2.3,

$$L_f(P) - U_f(P) \le I - U_f(P) \le 0$$

Now multiply by -1 and the result follows.

23. (a) $f'(x) = \dfrac{x}{\sqrt{1+x^2}} > 0$ for $x \in (0, 2)$. Thus, f is increasing on $[0, 2]$.

(b) Let $P = \{x_0, x_1, \ldots, x_n\}$ be a regular partition of $[0, 2]$ and let $\Delta x = 2/n$

By Exercise 22 (a),

$$\int_0^2 f(x)\,dx - L_f(P) \le |f(2) - f(0)|\,\frac{2}{n} = \frac{2(\sqrt{5}-1)}{n} \cong \frac{2.47}{n}$$

It now follows that $\int_0^2 f(x)\,dx - L_f(P) < 0.1$ if $n > 25$.

(c) $\int_0^2 f(x)\,dx \cong 2.96$

25. Let S be the set of positive integers for which the statement is true. Since $1 = \dfrac{1(2)}{2} = 1$, $1 \in S$. Assume that $k \in S$. Then

$$1 + 2 + \cdots + k + k + 1 = (1 + 2 + \cdots + k) + k + 1 = \frac{k(k+1)}{2} + k + 1$$
$$= \frac{k+1)(k+2)}{2}$$

Thus, $k + 1 \in S$ and so S is the set of positive integers.

27. Let $f(x) = x$ and let $P = \{x_0, x_1, x_2, \ldots, x_n\}$ be a regular partition of $[0, b]$. Then $\Delta x = b/n$ and $x_i = \dfrac{ib}{n}$, $i = 0, 1, 2, \ldots, n$.

(a) Since f is increasing on $[0, b]$,

$$L_f(P) = \left[f(0) + f\left(\frac{b}{n}\right) + f\left(\frac{2b}{n}\right) + \cdots + f\left(\frac{(n-1)b}{n}\right)\right]\frac{b}{n}$$

$$= \left[0 + \frac{b}{n} + \frac{2b}{n} + \cdots + \frac{(n-1)b}{n}\right]\frac{b}{n}$$

$$= \frac{b^2}{n^2}[1 + 2 + \cdots + (n-1)]$$

(b)

$$U_f(P) = \left[f\left(\frac{b}{n}\right) + f\left(\frac{2b}{n}\right) + \cdots + f\left(\frac{(n-1)b}{n}\right) + f(b)\right]\frac{b}{n}$$

$$= \left[\frac{b}{n} + \frac{2b}{n} + \cdots + \frac{(n-1)b}{n} + b\right]\frac{b}{n}$$

$$= \frac{b^2}{n^2}[1 + 2 + \cdots + (n-1) + n]$$

(c) By Exercise 25,

$$L_f(P) = \frac{b^2}{n^2} \cdot \frac{(n-1)n}{2} \quad \text{and} \quad U_f(P) = \frac{b^2}{n^2} \cdot \frac{n(n+1)}{2}.$$

As $n \to \infty$, $L_f(P) \to \dfrac{b^2}{2}$ and $U_f(P) \to \dfrac{b^2}{2}$. Therefore, $\int_0^2 x\, dx = \dfrac{b^2}{2}$.

29. Let P be an arbitrary partition of $[0, 4]$. Since each $m_i = 2$ and each $M_i \geq 2$,

$$L_g(P) = 2\Delta x_1 + \cdots + 2\Delta x_n = 2(\Delta x_1 + \cdots + \Delta x_n) = 2 \cdot 4 = 8$$

and

$$U_g(P) \geq 2\Delta x_1 + \cdots + 2\Delta x_n = 2(\Delta x_1 + \cdots + \Delta x_n) = 2 \cdot 4 = 8.$$

Thus

$$L_g(P) \leq 8 \leq U_g(P) \quad \text{for all partitions } P \text{ of } [0, 4].$$

Uniqueness: Suppose that

(*) $$L_g(P) \leq I \leq U_g(P) \text{ for all partitions } P \text{ of } [0,4].$$

Since $L_g(P) = 8$ for all P, I is at least 8. Suppose now that $I > 8$ and choose a partition P of $[0,4]$ with max $\Delta x_i < \frac{1}{5}(I - 8)$ and

$$0 = x_1 < \cdots < x_{i-1} < 3 < x_i < \cdots < x_n = 4.$$

Then

$$U_g(P) = 2\Delta x_1 + \cdots + 2\Delta x_{i-1} + 7\Delta x_i + 2\Delta x_{i+1} + \cdots + 2\Delta x_n$$

$$= 2(\Delta x_1 + \cdots + \Delta x_n) + 5\Delta x_i$$

$$= 8 + 5\Delta x_i < 8 + \tfrac{5}{5}(I - 8) = I$$

and I does not satisfy (*). This contradiction proves that I is not greater than 8 and therefore $I = 8$.

31.   Let $P = \{x_0, x_1, x_2, \ldots, x_n\}$ be any partition of $[2, 10]$.

(a) Since each subinterval $[x_{i-1}, x_i]$ contains both rational and irrational numbers, $m_i = 4$

and $M_i = 7$. Thus,

$$L_f(P) = 4\Delta x_1 + 4\Delta x_2 + \cdots + 4\Delta x_n = 4(\Delta x_1 + \Delta x_2 + \cdots + \Delta x_n) = 4(10 - 2) = 32$$

and

$$U_f(P) = 7\Delta x_1 + 7\Delta x_2 + \cdots + 7\Delta x_n = 7(\Delta x_1 + \Delta x_2 + \cdots + \Delta x_n) = 7(10 - 2) = 56$$

Therefore, $L_f(P) \leq 40 \leq U_f(P)$.

(b) Every number $I \in [32, 56]$ satisfies the inequalities

$$L_f(P) \leq I \leq U_f(P) \quad \text{for all partitions } P$$

(c) See part (a).

SECTION 6.3

1.   (a) $\displaystyle\int_0^5 f(x)\,dx = \int_0^2 f(x)\,dx + \int_2^5 f(x)\,dx = 4 + 1 = 5$

(b) $\displaystyle\int_1^2 f(x)\,dx = \int_0^2 f(x)\,dx - \int_0^1 f(x)\,dx = 4 - 6 = -2$

(c) $\displaystyle\int_1^5 f(x)\,dx = \int_0^5 f(x)\,dx - \int_0^1 f(x)\,dx = 5 - 6 = -1$

(d) 0 (e) $\displaystyle\int_2^0 f(x)\,dx = -\int_0^2 f(x)\,dx = -4$

(f) $\displaystyle\int_5^1 f(x)\,dx = -\int_1^5 f(x)\,dx = 1$

3.   With $P = \left\{1, \dfrac{3}{2}, 2\right\}$ and $f(x) = \dfrac{1}{x}$, we have

$$0.5 < \frac{7}{12} = L_f(P) \leq \int_1^2 \frac{dx}{x} \leq U_f(P) = \frac{5}{6} < 1.$$

5.   (a) $F(0) = 0$ (b) $F'(x) = x\sqrt{x+1}$ (c) $F'(2) = 2\sqrt{3}$

(d) $F(2) = \displaystyle\int_0^2 t\sqrt{t+1}\,dt$ (e) $-F(x) = \displaystyle\int_x^0 t\sqrt{t+1}\,dt$

7.   (a) $F'(x) = x \ln x = 0$: $F'(x) = 0$ at $x = 1$; $F'(x) < 0$ for $0 < x < 1$ and $F'(x) > 0$ for $x > 1$.

Thus, F is decreasing on $(0, 1)$ and increasing on $(1, \infty)$.

(b) $F''(x) = \ln x + 1$: $F''(x) = 0$ when $\ln x = -1$, that is, when $x = 1/e$; the graph of F

is concave down for $0 < x < 1/e$, and concave up for $x > 1/e$; $1/e$ is a point of inflection.

(c)

9. $F'(x) = \dfrac{1}{x^2 + 9}$;   (a) $\dfrac{1}{10}$ (b) $\dfrac{1}{9}$ (c) $\dfrac{4}{37}$ (d) $\dfrac{-2x}{(x^2+9)^2}$

11. $F'(x) = -\sqrt{x^2 + 1}$;   (a) $-\sqrt{2}$ (b) -1 (c) $-\frac{1}{2}\sqrt{5}$ (d) $-\dfrac{x}{\sqrt{x^2+1}}$

13. $F'(x) = \cos \pi x$;   (a) -1 (b) 1 (c) 0 (d) $-\pi \sin \pi x$

15. (a) Since $P_1 \subseteq P_2$, $U_f(P_2) \le U_f(P_1)$ but $5 \not\le 4$.

   (b) Since $P_1 \subseteq P_2$, $L_f(P_1) \le L_f(P_2)$ but $5 \not\le 4$.

17. Let $u = x^3$. Then $F(u) = \displaystyle\int_1^u t \cos t \, dt$ and

$$\frac{dF}{dx} = \frac{dF}{du}\frac{du}{dx} = u \cos u \,(3x^2) = 3x^5 \cos x^3.$$

19. $F(x) = \displaystyle\int_{x^2}^1 (t - \sin^2 t)\, dt = -\int_1^{x^2} (t - \sin^2 t)\, dt$. Let $u = x^2$. Then

$$\frac{dF}{dx} = \frac{dF}{du}\frac{du}{dx} = -(u - \sin^2 u)(2x) = 2x\left[\sin^2(x^2) - x^2\right].$$

21.

$$F(x) = \int_{x^2}^{x^4} \sin\sqrt{t}\, dt = \int_{x^2}^0 \sin\sqrt{t}\, dt + \int_0^{x^4} \sin\sqrt{t}\, dt$$

$$= -\int_0^{x^2} \sin\sqrt{t}\, dt + \int_0^{x^4} \sin\sqrt{t}\, dt$$

Now, $F'(x) = -\sin x\,(2x) + \sin(x^2)4x^3 = 4x^3 \sin x^2 - 2x \sin x$ (assuming $x > 0$).

23. $f(x) = \dfrac{d}{dx}\left(\dfrac{2x}{4 + x^2}\right) = \dfrac{8 - 2x^2}{(4 + x^2)^2}$

   (a) $f(0) = \dfrac{1}{2}$ (b) $f(x) = 0$ at $x = -2,\ 2$

25. By the hint $\dfrac{F(b) - F(a)}{b - a} = F'(c)$ for some c in (a, b). The result follows by observing that

$$F(b) = \int_a^b f(t)\, dt, \quad F(a) = 0, \quad \text{and} \quad F'(c) = f(c).$$

27. Set $G(x) = \displaystyle\int_a^x f(t)\, dt$. Then $F(x) = \displaystyle\int_c^a f(t)\, dt + G(x)$. First, note that $\displaystyle\int_c^a f(t)\, dt$ is a constant. By (5.3.5) G, and thus F, is continuous on $[a, b]$, is differentiable on (a, b), and $F'(x) = G'(x) = f(x)$ for all x in (a, b).

SECTION 6.4

1. $\displaystyle\int_0^1 (2x-3)\,dx = [x^2-3x]_0^1 = (-2)-(0) = -2$

3. $\displaystyle\int_{-1}^0 5x^4\,dx = [x^5]_{-1}^0 = (0)-(-1) = 1$

5. $\displaystyle\int_1^4 2\sqrt{x}\,dx = 2\int_1^4 x^{1/2}\,dx = 2\left[\frac{2}{3}x^{3/2}\right]_1^4 = \frac{4}{3}\left[x^{3/2}\right]_1^4 = \frac{4}{3}(8-1) = \frac{28}{3}$

7. $\displaystyle\int_1^5 2\sqrt{x-1}\,dx = \int_1^5 2(x-1)^{1/2}\,dx = \left[\frac{4}{3}(x-1)^{3/2}\right]_1^5 = \frac{4}{3}[4^{3/2}-0] = \frac{32}{3}$

9. $\displaystyle\int_{-2}^0 (x+1)(x-2)\,dx = \int_{-2}^0 (x^2-x-2)\,dx = \left[\frac{x^3}{3}-\frac{x^2}{2}-2x\right]_{-2}^0 = \left[0-\left(\frac{-8}{3}-2+4\right)\right] = \frac{2}{3}$

11. $\displaystyle\int_3^3 \sqrt{x}\,dx = 0$

13. $\displaystyle\int_0^1 \frac{dt}{(t+2)^3} = \int_0^1 (t+2)^{-3}\,dt = \left[-\frac{1}{2}(t+2)^{-2}\right]_0^1 = -\frac{1}{2}[3^{-2}-2^{-2}] = \frac{5}{72}$

15. $\displaystyle\int_1^2 \left(3t+\frac{4}{t^2}\right)dt = \int_1^2 (3t+4t^{-2})\,dt = \left[\frac{3}{2}t^2-4t^{-1}\right]_1^2 = \left[(6-2)-\left(\frac{3}{2}-4\right)\right] = \frac{13}{2}$

17. $\displaystyle\int_0^1 (x^{3/2}-x^{1/2})\,dx = \left[\frac{2}{5}x^{5/2}-\frac{2}{3}x^{3/2}\right]_0^1 = \left[\left(\frac{2}{5}-\frac{2}{3}\right)-0\right] = -\frac{4}{15}$

19. $\displaystyle\int_0^1 (x+1)^{17}\,dx = \left[\frac{1}{18}(x+1)^{18}\right]_0^1 = \frac{1}{18}(2^{18}-1)$

21. $\displaystyle\int_0^a (\sqrt{a}-\sqrt{x})^2\,dx = \int_0^a (a-2\sqrt{a}\,x^{1/2}+x)\,dx = \left[ax-\frac{4}{3}\sqrt{a}\,x^{3/2}+\frac{x^2}{2}\right]_0^a = a^2-\frac{4}{3}a^2+\frac{a^2}{2} = \frac{1}{6}a^2$

23. $\displaystyle\int_1^2 \frac{6-t}{t^3}\,dt = \int_1^2 (6t^{-3}-t^{-2})\,dt = [-3t^{-2}+t^{-1}]_1^2 = \left[\frac{-3}{4}+\frac{1}{2}\right]-[-3+1] = \frac{7}{4}$

25. $\displaystyle\int_0^1 x^2(x-1)\,dx = \int_0^1 (x^3-x^2)\,dx = \left[\frac{x^4}{4}-\frac{x^3}{3}\right]_0^1 = -\frac{1}{12}$

27. $\displaystyle\int_1^2 2x(x^2+1)\,dx = \int_1^2 (2x^3+2x)\,dx = \left[\frac{x^4}{2}+x^2\right]_1^2 = 12-\frac{3}{2} = \frac{21}{2}$

29. $\displaystyle\int_0^{\pi/2} \cos x\,dx = [\sin x]_0^{\pi/2} = 1$

31. $\displaystyle\int_0^{\pi/4} 2\sec^2 x\,dx = 2[\tan x]_0^{\pi/4} = 2$

33. $\displaystyle\int_{\pi/6}^{\pi/4} \csc x\cot x\,dx = [-\csc x]_{\pi/6}^{\pi/4} = -\sqrt{2}-(-2) = 2-\sqrt{2}$

35. $\displaystyle\int_0^{2\pi} \sin x\,dx = [-\cos x]_0^{2\pi} = -1-(-1) = 0$

37. $\int_0^{\pi/3} \left(\dfrac{2}{\pi}x - 2\sec^2 x \right) dx = \left[\dfrac{1}{\pi}x^2 - 2\tan x \right]_0^{\pi/3} = \dfrac{\pi}{9} - 2\sqrt{3}$

39. $\int_0^3 \left[\dfrac{d}{dx}\left(\sqrt{4+x^2} \right) \right] dx = \left[\sqrt{4+x^2} \right]_0^3 = \sqrt{13} - 2$

41. $\int_2^4 \dfrac{1}{3+x}\, dx = \left[\, \ln|3+x| \,\right]_2^4 = \ln 7 - \ln 5 = \ln(7/5)$

43. $\int_{-1}^0 e^{3x}\, dx = \left[\tfrac{1}{3} e^{3x} \right]_{-1}^0 = \tfrac{1}{3}\left(1 - e^{-3} \right)$

45. $\int_0^{1/\sqrt{2}} \dfrac{1}{(1-x^2)^{1/2}}\, dx = \left[\sin^{-1} x \right]_0^{1/\sqrt{2}} = \dfrac{\pi}{4}$

47. (a) $\ F(x) = \int_2^x \dfrac{dt}{t}$ (b) $\ F(x) = -3 + \int_2^x \dfrac{dt}{t}$

49. $\text{Area} = \int_0^4 (4x - x^2)\, dx = \left[2x^2 - \dfrac{x^3}{3} \right]_0^4 = \dfrac{32}{3}$

51. $\text{Area} = \int_{-\pi/2}^{\pi/4} 2\cos x\, dx = 2\left[\sin x \right]_{-\pi/2}^{\pi/4} = \sqrt{2} + 2$

53. (a) $\int_2^5 (x-3)\, dx = \left[\dfrac{x^2}{2} - 3x \right]_2^5 = \dfrac{3}{2}$ (b) $\int_2^5 |x-3|\, dx = \int_2^3 (3-x)\, dx + \int_3^5 (x-3)\, dx$

$$= \left[3x - \dfrac{x^2}{2} \right]_2^3 + \left[\dfrac{x^2}{2} - 3x \right]_3^5 = \dfrac{5}{2}$$

55. (a) $\int_{-2}^2 (x^2 - 1)\, dx = \left[\dfrac{x^3}{3} - x \right]_{-2}^2 = \dfrac{4}{3}$

(b)

$$\int_{-2}^2 |x^2 - 1|\, dx = \int_{-2}^{-1} (x^2 - 1)\, dx + \int_{-1}^1 (1 - x^2)\, dx + \int_1^2 (x^2 - 1)\, dx$$

$$= \left[\dfrac{x^3}{3} - x \right]_{-2}^{-1} + \left[x - \dfrac{x^3}{3} \right]_{-1}^1 + \left[\dfrac{x^3}{3} - x \right]_1^2 = 4$$

57. (a) $x(t) = \int_0^t (10u - u^2)\, du = \left[5u^2 - \dfrac{u^3}{3} \right]_0^t = 5t^2 - \dfrac{t^3}{3}, \quad 0 \le t \le 10$

(b) $v'(t) = 10 - 2t;$ v has an absolute maximum at $t = 5$. The object's position at $t = 5$ is

    $x(5) = \dfrac{250}{3}.$

59. $\int_0^4 f(x)\, dx = \int_0^1 (2x+1)\, dx + \int_1^4 (4-x)\, dx = \left[x^2 + x \right]_0^1 + \left[4x - \dfrac{x^2}{2} \right]_1^4 = \dfrac{13}{2}$

61. $\int_{-\pi/2}^{\pi} f(x)\, dx = \int_{-\pi/2}^{\pi/3} \cos x\, dx + \int_{\pi/3}^{\pi} \left[\dfrac{3}{\pi}x + 1 \right] dx = \left[\sin x \right]_{-\pi/2}^{\pi/3} + \left[\dfrac{3x^2}{2\pi} + x \right]_{\pi/3}^{\pi}$

$$= \dfrac{2 + \sqrt{3}}{2} + 2\pi$$

63. (a) f is continuous on $[-2, 2]$.

For $x \in [-2, 0]$, $\quad g(x) = \int_{-2}^{x} (t+2)dt = \left[\frac{1}{2}t^2 + 2t\right]_{-2}^{x} = \frac{1}{2}x^2 + 2x + 2$.

For $x \in [0, 1]$, $\quad g(x) = \int_{-2}^{0} (t+2)\,dt + \int_{0}^{x} 2\,dt = 2 + [2t]_0^x = 2 + 2x$.

For $x \in [1, 2]$, $\quad g(x) = \int_{-2}^{0} (t+2)\,dt + \int_{0}^{1} 2\,dt + \int_{1}^{x} (4-2x)\,dx = 2+2+\left[4t - t^2\right]_1^x = 1 + 4x - x^2$.

Thus $\quad g(x) = \begin{cases} \frac{1}{2}x^2 + 2x + 2, & -2 \le x \le 0 \\ 2x + 2, & 0 \le x \le 1 \\ 1 + 4x - x^2, & 1 \le x \le 2 \end{cases}$

(b)

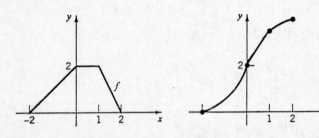

(c) f is continuous on $[-2, 2]$; f is differentiable on $(-2, 0)$, $(0, 1)$, and $(1, 2)$.

g is differentiable on $(-2, 2)$.

65. (a) $f'(x) = \sqrt{1 + x^2} > 0$, so f is always increasing, hence one-to-one.

(b) $\left(f^{-1}\right)'(0) = \dfrac{1}{f'\left(f^{-1}(0)\right)} = \dfrac{1}{f'(2)} = \dfrac{1}{\sqrt{5}}$.

67. $\dfrac{d}{dx}\left[\int_a^x f(t)\,dt\right] = f(x); \quad \int_a^x \dfrac{d}{dt}\left[f(t)\right]dt = f(x) - f(a)$

SECTION 6.5

1. $A = \int_0^1 (2 + x^3)\,dx = \left[2x + \dfrac{x^4}{4}\right]_0^1 = \dfrac{9}{6}$

3. $A = \int_3^8 \sqrt{x+1}\,dx = \int_3^8 (x+1)^{1/2}\,dx = \left[\dfrac{2}{3}(x+1)^{3/2}\right]_3^8 = \dfrac{2}{3}[27 - 8] = \dfrac{38}{3}$

5. $A = \int_0^1 (2x^2 + 1)^2\,dx = \int_0^1 (4x^4 + 4x^2 + 1)\,dx = \left[\dfrac{4}{5}x^5 + \dfrac{4}{3}x^3 + x\right]_0^1 = \dfrac{47}{15}$

7. $A = \int_1^2 [0 - (x^2 - 4)]\,dx = \int_1^2 (4 - x^2)\,dx = \left[4x - \dfrac{x^3}{3}\right]_1^2 = \left[8 - \dfrac{8}{3}\right] - \left[4 - \dfrac{1}{3}\right] = \dfrac{5}{3}$

9. $\quad A = \displaystyle\int_{\pi/3}^{\pi/2} \sin x \, dx = [-\cos x]_{\pi/3}^{\pi/2} = (0) - \left(-\frac{1}{2}\right) = \frac{1}{2}$

11.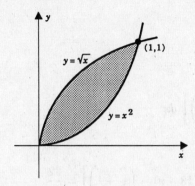

$$A = \int_0^1 [x^{1/2} - x^2] \, dx$$

$$= \left[\frac{2}{3}x^{3/2} - \frac{1}{3}x^3\right]_0^1 = \frac{1}{3}$$

13.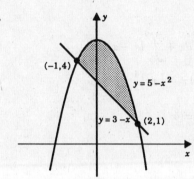

$$A = \int_{-1}^2 [(5 - x^2) - (3 - x)] \, dx$$

$$= \int_{-1}^2 (2 + x - x^2) \, dx$$

$$= \left[2x + \frac{x^2}{2} - \frac{x^3}{3}\right]_{-1}^2$$

$$= \left[4 + 2 - \frac{8}{3}\right] - \left[-2 + \frac{1}{2} + \frac{1}{3}\right] = \frac{9}{2}$$

15.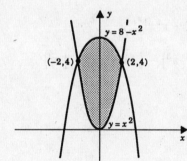

$$A = \int_{-2}^2 [(8 - x^2) - (x^2)] \, dx$$

$$= \int_{-2}^2 (8 - 2x^2) \, dx$$

$$= \left[8x - \frac{2}{3}x^3\right]_{-2}^2$$

$$= \left[16 - \frac{16}{3}\right] - \left[-16 + \frac{16}{3}\right] = \frac{64}{3}$$

17.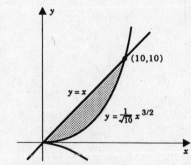

$$A = \int_0^{10} \left[x - \frac{1}{\sqrt{10}}\, x^{3/2}\right] dx$$

$$= \left[\frac{x^2}{2} - \frac{2\sqrt{10}}{50}\, x^{5/2}\right]_0^{10}$$

$$= 50 - \frac{2\sqrt{10}}{50}(10)^{5/2} = 10$$

19.

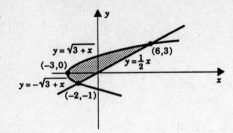

$$A = \int_{-3}^{-2} [(\sqrt{3+x}) - (-\sqrt{3+x})]\, dx + \int_{-2}^{6} \left[(\sqrt{3+x}) - \left(\frac{1}{2}x \right) \right] dx$$

$$= \int_{-3}^{-2} 2(3+x)^{1/2}\, dx + \int_{-2}^{6} \left[(3+x)^{1/2} - \frac{1}{2}x \right] dx$$

$$= \left[\frac{4}{3}(3+x)^{3/2} \right]_{-3}^{-2} + \left[\frac{2}{3}(3+x)^{3/2} - \frac{x^2}{4} \right]_{-2}^{6} = \left[\frac{4}{3} - 0 \right] + \left[(18-9) - \left(\frac{2}{3} - 1 \right) \right] = \frac{32}{3}$$

21.

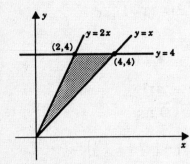

$$A = \int_{0}^{2} [2x - x]\, dx + \int_{2}^{4} [4 - x]\, dx$$

$$= \left[\tfrac{1}{2}x^2 \right]_{0}^{2} + \left[4x - \tfrac{1}{2}x^2 \right]_{2}^{4}$$

$$= 2 + [8 - 6] = 4$$

23.

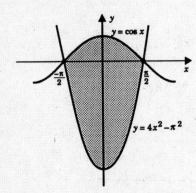

$$A = \int_{-\pi/2}^{\pi/2} [\cos x - (4x^2 - \pi^2)]\, dx$$

$$= \left[\sin x - \tfrac{4}{3}x^3 + \pi^2 x \right]_{-\pi/2}^{\pi/2}$$

$$= \left[1 - \tfrac{1}{6}\pi^3 + \tfrac{1}{2}\pi^3 \right] - \left[-1 + \tfrac{1}{6}\pi^3 - \tfrac{1}{2}\pi^3 \right]$$

$$= 2 + \tfrac{2}{3}\pi^3$$

25.

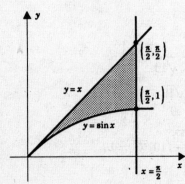

$$A = \int_{0}^{\pi/2} [x - \sin x]\, dx$$

$$= \left[\frac{x^2}{2} + \cos x \right]_{0}^{\pi/2}$$

$$= \frac{\pi^2}{8} - 1$$

27.

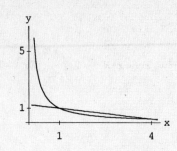

$$A = \int_1^4 \left[\frac{5-x}{4} - \frac{1}{x} \right] dx$$

$$= \left[\frac{5}{4}x - \frac{1}{8}x^2 - \ln x \right]_1^4$$

$$= \frac{15}{8} - \ln 4$$

29. (a) $\displaystyle\int_{-3}^4 (x^2 - x - 6)\, dx = \left[\frac{1}{3}x^3 - \frac{1}{2}x^2 - 6x \right]_{-3}^4 = -\frac{91}{6};$

the area of the region bounded by the graph of f and the x-axis for $x \in [-3,-2] \cup [3,4]$

minus the area of the region bounded the graph of f and the x-axis for $x \in [-2,3]$.

  (b) $\displaystyle A = \int_{-3}^{-2}(x^2 - x - 6)\,dx + \int_{-2}^{3}3(-x^2 + x + 6)\,dx + \int_3^4 (x^2 - x - 6)\,dx$

$$= \left[\tfrac{1}{3}x^3 - \tfrac{1}{2}x^2 - 6 \right]_{-3}^{-2} + \left[-\tfrac{1}{3}x^3 + \tfrac{1}{2}x^2 + 6 \right]_{-2}^{3} + \left[\tfrac{1}{3}x^3 - \tfrac{1}{2}x^2 - 6 \right]_3^4 = \frac{17}{6} + \frac{125}{6} + \frac{17}{6} = \frac{53}{2}$$

  (c) $\displaystyle A = -\int_{-2}^{3}(x^2 - x - 6)\,dx = \frac{125}{6}$

31. $\displaystyle A = \int_x^1 \frac{1}{t}\, dt = -\int_1^x \frac{1}{t}\, dt$

$$= -\ln x = \ln \frac{1}{x}$$

33. $\displaystyle \ln 2.5 \cong \frac{1}{2}\left[L_f(P) + U_f(P) \right] \cong 0.921$

35. (a) $\displaystyle\int_{-\pi/2}^{3\pi/4} 2 \sin x\, dx = [-2\cos x]_{-\pi/2}^{3\pi/4} = \sqrt{2} = $ area above $-$ area below

  (b) $\displaystyle A = \int_{-\pi/2}^{0} -2\sin x\, dx + \int_0^{3\pi/4} 2\sin x\, dx = [2\cos x]_{-\pi/2}^{0} + [-2\cos x]_0^{3\pi/4} = \sqrt{2} + 4$

  (c) $\displaystyle A = \int_{-\pi/2}^{0} -2\sin x\, dx = [2\cos x]_{-\pi/2}^{0} = 2$

37. (a) $\displaystyle\int_{-\pi}^{\pi} (\cos x + \sin x)\, dx = [\sin x - \cos x]_{-\pi}^{\pi} = 0$

  (b)

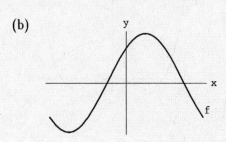

$$A = -\int_{-\pi}^{-\pi/4} f(x)\, dx + \int_{-\pi/4}^{3\pi/4} f(x)\, dx - \int_{3\pi/4}^{\pi} f(x)\, dx$$

$$= 4\sqrt{2}$$

39. (a) $\int_{-\pi/2}^{\pi/2} (3x^2 - 2\cos x)\, dx = \left[x^3 - 2\sin x\right]_{-\pi/2}^{\pi/2} = \frac{\pi^3}{4} - 4$

(b)

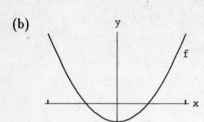

$A \cong -\int_{-0.72}^{0.72} (3x^2 - 2\cos x)\, dx + 2\int_{0.72}^{\pi/2} (3x^2 - 2\cos x)\, dx$

$\cong 7.45$

41.

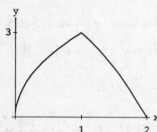

$A = \int_0^1 3\sqrt{x}\, dx + \int_1^2 (4 - x^2)\, dx$

$= \left[2x^{3/2}\right]_0^1 + \left[4x - \frac{x^3}{3}\right]_1^2$

$= 2 + \frac{5}{3} = \frac{11}{3}$

43.

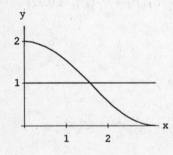

$A = 2\int_0^{\pi/2} (1 + \cos x - 1)\, dx$

$= \left[2\sin x\right]_0^{\pi/2}$

$= 2$

45. By symmetry

$A = 2\left(\int_0^{1.60} \left[(4 - x^2) - (x^4 - 2x^2)\right]\, dx + \right.$

$\left. \int_{1.60}^2 \left[(x^4 - 2x^2) - (4 - x^2)\right]\, dx\right)$

$= 2\left(\int_0^{1.60} \left[4 + x^2 - x^4\right]\, dx + \int_{1.60}^2 \left[x^4 - x^2 - 4\right]\, dx\right)$

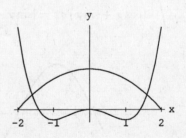

$\cong 14.14$

SECTION 6.6

1. $\displaystyle \int \frac{dx}{x^4} = \int x^{-4}\,dx = -\frac{1}{3}x^{-3} + C$ 3. $\displaystyle \int (ax+b)\,dx = \frac{1}{2}ax^2 + bx + C$

5. $\displaystyle \int \frac{dx}{\sqrt{1+x}} = \int (1+x)^{-1/2}\,dx = 2(1+x)^{1/2} + C$

7. $\displaystyle \int \left(\frac{x^3-1}{x^2}\right) dx = \int (x - x^{-2})\,dx = \frac{1}{2}x^2 + x^{-1} + C$

9. $\displaystyle \int (t-a)(t-b)\,dt = \int [t^2 - (a+b)t + ab]\,dt = \frac{1}{3}t^3 - \frac{a+b}{2}t^2 + abt + C$

11. $\displaystyle \int \frac{(t^2-a)(t^2-b)}{\sqrt{t}}\,dt = \int [t^{7/2} - (a+b)t^{3/2} + abt^{-1/2}]\,dt$

$$= \tfrac{2}{9}t^{9/2} - \tfrac{2}{5}(a+b)t^{5/2} + 2abt^{1/2} + C$$

13. $\displaystyle \int g(x)g'(x)\,dx = \frac{1}{2}[g(x)]^2 + C$

15. $\displaystyle \int \tan x \sec^2 x\,dx = \int \sec x \frac{d}{dx}[\sec x]\,dx = \frac{1}{2}\sec^2 x + C$

$\displaystyle \int \tan x \sec^2 x\,dx = \int \tan x \frac{d}{dx}[\tan x]\,dx = \frac{1}{2}\tan^2 x + C$

17. $\displaystyle \int \frac{4}{(4x+1)^2}\,dx = \int 4(4x+1)^{-2}\,dx = -(4x+1)^{-1} + C$

19. $\displaystyle \frac{d}{dx}(\ln|\csc x - \cot x| + C) = \frac{1}{\csc x - \cot x}\left(-\csc x \cot x + \csc^2 x\right) = \csc x.$

Thus, $\ln|\csc x - \cot x| + C$ is an antiderivative of $\csc x$ and the result follows.

21. $\displaystyle f(x) = \int (3-4x)\,dx = 3x - 2x^2 + C, \quad f(1) = 6 \implies f(x) = -2x^2 + 3x + 5$

23. $\displaystyle f(x) = \int f'(x)\,dx = \int (\sin x + e^x)\,dx = -\cos x + e^x + C.$

$f(0) = 0 \implies C = 0$ and $f(x) = e^x - \cos x.$

25. $\displaystyle f(x) = \int \cos x\,dx = \sin x + C, \quad f(\pi) = 3 \implies f(x) = 3 + \sin x$

27. $\displaystyle f'(x) = \int -12x^2\,dx = -4x^3 + C, \quad f'(0) = 1 \implies f'(x) = -4x^3 + 1$

$$f(x) = \int (-4x^3 + 1)\,dx = -x^4 + x + K, \quad f(0) = 2 \implies f(x) = -x^4 + x + 2$$

29. $\quad f'(x) = \int (1-x)\,dx = x - \dfrac{x^2}{2} + C, \quad f'(2) = 1 \implies f'(x) = x - \dfrac{x^2}{2} + 1$

$$f(x) = \int \left(x - \dfrac{x^2}{2} + 1\right) dx = \dfrac{x^2}{2} - \dfrac{x^3}{6} + x + K, \quad f(2) = 0 \implies f(x) = -\dfrac{x^3}{6} + \dfrac{x^2}{2} + x - \dfrac{8}{3}$$

31. $\quad f'(x) = \int \sin x\,dx = -\cos x + C, \quad f'(0) = -2 \implies f'(x) = -\cos x - 1$

$$f(x) = \int (-\cos x - 1)\,dx = -\sin x - x + K, \quad f(0) = 1 \implies f(x) = 1 - \sin x - x$$

33. $\quad f'(x) = \int (5 - 4x)\,dx = 5x - 2x^2 + C,$

$$f(x) = \int (5x - 2x^2 + C)\,dx = \dfrac{5}{2}x^2 - \dfrac{2}{3}x^3 + Cx + K$$

$$f(1) = \dfrac{5}{2} - \dfrac{2}{3} + C + K = 1, \quad f(0) = k = -2 \implies f(x) = -\dfrac{2}{3}x^3 + \dfrac{5}{2}x^2 + \dfrac{7}{6}x - 2$$

35. $\quad \displaystyle\int [f(x)g''(x) - g(x)f''(x)]\,dx = \int [f(x)g''(x) + f'(x)g'(x) - f'(x)g'(x) - g(x)f''(x)]\,dx$

$$= \int \left(\dfrac{d}{dx}[f(x)g'(x)] - \dfrac{d}{dx}[f'(x)g(x)]\right) dx = f(x)g'(x) - g(x)f'(x) + C$$

37. (a) $\quad v(t) = \displaystyle\int a(t)\,dt = \int (t+2)^3\,dt = \dfrac{1}{4}(t+2)^4 + C,$

$$v(0) = 3 \implies v(t) = \dfrac{1}{4}(t+2)^4 - 1$$

(b) $\quad x(t) = \displaystyle\int \left[\dfrac{(t+2)^4}{4} - 1\right] dt = \dfrac{(t+2)^5}{20} - t + K, \quad x(0) = 0 \implies x(t) = \dfrac{(t+2)^5}{20} - t - \dfrac{8}{5}$

39. (a) $\quad x(t) = \displaystyle\int t(1-t)\,dt = \dfrac{t^2}{2} - \dfrac{t^3}{3} + C, \quad x(0) = -2 \implies x(t) = \dfrac{t^2}{2} - \dfrac{t^3}{3} - 2$

$$x(0) = -\dfrac{856}{3} : \quad 285\dfrac{1}{3} \text{ units to the left of the origin.}$$

(b) $\quad s = \displaystyle\int_0^{10} |v(t)|\,dt = \int_0^1 t(1-t)\,dt + \int_1^{10} -t(1-t)\,dt = \left[\dfrac{t^2}{2} - \dfrac{t^3}{3}\right]_0^1 - \left[\dfrac{t^2}{2} - \dfrac{t^3}{3}\right]_1^{10}$

$$= \frac{851}{3} = 283\frac{2}{3} \text{ units.}$$

41. Let acceleration $= a$. Then $v(t) = \int a \, dt = at + v_0$.

$$x(t) = \int v(t) \, dt = \int (at + v_0) \, dt = \frac{1}{2}at^2 + v_0 t + x_0 = \frac{1}{2}\left[v(t) + v_0\right]t + x_0$$

43. (a) $v(t) = at + v_0$, and by Exercise 42 $x(t) = \frac{1}{2}\left[v(t) + v_0\right]t$, so

$$a = \frac{v(t) - v_0}{t} = \frac{v(t) - v_0}{2x(t)}(v(t) + v_0) = \frac{v(t)^2 - v_0^2}{2x(t)} = \frac{58.7^2 - 88^2}{2.264} = -8.15 \text{ ft/sec}^2$$

[Note $60 \text{ mph} = 88 \text{ ft/sec}$, $40 \text{ mph} = 58\frac{2}{3}\text{ft/sec.}$]

(b) $t = \dfrac{2x(t)}{v(t) + v_0} = \dfrac{2.264}{58\frac{2}{3} + 88} = 3.6 \text{ sec}$

(c) We don't know $x(t)$, so we will use $t = \dfrac{v(t) - v_0}{a} = \dfrac{0 - 88}{-8.15} = 10.8 \text{ sec}$

(d) $x(t) = \dfrac{1}{2}\left[v(t) + v_0\right]t = \dfrac{1}{2}[0 + 88]\,10.8 = 475.2 \text{ ft}$

45. $v(t) = \int \sin t \, dt = -\cos t + C$, $v(0) = v_0 \implies v(t) = -\cos t + v_0 + 1$

$$x(t) = \int (-\cos t + v_0 + 1) \, dt = -\sin t + (v_0 + 1)t + K, \quad x(0) = x_0 \implies x(t) = x_0 + (v_0 + 1)t - \sin t$$

47. $v(t) = \int \cos t \, dt = \sin t + C$, $v(0) = v_0 \implies v(t) = \sin t + v_0$

$$x(t) = \int (\sin t + v_0) \, dt = -\cos t + v_0 t + K, \quad x(0) = x_0 \implies x(t) = x_0 + 1 + v_0 t - \cos t$$

49.

$$x'(t) = t^2 - 5, \qquad\qquad y'(t) = 3t,$$

$$x(t) = \tfrac{1}{3}t^3 - 5t + C. \qquad\qquad y(t) = \tfrac{3}{2}t^2 + K.$$

When $t = 2$, the particle is at $(4, 2)$. Thus, $x(2) = 4$ and $y(2) = 2$.

$$4 = \tfrac{8}{3} - 10 + C \implies C = \tfrac{34}{3}. \qquad\qquad 2 = 6 + K \implies K = -4.$$

$$x(t) = \tfrac{1}{3}t^3 - 5t + \tfrac{34}{3}, \qquad\qquad y(t) = \tfrac{3}{2}t^2 - 4.$$

Four seconds later the particle is at $(x(6), y(6)) = \left(\tfrac{160}{3}, 50\right)$.

51. Since $v(0) = 2$, we have $2 = A \cdot 0 + B$ so that $B = 2$. Therefore

$$x(t) = \int v(t) \, dt = \int (At + 2) \, dt = \frac{1}{2}At^2 + 2t + C.$$

Since $x(2) = x(0) - 1$, we have

$$2A + 4 + C = C - 1 \quad \text{so that} \quad A = -\tfrac{5}{2}.$$

53. $x(t) = \int v(t) \, dt = \int \sin t \, dt = -\cos t + C$

Since $x(\pi/6) = 0$, we have $0 = -\dfrac{\sqrt{3}}{2} + C$ so that $C = \dfrac{\sqrt{3}}{2}$ and $x(t) = \dfrac{\sqrt{3}}{2} - \cos t$.

(a)   At $t = 11\pi/6$ sec.

(b)   We want to find the smallest $t_0 > \pi/6$ for which $x(t_0) = 0$ and $v(t_0) > 0$. We get

$$t_0 = 13\pi/6 \text{ seconds.}$$

55.   The mean-value theorem. With obvious notation

$$\frac{x(1/12) - x(0)}{1/12} = \frac{4}{1/12} = 48.$$

By the mean-value theorem there exists some time t_0 at which

$$x'(t_0) = \frac{x(1/12) - x(0)}{1/12} = 48.$$

57. $\dfrac{v'(t)}{[v(t)]^2} = 2 \implies -[v(t)]^{-1} = 2t - v_0^{-1}.$

$\implies [v(t)]^{-1} = v_0^{-1} - 2t \implies v(t) = \dfrac{1}{v_0^{-1} - 2t} = \dfrac{v_0}{1 - 2tv_0}.$

59. $V(t) = \displaystyle\int a(t)\,dt = \int -(t+1)^{-2}\,dt = \dfrac{1}{t+1} + C, \quad v(0) = 2 \implies v(t) = \dfrac{1}{t+1} + 1$

Then $s = \displaystyle\int_0^4 v(t)\,dt = \int_0^4 \left(\dfrac{1}{1+t} + 1\right)\,dt = \left[\, \ln(t+1) + t \,\right]_0^4 = 4 + \ln 5 \ \text{ft}$

SECTION 6.7

1. $\left\{ \begin{array}{l} u = 2 - 3x \\ du = -3\,dx \end{array} \right\};\quad \displaystyle\int \frac{dx}{(2-3x)^2} = \int (2-3x)^{-2}\,dx = -\frac{1}{3}\int u^{-2}\,du = \frac{1}{3}u^{-1} + C$

$$= \tfrac{1}{3}(2-3x)^{-1} + C$$

3. $\left\{ \begin{array}{l} u = 2x + 1 \\ du = 2\,dx \end{array} \right\};\quad \displaystyle\int \sqrt{2x+1}\,dx = \int (2x+1)^{1/2}\,dx = \frac{1}{2}\int u^{1/2}\,du = \frac{1}{3}u^{3/2} + C$

$$= \tfrac{1}{3}(2x+1)^{3/2} + C$$

5. $\left\{ \begin{array}{l} u = ax + b \\ du = a\,dx \end{array} \right\};\quad \displaystyle\int (ax+b)^{3/4}\,dx = \frac{1}{a}\int u^{3/4}\,du = \frac{4}{7a}u^{7/4} + C$

$$= \frac{4}{7a}(ax+b)^{7/4} + C$$

7. $\left\{ \begin{array}{l} u = 4t^2 + 9 \\ du = 8t\,dt \end{array} \right\}$; $\displaystyle\int \frac{t}{(4t^2+9)^2}\,dt = \frac{1}{8}\int \frac{du}{u^2} = -\frac{1}{8}u^{-1} + C = -\frac{1}{8}\left(4t^2+9\right)^{-1} + C$

9. $\left\{ \begin{array}{l} u = 5t^3 + 9 \\ du = 15t^2\,dt \end{array} \right\}$; $\displaystyle\int t^2\left(5t^3+9\right)^4\,dt = \frac{1}{15}\int u^4\,du = \frac{1}{75}u^5 + C$

$$= \tfrac{1}{75}(5t^3+9)^5 + C$$

11. $\left\{ \begin{array}{l} u = 1 + x^3 \\ du = 3x^2\,dx \end{array} \right\}$; $\displaystyle\int x^2\left(1+x^3\right)^{1/4}\,dx = \frac{1}{3}\int u^{1/4}\,du = \frac{4}{15}u^{5/4} + C$

$$= \tfrac{4}{15}(1+x^3)^{5/4} + C$$

13. $\left\{ \begin{array}{l} u = 1 + s^2 \\ du = 2s\,ds \end{array} \right\}$; $\displaystyle\int \frac{s}{(1+s^2)^3}\,ds = \frac{1}{2}\int \frac{du}{u^3} = -\frac{1}{4}u^{-2} + C = -\frac{1}{4}(1+s^2)^{-2} + C$

15. $\left\{ \begin{array}{l} u = x^2 + 1 \\ du = 2x\,dx \end{array} \right\}$; $\displaystyle\int \frac{x}{\sqrt{x^2+1}}\,dx = \int \left(x^2+1\right)^{-1/2} x\,dx = \frac{1}{2}\int u^{-1/2}\,du$

$$= u^{1/2} + C = \sqrt{x^2+1} + C$$

17. $\left\{ \begin{array}{l} u = 1 - x^3 \\ du = -3x^2\,dx \end{array} \right\}$; $\displaystyle\int x^2\left(1-x^3\right)^{2/3}\,dx = -\frac{1}{3}\int u^{2/3}\,du$

$$= -\tfrac{1}{5}u^{5/3} + C = -\tfrac{1}{5}(1-x^3)^{5/3} + C$$

19. $\left\{ \begin{array}{l} u = x^2 + 1 \\ du = 2x \end{array} \right\}$; $\displaystyle\int 5x\left(x^2+1\right)^{-3}\,dx = \frac{5}{2}\int u^{-3}\,du = -\frac{5}{4}u^{-2} + C$

$$= -\tfrac{5}{4}(x^2+1)^{-2} + C$$

21. $\left\{ \begin{array}{l} u = x^{1/4} + 1 \\ du = \frac{1}{4}x^{-3/4}\,dx \end{array} \right\}$; $\displaystyle\int x^{-3/4}\left(x^{1/4}+1\right)^{-2}\,dx = 4\int u^{-2}\,du = -4u^{-1} + C$

$$= -4(x^{1/4}+1)^{-1} + C$$

23. $\left\{ \begin{array}{l} u = 1 - a^4x^4 \\ du = -4a^4x^3\,dx \end{array} \right\}$; $\displaystyle\int \frac{b^3x^3}{\sqrt{1-a^4x^4}}\,dx = -\frac{b^3}{4a^4}\int u^{-1/2}\,du = -\frac{b^3}{2a^4}u^{1/2} + C$

$$= -\frac{b^3}{2a^4}\sqrt{1-a^4x^4} + C$$

25.
$$\left\{ \begin{array}{l} u = x^2 + 1 \\ du = 2x\,dx \end{array} \middle| \begin{array}{l} x = 0 \implies u = 1 \\ x = 1 \implies u = 2 \end{array} \right\}; \qquad \int_0^1 x\left(x^2+1\right)^3 dx = \frac{1}{2}\int_1^2 u^3\,du$$

$$= \tfrac{1}{8}\left[u^4\right]_1^2 = \tfrac{15}{8}$$

27.
$$\left\{ \begin{array}{l} u = 1 + x^2 \\ du = 2x\,dx \end{array} \middle| \begin{array}{l} x = 0 \implies u = 1 \\ x = 1 \implies u = 2 \end{array} \right\}; \qquad \int_0^1 5x\left(1+x^2\right)^4 dx = \frac{5}{2}\int_1^2 u^4\,du$$

$$= \tfrac{1}{2}\left[u^5\right]_1^2 = \tfrac{31}{2}$$

29.   0;   integrand is an odd function

31.
$$\left\{ \begin{array}{l} u = a^2 - y^2 \\ du = -2y\,dy \end{array} \middle| \begin{array}{l} y = 0 \implies u = a^2 \\ y = a \implies u = 0 \end{array} \right\}; \qquad \int_0^a y\sqrt{a^2 - y^2}\,dy = -\frac{1}{2}\int_{a^2}^0 u^{1/2}\,du$$

$$= -\tfrac{1}{3}\left[u^{3/2}\right]_{a^2}^0 = \tfrac{1}{3}\left(a^2\right)^{3/2} = \tfrac{1}{3}|a|^3$$

33.
$$\left\{ \begin{array}{l} u = x + 1 \\ du = dx \end{array} \right\}; \qquad \int x\sqrt{x+1}\,dx = \int (u-1)\sqrt{u}\,du = \int \left(u^{3/2} - u^{1/2}\right)du$$

$$= \tfrac{2}{5}u^{5/2} - \tfrac{2}{3}u^{3/2} + C = \tfrac{2}{5}(x+1)^{5/2} - \tfrac{2}{3}(x+1)^{3/2} + C$$

35.
$$\left\{ \begin{array}{l} u = 2x - 1 \\ du = 2\,dx \end{array} \right\}; \qquad \int x\sqrt{2x-1}\,dx = \frac{1}{2}\int \frac{u+1}{2}\sqrt{u}\,du = \frac{1}{4}\int \left(u^{3/2} + u^{1/2}\right)du$$

$$= \tfrac{1}{10}u^{5/2} + \tfrac{1}{6}u^{3/2} + C = \tfrac{1}{10}(2x-1)^{5/2} + \tfrac{1}{6}(2x-1)^{3/2} + C$$

37.
$$\left\{ \begin{array}{l} u = y + 1 \\ du = dy \end{array} \right\}; \qquad \int y(y+1)^{12}\,dy = \int (u-1)u^{12}\,du = \int \left(u^{13} - u^{12}\right)du$$

$$= \tfrac{1}{14}u^{14} - \tfrac{1}{13}u^{13} + C = \tfrac{1}{14}(y+1)^{14} - \tfrac{1}{13}(y+1)^{13} + C$$

39.
$$\left\{ \begin{array}{l} u = t - 2 \\ du = dt \end{array} \right\}; \qquad \int t^2 (t-2)^{-5}\,dt = \int (u+2)^2 u^{-5}\,du = \int \left(u^2 + 4u + 4\right)u^{-5}\,du$$

$$= \int \left(u^{-3} + 4u^{-4} + 4u^{-5}\right)du = -\tfrac{1}{2}u^{-2} - \tfrac{4}{3}u^{-3} - u^{-4} + C$$

$$= -\tfrac{1}{2}(t-2)^{-2} - \tfrac{4}{3}(t-2)^{-3} - (t-2)^{-4} + C$$

41.
$$\left\{ \begin{array}{l} u = x + 1 \\ du = dx \end{array} \middle| \begin{array}{l} x = 0 \implies u = 1 \\ x = 1 \implies u = 2 \end{array} \right\}; \qquad \int_0^1 \frac{x+3}{\sqrt{x+1}}\,dx = \int_1^2 \frac{u+2}{\sqrt{u}}\,du$$

$$= \int_1^2 \left(u^{1/2} + 2u^{-1/2}\right)du$$

$$= \left[\tfrac{2}{3}u^{3/2} + 4u^{1/2}\right]_1^2 = \tfrac{2}{3}\sqrt{8} + 4\sqrt{2} - \tfrac{2}{3} - 4 = \tfrac{16}{3}\sqrt{2} - \tfrac{14}{3}$$

43. $\left\{\begin{array}{l} u = x^2 + 1 \\ du = 2x\,dx \end{array} \,\middle|\, \begin{array}{l} x = -1 \implies u = 2 \\ x = 0 \implies u = 1 \end{array}\right\}$; $\displaystyle\int_{-1}^{0} x^3 \left(x^2 + 1\right)^6 dx = \frac{1}{2} \int_{2}^{1} (u-1) u^6 \, du$

$$= \frac{1}{2} \int_{2}^{1} (u^7 - u^6)\,du = \left[\frac{1}{16} u^8 - \frac{1}{14} u^7\right]_{2}^{1} = -\frac{255}{16} + \frac{127}{14} = -\frac{769}{112}$$

45. $\displaystyle\int \cos(3x - 1)\,dx = \frac{1}{3} \sin(3x - 1) + C$ **47.** $\displaystyle\int \csc^2 \pi x \, dx = -\frac{1}{\pi} \cot \pi x + C$

49. $\left\{\begin{array}{l} u = 3 - 2x \\ du = -2\,dx \end{array}\right\}$; $\displaystyle\int \sin(3 - 2x)\,dx = \int -\frac{1}{2} \sin u \, du = \frac{1}{2} \cos u + C = \frac{1}{2} \cos(3 - 2x) + C$

51. $\left\{\begin{array}{l} u = \cos x \\ du = -\sin x\,dx \end{array}\right\}$; $\displaystyle\int \cos^4 x \sin x \, dx = \int -u^4 \, du = -\frac{1}{5} u^5 + C = -\frac{1}{5} \cos^5 x + C$

53. $\left\{\begin{array}{l} u = x^{1/2} \\ du = \frac{1}{2} x^{-1/2}\,dx \end{array}\right\}$; $\displaystyle\int x^{-1/2} \sin x^{1/2}\,dx = \int 2 \sin u \, du = -2 \cos u + C$

$$= -2 \cos x^{1/2} + C$$

55. $\left\{\begin{array}{l} u = 1 + \sin x \\ du = \cos x\,dx \end{array}\right\}$; $\displaystyle\int \sqrt{1 + \sin x}\, \cos x \, dx = \int u^{1/2}\,du = \frac{2}{3} u^{3/2} + C$

$$= \tfrac{2}{3}(1 + \sin x)^{3/2} + C$$

57. $\displaystyle\int \frac{1}{\cos^2 x}\,dx = \int \sec^2 x \, dx = \tan x + C$

59. $\left\{\begin{array}{l} u = x^2 \\ du = 2x\,dx \end{array}\right\}$; $\displaystyle\int x \sin^3 x^2 \cos x^2 \, dx = \int \frac{1}{2} \sin^3 u \cos u \, du = \frac{1}{8} \sin^4 u + C$

$$= \tfrac{1}{8} \sin^4 x^2 + C$$

61. $\left\{\begin{array}{l} u = \cot x \\ du = -\csc^2 x\,dx \end{array}\right\}$; $\displaystyle\int \left(1 + \cot^2 x\right) \csc^2 x \, dx = \int -\left(1 + u^2\right)\,du = -u - \frac{1}{3} u^3 + C$

$$= -\cot x - \tfrac{1}{3} \cot^3 x + C$$

63. $\left\{\begin{array}{l} u = 1 + \tan x \\ du = \sec^2 x\,dx \end{array}\right\}$; $\displaystyle\int \frac{\sec^2 x}{\sqrt{1 + \tan x}}\,dx = \int u^{-1/2}\,du = 2u^{1/2} + C$

$$= 2(1 + \tan x)^{1/2} + C$$

65. 0;   the sine is an odd function

67. $\int_{1/4}^{1/3} \sec^2 \pi x \, dx = \frac{1}{\pi} \left[\tan \pi x\right]_{1/4}^{1/3} = \frac{1}{\pi}(\sqrt{3} - 1)$

69. $\int_0^{\pi/2} \sin^3 x \cos x \, dx = \frac{1}{4} \left[\sin^4 x\right]_0^{\pi/2} = \frac{1}{4}$

71. $\int_{\pi/6}^{\pi/4} \csc x \cot x \, dx = \left[-\csc x\right]_{\pi/6}^{\pi/4} = 2 - \sqrt{2}$

73. $\int \sin^2 x \, dx = \int \frac{1 - \cos 2x}{2} \, dx = \frac{1}{2} x - \frac{1}{4} \sin 2x + C$

75. $\int \cos^2 5x \, dx = \int \frac{1 + \cos 10x}{2} \, dx = \frac{1}{2} x + \frac{1}{20} \sin 10x + C$

77. $\int_0^{\pi/2} \cos^2 2x \, dx = \int_0^{\pi/2} \frac{1 + \cos 4x}{2} \, dx = \left[\frac{1}{2} x + \frac{1}{8} \sin 4x\right]_0^{\pi/2} = \frac{\pi}{4}$

79.
$$A = \int_0^{1/4} \left(\cos^2 \pi x - \sin^2 \pi x\right) dx = \int_0^{1/4} \cos 2\pi x \, dx$$

$$= \frac{1}{2\pi} \left[\sin 2\pi x\right]_0^{1/4} = \frac{1}{2\pi}$$

81.
$$A = \int_{1/6}^{1/4} \left(\csc^2 \pi x - \sec^2 \pi x\right) dx = \frac{1}{\pi} \left[-\cot \pi x - \tan \pi x\right]_{1/6}^{1/4}$$

$$= \frac{1}{\pi} \left(-2 + \cot \frac{\pi}{6} + \tan \frac{\pi}{6}\right)$$

$$= \frac{1}{\pi} \left(-2 + \sqrt{3} + \frac{1}{\sqrt{3}}\right) = \frac{1}{3\pi}(4\sqrt{3} - 6)$$

83. (a)
$$\left.\begin{matrix} u = \sec x \\[1mm] du = \sec x \tan x \, dx \end{matrix}\right\}; \quad \int \sec^2 x \tan x \, dx = \int u \, du = \frac{1}{2} u + C$$

$$= \tfrac{1}{2} \sec^2 x + C$$

(b)
$$\left.\begin{matrix} u = \tan x \\[1mm] du = \sec^2 x \, dx \end{matrix}\right\}; \quad \int \sec^2 x \tan x \, dx = \int u \, du = \frac{1}{2} u^2 + C'$$

$$= \tfrac{1}{2} \tan^2 x + C'$$

(c) $C' = C + \tfrac{1}{2}$

85.
$$A = \frac{4b}{a} \int_0^a \sqrt{a^2 - x^2} \, dx = \frac{4b}{a} \left(\frac{\text{area of circle of radius } a}{4}\right)$$

$$= \frac{4b}{a} \left(\frac{\pi a^2}{4}\right) = \pi ab$$

87. (a) $v(t) = \int a(t)\,dt = \int \left[2\cos 2(t+1) + \dfrac{2}{t+1} \right] dt = \sin 2(t+1) + 2\ln(t+1) + C$

$v(0) = 2 \implies v(t) = \sin 2(t+1) + 2\ln(t+1) + 2 - \sin 2$

(b)

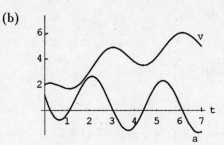

(c) max at $t \cong 6.1389$; min at $t \cong 1.1092$

SECTION 6.8

1. $\displaystyle\int \frac{dx}{x+1} = \ln|x+1| + C$

3. $\left\{ \begin{array}{l} u = 3 - x^2 \\ du = -2x\,dx \end{array} \right\};\quad \displaystyle\int \frac{x}{3-x^2}\,dx = -\frac{1}{2}\int \frac{du}{u} = -\frac{1}{2}\ln|u| + C = -\frac{1}{2}\ln|3 - x^2| + C$

5. $\left\{ \begin{array}{l} u = 3 - x^2 \\ du = -2x\,dx \end{array} \right\};\quad \displaystyle\int \frac{x}{(3-x^2)^2}\,dx = -\frac{1}{2}\int \frac{du}{u^2} = \frac{1}{2u} + C = \frac{1}{2(3-x^2)} + C$

7. $\left\{ \begin{array}{l} u = 2 + \cos x \\ du = -\sin x\,dx \end{array} \right\};\quad \displaystyle\int \frac{\sin x}{2+\cos x}\,dx = -\int \frac{1}{u}\,du = -\ln|u| + C = -\ln|2 + \cos x| + C$

9. $\displaystyle\int e^{2x}\,dx = \frac{1}{2}e^{2x} + C$

11. $\left\{ u = x^2,\quad du = 2x\,dx \right\};\quad \displaystyle\int x e^{x^2}\,dx = \frac{1}{2}\int e^u\,du = \frac{1}{2}e^u + C = \frac{1}{2}e^{x^2} + C$

13. $\left\{ u = \dfrac{1}{x},\quad du = -\dfrac{1}{x^2}\,dx \right\};\quad \displaystyle\int \frac{e^{1/x}}{x^2}\,dx = -\int e^u\,du = -e^u + C = -e^{1/x} + C$

15. $\displaystyle\int \left(e^x + e^{-x} \right)^2 dx = \int \left(e^{2x} + 2 + e^{-2x} \right) dx = \frac{1}{2}e^{2x} + 2x - \frac{1}{2}e^{-2x} + C$

17. $\displaystyle\int \ln e^x\,dx = \int x\,dx = \frac{1}{2}x^2 + C$ 　　　　**19.** $\displaystyle\int \left(x^3 + 3^{-x} \right) dx = \frac{1}{4}x^4 - \frac{3^{-x}}{\ln 3} + C$

21. $\displaystyle\int \frac{\log_2 x^3}{x}\,dx = \frac{1}{\ln 2}\int \frac{\ln x^3}{x}\,dx = \frac{3}{\ln 2}\int \frac{\ln x}{x}\,dx$

$$= \frac{3}{\ln 2}\left[\frac{1}{2}\,(\ln x)^2\right] + C = \frac{3}{\ln 4}\,(\ln x)^2 + C$$

23. $\left\{\begin{array}{l} u = 3x \\ du = 3dx \end{array}\right\}$; $\displaystyle\int \tan 3x\,dx = \frac{1}{3}\int \tan u\,du = \frac{1}{3}\ln|\sec u| + C = \frac{1}{3}\ln|\sec 3x| + C$

25. $\left\{\begin{array}{l} u = e^x \\ du = e^x\,dx \end{array}\right\}$; $\displaystyle\int e^x \cot e^x\,dx = \int \cot u\,du = \ln|\sin u| + C = \ln|\sin e^x| + C$

27. $\left\{\begin{array}{l} u = 3 - 2\cos 2x \\ du = 4\sin 2x\,dx \end{array}\right\}$; $\displaystyle\int \frac{\sin 2x}{3 - 2\cos 2x}\,dx = \frac{1}{4}\int \frac{du}{u} = \frac{1}{4}\ln|u| + C$

$$= \tfrac{1}{4}\ln|3 - 2\cos 2x| + C$$

29. $\left\{\begin{array}{l} u = \ln|\sin x| \\ du = \cot x\,dx \end{array}\right\}$; $\displaystyle\int \cot x \ln|\sin x|\,dx = \int u\,du = \frac{1}{2}u^2 + C$

$$= \tfrac{1}{2}\left(\ln|\sin x|\right)^2 + C$$

31. $\left\{\begin{array}{l} u = 1 + \cot x \\ du = -\csc^2 x\,dx \end{array}\right\}$; $\displaystyle\int \left(\frac{\csc x}{1 + \cot x}\right)^2 dx = -\int \frac{du}{u^2} = \frac{1}{u} + C = \frac{1}{1 + \cot x} + C$

33. $\left\{u = \ln x,\ du = \dfrac{dx}{x}\right\}$; $\displaystyle\int \frac{dx}{x\,(\ln x)^2} = \int \frac{du}{u^2} = -\frac{1}{u} + C = -\frac{1}{\ln x} + C$

35. $\displaystyle\int \frac{4}{\sqrt{e^x}}\,dx = \int 4e^{-x/2}\,dx = -8e^{-x/2} + C$

37. $\left\{\begin{array}{l} u = e^x + 1 \\ du = e^x\,dx \end{array}\right\}$; $\displaystyle\int \frac{e^x}{\sqrt{e^x + 1}}\,dx = \int \frac{du}{\sqrt{u}} = \int u^{-1/2}\,du = 2u^{1/2} + C = 2\sqrt{e^x + 1} + C$

39. $\left\{\begin{array}{l} u = x^2 \\ du = 2x\,dx \end{array}\right\}$; $\displaystyle\int \frac{x}{\sqrt{1 - x^4}}\,dx = \frac{1}{2}\int \frac{du}{\sqrt{1 - u^2}} = \frac{1}{2}\sin^{-1} u + C = \frac{1}{2}\sin^{-1} x^2 + C$

41. $\left\{ \begin{array}{l} u = x^2 \\ du = 2x\,dx \end{array} \right\}$; $\displaystyle \int \frac{x}{1+x^4}\,dx = \frac{1}{2}\int \frac{du}{1+u^2} = \frac{1}{2}\tan^{-1}u + C = \frac{1}{2}\tan^{-1}x^2 + C$

43. $\left\{ \begin{array}{l} u = \sin^{-1}x \\ du = \dfrac{1}{\sqrt{1-x^2}}\,dx \end{array} \right\}$; $\displaystyle \int \frac{\sin^{-1}x}{\sqrt{1-x^2}}\,dx = \int u\,du = \frac{1}{2}u^2 + C = \frac{1}{2}\left(\sin^{-1}x\right)^2 + C$

45. $\left\{ \begin{array}{l} u = \ln x \\ du = \dfrac{1}{x}\,dx \end{array} \right\}$; $\displaystyle \int \frac{dx}{x\sqrt{1-(\ln x)^2}} = \int \frac{du}{\sqrt{1-u^2}} = \sin^{-1}u + C = \sin^{-1}(\ln x) + C$

47. $\left\{ \begin{array}{l} u = \ln|\sec x + \tan x| \\ du = \sec x\,dx \end{array} \right\}$;

$$\int \frac{\sec x}{\sqrt{\ln|\sec x + \tan x|}}\,dx = \int \frac{du}{\sqrt{u}} = 2\sqrt{u} + C = 2\sqrt{\ln|\sec x + \tan x|} + C$$

49. $\left\{ \begin{array}{l} u = \sin x + \cos x \\ du = (\cos x - \sin x)\,dx \end{array} \right\}$; $\displaystyle \int \frac{\sin x - \cos x}{\sin x + \cos x}\,dx = -\int \frac{1}{u}\,du = -\ln|u| + C$

$$= -\ln|\sin x + \cos x| + C$$

51. $\left\{ \begin{array}{l} u = 2e^{2x} + 3 \\ du = 4e^{2x}\,dx \end{array} \right\}$; $\displaystyle \int \frac{e^{2x}}{2e^{2x}+3}\,dx = \frac{1}{4}\int \frac{du}{u} = \frac{1}{4}\ln u + C = \frac{1}{4}\ln\left(2e^{2x}+3\right) + C$

53. $\{u = \sin x, \quad du = \cos x\,dx\}$; $\displaystyle \int \cos x\, e^{\sin x}\,dx = \int e^u\,du = e^u + C = e^{\sin x} + C$

55. $\displaystyle \int_e^{e^2} \frac{dx}{x} = \left[\ln x\right]_e^{e^2} = \ln e^2 - \ln e = 2 - 1 = 1$

57. $\left\{ \begin{array}{l|l} u = \ln(x+1) & x = 0 \implies u = 0 \\ du = \dfrac{dx}{x+1} & x = 1 \implies u = \ln 2 \end{array} \right\}$;

$$\int_0^1 \frac{\ln(x+1)}{x+1}\,dx = \int_0^{\ln 2} u\,du = \frac{1}{2}\left[u^2\right]_0^{\ln 2} = \frac{1}{2}(\ln 2)^2$$

59. $\displaystyle \int_0^{\ln \pi} e^{-6x}\,dx = \left[-\frac{1}{6}e^{-6x}\right]_0^{\ln \pi} = -\frac{1}{6}e^{-6\ln \pi} + \frac{1}{6}e^0 = \frac{1}{6}\left(1 - \pi^{-6}\right)$

61. $\displaystyle\int_0^{\ln 2} \frac{e^x}{e^x+1}\,dx = \left[\ln\left(e^x+1\right)\right]_0^{\ln 2} = \ln\left(e^{\ln 2}+1\right) - \ln\left(e^0+1\right) = \ln 3 - \ln 2 = \ln\frac{3}{2}$

63.

$$\left\{ \begin{array}{l} u = \log_{10} x \\[2mm] du = \dfrac{dx}{x\ln 10} \end{array} \;\middle|\; \begin{array}{lll} x=10 & \Longrightarrow & u=1 \\[2mm] x=100 & \Longrightarrow & u=2 \end{array} \right\}; \quad \int_{10}^{100} \frac{dx}{x\log_{10}x} = \ln 10 \int_1^2 \frac{du}{u}$$

$$= \ln 10\,[\ln u]_1^2 = (\ln 2)(\ln 10)$$

65. $\displaystyle\int_0^1 \left(2^x + x^2\right) dx = \left[\frac{2^x}{\ln 2} + \frac{x^3}{3}\right]_0^1 = \frac{1}{3} + \frac{1}{\ln 2}$

67.

$$\left\{ \begin{array}{l} u = e^x \\[2mm] du = e^x\,dx \end{array} \;\middle|\; \begin{array}{lll} x=0 & \Longrightarrow & u=1 \\[2mm] x=\ln\dfrac{\pi}{4} & \Longrightarrow & u=\dfrac{\pi}{4} \end{array} \right\};$$

$$\int_0^{\ln \pi/4} e^x \sec e^x\,dx = \int_1^{\pi/4} \sec u\,du = \left[\ln|\sec u + \tan u|\right]_1^{\pi/4} = \ln\left|\sqrt{2}+1\right| - \ln|\sec 1 + \tan 1|$$

$$= \ln\left|\frac{1+\sqrt{2}}{\sec 1 + \tan 1}\right| \cong -0.345$$

69. $\displaystyle\int_0^5 \frac{dx}{25+x^2} = \left[\frac{1}{5}\tan^{-1}\frac{x}{5}\right]_0^5 = \frac{\pi}{20}$

$\qquad\qquad\qquad$ —— (6.8.4)

71. $\left\{ \begin{array}{l} 3u=2x \\[2mm] 3\,du = 2\,dx \end{array} \;\middle|\; \begin{array}{lll} x=0 & \Longrightarrow & u=0 \\[2mm] x=3/2 & \Longrightarrow & u=1 \end{array} \right\};$ $\displaystyle\int_0^{3/2} \frac{dx}{9+4x^2} = \frac{1}{6}\int_0^1 \frac{du}{1+u^2} = \frac{1}{6}\left[\tan^{-1} u\right]_0^1 = \frac{\pi}{24}$

73. $\displaystyle\int_{\ln 2}^{\ln 3} \left(e^x - e^{-x}\right)^2 dx = \int_{\ln 2}^{\ln 3} \left(e^{2x} - 2 + e^{-2x}\right) dx$

$$= \left[\tfrac{1}{2}e^{2x} - 2x - \tfrac{1}{2}e^{-2x}\right]_{\ln 2}^{\ln 3}$$

$$= \left[\tfrac{1}{2}(9) - 2\ln 3 - \tfrac{1}{2}\left(\tfrac{1}{9}\right)\right] - \left[\tfrac{1}{2}(4) - 2\ln 2 - \tfrac{1}{2}\left(\tfrac{1}{4}\right)\right]$$

$$= \tfrac{185}{72} + \ln\tfrac{4}{9}$$

75.

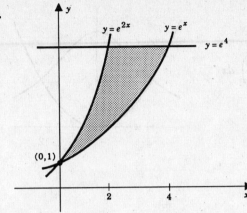

$$A = \int_0^2 \left(e^{2x} - e^x\right) dx + \int_2^4 \left(e^4 - e^x\right) dx$$

$$= \left[\tfrac{1}{2}e^{2x} - e^x\right]_0^2 + \left[e^4 x - e^x\right]_2^4$$

$$= \left(\tfrac{1}{2}e^4 - e^2 - \tfrac{1}{2} + 1\right) + \left(4e^4 - e^4 - 2e^4 + e^2\right)$$

$$= \tfrac{1}{2}\left(3e^4 + 1\right)$$

77.

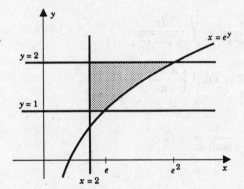

$$A = \int_1^2 \left(e^y - 2\right) dy$$

$$= \left[e^y - 2y\right]_1^2 = e^2 - e - 2$$

79.

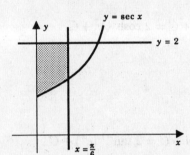

$$A = \int_0^{\pi/6} \left(2 - \sec x\right) dx$$

$$= \left[2x - \ln|\sec x + \tan x|\right]_0^{\pi/6}$$

$$= \frac{\pi}{3} - \ln\left|\frac{2}{\sqrt{3}} + \frac{1}{\sqrt{3}}\right| = \frac{\pi}{3} - \frac{1}{2}\ln 3$$

81.

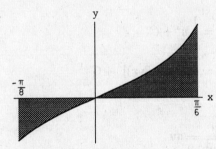

$$A = \int_{-\pi/8}^0 \left(-\tan 2x\right) dx + \int_0^{\pi/6} \tan 2x \, dx$$

$$= \left[-\frac{1}{2}\ln|\sec 2x|\right]_{-\pi/8}^0 + \left[\frac{1}{2}\ln|\sec 2x|\right]_0^{\pi/6}$$

$$= \frac{3}{4}\ln 2$$

83. $\quad A = \displaystyle\int_{-1}^1 \frac{1}{\sqrt{4 - x^2}} \, dx = 2\int_0^1 \frac{1}{\sqrt{4 - x^2}} \, dx = 2\left[\sin^{-1}\left(\frac{x}{2}\right)\right]_0^1 = \frac{\pi}{3}$

85. $\dfrac{8}{x^2+4} = \dfrac{1}{4}x^2 \;\Rightarrow\; x = \pm 2$

$$A = \int_{-2}^{2}\left(\frac{8}{x^2+4}-\frac{1}{4}x^2\right)dx = 2\int_{0}^{2}\left(\frac{8}{x^2+4}-\frac{1}{4}x^2\right)dx$$

$$= 2\left[8\cdot\frac{1}{2}\tan^{-1}\left(\frac{x}{2}\right)-\frac{1}{12}x^3\right]_{0}^{2} = 2\pi - \frac{4}{3}$$

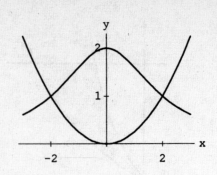

87. $\left\{\begin{array}{l} au = x+b \\ a\,du = dx \end{array}\right\}; \quad \displaystyle\int \frac{dx}{\sqrt{a^2-(x+b)^2}} = \int \frac{a\,du}{\sqrt{a^2-a^2u^2}} = \int \frac{du}{\sqrt{1-u^2}}$

$$= \sin^{-1}u + C = \sin^{-1}\left(\frac{x+b}{a}\right)+C$$

89. $\dfrac{1}{a}\ln\left(\cosh ax\right)+C$ 　　　　　　　　**91.** $-\dfrac{1}{a\cosh ax}+C$

93. $\left\{\begin{array}{l} u = \sqrt{x} \\ du = dx/2\sqrt{x} \end{array}\right\}; \quad \displaystyle\int \frac{\sinh\sqrt{x}}{\sqrt{x}}\,dx = 2\int \sinh u\,du = 2\cosh u + C = 2\cosh\sqrt{x}+C$

95. $\displaystyle\int \operatorname{sech} x\,dx = \int \frac{1}{\cosh x}\,dx = \int \frac{2}{e^x+e^{-x}}\,dx = \int \frac{2e^x}{e^{2x}+1}\,dx$

$\left\{\begin{array}{l} u = e^x \\ du = e^x\,dx \end{array}\right\}; \quad \displaystyle\int \frac{2e^x}{e^{2x}+1}\,dx = 2\int \frac{1}{u^2+1}\,du = 2\tan^{-1}u + C = 2\tan^{-1}(e^x)+C$

97. $\left\{\begin{array}{l} u = \ln\left(\cosh x\right) \\ du = \tanh x\,dx \end{array}\right\}; \quad \displaystyle\int \tanh x\,\ln\left(\cosh x\right)dx = \int u\,du = \frac{1}{2}u^2 + C$

$$= \frac{1}{2}\left[\ln\left(\cosh x\right)\right]^2 + C$$

99. $\left\{\begin{array}{l} x = a\sinh u \\ dx = a\cosh u\,du \end{array}\right\}; \quad \displaystyle\int \frac{dx}{\sqrt{a^2+x^2}}\,dx = \int \frac{a\cosh u}{\sqrt{a^2+a^2\sinh^2 u}}\,du$

$$= \int du = u + C = \sinh^{-1}\left(\frac{x}{a}\right)+C$$

101. Suppose $|x| < a$.

$$\left\{\begin{array}{l} x = a\tanh u \\ dx = a\,\text{sech}^2 u\,du \end{array}\right\};\quad \int \frac{dx}{a^2 - x^2}\,dx = \int \frac{a\,\text{sech}^2 u}{a^2 - a^2\tanh^2 u}\,du$$

$$= \frac{1}{a}\int du = \frac{u}{a} + C = \frac{1}{a}\tanh^{-1}\left(\frac{x}{a}\right) + C$$

The other case is done in the same way.

SECTION 6.9

1. Yes; $\displaystyle\int_a^b [f(x) - g(x)]\,dx = \int_a^b f(x)\,dx - \int_a^b g(x)\,dx > 0.$

3. Yes; otherwise we would have $f(x) \le g(x)$ for all $x \in [a, b]$ and it would follow that

$$\int_a^b f(x)\,dx \le \int_a^b g(x)\,dx.$$

5. No; take $f(x) = 0$, $g(x) = -1$ on $[0,1]$.

7. No; take, for example, any odd function on an interval of the form $[-c, c]$.

9. No; $\displaystyle\int_{-1}^1 x\,dx = 0$ but $\displaystyle\int_{-1}^1 |x|\,dx \ne 0.$

11. Yes; $U_f(P) \ge \displaystyle\int_a^b f(x)\,dx = 0.$ **13.** No; $L_f(P) \le \displaystyle\int_a^b f(x)\,dx = 0.$

15. Yes; $\displaystyle\int_a^b [f(x) + 1]\,dx = \int_a^b f(x)\,dx + \int_a^b 1\,dx = 0 + b - a = b - a.$

17. $\dfrac{d}{dx}\left[\displaystyle\int_0^{1+x^2} \frac{dt}{\sqrt{2t+5}}\right] = \dfrac{1}{\sqrt{2(1+x^2)+5}}\dfrac{d}{dx}(1+x^2) = \dfrac{2x}{\sqrt{2x^2+7}}$

19. $\dfrac{d}{dx}\left[\displaystyle\int_x^a f(t)\,dt\right] = \dfrac{d}{dx}\left[-\displaystyle\int_a^x f(t)\,dt\right] = -f(x)$

21. $\dfrac{d}{dx}\left[\displaystyle\int_{x^2}^3 \frac{\sin t}{t}\,dt\right] = -\dfrac{d}{dx}\left[\displaystyle\int_3^{x^2} \frac{\sin t}{t}\,dt\right] = -\dfrac{\sin(x^2)}{x^2}(2x) = -\dfrac{2\sin(x^2)}{x}$

23. $\dfrac{d}{dx}\left[\displaystyle\int_1^{\sqrt{x}} \frac{t^2}{1+t^2}\,dt\right] = \dfrac{x}{1+x}\cdot\dfrac{1}{2\sqrt{x}} = \dfrac{\sqrt{x}}{2(1+x)}$

25. $\dfrac{d}{dx}\left[\displaystyle\int_x^{x^2} \frac{dt}{t}\right] = \dfrac{1}{x^2}\dfrac{d}{dx}(x^2) - \dfrac{1}{x}\dfrac{d}{dx}(x) = \dfrac{2x}{x^2} - \dfrac{1}{x} = \dfrac{1}{x}$

27.
$$\frac{d}{dx}\left[\int_{x^{1/3}}^{2+3x} \frac{dt}{1+t^{3/2}}\right] = \frac{1}{1+(2+3x)^{3/2}}\frac{d}{dx}(2+3x) - \frac{1}{1+(x^{1/3})^{3/2}}\frac{d}{dx}\left(x^{1/3}\right)$$

$$= \frac{3}{1+(2+3x)^{3/2}} - \frac{1}{3x^{2/3}(1+x^{1/2})}$$

29.
$$\frac{d}{dx}\left[\int_{\sqrt{x}}^{x^2} t\,\cos t^2\,dt\right] = x^2\cos(x^4)(2x) - \sqrt{x}\,\cos x\left(\frac{1}{2\sqrt{x}}\right) = 2x^3\cos(x^4) - \frac{1}{2}\cos x$$

31.
$$\frac{d}{dx}\left[\int_{\tan x}^{2x} t\sqrt{1+t^2}\,dt\right] = 2x\sqrt{1+(2x)^2}\,(2) - \tan x\,\sqrt{1+\tan^2 x}(\sec^2 x)$$

$$= 4x\sqrt{1+4x^2} - \tan x\,\sec^2 x|\sec x|$$

33.   (a)   With P a partition of $[a, b]$

$$L_f(P) \le \int_a^b f(x)\,dx.$$

If f is nonnegative on $[a, b]$, then $L_f(P)$ is nonnegative and, consequently, so is the integral.

If f is positive on $[a, b]$, then $L_f(P)$ is positive and, consequently, so is the integral.

(b)   Take F as an antiderivative of f on $[a, b]$. Observe that

$$F'(x) = f(x) \text{ on } (a, b) \quad \text{and} \quad \int_a^b f(x)\,dx = F(b) - F(a).$$

If $f(x) \ge 0$ on $[a, b]$, then F is nondecreasing on $[a, b]$ and $F(b) - F(a) \ge 0$.

If $f(x) > 0$ on $[a, b]$, then F is increasing on $[a, b]$ and $F(b) - F(a) > 0$.

35.   For all $x \in [a, b]$

$$-f(x) \le |f(x)| \quad \text{and} \quad f(x) \le |f(x)|.$$

It follows from II that

$$-\int_a^b f(x)\,dx = \int_a^b -f(x)\,dx \le \int_a^b |f(x)|\,dx \quad \text{and} \quad \int_a^b f(x)\,dx \le \int_a^b |f(x)|\,dx,$$

and thus

$$\left|\int_a^b f(x)\,dx\right| \le \int_a^b |f(x)|\,dx.$$

37.
$$H(x) = \int_{2x}^{x^3-4} \frac{x\,dt}{1+\sqrt{t}} = x\int_{2x}^{x^3-4} \frac{dt}{1+\sqrt{t}},$$

$$H'(x) = x\cdot\left[\frac{3x^2}{1+\sqrt{x^3-4}} - \frac{2}{1+\sqrt{2x}}\right] + 1\cdot\int_{2x}^{x^3-4} \frac{dt}{1+\sqrt{t}},$$

$$H'(2) = 2\left[\frac{12}{3} - \frac{2}{3}\right] + \underbrace{\int_4^4 \frac{dt}{1+\sqrt{t}}}_{=0} = \frac{20}{3}$$

39.   (a) Let $u = -x$. Then $du = -dx$; and $u = 0$ when $x = 0$, $u = a$ when $x = -a$.

$$\int_{-a}^{0} f(x)\, dx = -\int_{a}^{0} f(-u)\, du = \int_{0}^{a} f(-u)\, du = \int_{0}^{a} f(-x)\, dx$$

41. $\displaystyle\int_{-\pi/4}^{\pi/4} (x + \sin 2x)\, dx = 0$ since $f(x) = x + \sin 2x$ is an odd function.

43. $\displaystyle\int_{-\pi/3}^{\pi/3} (1 + x^2 - \cos x)\, dx = 2\int_{0}^{\pi/3} (1 + x^2 - \cos x)\, dx$ since $f(x) = 1 + x^2 - \cos x$ is an even function.

$$2\int_{0}^{\pi/3} (1 + x^2 - \cos x)\, dx = 2\left[x + \frac{1}{3}x^3 - \sin x\right]_{0}^{\pi/3} = \frac{2}{3}\pi + \frac{2}{81}\pi^3 - \sqrt{3}$$

45.   (a) for $t \in (1, x)$ (b) for $x > 1$

$$\frac{1}{t} < \frac{1}{\sqrt{t}}$$

$$0 < \ln x = \int_{1}^{x} \frac{dt}{t} < \int_{1}^{x} \frac{dt}{\sqrt{t}} = \left[2\sqrt{t}\right]_{1}^{x} = 2\left(\sqrt{x} - 1\right)$$

(c) for $0 < x < 1$

$$0 < \ln \frac{1}{x} < 2\left(\sqrt{\frac{1}{x}} - 1\right) \qquad \text{by (b)}$$

$$0 < -\ln x < 2\left(\frac{1}{\sqrt{x}} - 1\right)$$

$$2\left(1 - \frac{1}{\sqrt{x}}\right) < \ln x < 0$$

$$2x\left(1 - \frac{1}{\sqrt{x}}\right) < x\ln x < 0$$

(d) Use part (c) and the pinching theorem for one-sided limits.

47.   By induction. True for $n = 0 : e^x > 1$ for $x > 0$.

Assume true for n. Then

$$e^x = 1 + \int_{0}^{x} e^t\, dt > 1 + \int_{0}^{x} \left(1 + t + \frac{t^2}{2!} + \cdots + \frac{t^n}{n!}\right) dt$$

$$= 1 + \left[t + \frac{t^2}{2} + \frac{t^3}{3!} + \cdots + \frac{t^{n+1}}{(n+1)!}\right]_{0}^{x}$$

$$= 1 + x + \frac{x^2}{2!} + \frac{x^3}{3!} + \cdots + \frac{x^{n+1}}{(n+1)!}$$

So the result is true for $n + 1$

SECTION 6.10

1.   A.V. $= \dfrac{1}{c}\displaystyle\int_{0}^{c} (mx + b)\, dx = \dfrac{1}{c}\left[\dfrac{m}{2}x^2 + bx\right]_{0}^{c} = \dfrac{mc}{2} + b;$ at $x = c/2$

3. A.V. $= \frac{1}{2} \int_{-1}^{1} x^3 \, dx = 0$ since the integrand is odd;   at $x = 0$

5. A.V. $= \frac{1}{4} \int_{-2}^{2} |x| \, dx = \frac{1}{2} \int_{0}^{2} |x| \, dx = \frac{1}{2} \int_{0}^{2} x \, dx = \frac{1}{2} \left[\frac{x^2}{2} \right]_{0}^{2} = 1$;   at $x = \pm 1$

7. A.V. $= \frac{1}{2} \int_{0}^{2} (2x - x^2) \, dx = \frac{1}{2} \left[x^2 - \frac{x^3}{3} \right]_{0}^{2} = \frac{2}{3}$;   at $x = 1 \pm \frac{1}{3}\sqrt{3}$

9. A.V. $= \frac{1}{9} \int_{0}^{9} \sqrt{x} \, dx = \frac{1}{9} \left[\frac{2}{3} x^{3/2} \right]_{0}^{9} = 2$;   at $x = 4$

11. A.V. $= \frac{1}{2\pi} \int_{0}^{2\pi} \sin x \, dx = \frac{1}{2\pi} \left[-\cos x \right]_{0}^{2\pi} = 0$;   at $x = 0, \pi, 2\pi$

13. $f_{\text{avg}} = \frac{1}{1 - (-1)} \int_{-1}^{1} \cosh x \, dx = \frac{1}{2} \left[\sinh x \right]_{-1}^{1} = \frac{e^2 - 1}{2e} \cong 1.175$

15. (a) A.V. $= \frac{1}{8} \int_{1}^{9} \sqrt{x} \, dx = \frac{1}{8} \left[\frac{2}{3} x^{2/3} \right]_{1}^{9} = \frac{13}{6}$

(b) $\sqrt{x} = \frac{13}{6} \implies x = 4.694$ (c)

17. (a) A.V. $= \frac{1}{\pi} \int_{0}^{\pi} \sin x \, dx = \frac{1}{\pi} \left[-\cos x \right]_{0}^{\pi} = \frac{2}{\pi}$

(b) $\sin x = \frac{2}{\pi} \implies x = 0.691$ (c)

19. $f_{avg} = \frac{1}{\frac{\pi}{2} - 0} \int_{0}^{\pi/2} \frac{\sin x}{1 + 2 \cos x} \, dx = \frac{2}{\pi} \int_{0}^{\pi/2} \frac{\sin x}{1 + 2 \cos x} \, dx$

$$\begin{cases} u = 1 + 2\cos x; & x = 0 \quad \Rightarrow \quad u = 3 \\ du = -2\sin x\, dx; & x = \pi/2 \quad \Rightarrow \quad u = 1 \end{cases};$$

$$\frac{2}{\pi}\int_0^{\pi/2}\frac{\sin x}{1+2\cos x}\, dx = \frac{2}{\pi}\left(-\frac{1}{2}\right)\int_3^1\frac{du}{u} = \frac{1}{\pi}\int_1^3\frac{du}{u} = \frac{1}{\pi}\left[\ln|u|\right]_1^3 = \frac{\ln 3}{\pi}$$

21. $0 = \int_a^b [f(x) - A]\, dx = \int_a^b f(x)\, dx - A\int_a^b dx = \int_a^b f(x)\, dx - A\,(b-a).$

Thus, $A = \dfrac{1}{b-a}\displaystyle\int_a^b f(x)\, dx = $ average value of f on $[a, b]$.

23. Average of f' on $[a, b] = \dfrac{1}{b-a}\displaystyle\int_a^b f'(x)\, dx = \dfrac{1}{b-a}\left[f(x)\right]_a^b = \dfrac{f(b)-f(a)}{b-a}.$

25. (a) A.V. $= \dfrac{1}{\sqrt{3}}\displaystyle\int_0^{\sqrt{3}} y\, dx = \dfrac{1}{\sqrt{3}}\int_0^{\sqrt{3}} x^2\, dx = \dfrac{1}{\sqrt{3}}\left[\dfrac{x^3}{3}\right]_0^{\sqrt{3}} = 1$

(b) A.V. $= \dfrac{1}{3}\displaystyle\int_0^3 x\, dy = \dfrac{1}{3}\int_0^3 \sqrt{y}\, dy = \dfrac{1}{3}\left[\dfrac{2}{3}y^{3/2}\right]_0^3 = \dfrac{2}{3}\sqrt{3}$

(c) $A.V. = \dfrac{1}{\sqrt{3}}\displaystyle\int_0^{\sqrt{3}}\sqrt{(x-0)^2 + (x^2-0)^2}\, dx = \dfrac{1}{\sqrt{3}}\int_0^{\sqrt{3}} x\sqrt{1+x^2}\, dx$

$$= \frac{1}{\sqrt{3}}\left[\frac{1}{3}\left(1+x^2\right)^{3/2}\right]_0^{\sqrt{3}} = \frac{7}{9}\sqrt{3}$$

27. The distance the stone has fallen after t seconds is given by $s(t) = 16t^2$.

(a) The terminal velocity after x seconds is $s'(x) = 32x$. The average velocity is $\dfrac{s(x) - s(0)}{x - 0} = 16x$. Thus the terminal velocity is twice the average velocity.

(b) For the first $\frac{1}{2}x$ seconds, aver. vel. $= \dfrac{s\left(\frac{1}{2}x\right) - s(0)}{\frac{1}{2}x - 0} = 8x$.

For the next $\frac{1}{2}x$ seconds, aver. vel. $= \dfrac{s(x) - s\left(\frac{1}{2}x\right)}{x - \frac{1}{2}x} = 24x$.

Thus, for the first $\frac{1}{2}x$ seconds the average velocity is one-third of the average velocity during the next $\frac{1}{2}x$ seconds.

29. (a) $v(t) - v(0) = \displaystyle\int_0^t a\, du; \quad v(0) = 0.$ Thus $v(t) = at.$

$x(t) - x(0) = \displaystyle\int_0^t v(u)\, du; \quad x(0) = x_0.$ Thus $x(t) = \displaystyle\int_0^t au\, du + x_0 = \frac{1}{2}at^2 + x_0.$

(b) $v_{avg} = \dfrac{1}{t_2 - t_1} \displaystyle\int_{t_1}^{t_2} at\, dt = \dfrac{1}{t_2 - t_1} \left[\dfrac{1}{2} at^2\right]_{t_1}^{t_2}$

$$= \dfrac{at_2^2 - at_1^2}{2(t_2 - t_1)} = \dfrac{v(t_1) + v(t_2)}{2}$$

31. (a) $\quad M = \displaystyle\int_0^L k\sqrt{x}\, dx = k\left[\dfrac{2}{3}x^{3/2}\right]_0^L = \dfrac{2}{3}kL^{3/2}$

$$x_M M = \int_0^L x\left(k\sqrt{x}\right)dx = \int_0^L kx^{3/2}\, dx = \left[\dfrac{2}{5}kx^{5/2}\right]_0^L = \dfrac{2}{5}kL^{5/2}$$

$$x_M = \left(\tfrac{2}{5}kL^{5/2}\right)\Big/\left(\tfrac{2}{3}kL^{3/2}\right) = \tfrac{3}{5}L$$

(b) $\quad M = \displaystyle\int_0^L k(L-x)^2\, dx = \left[-\dfrac{1}{3}k(L-x)^3\right]_0^L = \dfrac{1}{3}kL^3$

$$x_M M = \int_0^L x\left[k(L-x)^2\right]dx = \int_0^L k\left(L^2 x - 2Lx^2 + x^3\right)dx$$

$$= k\left[\tfrac{1}{2}L^2 x^2 - \tfrac{2}{3}Lx^3 + \tfrac{1}{4}x^4\right]_0^L = \tfrac{1}{12}kL^4$$

$$x_M = \left(\tfrac{1}{12}kL^4\right)\Big/\left(\tfrac{1}{3}kL^3\right) = \tfrac{1}{4}L$$

33. $\quad \tfrac{1}{4}LM = \tfrac{1}{8}LM_1 + x_{M_2}M_2$

$$x_{M_2} = \dfrac{1}{M_2}\left(\dfrac{1}{4}LM - \dfrac{1}{8}LM_1\right) = \dfrac{L}{8M_2}(2M - M_1)$$

35. If f is continuous on $[a, b]$, then, by Theorem 6.3.5, F satisfies the conditions of the mean-value theorem of differential calculus (Theorem 4.1.1). Therefore, by that theorem, there is at least one number c in (a, b) for which

$$F'(c) = \dfrac{F(b) - F(a)}{b - a}.$$

Then

$$\int_a^b f(x)\, dx = F(b) - F(a) = F'(c)(b - a) = f(c)(b - a).$$

37. If f and g take on the same average value on every interval $[a, x]$, then

$$\dfrac{1}{x - a}\int_a^x f(t)\, dt = \dfrac{1}{x - a}\int_a^x g(t)\, dt.$$

Multiplication by $(x - a)$ gives

$$\int_a^x f(t)\, dt = \int_a^x g(t)\, dt.$$

Differentiation with respect to x gives $f(x) = g(x)$. This shows that, if the averages are everywhere the same, then the functions are everywhere the same.

SECTION 6.11

1.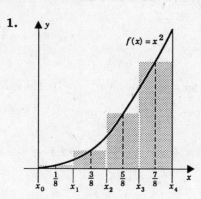

3. (a) $\Delta x_1 = \Delta x_2 = \frac{1}{8}, \quad \Delta x_3 = \Delta x_4 = \Delta x_5 = \frac{1}{4}$

   (b) $\|P\| = \frac{1}{4}$

   (c) $m_1 = 0, \quad m_2 = \frac{1}{4}, \quad m_3 = \frac{1}{2}, \quad m_4 = 1, \quad m_5 = \frac{3}{2}$

   (d) $f(x_1^*) = \frac{1}{8}, \quad f(x_2^*) = \frac{3}{8}, \quad f(x_3^*) = \frac{3}{4}, \quad f(x_4^*) = \frac{5}{4},$

   $f(x_5^*) = \frac{3}{2}$

   (e) $M_1 = \frac{1}{4}, \quad M_2 = \frac{1}{2}, \quad M_3 = 1, \quad M_4 = \frac{3}{2}, \quad M_5 = 2$

   (f) $L_f(P) = \frac{25}{32}$ (g) $S^*(P) = \frac{15}{16}$

   (h) $U_f(P) = \frac{39}{32}$ (i) $\int_a^b f(x)\,dx = 1$

5. (a) $\dfrac{1}{n^2}(1 + 2 + \cdots + n) = \dfrac{1}{n^2}\left[\dfrac{n(n+1)}{2}\right] = \dfrac{1}{2} + \dfrac{1}{2n}$

   (b) $S_n^* = \dfrac{1}{2} + \dfrac{1}{2n}, \quad \displaystyle\int_0^1 x\,dx = \left[\dfrac{1}{2}x^2\right]_0^1 = \dfrac{1}{2}$

   $\left| S_n^* - \displaystyle\int_0^1 x\,dx \right| = \dfrac{1}{2n} < \dfrac{1}{n} < \epsilon \quad \text{if} \quad n > \dfrac{1}{\epsilon}$

7. (a) $\dfrac{1}{n^4}\left(1^3 + 2^3 + \cdots + n^3\right) = \dfrac{1}{n^4}\left[\dfrac{n^2(n+1)^2}{4}\right] = \dfrac{1}{4} + \dfrac{1}{2n} + \dfrac{1}{4n^2}$

   (b) $S_n^* = \dfrac{1}{4} + \dfrac{1}{2n} + \dfrac{1}{4n^2}, \quad \displaystyle\int_0^1 x^3\,dx = \left[\dfrac{1}{4}x^4\right]_0^1 = \dfrac{1}{4}$

   $\left| S_n^* - \displaystyle\int_0^1 x^3\,dx \right| = \dfrac{1}{2n} + \dfrac{1}{4n^2} < \dfrac{1}{n} < \epsilon \quad \text{if} \quad n > \dfrac{1}{\epsilon}$

9. $S^*(P) = \frac{1}{3}\left[\frac{1}{6}\cos\left(\frac{1}{6}\right)^2 + \frac{3}{6}\cos\left(\frac{3}{6}\right)^2 + \frac{5}{6}\cos\left(\frac{5}{6}\right)^2 + \frac{7}{6}\cos\left(\frac{7}{6}\right)^2 + \frac{9}{6}\cos\left(\frac{9}{6}\right)^2 + \frac{11}{6}\cos\left(\frac{11}{6}\right)^2\right]$

   $\cong -0.3991$

   $\int_0^2 x\cos x^2\,dx = \left[\frac{1}{2}\sin x^2\right]_0^2 = \frac{1}{2}\sin 4 \cong -0.3784$

11. $I = \displaystyle\int_0^{0.5} \dfrac{1}{\sqrt{1 - x^2}}\,dx \cong \dfrac{1}{10}\left[f(0.05) + f(0.15) + f(0.25) + f(0.35) + f(0.45)\right] \cong 0.523;$

   and $\sin(0.523) \cong 0.499$. Explanation: $I = \sin^{-1}(0.5)$ and $\sin\left[\sin^{-1}(0.5)\right] = 0.5$.

PROJECTS AND EXPLORATIONS

6.1. (a) Dom $(F) = (-\infty, \infty)$ (b) $F'(x) = \sin^2(x^2)$ (by Theorem 6.3.5)

$$F''(x) = 2\sin(x^2)\cos(x^2)2x = 2x\sin(2x^2)$$

(c) Again by Theorem 5.3.5, F is continuous on (∞, ∞).

(d) $F'(x) \geq 0$ for all x, and $F'(x) = 0$ only when $x = \pm\sqrt{n\pi}$ for some integer n, $n \geq 0$. Thus, F is increasing on (∞, ∞). Since F is increasing, it is one-to-one.

(e) Since F' is even, we would expect F to be odd (see Exercises 59 and 60, Section 3.1.)

$$F(-x) = \int_0^{-x} \sin^2(t^2)\, dt$$

Set $u = -t$. Then $du = -dt$, and $u = 0$ at $t = 0$, $u = x$ at $t = -x$. Then

$$F(-x) = \int_0^x \sin^2(-u)^2\,(-du) = -\int_0^x \sin^2(u^2)\, du = -F(x).$$

(f) $F''(x) = 0$ when $x^2 = \dfrac{n}{2}\pi$, $n = 0, 1, 2, \ldots$. That is, when $x = \pm\sqrt{n\pi/2}$. Since F'' changes signs at each of these values, the points $\left(\pm\sqrt{n\pi/2},\, F\left(\pm\sqrt{n\pi/2}\right)\right)$ are points of inflection. The graph of F is concave up on $\left(0, \sqrt{\pi/2}\right)$, concave down on $\left(\sqrt{\pi/2}, \sqrt{\pi}\right)$, and so on.

(g)

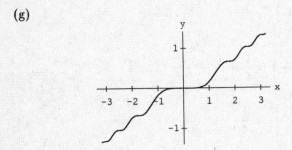

6.3. (a) We would expect $v(t) \geq 0$ and $a(t) \geq 0$, so $T = 20$.

(b) Let s denote the distance traveled. Then

$$s = \int_0^{20} (40t - t^2)\, dt = \left[20t^2 - \frac{1}{3}t^3\right]_0^{20} = \frac{16,000}{3} \cong 5333.33 \text{ ft.}$$

(c) $V_{avg} = \dfrac{1}{3.5}\displaystyle\int_2^{5.5} (40t - t^2)\, dt = \dfrac{1}{3.5}\left[20t^2 - \dfrac{1}{3}t^3\right]_2^{5.5} \cong 134.917$

(d) $\dfrac{v(2)+v(5.5)}{2} = \dfrac{189.75+76}{2} = 132.875.$ (e) $\dfrac{v(2)+v(2.5)\cdots+v(5.5)}{8} \cong 134.625$

(f) $\dfrac{v(2)+v(2.1)+v(2.2)+\cdots+v(5.5)}{36} \cong 134.858$

(g) The function v is increasing on $[2,5.5]$ and its graph is concave down on $(2,5.5)$. As a result the area under the graph is greater than the area under any "polygonal" curve (see the figure).

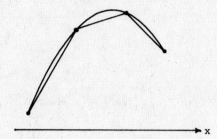

SECTION 7.1

1.

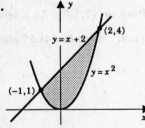

(a) $\displaystyle\int_{-1}^{2} [(x+2) - x^2]\, dx$

(b) $\displaystyle\int_{0}^{1} [(\sqrt{y}) - (-\sqrt{y})]\, dy + \int_{1}^{4} [(\sqrt{y}) - (y-2)]\, dy$

3.

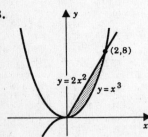

(a) $\displaystyle\int_{0}^{2} [(2x^2) - (x^3)]\, dx$

(b) $\displaystyle\int_{0}^{8} \left[(y^{1/3}) - \left(\frac{1}{2}y\right)^{1/2} \right] dy$

5.

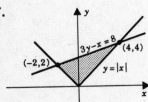

(a) $\displaystyle\int_{0}^{4} [(0) - (-\sqrt{x})]\, dx + \int_{4}^{6} [(0) - (x-6)]\, dx$

(b) $\displaystyle\int_{-2}^{0} [(y+6) - (y^2)]\, dy$

7.

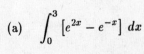

(a) $\displaystyle\int_{-2}^{0} \left[\left(\frac{8+x}{3}\right) - (-x) \right] dx + \int_{0}^{4} \left[\left(\frac{8+x}{3}\right) - (x) \right] dx$

(b) $\displaystyle\int_{0}^{2} [(y) - (-y)]\, dy + \int_{2}^{4} [(y) - (3y-8)]\, dy$

9.

(a) $\displaystyle\int_{0}^{3} [e^{2x} - e^{-x}]\, dx$

(b) $\displaystyle\int_{e^{-3}}^{1} [3 - (-\ln y)]\, dy + \int_{1}^{e^6} [3 - \tfrac{1}{2}\ln y]\, dy$

11.

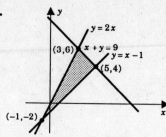

(a) $\displaystyle\int_{-1}^{3} [(2x) - (x-1)] \, dx + \int_{3}^{5} [(9-x) - (x-1)] \, dx$

(b) $\displaystyle\int_{-2}^{4} \left[(y+1) - \left(\frac{1}{2}y\right)\right] dy + \int_{4}^{6} \left[(9-y) - \left(\frac{1}{2}y\right)\right] dy$

13.

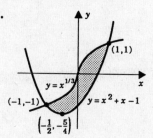

(a) $\displaystyle\int_{-1}^{1} \left[\left(x^{1/3}\right) - (x^2 + x - 1)\right] dx$

(b) $\displaystyle\int_{-5/4}^{-1} \left[\left(-\frac{1}{2} + \frac{1}{2}\sqrt{4y+5}\right) - \left(-\frac{1}{2} - \frac{1}{2}\sqrt{4y+5}\right)\right] dy$

$\displaystyle + \int_{-1}^{1} \left[\left(-\frac{1}{2} + \frac{1}{2}\sqrt{4y+5}\right) - (y^3)\right] dy$

15.

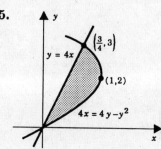

$\displaystyle A = \int_{0}^{3} \left[\left(\frac{4y - y^2}{4}\right) - \left(\frac{y}{4}\right)\right] dy$

$\displaystyle = \int_{0}^{3} \left(\frac{3}{4}y - \frac{1}{4}y^2\right) dy$

$\displaystyle = \left[\frac{3}{8}y^2 - \frac{1}{12}y^3\right]_{0}^{3} = \frac{9}{8}$

17.

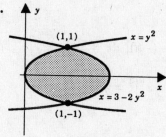

$\displaystyle A = 2\int_{0}^{1} \left[(3 - 2y^2) - (y^2)\right] dy$

$\displaystyle = 2\int_{0}^{1} (3 - 3y^2) \, dy$

$\displaystyle = 2\left[3y - y^3\right]_{0}^{1} = 2(2) = 4$

19.

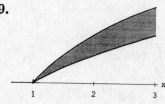

$y = \ln x \implies x = e^y$

$\displaystyle A = \int_{0}^{\ln 2} [2e^y - 1 - e^y] \, dy + \int_{\ln 2}^{\ln 3} [3 - e^y] \, dy$

$\displaystyle = [e^y - y]_{0}^{\ln 2} + [3y - e^y]_{\ln 2}^{\ln 3}$

$(2 - \ln 2) - 1 + (3\ln 3 - 3) - (3\ln 2 - 2) = \ln(27/16)$

21.

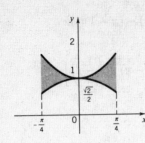

$$A = \int_{-\pi/4}^{\pi/4} \left[\sec^2 x - \cos x\right] dx$$

$$= 2 \int_0^{\pi/4} \left[\sec^2 x - \cos x\right] dx$$

$$= 2 \left[\tan x + \sin x\right]_0^{\pi/4} = 2 \left[1 + \sqrt{2}/2\right] = 2 + \sqrt{2}$$

23.

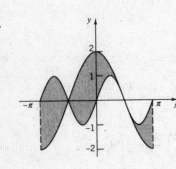

$$A = \int_{-\pi}^{-\pi/2} \left[\sin 2x - 2\cos x\right] dx + \int_{-\pi/2}^{\pi/2} \left[2\cos x - \sin 2x\right] dx$$

$$+ \int_{\pi/2}^{\pi} \left[\sin 2x - 2\cos x\right] dx$$

$$= \left[-\tfrac{1}{2}\cos 2x - 2\sin x\right]_{\pi}^{\pi/2} + \left[2\sin x + \tfrac{1}{2}\cos 2x\right]_{-\pi/2}^{\pi/2}$$

$$+ \left[-\tfrac{1}{2}\cos 2x - 2\cos x\right]_{\pi/2}^{\pi}$$

$$= 8$$

25.

$$A = \int_0^1 \left[3x - \tfrac{1}{3}x\right] dx + \int_1^3 \left[-x + 4 - \tfrac{1}{3}x\right] dx$$

$$= \left[\tfrac{4}{3}x^2\right]_0^1 + \left[-\tfrac{2}{3}x^2 + 4x\right]_1^3 = 4$$

27.

$$A = \int_{-2}^1 \left[x - (-2)\right] dx + \int_1^5 \left[1 - (-2)\right] dx + \int_5^7 \left[-\tfrac{3}{2}x + \tfrac{17}{2} - (-2)\right] dx$$

$$= \left[\tfrac{1}{2}x^2 + 2x\right]_{-2}^1 + \left[3x\right]_1^5 \left[-\tfrac{3}{4}x^2 + \tfrac{21}{2}x\right]_5^7 = \frac{39}{2}$$

29.

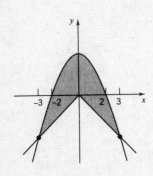

$$A = \int_{-3}^0 \left[6 - x^2 - x\right] dx + \int_0^3 \left[6 - x^2 - (-x)\right] dx$$

$$= \left[6x - \tfrac{1}{3}x^3 - \tfrac{x^2}{2}\right]_{-3}^0 + \left[6x - \tfrac{1}{3}x^3 + \tfrac{1}{2}x^2\right]_0^3 = 27$$

31.

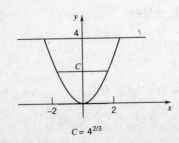

$C = 4^{2/3}$

$$\int_0^{\sqrt{c}} \left[c - x^2\right] dx = \frac{1}{2} \int_0^2 \left[4 - x^2\right] dx$$

$$\left[cx - \tfrac{1}{3}x^3\right]_0^{\sqrt{c}} = \tfrac{1}{2}\left[4x - \tfrac{1}{3}x^3\right]_0^2$$

$$\tfrac{2}{3}c^{3/2} = \tfrac{8}{3} \quad \text{and} \quad c = 4^{2/3}$$

33.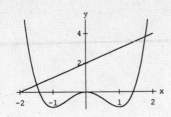

$$A \cong \int_{-1.49}^{1.79} \left[x + 2 - (x^4 - 2x^2) \right] \, dx$$

$$= \left[\tfrac{1}{2} x^2 - 2x - \tfrac{1}{5} x^5 + \tfrac{2}{3} x^3 \right]_{-1.49}^{1.78} \cong 7.93$$

35.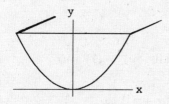

$$V = 8 \cdot 12 \int_{-3}^{3} \left[4 - \frac{4}{9} x^2 \right] \, dx$$

$$= 96 \cdot 2 \int_{0}^{3} \left[4 - \frac{4}{9} x^2 \right] \, dx$$

$$= 192 \left[4x - \tfrac{4}{27} x^3 \right]_{0}^{3}$$

$$= 1536 \text{ cu. in.} \cong 0.89 \text{ cu. ft.}$$

37. (a) $L(50) = \tfrac{7}{12} \left(\tfrac{1}{2} \right)^2 + \tfrac{5}{12} \left(\tfrac{1}{2} \right) = 35.4$

(b) $C = 2 \int_{0}^{1} \left[x - \tfrac{7}{12} x^2 - \tfrac{5}{12} x \right] \, dx = 2 \int_{0}^{1} \left[\tfrac{7}{12} x - \tfrac{7}{12} x^2 \right] \, dx$

$$= 2 \left[\tfrac{7}{24} x^2 - \tfrac{7}{36} x^3 \right]_{0}^{1} = \tfrac{7}{36}$$

SECTION 7.2

1.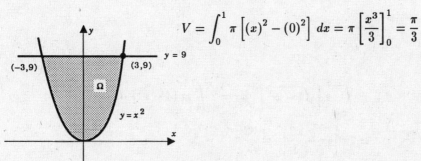

$$V = \int_{0}^{1} \pi \left[(x)^2 - (0)^2 \right] \, dx = \pi \left[\frac{x^3}{3} \right]_{0}^{1} = \frac{\pi}{3}$$

3.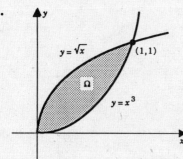

$$V = \int_{-3}^{3} \pi \left[(9)^2 - (x^2)^2 \right] \, dx = 2 \int_{0}^{3} \pi \left(81 - x^4 \right) \, dx$$

$$= 2\pi \left[81x - \frac{x^5}{5} \right]_{0}^{3} = \frac{1944\pi}{5}$$

5.

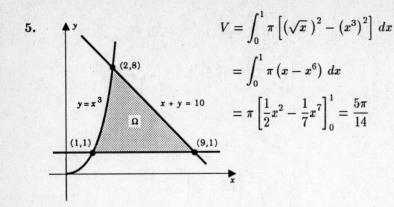

$$V = \int_0^1 \pi \left[\left(\sqrt{x} \right)^2 - \left(x^3 \right)^2 \right] dx$$

$$= \int_0^1 \pi \left(x - x^6 \right) dx$$

$$= \pi \left[\frac{1}{2} x^2 - \frac{1}{7} x^7 \right]_0^1 = \frac{5\pi}{14}$$

7.

$$V = \int_1^2 \pi \left[\left(x^3 \right)^2 - (1)^2 \right] dx + \int_2^9 \pi \left[(10 - x)^2 - (1)^2 \right] dx$$

$$= \int_1^2 \pi \left(x^6 - 1 \right) dx + \int_2^9 \pi \left(99 - 20x + x^2 \right) dx$$

$$= \pi \left[\frac{1}{7} x^7 - x \right]_1^2 + \pi \left[99x - 10x^2 + \frac{1}{3} x^3 \right]_2^9 = \frac{3790\pi}{21}$$

9.

$$V = \int_{-1}^2 \pi \left[(x + 2)^2 - \left(x^2 \right)^2 \right] dx$$

$$= \int_{-1}^2 \pi \left(x^2 + 4x + 4 - x^4 \right) dx$$

$$= \pi \left[\tfrac{1}{3} x^3 + 2x^2 + 4x - \tfrac{1}{5} x^5 \right]_{-1}^2 = \tfrac{72}{5}\pi$$

11.

$$V = \int_{-2}^2 \pi \left[\sqrt{4 - x^2} \right]^2 dx = 2 \int_0^2 \pi \left(4 - x^2 \right) dx$$

$$= 2\pi \left[4x - \frac{x^3}{3} \right]_0^2 = \frac{32}{3}\pi$$

13.

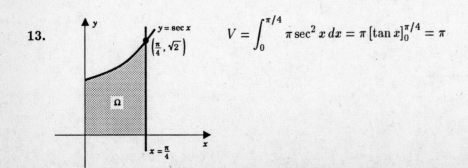

$$V = \int_0^{\pi/4} \pi \sec^2 x \, dx = \pi \left[\tan x \right]_0^{\pi/4} = \pi$$

15.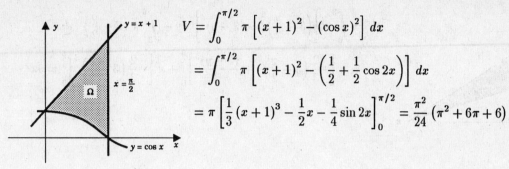

$$V = \int_0^{\pi/2} \pi \left[(x+1)^2 - (\cos x)^2 \right] dx$$

$$= \int_0^{\pi/2} \pi \left[(x+1)^2 - \left(\frac{1}{2} + \frac{1}{2} \cos 2x \right) \right] dx$$

$$= \pi \left[\frac{1}{3}(x+1)^3 - \frac{1}{2}x - \frac{1}{4} \sin 2x \right]_0^{\pi/2} = \frac{\pi^2}{24} \left(\pi^2 + 6\pi + 6 \right)$$

17.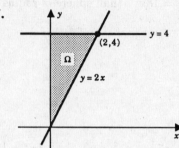

$$V = \int_0^4 \pi \left(\frac{y}{2} \right)^2 dy = \frac{\pi}{12} \left[y^3 \right]_0^4 = \frac{16\pi}{3}$$

19.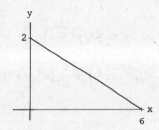

$$V = (\text{Volume of cone of radius 6, height 2}) = \frac{1}{3} \pi 6^2 \cdot 2 = 24\pi$$

21.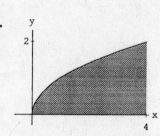

$$V = \int_{-2}^2 \pi \left[4^2 - y^4 \right] dy = \pi \left[16y - \frac{y^5}{5} \right]_{-2}^2 = \frac{256}{5}\pi$$

23.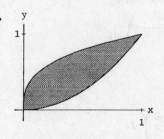

$$V = \int_0^1 \pi \left[(\sqrt{y})^2 - (y^3)^2 \right] dy = \int_0^1 \pi (y - y^6)\, dy$$

$$= \pi \left[\frac{y^2}{2} - \frac{y^7}{7} \right]_0^1 = \frac{5}{14}\pi$$

25.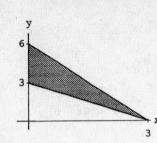

$$V = \int_0^3 \pi \left[\left(3 - \frac{y}{2}\right)^2 - (3-y)^2 \right] dy + \int_3^6 \pi \left(3 - \frac{y}{2}\right)^2 dy$$

$$= \pi \left[-\frac{2}{3}\left(3 - \frac{y}{2}\right)^3 + \frac{(3-y)^3}{3} \right]_0^3 + \pi \left[-\frac{2}{3}\left(3 - \frac{y}{2}\right)^3 \right]_3^6 = 9\pi$$

27.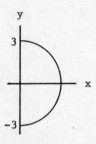

$$V = \int_0^3 \pi(9 - y^2)\, dy = \pi \left[9y - \frac{y^3}{3} \right]_0^3 = 18\pi \quad \text{(half sphere of radius 3.)}$$

29. Rotate about x-axis: $V = \int_{-1}^{1} \pi \operatorname{sech}^2 x\, dx = [\pi \tanh x]_{-1}^{1} = \pi \left(\frac{e - e^{-1}}{e + e^{-1}} - \frac{e^{-1} - e}{e^{-1} + e} \right) = 2\pi \left(\frac{e^2 - 1}{e^2 + 1} \right)$

31. For each $x \in [-a, a]$, the length of the base of the cross section at x is $2y = \frac{2b}{a}\sqrt{a^2 - x^2}$

(a) height $= \frac{1}{2}$ base, so area of cross section is $\frac{1}{2} \cdot \left(\frac{2b}{a}\sqrt{a^2 - x^2} \right) \left(\frac{b}{a}\sqrt{a^2 - x^2} \right) = \frac{b^2}{a^2}(a^2 - x^2)$.

$$\implies V = \int_{-a}^{a} \frac{b^2}{a^2}(a^2 - x^2)\, dx = \frac{b^2}{a^2}\left[a^2 x - \frac{x^3}{3} \right]_{-a}^{a} = \frac{4}{3}ab^2$$

(b) Area of cross section is $\left(\frac{2b}{a}\sqrt{a^2 - x^2} \right)^2$, so $V = \int_{-a}^{a} \frac{4b^2}{a^2}(a^2 - x^2)\, dx = \frac{16}{3}ab^2$

(c) Area of cross section is $\frac{1}{2}\left(\frac{2b}{a}\sqrt{a^2 - x^2} \right) \cdot 2 = \frac{2b}{a}\sqrt{a^2 - x^2}$, so

$$V = \frac{2b}{a}\int_{-a}^{a}\sqrt{a^2 - x^2}\, dx = \frac{2b}{a}\,[\text{Area of half circle of radius } a] = \frac{2b}{a} \cdot \frac{1}{2}\pi a^2 = \pi ab.$$

33. (a) $V = \int_0^1 2\sqrt{x}\, h\, dx + \int_1^3 2 \cdot \frac{1}{\sqrt{2}}\sqrt{3 - x}\, h\, dx = \frac{4h}{3}\left[x^{3/2} \right]_0^1 + \left[\frac{-2\sqrt{2}h}{3}(3 - x)^{3/2} \right]_1^3 = 4h$

(b) $V = \int_0^1 \left(\frac{1}{2} \cdot 2\sqrt{x} \cdot \sqrt{3}\sqrt{x} \right) dx + \int_1^3 \left(\frac{1}{2} \cdot \sqrt{2}\sqrt{3 - x} \cdot \frac{\sqrt{3}}{\sqrt{2}} \cdot \sqrt{3 - x} \right) dx$

$$= \sqrt{3}\int_0^1 x\, dx + \frac{\sqrt{3}}{2}\int_1^3 (3 - x)\, dx = \frac{3\sqrt{3}}{2}$$

(c) $V = \int_0^1 \left(\frac{1}{2} \cdot 2\sqrt{x} \cdot \sqrt{x} \right) dx + \int_1^3 \left(\frac{1}{2} \cdot \sqrt{2}\sqrt{3 - x} \cdot \frac{\sqrt{2}}{2} \cdot \sqrt{3 - x} \right) dx$

$$= \int_0^1 x\,dx + \frac{1}{2}\int_1^3 (3-x)\,dx = \frac{3}{2}$$

35. (a) $\quad V = \int_{-1}^1 (3-3y^2)h\,dy = h\left[3y - y^3\right]_{-1}^1 = 4h$

(b) $\quad V = \int_{-1}^1 \frac{1}{2}(3-3y^2)\frac{\sqrt{3}}{2}(3-3y^2)\,dy = \frac{9\sqrt{3}}{4}\int_{-1}^1 (1-2y^2+y^4)\,dy$

$$= \frac{9\sqrt{3}}{4}\left[y - \frac{2}{3}y^3 + \frac{y^5}{5}\right]_{-1}^1 = \frac{12}{5}\sqrt{3}$$

(c) $\quad V = \int_{-1}^1 \frac{1}{2}(3-3y^2)\frac{1}{2}(3-3y^2)\,dy = \frac{1}{\sqrt{3}}\cdot[\text{Volume in (b)}] = \frac{12}{5}$

37. $\quad V = \int_{-b}^b \pi\left(\frac{a}{b}\sqrt{b^2-y^2}\right)^2 dy = \frac{\pi a^2}{b^2}\int_{-b}^b (b^2-y^2)\,dy$

$$= \frac{\pi a^2}{b^2}\left[b^2 y - \frac{y^3}{3}\right]_{-b}^b = \frac{4}{3}\pi a^2 b$$

39. $\quad V = 2\int_0^{a/2} \pi\left(\sqrt{3}x\right)^2 dx = 6\pi\left[\frac{x^3}{3}\right]_0^{a/2} = \frac{1}{4}\pi a^3$ (twice the volume of cone of radius $\frac{\sqrt{3}}{2}a$, height $\frac{a}{2}$)

41. $\quad V = \int_a^b \pi\left(\sqrt{r^2-x^2}\right)^2 dx = \pi\int_a^b (r^2-x^2)\,dx = \pi\left[r^2 x - \frac{x^3}{3}\right]_a^b = \pi r^2 (b-a) - \frac{1}{3}\pi(b^3-a^3).$

43.

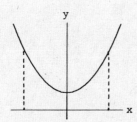

$$V = \int_{-\pi/3}^{\pi/3} \pi\left(\sqrt{\sec x}\right)^2 dx$$

$$= 2\pi\int_0^{\pi/3} \sec x\,dx$$

$$= 2\pi\left[\ln|\sec x + \tan x|\right]_0^{\pi/3} = 2\pi\ln(2+\sqrt{3})$$

45. $\quad V = \int_0^1 \pi\left(\cosh^2 x - \sinh^2 x\right)dx = \int_0^1 \pi\,dx = \pi$

47. The cross section with coordinate x is a washer with outer radius k, inner radius $k - f(x)$, and area

$$A(x) = \pi k^2 - \pi[k - f(x)]^2 = \pi\left(2kf(x) - [f(x)]^2\right)$$

Thus

$$V = \int_a^b \pi\left(2kf(x) - [f(x)]^2\right)dx$$

49. $V = \int_0^4 \pi \left[4\sqrt{x} - x \right] dx$

$= \pi \left[\frac{8}{3} x^{3/2} - \frac{1}{2} x^2 \right]_0^4 = \frac{40\pi}{3}$

51. $V = \int_0^\pi \pi \left[2 \sin x - \sin^2 x \right] dx$

$= \pi \left[-2 \cos x \right]_0^\pi - \frac{\pi}{2} \int_0^\pi (1 - \cos 2x) \, dx$

$= 4\pi - \frac{\pi}{2} \left[1 - \frac{1}{2} \sin 2x \right]_0^\pi = 4\pi - \frac{1}{2} \pi^2$

53.

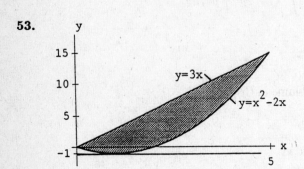

$V = \int_0^5 \pi \left([(3x-) - 1)]^2 - [x^2 - 2x - (-1)]^2 \right) dx$

$= \pi \int_0^5 [3x + 1]^2 \, dx - \int_0^5 [x - 1]^4 \, dx$

$= \pi \left[\frac{1}{9}(3x + 1)^3 \right]_0^5 - \frac{1}{5} [x - 1]_0^5$

$= 250\pi$

55.

(a) $V = \int_0^4 \pi \left[\left(\sqrt{4x} \right)^2 - x^2 \right] dx$

$= \pi \int_0^4 [4x - x^2] \, dx$

$= \pi \left[2x^2 - \frac{1}{3} x^3 \right]_0^4 = \frac{32\pi}{3}$

(b) $V = \int_0^4 \pi \left[\left(\frac{1}{4} y^2 - 4 \right)^2 - (y - 4)^2 \right] dy$

$= \pi \int_0^4 \left[\frac{1}{16} y^4 - 3y^2 + 8y \right] dy$

$= \pi \left[\frac{1}{80} y^5 - y^3 + 4y^2 \right]_0^4 = \frac{64\pi}{5}$

57. (a) $V = \int_0^4 \pi \left(x^{3/2} \right)^2 dx = \pi \int_0^4 x^3 \, dx = \pi \left[\frac{1}{4} x^4 \right]_0^4 = 64\pi$

(b) $V = \int_0^8 \pi \left(4 - y^{3/2} \right)^2 dy = \pi \int_0^8 \left(16 - 8y^{2/3} + y^{4/3} \right) dy$

$= \pi \left[16y - \frac{24}{5} y^{5/3} + \frac{3}{7} y^{7/3} \right]_0^8 = \frac{1024}{35} \pi$

(c) $V = \int_0^4 \pi \left[(8)^2 - \left(8 - x^{3/2} \right)^2 \right] dx = \pi \int_0^4 \left(16x^{3/2} - x^3 \right) dx$

$= \pi \left[\frac{32}{5} x^{5/2} - \frac{1}{4} x^4 \right]_0^4 = \frac{704}{5} \pi$

(d) $V = \int_0^8 \pi \left[(4)^2 - \left(y^{2/3} \right)^2 \right] dy = \pi \int_0^8 \left(16 - y^{4/3} \right) dy = \pi \left[16y - \frac{3}{7} y^{7/3} \right]_0^8 = \frac{512}{7} \pi$

SECTION 7.3

1.
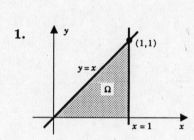

$V = \int_0^1 2\pi x \left[x - 0 \right] dx = 2\pi \int_0^1 x^2 \, dx$

$\quad = 2\pi \left[\frac{1}{3} x^3 \right]_0^1 = \frac{2\pi}{3}$

3.
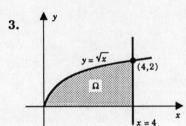

$V = \int_0^4 2\pi x \left[\sqrt{x} - 0 \right] dx = 2\pi \int_0^4 x^{3/2} \, dx$

$\quad = 2\pi \left[\frac{2}{5} x^{5/2} \right]_0^4 = \frac{128}{5} \pi$

5.
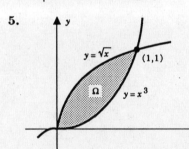

$V = \int_0^1 2\pi x \left[\sqrt{x} - x^3 \right] dx$

$\quad = 2\pi \int_0^1 \left(x^{3/2} - x^4 \right) dx$

$\quad = 2\pi \left[\frac{2}{5} x^{5/2} - \frac{1}{5} x^5 \right]_0^1 = \frac{2\pi}{5}$

7.
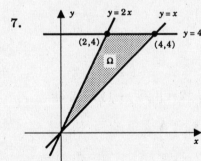

$V = \int_0^2 2\pi x \left[2x - x \right] dx + \int_2^4 2\pi x \left[4 - x \right] dx$

$\quad = 2\pi \int_0^2 x^2 \, dx + 2\pi \int_2^4 \left(4x - x^2 \right) dx$

$\quad = 2\pi \left[\frac{1}{3} x^3 \right]_0^2 + 2\pi \left[2x^2 - \frac{1}{3} x^3 \right]_2^4 = 16\pi$

9.
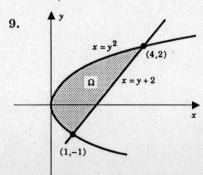

$V = \int_0^1 2\pi x \left[(\sqrt{x}) - (-\sqrt{x}) \right] dx + \int_1^4 2\pi x \left[(\sqrt{x}) - (x - 2) \right] dx$

$\quad = 4\pi \int_0^1 x^{3/2} \, dx + 2\pi \int_1^4 \left(x^{3/2} - x^2 + 2x \right) dx$

$\quad = 4\pi \left[\frac{2}{5} x^{5/2} \right]_0^1 + 2\pi \left[\frac{2}{5} x^{5/2} - \frac{1}{3} x^3 + x^2 \right]_1^4 = \frac{72}{5} \pi$

11.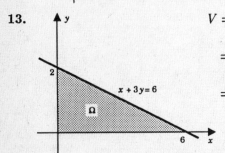

$$V = \int_0^3 2\pi x \left[\sqrt{9 - x^2} - \left(-\sqrt{9 - x^2} \right) \right] dx$$

$$= 4\pi \int_0^3 x \left(9 - x^2 \right)^{1/2} dx$$

$$= 4\pi \left[-\tfrac{1}{3} \left(9 - x^2 \right)^{3/2} \right]_0^3 = 36\pi$$

13.

$$V = \int_0^2 2\pi y \left[6 - 3y \right] dy$$

$$= 6\pi \int_0^2 \left(2y - y^2 \right) dy$$

$$= 6\pi \left[y^2 - \tfrac{1}{3} y^3 \right]_0^2 = 8\pi$$

15.

$$V = \int_0^9 2\pi y \left[(\sqrt{y}) - (-\sqrt{y}) \right] dy$$

$$= 4\pi \int_0^9 y^{3/2} dy$$

$$= 4\pi \left[\tfrac{2}{5} y^{5/2} \right]_0^9 = \tfrac{1944}{5}\pi$$

17.

$$V = \int_0^1 2\pi y \left[y^{1/3} - y^2 \right] dy$$

$$= 2\pi \int_0^1 \left(y^{4/3} - y^3 \right) dy$$

$$= 2\pi \left[\tfrac{3}{7} y^{7/3} - \tfrac{1}{4} y^4 \right]_0^1 = \tfrac{5}{14}\pi$$

19.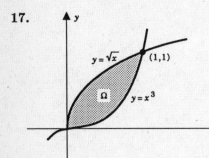

$$V = \int_0^1 2\pi y \left[(\sqrt{y}) - (-\sqrt{y}) \right] dy + \int_1^4 2\pi y \left[(\sqrt{y}) - (y - 2) \right] dy$$

$$= 4\pi \int_0^1 y^{3/2} dy + 2\pi \int_1^4 \left(y^{3/2} - y^2 + 2y \right) dy$$

$$= 4\pi \left[\tfrac{2}{5} y^{5/2} \right]_0^1 + 2\pi \left[\tfrac{2}{5} y^{5/2} - \tfrac{1}{3} y^3 + y^2 \right]_1^4 = \tfrac{72}{5}\pi$$

21.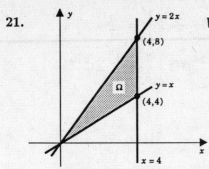

$$V = \int_0^4 2\pi y \left[y - \frac{y}{2}\right] dy + \int_4^8 2\pi y \left[4 - \frac{y}{2}\right] dy$$

$$= \pi \int_0^4 y^2 \, dy + \pi \int_4^8 (8y - y^2) \, dy$$

$$= \pi \left[\tfrac{1}{3} y^3\right]_0^4 + \pi \left[4y^2 - \tfrac{1}{3} y^3\right]_4^8 = 64\pi$$

23.

$$V = \int_0^1 2\pi y \left[\sqrt{1 - y^2} - (1 - y)\right] dy$$

$$= 2\pi \int_0^1 \left[y \left(1 - y^2\right)^{1/2} - y + y^2\right] dy$$

$$= 2\pi \left[-\frac{1}{3} \left(1 - y^2\right)^{3/2} - \frac{1}{2} y^2 + \frac{1}{3} y^3\right]_0^1 = \frac{\pi}{3}$$

25. (a) $\quad V = \int_0^1 2\pi x \left[1 - \sqrt{x}\right] dx$

(b) $\quad V = \int_0^1 \pi \, y^4 \, dy$

$$= \pi \left[\tfrac{1}{5} y^5\right]_0^1 = \tfrac{1}{5}\pi$$

27. (a) $\quad V = \int_0^1 \pi \left(x - x^4\right) dx$

(b) $\quad V = \int_0^1 2\pi y \left(\sqrt{y} - y^2\right) dy$

$$= \pi \left[\tfrac{1}{2} x^2 - \tfrac{1}{5} x^5\right]_0^1 = \frac{3\pi}{10}$$

29. (a) $\quad V = \int_0^1 2\pi x \cdot x^2 \, dx$

(b) $V = \int_0^1 \pi (1 - y) \, dy$

$$= 2\pi \left[\tfrac{1}{4} x^4\right]_0^1 = \frac{\pi}{2}$$

31. $\quad V = \int_{2\sqrt{3}}^6 2\pi x \cdot \frac{1}{x^2 \sqrt{x^2 - 9}} \, dx = 2\pi \int_{2\sqrt{3}}^6 \frac{dx}{x \sqrt{x^2 - 9}} = 2\pi \left[\frac{1}{3} \sec^{-1}\left(\frac{x}{3}\right)\right]_{2\sqrt{3}}^6 = \frac{\pi^2}{9}$

33. $\quad V = \int_0^b 2\pi y \frac{2a}{b} \sqrt{b^2 - y^2} \, dy = \frac{4\pi a}{b} \int_0^b y (b^2 - y^2)^{1/2} \, dy$

$$= \frac{4\pi a}{b} \left[-\frac{1}{3} (b^2 - y^2)^{3/2}\right]_0^b = \frac{4}{3} \pi a b^2$$

35. $\quad V = \int_0^{\sqrt{r^2 - a^2}} 2\pi x \left(\sqrt{r^2 - x^2} - a\right) dx = 2\pi \int_0^{\sqrt{r^2 - a^2}} \left[x(r^2 - x^2)^{1/2} - ax\right] dx$

$$= 2\pi \left[-\frac{1}{3} (r^2 - x^2)^{3/2} - \frac{a}{2} x^2\right]_0^{\sqrt{r^2 - a^2}} = \frac{1}{3} \pi (2r^3 + a^3 - 3ar^2)$$

37. (a) $V = \int_0^4 2\pi x(8 - x^{3/2})\,dx = 2\pi \int_0^4 (8x - x^{5/2})\,dx = 2\pi \left[4x^2 - \frac{2}{7}x^{7/2}\right]_0^4 = \frac{384}{7}\pi$

(b) $V = \int_0^8 2\pi(8 - y)y^{2/3}\,dy = 2\pi \int_0^8 (8y^{2/3} - y^{5/3})\,dy = 2\pi \left[\frac{24}{5}y^{5/3} - \frac{3}{8}y^{8/3}\right]_0^8 = \frac{576}{5}\pi$

(c) $V = \int_0^4 2\pi(4 - x)(8 - x^{3/2}\,dx = 2\pi \int_0^8 (8y^{2/3} - y^{5/3})\,dy$

$= 2\pi \left[32x - 4x^2 - \frac{8}{5}x^{5/2} + \frac{2}{7}x^{7/2}\right]_0^4 = \frac{3456}{35}\pi$

(d) $V = \int_0^8 2\pi y\, y^{2/3}\,dy = 2\pi \int_0^8 y^{5/3}\,dy = 2\pi \left[\frac{3}{8}y^{8/3}\right]_0^8 = 192\pi$

39. (a)

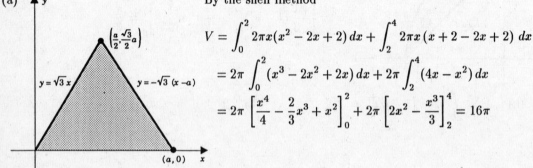

By the shell method

$V = \int_0^2 2\pi x(x^2 - 2x + 2)\,dx + \int_2^4 2\pi x\,(x + 2 - 2x + 2)\,dx$

$= 2\pi \int_0^2 (x^3 - 2x^2 + 2x)\,dx + 2\pi \int_2^4 (4x - x^2)\,dx$

$= 2\pi \left[\frac{x^4}{4} - \frac{2}{3}x^3 + x^2\right]_0^2 + 2\pi \left[2x^2 - \frac{x^3}{3}\right]_2^4 = 16\pi$

41. (a) $V = \int_0^1 \pi(\sqrt{3}x)^2\,dx + \int_1^2 \pi\left(\sqrt{4 - x^2}\right)^2\,dx$

(b) $V = \int_0^{\sqrt{3}} 2\pi y\left[\sqrt{4 - y^2} - \frac{y}{\sqrt{3}}\right]\,dy$

(c) use (a): $V = 3\pi \int_0^1 x^2\,dx + \pi \int_1^2 (4 - x^2)\,dx = \pi\left[x^3\right]_0^1 + \pi\left[4x - \frac{x^3}{3}\right]_1^2 = \frac{8}{3}\pi$

43. (a) $V = \int_0^1 \pi\left[(\sqrt{3} + 1)^2 - 1^2\right]\,dx + \int_1^2 \pi\left[\left(\sqrt{4 - x^2} + 1\right)^2 - 1^2\right]\,dx$

(b) $V = \int_0^{\sqrt{3}} 2\pi(y + 1)\left(\sqrt{4 - y^2} - \frac{y}{\sqrt{3}}\right)\,dy$

45. $V = \int_{-a}^a 2\pi(a - x)2\sqrt{a^2 - x^2}\,dx = 4\pi a \int_{-a}^a \sqrt{a^2 - x^2}\,dx - 4\pi \int_{-a}^a x\sqrt{a^2 - x^2}\,dx$

$= 4\pi a(\text{ Area of half circle}) - 4\pi \left[-\frac{1}{3}(a^2 - x^2)^{3/2}\right]_{-a}^a$

$= 4\pi a \cdot \pi a^2 - 0 = 4\pi^2 a^3$

SECTION 7.4

1.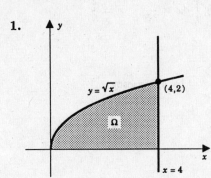

$$A = \int_0^4 \sqrt{x}\, dx = \frac{16}{3}$$

$$\overline{x}A = \int_0^4 x\sqrt{x}\, dx = \frac{64}{5}, \quad \overline{x} = \frac{12}{5}$$

$$\overline{y}A = \int_0^4 \frac{1}{2}\left(\sqrt{x}\right)^2 dx = 4, \quad \overline{y} = \frac{3}{4}$$

$$V_x = 2\pi\overline{y}A = 8\pi, \quad V_y = 2\pi\overline{x}A = \frac{128}{5}\pi$$

3.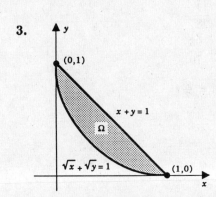

$$A = \int_0^1 \left(x^{1/3} - x^2\right) dx = \frac{5}{12}$$

$$\overline{x}A = \int_0^1 x\left(x^{1/3} - x^2\right) dx = \frac{5}{28}, \quad \overline{x} = \frac{3}{7}$$

$$\overline{y}A = \int_0^1 \frac{1}{2}\left[\left(x^{1/3}\right)^2 - \left(x^2\right)^2\right] dx = \frac{1}{5}, \quad \overline{y} = \frac{12}{25}$$

$$V_x = 2\pi\overline{y}A = \frac{2}{5}\pi, \quad V_y = 2\pi\overline{x}A = \frac{5}{14}\pi$$

5.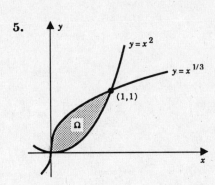

$$A = \int_1^3 (2x - 2)\, dx = 4$$

$$\overline{x}A = \int_1^3 x(2x - 2)\, dx = \frac{28}{3}, \quad \overline{x} = \frac{7}{3}$$

$$\overline{y}A = \int_1^3 \frac{1}{2}\left[(2x)^2 - (2)^2\right] dx = \frac{40}{3}, \quad \overline{y} = \frac{10}{3}$$

$$V_x = 2\pi\overline{y}A = \frac{80}{3}\pi, \quad V_y = 2\pi\overline{x}A = \frac{56}{3}\pi$$

7.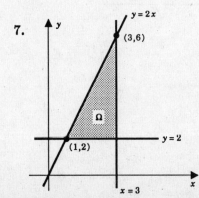

$$A = \int_0^2 \left[6 - \left(x^2 + 2x\right)\right] dx = \frac{16}{3}$$

$$\overline{x}A = \int_0^2 x\left[6 - \left(x^2 + 2\right)\right] dx = 4, \quad \overline{x} = \frac{3}{4}$$

$$\overline{y}A = \int_0^2 \frac{1}{2}\left[(6)^2 - \left(x^2 + 2\right)^2\right] dx = \frac{352}{15}, \quad \overline{y} = \frac{22}{5}$$

$$V_x = 2\pi\overline{y}A = \frac{704}{15}\pi, \quad V_y = 2\pi\overline{x}A = 8\pi$$

9.

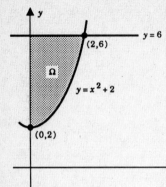

$$A = \int_0^1 \left[(1-x) - (1-\sqrt{x})^2 \right] dx = \frac{1}{3}$$

$$\overline{x}A = \int_0^1 x \left[(1-x) - (1-\sqrt{x})^2 \right] dx = \frac{2}{15}, \quad \overline{x} = \frac{2}{5}$$

$$\overline{y} = \frac{2}{5} \qquad \text{by symmetry}$$

$$V_x = 2\pi\overline{y}A = \frac{4}{15}\pi, \quad V_y = \frac{4}{15}\pi \qquad \text{by symmetry}$$

11.

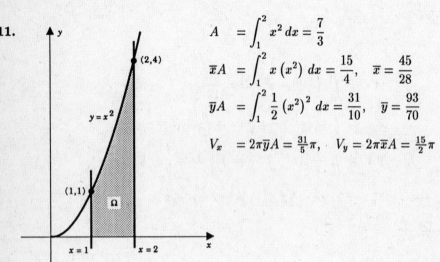

$$A = \int_1^2 x^2 \, dx = \frac{7}{3}$$

$$\overline{x}A = \int_1^2 x \left(x^2 \right) dx = \frac{15}{4}, \quad \overline{x} = \frac{45}{28}$$

$$\overline{y}A = \int_1^2 \frac{1}{2} \left(x^2 \right)^2 dx = \frac{31}{10}, \quad \overline{y} = \frac{93}{70}$$

$$V_x = 2\pi\overline{y}A = \frac{31}{5}\pi, \quad V_y = 2\pi\overline{x}A = \frac{15}{2}\pi$$

13.

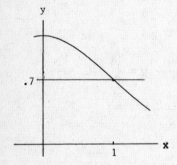

$$A = \int_0^1 \left[\frac{1}{\sqrt{1+x^2}} - \frac{1}{\sqrt{2}} \right] dx = \left[\ln |x + \sqrt{1+x^2}| - \frac{1}{\sqrt{2}} x \right]_0^1 = \frac{\sqrt{2}\ln\left(1+\sqrt{2}\right) - 1}{\sqrt{2}}$$

$$\overline{x}A = \int_0^1 x \left[\frac{1}{\sqrt{1+x^2}} - \frac{1}{\sqrt{2}} \right] dx = \left[\sqrt{1+x^2} - \frac{1}{2\sqrt{2}} \right]_0^1 = \frac{3 - 2\sqrt{2}}{2\sqrt{2}}$$

Therefore, $\quad \overline{x} = \dfrac{\dfrac{3-2\sqrt{2}}{2\sqrt{2}}}{\dfrac{\sqrt{2}\ln\left(1+\sqrt{2}\right)-1}{\sqrt{2}}} = \dfrac{3-2\sqrt{2}}{2\sqrt{2}\ln\left(1+\sqrt{2}\right)-2}$

$$\overline{y}A = \int_0^1 \frac{1}{2}\left[\frac{1}{1+x^2} - \frac{1}{2}\right] dx = \left[\frac{1}{2}\tan^{-1}x - \frac{1}{4}x\right]_0^1 = \frac{\pi-2}{8}$$

Therefore, $\overline{y} = \dfrac{\dfrac{\pi-2}{8}}{\dfrac{\sqrt{2}\ln(1+\sqrt{2})-1}{\sqrt{2}}} = \dfrac{\sqrt{2}(\pi-2)}{8[\sqrt{2}\ln(1+\sqrt{2})-1]}$

15.

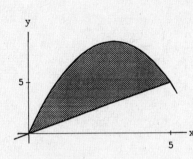

$$A = \int_0^5 \left[6x - x^2 - x\right] dx = \int_0^5 \left[5x - x^2\right] dx$$
$$= \left[\frac{5}{2}x^2 - \frac{1}{3}x^3\right]_0^5 = \frac{125}{6}$$

$$\overline{x}A = \int_0^5 x\left[6x - x^2 - x\right] dx = \int_0^5 \left[5x^2 - x^3\right] dx = \left[\frac{5}{3}x^3 - \frac{1}{4}x^4\right]_0^5 = \frac{625}{12}$$

Therefore, $\overline{x} = \dfrac{625/12}{125/6} = \dfrac{5}{2}$.

$$\overline{y}A = \int_0^5 \frac{1}{2}\left[(6x-x^2)^2 - x^2\right] dx = \frac{1}{2}\int_0^5 \left[x^4 - 12x^3 + 35x^2\right] dx == \frac{1}{2}\left[\frac{1}{5}x^5 - 3x^4 + \frac{35}{3}x^3\right]_0^5 = \frac{625}{6}$$

Therefore, $\overline{y} = \dfrac{625/6}{125/6} = 5$. The centroid is $\left(\frac{5}{2}, 5\right)$

17.

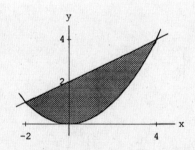

$$A = \int_{-2}^4 \left[\frac{1}{2}x + 2 - \frac{1}{4}x^2\right] dx = \left[\frac{1}{4}x^2 + 2x - \frac{1}{12}x^3\right]_{-2}^4 = 9$$

$$\overline{x}A = \int_{-2}^4 x\left[\frac{1}{2}x + 2 - \frac{1}{4}x^2\right] dx \int_{-2}^4 \left[\frac{1}{2}x^2 + 2x - \frac{1}{4}x^3\right] dx = \left[\frac{1}{6}x^3 + x^2 - \frac{1}{16}x^4\right]_{-2}^4 = 9$$

Therefore, $\overline{x} = \dfrac{9}{9} = 1$.

$$\overline{y}A = \int_{-2}^4 \frac{1}{2}\left[\left(\frac{1}{2}x + 2\right)^2 - \frac{1}{16}x^4\right] dx \frac{1}{2}\int_{-2}^4 \left[\frac{1}{4}x^2 + 2x + 4 - \frac{1}{16}x^4\right] dx$$

$$= \frac{1}{2}\left[\frac{1}{12}x^3 + x^2 + 4x - \frac{1}{80}x^5\right]_{-2}^4 = \frac{72}{5}$$

Therefore, $\overline{y} = \dfrac{72/5}{9} = \dfrac{8}{5}$. The centroid is $\left(1, \frac{8}{5}\right)$

19.

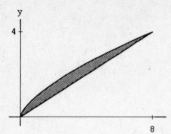

$$A = \int_0^8 \left[x^{2/3} - \tfrac{1}{2}\,x \right] dx = \left[\tfrac{3}{5}\,x^{5/3} - \tfrac{1}{4}\,x^2 \right]_0^8 = \frac{16}{5}$$

$$\overline{x}A = \int_0^8 4\,x \left[x^{2/3} - \tfrac{1}{2}\,x \right] dx \int_0^8 \left[x^{5/3} - \tfrac{1}{2}\,x^2 \right] dx = \left[\tfrac{3}{8}\,x^{8/3} - \tfrac{1}{6}\,x^3 \right]_0^8 = \frac{32}{3}$$

Therefore, $\ \overline{x} = \dfrac{32/3}{16/5} = \dfrac{10}{3}$.

$$\overline{y}A = \int_0^8 \tfrac{1}{2} \left[x^{4/3} - \tfrac{1}{4}\,x^2 \right] dx \, \tfrac{1}{2} \int_0^8 \left[\tfrac{3}{7}\,x^{7/3} - \tfrac{1}{12}\,x^3 \right]_0^8 = \frac{128}{21}$$

Therefore, $\ \overline{y} = \dfrac{128/21}{16/5} = \dfrac{40}{21}$. The centroid is $\left(\tfrac{10}{3}, \tfrac{40}{21} \right)$

21.

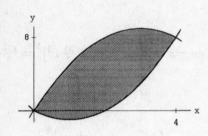

$$A = \int_0^4 \left[6x - x^2 - (x^2 - 2x) \right] dx = \int_0^4 \left[8x - 2x^2 \right] dx$$
$$= \left[4x^2 - \tfrac{2}{3}\,x^3 \right]_0^4 = \frac{64}{3}$$

$$\overline{x}A = \int_0^4 x \left[8x - 2x^2 \right] dx = \int_0^4 \left[8x^2 - 2x^3 \right] dx = \left[\tfrac{8}{3}\,x^3 - \tfrac{1}{2}\,x^4 \right]_0^4 = \frac{128}{3}$$

Therefore, $\ \overline{x} = \dfrac{128/3}{64/3} = 2$.

$$\overline{y}A = \int_0^4 \tfrac{1}{2} \left[(6x - x^2)^2 - (x^2 - 2x)^2 \right] dx = \int_0^4 \left[16x^2 - 4x^3 \right] dx == \left[\tfrac{16}{3}\,x^3 - x^4 \right]_0^4 = \frac{256}{3}$$

Therefore, $\ \overline{y} = \dfrac{256/3}{64/3} = 4$. The centroid is $(2, 4)$

23.

$$A = 2 \int_{-1}^0 \left[(-x)^{1/2} \right] dx = \left[-\tfrac{4}{3}\,(-x)^{3/2} \right]_{-1}^0 = \frac{4}{3}$$

$$\overline{x}A = \int_{-1}^0 (-x) \left\{ (-x)^{1/2} - [-(-x)^{1/2}] \right\} dx = \int_{-1}^0 -2(-x)^{3/2}\, dx = \left[\tfrac{4}{5}\,x^{5/2} \right]_{-1}^0 = -\frac{4}{5}$$

Therefore, $\bar{x} = \dfrac{-4/5}{4/3} = -\dfrac{3}{5}$.

By symmetry, $\bar{y} = 0$. The centroid is $\left(-\frac{3}{5}, 0\right)$.

25.   (a) $(0,0)$ by symmetry

(b) Ω_1 smaller quarter disc, Ω_2 the larger quarter disc

$$A_1 = \frac{1}{16}\pi, \quad A_2 = \pi; \quad \bar{x}_1 = \bar{y}_1 = \frac{2}{3\pi}, \quad \bar{x}_2 = \bar{y}_2 = \frac{8}{3\pi} \quad \text{(Problem 1)}$$

$$\bar{x}A = \left(\frac{8}{3\pi}\right)(\pi) - \frac{2}{3\pi}\left(\frac{1}{16}\pi\right)\frac{63}{24}, \quad A = \frac{15}{16}\pi$$

$$\bar{x} = \left(\frac{63}{24}\right)\Big/\left(\frac{15\pi}{16}\right) = \frac{14}{5\pi}, \quad \bar{y} = \bar{x} = \frac{14}{5\pi} \quad \text{(symmetry)}$$

(c) $\bar{x} = 0, \quad \bar{y} = \dfrac{14}{5\pi}$

27.   Use theorem of Pappus. Centroid of rectangle is located

$$c + \sqrt{\left(\frac{a}{2}\right)^2 + \left(\frac{b}{2}\right)^2} \text{ units}$$

from line l. The area of the rectangle is ab. Thus,

$$\text{volume } = 2\pi \left[c + \sqrt{\left(\frac{a}{2}\right)^2 + \left(\frac{b}{2}\right)^2}\right](ab) = \pi ab \left(2c + \sqrt{a^2 + b^2}\right).$$

29.   (a) $\left(\frac{2}{3}a, \frac{1}{3}h\right)$ (b) $\left(\frac{2}{3}a + \frac{1}{3}b, \frac{1}{3}h\right)$ (c) $\left(\frac{1}{3}a + \frac{1}{3}b, \frac{1}{3}h\right)$

31.   (a) $V = \frac{2}{3}\pi R^3 \sin^3\theta + \frac{1}{3}\pi R^3 \sin^2\theta\cos\theta = \frac{1}{3}\pi R^3 \sin^2\theta\,(2\sin\theta + \cos\theta)$

(b) $\bar{x} = \dfrac{V}{2\pi A} = \dfrac{\frac{1}{3}\pi R^3 \sin^2\theta\,(2\sin\theta + \cos\theta)}{2\pi\left(\frac{1}{2}R^2\sin\theta\cos\theta + \frac{1}{4}\pi R^2\sin^2\theta\right)} = \dfrac{2R\sin\theta\,(2\sin\theta + \cos\theta)}{3\,(\pi\sin\theta + 2\cos\theta)}$

33.   (a)   The mass contributed by $[x_{i-1}, x_i]$ is approximately $\lambda\,(x_i^*)\,\Delta x_i$ where x_i^* is the midpoint of $[x_{i-1}, x_i]$. The sum of these contributions,

$$\lambda\,(x_1^*)\,\Delta x_1 + \cdots + \lambda\,(x_n^*)\,\Delta x_n,$$

is a Riemann sum, which as $\|P\| \to 0$, tends to the given integral.

(b)   Take M_i as the mass contributed by $[x_{i-1}, x_i]$. Then $x_{M_i}M_i \cong x_i^*\lambda\,(x_i^*)\,\Delta x_i$ where x_i^* is the midpoint of $[x_{i-1}, x_i]$. Therefore

$$x_M M = x_{M_1} M_1 + \cdots + x_{M_n} M_n \cong x_1^* \lambda (x_1^*) \, \Delta x_1 + \cdots + x_n^* \lambda (x_n^*) \, \Delta x_n.$$

As $\|P\| \to 0$, the sum on the right converges to the given integral.

35.

$$\overline{x} \left(\tfrac{1}{3} \pi r^2 h \right) = \int_0^h \pi x \left(\frac{r}{h} x \right)^2 dx = \frac{1}{4} \pi r^2 h^2$$

$$\overline{x} = \left(\tfrac{1}{4} \pi r^2 h^2 \right) / \left(\tfrac{1}{3} \pi r^2 h \right) = \tfrac{3}{4} h.$$

The centroid of the cone lies on the axis of the cone at a distance $\tfrac{3}{4} h$ from the vertex.

37. (a) $V_x = \displaystyle\int_0^1 \pi \left(\sqrt{x} \right)^2 dx = \frac{1}{2} \pi, \quad \overline{x} V_x = \int_0^1 \pi x \left(\sqrt{x} \right)^2 dx = \frac{1}{3} \pi$

$\overline{x} = \left(\tfrac{1}{3} \pi \right) / \left(\tfrac{1}{2} \pi \right) = \tfrac{2}{3};$ centroid $\left(\tfrac{2}{3}, 0 \right)$

(b) $V_y = \displaystyle\int_0^1 2 \pi x \sqrt{x} \, dx = \frac{4}{5} \pi, \quad \overline{y} V_y = \int_0^1 \pi x \left(\sqrt{x} \right)^2 dx = \frac{1}{3} \pi$

$\overline{y} = \left(\tfrac{1}{3} \pi \right) / \left(\tfrac{4}{5} \pi \right) = \tfrac{5}{12};$ centroid $\left(0, \tfrac{5}{12} \right)$

39. $V_x = \displaystyle\int_0^a \pi \frac{b^2}{a^2} \left(a^2 - x^2 \right) dx = \frac{2}{3} \pi a b^2, \quad \overline{x} V_x = \int_0^a \pi x \frac{b^2}{a^2} \left(a^2 - x^2 \right) dx = \frac{1}{4} \pi a^2 b^2$

$\overline{x} = \left(\tfrac{1}{4} \pi a^2 b^2 \right) / \left(\tfrac{2}{3} \pi a b^2 \right) = \tfrac{3}{8} a;$ centroid $\left(\tfrac{3}{8} a, 0 \right)$

SECTION 7.5

1. $W = \displaystyle\int_1^4 x \left(x^2 + 1 \right)^2 dx$

$= \tfrac{1}{6} \left[\left(x^2 + 1 \right)^3 \right]_1^4 = 817.5$ ft-lb

3. $W = \displaystyle\int_{\pi/6}^{pi} \left(x + \sin 2x \right) dx$

$= \left[\tfrac{1}{2} x^2 - \tfrac{1}{2} \cos 2x \right]_{\pi/6}^{\pi}$

$= \tfrac{35}{72} \pi^2 - \tfrac{1}{4}$ Newton-meters

5. By Hooke's law, we have $600 = -k(-1)$. Therefore $k = 600$.

The work required to compress the spring to 5 inches is given by

$$W = \int_{10}^5 600(x - 10) \, dx = 600 \left[\tfrac{1}{2} x^2 - 10x \right]_{10}^5$$

$= 7500$ in-lb, or 625 ft-lb

7.   To counteract the restoring force of the spring we must apply a force $F(x) = kx$.

Since $F(4) = 200$, we see that $k = 50$ and therefore $F(x) = 50x$.

(a) $W = \displaystyle\int_0^1 50x \, dx = 25$ ft-lb   (b) $W = \displaystyle\int_0^{3/2} 50x \, dx = \dfrac{225}{4}$ ft-lb

9.   Let L be the natural length of the spring.

$$\int_{2-L}^{2.1-L} kx \, dx = \frac{1}{2} \int_{2.1-L}^{2.2-L} kx \, dx$$

$$\left[\tfrac{1}{2}kx^2\right]_{2-L}^{2.1-L} = \tfrac{1}{2}\left[\tfrac{1}{2}kx^2\right]_{2.1-L}^{2.2-L}$$

$$(2.1 - L)^2 - (2 - L)^2 = \tfrac{1}{2}\left[(2.2 - L)^2 - (2.1 - L)^2\right].$$

Solve this equation for L and you will find that $L = 1.95$.

Answer: 1.95 ft

11.

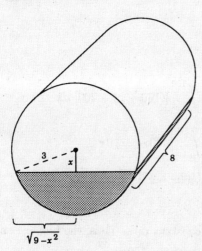

(a) $W = \displaystyle\int_0^3 (x+3)(60)(8)\left(2\sqrt{9-x^2}\right) dx$

$= 960 \displaystyle\int_0^3 x\left(9-x^2\right)^{1/2} dx$

$+ 2880 \underbrace{\displaystyle\int_0^3 \sqrt{9-x^2} \, dx}_{\substack{\text{area of quarter} \\ \text{circle of radius 3}}}$

$= 960\left[-\tfrac{1}{3}\left(9-x^2\right)^{3/2}\right]_0^3 + 2880\left[\tfrac{9}{4}\pi\right]$

$= (8640 + 6480\pi)$ ft-lb

(b) $W = \displaystyle\int_0^3 (x+7)(60)(8)\left(2\sqrt{9-x^2}\right) dx = 960 \displaystyle\int_0^3 x\left(9-x^2\right)^{1/2} dx + 6720 \displaystyle\int_0^3 \sqrt{9-x^2} \, dx$

$= (8640 + 15120\pi)$ ft-lb

13.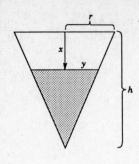

By similar triangles

$$\frac{h}{r} = \frac{h-x}{y} \quad \text{so that} \quad y = \frac{r}{h}(h-x).$$

Thus, the area of a cross section of the fluid at a depth of x feet is

$$\pi y^2 = \pi \frac{r^2}{h^2}(h-x)^2.$$

(a) $W = \displaystyle\int_0^{h/2} x\sigma\left[\pi\frac{r^2}{h^2}(h-x)^2\right] dx = \frac{\sigma\pi r^2}{h^2}\int_0^{h/2}(h^2 x - 2hx^2 + x^3)\,dx = \frac{11}{192}\sigma\pi r^2 h^2$ ft-lb

(b) $W = \displaystyle\int_0^{h/2}(x+k)\sigma\left[\pi\frac{r^2}{h^2}(h-x)^2\right] dx = \frac{11}{192}\pi r^2 h^2\sigma + \frac{7}{24}\pi r^2 hk\sigma$ ft-lb

15. $W = \displaystyle\int_{r_1}^{r_2} F\,dr = \int_{r_1}^{r_2} -\frac{GmM}{r^2}\,dr = \left[\frac{GmM}{r}\right]_{r_1}^{r_2} = GmM\left[\frac{1}{r_2} - \frac{1}{r_1}\right]$

17. $W = \displaystyle\int_0^{80}(80-x)15\,dx = 15\left[80x - \frac{1}{2}x^2\right]_0^{80} = 48{,}000$ ft-lb

19. (a) $W = 200 \cdot 100 = 20{,}000$ ft-lb

(b) $W = \displaystyle\int_0^{100}[(100-x)+200]\,dx$

$= \displaystyle\int_0^{100}(400-2x)\,dx$

$= \left[400x - x^2\right]_0^{100} = 30{,}000$ ft-lb

21. The bag is raised 8 feet and loses a total of 3 pounds at a constant rate. Thus, the bag loses sand at the rate of 3/8 lb/ft. After the bag has been raised x feet it weighs $100 - \dfrac{3x}{8}$ pounds.

$$W = \int_0^8\left(100 - \frac{3x}{8}\right) dx = \left[100 - \frac{3x^2}{16}\right]_0^8 = 788 \text{ ft-lb.}$$

23. (a) $W = \displaystyle\int_0^l x\sigma\,dx = \frac{1}{2}\sigma l^2$ ft-lb

(b) $W = \displaystyle\int_0^l (x+l)\sigma\,dx = \frac{3}{2}\sigma l^2$ ft-lb

25. Thirty feet of cable and the steel beam weighing a total of

$$800 + 30(6) = 980 \text{ lb}$$

are raised 20 feet. The work requires $(20)(980)$ ft-lb.

Next, the remaining 20 feet of cable is raised a varying distance and wound onto the steel drum. Thus the total work is given by

$$W = (20)(980) + \int_0^{20} 6x \, dx = 19600 + 1200 = 20800 \text{ ft-lb.}$$

27.   By the hint

$$W = \int_a^b F(x) \, dx = \int_a^b ma \, dx = \int_a^b mv \frac{dv}{dx} \, dx = \int_{v_a}^{v_b} mv \, dv = \left[\frac{1}{2} mv^2 \right]_{v_a}^{v_b} = \frac{1}{2} v_b^2 - \frac{1}{2} v_a^2$$

29.   (a)   The work required to pump the water out of the tank is given by

$$W = \int_5^{10} (62.5) \pi \, 5^2 x \, dx = 1562.5\pi \left[\frac{1}{2} x^2 \right]_5^{10} \cong 184,078 \text{ ft-lb}$$

A $\frac{1}{2}$-horsepower pump can do 275 ft-lb of work per second. Therefore it will take

$$\frac{184,078}{275} \cong 669 \text{ seconds} \cong 11 \text{ min, } 10 \text{ sec, to empty the tank.}$$

(b)   The work required to pump the water to a point 5 feet above the top of the tank

is given by

$$W = \int_5^{10} (62.5) \pi \, 5^2 (x + 5) \, dx = \int_5^{10} (62.5) \pi \, 5^2 x \, dx + \int_5^{10} (62.5) \pi \, 5^3 \, dx \cong 306796 \text{ ft-lb}$$

It will take a $\frac{1}{2}$-horsepower pump approximately $1,116$ sec, or 18 min, 36 sec, in this case.

SECTION 7.6

1. $$F = \int_0^6 (62.5) \cdot x \cdot 8 \, dx = 250 \left[x^2 \right]_0^6 = 9000 \text{ lb}$$

3.   The width of the plate x meters below the surface is given by $w(x) = 60 + 2(20 - x) = 100 - 2x$

(see the figure). The force against the dam is

$$F = \int_0^{20} 9800x(100 - 2x) \, dx$$

$$= 9800 \int_0^{20} (100x - 2x^2) \, dx$$

$$= 9800 \left[50x^2 - \tfrac{2}{3} x^3 \right]_0^{20}$$

$$\cong 1.437 \times 10^8 \text{ Newtons}$$

5.   The width of the gate x meters below its top is given by $w(x) = 4 + \frac{2}{3}x$ (see the figure).

The force of the water against the gate is

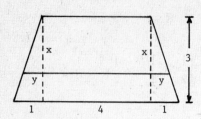

$$F = \int_0^3 9800(10 + x)\left(4 + \frac{2}{3}x\right) dx$$

$$= 9800 \int_0^3 \left[\frac{2}{3}x^2 + \frac{32}{3} + 40\right] dx$$

$$= 9800 \left[\frac{2}{9}x^3 + \frac{16}{3}x^2 + 40x\right]_0^3$$

$$\cong 1.7052 \times 10^6 \text{ Newtons}$$

7.

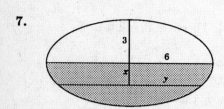

ellipse: $\dfrac{x^2}{3^2} + \dfrac{y^2}{6^2} = 1$

$$F = \int_0^3 (60)x \left[12\sqrt{1 - \frac{x^2}{9}}\right] dx$$

$$= 240 \int_0^3 x\left(9 - x^2\right)^{1/2} dx$$

$$= 240 \left[-\frac{1}{3}\left(9 - x^2\right)^{3/2}\right]_0^3 = 2160 \text{ lb}$$

9.

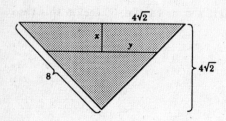

By similar triangles

$$\frac{4\sqrt{2}}{4\sqrt{2}} = \frac{y}{4\sqrt{2} - x} \quad \text{so} \quad y = 4\sqrt{2} - x.$$

$$F = \int_0^{4\sqrt{2}} (62.5)x \left[2\left(4\sqrt{2} - x\right)\right] dx$$

$$= 125 \int_0^{4\sqrt{2}} \left(4\sqrt{2}\,x - x^2\right) dx = \frac{8000}{3}\sqrt{2} \text{ lb}$$

11.

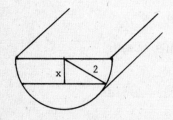

$$F = \int_0^2 (62.5)\cdot x \cdot 2\sqrt{4 - x^2}\, dx$$

$$= 125 \int_0^2 x\sqrt{4 - x^2}\, dx$$

$$= -\frac{125}{3}\left[\left(4 - x^2\right)^{3/2}\right]_0^2 = 333.33 \text{ lb}$$

13.

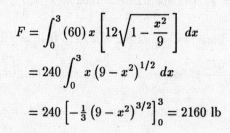

$$F = \int_0^4 60x \left(2\sqrt{16 - x^2}\right) dx$$

$$= 120 \int_0^4 x\left(16 - x^2\right)^{1/2} dx$$

$$= 120 \left[-\frac{1}{3}\left(16 - x^2\right)^{3/2}\right]_0^4 = 2560 \text{ lb}$$

15. (a) The width of the plate is 10 feet and the depth of the plate ranges from 8 feet to 14 feet. Thus

$$F = \int_8^{14} 62.5x\,(10)\,dx = 41{,}250 \text{ lb.}$$

(b) The width of the plate is 6 feet and the depth of the plate ranges from 6 feet to 16 feet. Thus

$$F = \int_6^{16} 62.5x\,(6)\,dx = 41{,}250 \text{ lb.}$$

17. (a) Force on the sides:

$$F = \int_0^1 (9800)\,x\,14\,dx + \int_0^2 (9800)(1+x)\,7(2-x)\,dx$$

$$= 68{,}600\left[x^2\right]_0^1 + 68{,}600\int_0^2 \left[2+x-x^2\right]\,dx$$

$$= 68{,}600 + 68{,}600\left[2x + \tfrac{1}{2}x^2 - \tfrac{1}{3}x^3\right]_0^2$$

$$\cong 297{,}267 \text{ Newtons}$$

(b) Force at the shallow end:

$$F = \int_0^1 (9800)\cdot x \cdot 8\,dx = 39{,}200\left[x^2\right]_0^1 = 39{,}200 \text{ Newtons}$$

Force at the deep end:

$$F = \int_0^3 (9800)\cdot x \cdot 8\,dx = 39{,}200\left[x^2\right]_0^3 = 352{,}800 \text{ Newtons}$$

19. $F = \int_0^{14} (9800)\left(1+\tfrac{1}{7}x\right)\cdot 8\tfrac{5\sqrt{2}}{7}\,dx = 392{,}000\tfrac{\sqrt{2}}{7}\left[x+\tfrac{1}{14}x^2\right]_0^1 4 \cong 2.217 \times 10^6 \text{ Newtons}$

21. $F = \int_a^b \sigma x w(x)\,dx = \sigma \int_a^b x w(x)\,dx = \sigma \overline{x} A$

where A is the area of the submerged surface and $\overline{x}$ is the depth of the centroid.

PROJECTS AND EXPLORATIONS

7.1. (a) Figure A shows the graph of f on $I = [-3,3]$. You can verify that f is defined and positive for all x, and that $\lim_{x\to\pm\infty} f(x) = 0$. You can also verify that f has only one critical number, $x \cong 1.2147$. Figure B shows the graphs of f and g_p for $p = 1, 2, \tfrac{1}{2}$; f and g_p intersect at two points for all $p > 0$. If $p < 0$ then the graphs of f and g_p do not intersect.

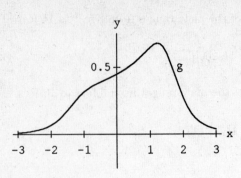

Figure A

Figure B

(b)

p	x_1	x_2	A
1	−0.6152	0.7702	0.4197
2	−0.4462	0.5167	0.2902
$\frac{1}{2}$	−0.8344	1.1623	0.6217

(c) A is continuous on $(0, \infty)$.

(d) We'll approximate $A'(2)$ using the symmetric difference quotient

$$\frac{A(2+h) - A(2-h)}{2h} \quad \text{with} \quad h = 0.1, 0.01, 0.001$$

(See Exercise 4.3 in the Projects and Explorations in Chapter 4.)

$$\frac{A(2.1) - A(1.9)}{0.2} \cong \frac{0.282889 - 0.298034}{0.2} \cong -0.07527$$

$$\frac{A(2.01) - A(1.99)}{0.02} \cong \frac{0.289412 - 0.290923}{0.02} \cong -0.07555$$

$$\frac{A(2.001) - A(1.999)}{0.002} \cong \frac{0.290089 - 0.290242}{0.002} \cong -0.0765$$

7.3. (a) $\frac{1}{3}(750)^2(400) = 75,000,000 = 7.5 \times 10^7$ cu. ft. (See Section 6.2, Example 2.)

(b) Weight $= 175 \times 7.5 \times 10^7 = 1.3125 \times 10^{10}$ lbs. $= 6,562,500$ tons.

(c) The cross-sectional area of the pyramid at height x above the ground is given by

$$A(x) = \left(\frac{750}{400}\right)^2 (400 - x)^2 = \left(\frac{15}{8}\right)^2 (400 - x)^2.$$

Therefore,

$$\text{Work} = \int_0^{400} 175 \left(\frac{15}{8}\right)^2 (400 - x)^2 \, x \, dx$$

$$= 175 \left(\frac{15}{8}\right)^2 \int_0^{400} \left[(400)^2 x - 800 x^2 + x^3\right] \, dx$$

$$= 175 \left(\frac{15}{8}\right)^2 \left[\frac{(400)^2}{2} x^2 - \frac{800}{3} x^3 + \frac{1}{4} x^4\right]_0^{400}$$

$$\cong 1.3125 \times 10^{12} \text{ ft-lbs}$$

(d) The facade of the pyramid is in two sections. The top section is a pyramid with a square base, 20 ft. on a side, and height 32/3 ft. The weight of the top section is:

$$\text{Weight} = 225 \left(\frac{1}{3}\right) (20)^2 \left(\frac{32}{3}\right) \cong 320,000 \text{ lbs or } 160 \text{ tons.}$$

The work required to construct the top section is given by:

$$\text{Work} = \int_0^{32/3} 225 \left[\frac{1168}{3} + x\right] \left(\frac{20}{32/3}\right)^2 \left(\frac{32}{3} - x\right)^2 \, dx$$

$$= \frac{225}{27} \left(\frac{15}{8}\right)^2 \int_0^{32/3} (1168 + 3x)(32 - 3x)^2 \, dx$$

$$= \frac{225}{27} \left(\frac{15}{8}\right)^2 \int_0^{32/3} \left[(1168)(1024) - 216(1024)x + 9936 x^2 + 27 x^3\right] \, dx$$

$$= \frac{225}{27} \left(\frac{15}{8}\right)^2 \left[(1168)(1024)x - 108(1024)x^2 + 3312 x^3 + \frac{27}{4} x^4\right]_0^{32/3}$$

$$\cong 1.33 \times 10^9 \text{ ft-lbs}$$

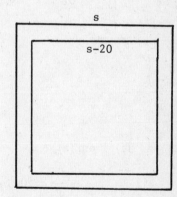

The bottom section is a truncated pyramid with a square base, 750 ft. on a side; a square top, 20 ft. on a side, and height $389\frac{1}{3}$ ft. A cross-section at height x is the region shown in the figure. Its area is

$$A = s^2 - (s - 20)^2 = 40s - 400, \quad \text{where} \quad s = \frac{750}{400}(400 - x) = \frac{15}{8}(400 - x).$$

Therefore,

$$A(x) = 40 \left(\frac{15}{8}\right) (400 - x) - 400 = 29,600 - 75x, \quad 0 \leq x \leq \frac{1168}{3}$$

The volume of stone in the facade is:

$$V = \int_0^{1168/3} (29,600 - 75x)\, dx = \left[29,600x - \frac{75}{2}x^2 \right]_0^{1168/3} \cong 5,840,000 \text{ cu. ft.},$$

and the weight of the facade is:

$$\text{Weight} = 5,840,000 \times 225 \cong 1.314 \times 10^9 \text{ lbs or } 657,000 \text{ tons.}$$

The work required to construct the facade on the bottom section is:

$$\text{Work} = \int_0^{1168/3} 225[29,600 - 75x]x\, dx = 225 \int_0^{1168/3} \left[29,600x - 75x^2 \right]\, dx$$

$$= 225 \left[14,800x^2 - 25x^3 \right]_0^{1168/3}$$

$$\cong 1.73 \times 10^{11} \text{ ft-lbs}$$

The total weight of the facade is: 657,160 tons.

The work required to construct the facade is: $1.33 \times 10^9 + 1.73 \times 10^{11} = 1.7433 \times 10^{11}$ ft-lbs.

7.5.   An equation for the parabola is: $y = \frac{3}{8}x^2.$

(a) The cross-sectional area with depth h is given by:

$$A(h) = 2 \int_0^{\sqrt{8h/3}} \left[h - \frac{3}{8}x^2 \right]\, dx = 2 \left[hx - \frac{1}{8}x^3 \right]_0^{\sqrt{8h/3}} = \frac{8}{9}\sqrt{6}\, h^{3/2}$$

The volume is: $V(h) = \frac{40}{3}\sqrt{6}\, h^{3/2}$

(b)

$$F = \int_0^h 9800\, x\, \frac{4}{3}\sqrt{6}\, \sqrt{h-x}\, dx = \frac{39,200}{3}\sqrt{6} \int_0^h x\sqrt{h-x}\, dx$$

$$= \frac{39,200}{3}\sqrt{6} \int_0^h \left[hu^{1/2} - u^{3/2} \right]\, du \quad (u = h - x)$$

$$= \frac{39,200}{3}\sqrt{6} \left[\frac{2}{3}h\, u^{3/2} - \frac{2}{5}u^{5/2} \right]_0^h$$

$$= \frac{31,360}{9}\sqrt{6}\, h^{5/2}$$

(c) From (a),

$$\frac{dV}{dt} = 20\sqrt{6}\, h^{1/2}\frac{dh}{dt} \quad \text{and since} \quad \frac{dV}{dt} = -\frac{1}{2}, \quad \frac{dh}{dt} = \frac{-1}{40\sqrt{6}\, h^{1/2}}$$

(d) From (b) and (c),

$$\frac{dF}{dt} = \frac{31,360}{9}\sqrt{6}\left(\frac{5}{2}\right)h^{3/2}\frac{dh}{dt} = -\frac{1960}{9}h$$

CHAPTER 8

SECTION 8.1

1. $\displaystyle\int e^{2-x}\,dx = -e^{2-x} + C$ **3.** $\displaystyle\int_0^1 \sin \pi x\,dx = \left[-\frac{1}{\pi}\cos \pi x\right]_0^1 = \frac{2}{\pi}$

5. $\displaystyle\int \sec^2 (1-x)\,dx = -\tan (1-x) + C$

7. $\displaystyle\int_{\pi/6}^{\pi/3} \cot x\,dx = \left[\ln (\sin x)\right]_{\pi/6}^{\pi/3} = \ln \frac{\sqrt{3}}{2} - \ln \frac{1}{2} = \frac{1}{2}\ln 3$

9. $\left\{\begin{array}{l} u = 1-x^2 \\ du = -2x\,dx \end{array}\right\};\quad \displaystyle\int \frac{x\,dx}{\sqrt{1-x^2}} = -\frac{1}{2}\int u^{-1/2}\,du = -u^{1/2} + C = -\sqrt{1-x^2} + C$

11. $\displaystyle\int_{-\pi/4}^{\pi/4} \frac{\sin x}{\cos^2 x}\,dx = \int_{-\pi/4}^{\pi/4} \sec x \tan x\,dx = \left[\sec x\right]_{-\pi/4}^{\pi/4} = 0$

13. $\left\{\begin{array}{l} u = 1/x \mid x = 1 \implies u = 1 \\ du = -\dfrac{dx}{x^2} \mid x = 2 \implies u = 1/2 \end{array}\right\};$

$\displaystyle\int_1^2 \frac{e^{1/x}}{x^2}\,dx = \int_1^{1/2} -e^u\,du = \left[-e^u\right]_1^{1/2} = e - \sqrt{e}$

15. $\displaystyle\int \frac{x\,dx}{x^2+1} = \frac{1}{2}\ln (x^2+1) + C$ **17.** $\displaystyle\int_0^c \frac{dx}{x^2+c^2} = \left[\frac{1}{c}\tan^{-1}\left(\frac{x}{c}\right)\right]_0^c = \frac{\pi}{4c}$

19. $\left\{\begin{array}{l} u \doteq 3\tan \theta + 1 \\ du = 3\sec^2 \theta\,d\theta \end{array}\right\};$

$\displaystyle\int \frac{\sec^2 \theta}{\sqrt{3\tan \theta + 1}}\,d\theta = \frac{1}{3}\int u^{-1/2}\,du = \frac{2}{3}u^{1/2} + C = \frac{2}{3}\sqrt{1+3+\tan \theta} + C$

21. $\displaystyle\int \frac{e^x}{ae^x - b}\,dx = \frac{1}{a}\ln |ae^x - b| + C$

23. $\displaystyle\int \frac{1+\cos 2x}{\sin^2 2x}\,dx = \int (\csc^2 2x + \csc 2x \cot 2x)\,dx = -\frac{1}{2}\cot 2x - \frac{1}{2}\csc 2x + C$

25. $\left\{\begin{array}{l} u = x+1 \\ du = dx \end{array}\right\};$

$\displaystyle\int \frac{x}{(x+1)^2 + 4}\,dx = \int \frac{u-1}{u^2+4}\,du = \int \frac{u}{u^2+4}\,du - \int \frac{du}{u^2+4}$

$\displaystyle\qquad = \frac{1}{2}\ln |u^2+4| - \frac{1}{2}\tan^{-1}\frac{u}{2} + C$

$\displaystyle\qquad = \frac{1}{2}\ln |(x+1)^2 + 4| - \frac{1}{2}\tan^{-1}\left(\frac{x+1}{2}\right) + C$

27. $\left\{ \begin{array}{l} u = x^2 \\ du = 2x\,dx \end{array} \right\};\ \int \dfrac{x}{\sqrt{1-x^4}}\,dx = \dfrac{1}{2}\int \dfrac{du}{\sqrt{1-u^2}} = \dfrac{1}{2}\sin^{-1}u + C$

$$= \tfrac{1}{2}\sin^{-1}(x^2) + C$$

29. $\left\{ \begin{array}{l} u = x + 3 \\ du = dx \end{array} \right\};\ \int \dfrac{dx}{x^2 + 6x + 10} = \int \dfrac{dx}{(x+3)^2 + 1} = \int \dfrac{du}{u^2 + 1}$

$$= \tan^{-1}u + C = \tan^{-1}(x+3) + C$$

31. $\displaystyle\int x\sin x^2\,dx = -\dfrac{1}{2}\cos x^2 + C$

33. $\displaystyle\int \tan^2 x\,dx = \int (\sec^2 x - 1)\,dx = \tan x - x + C$

35. $\left\{ \begin{array}{ll} u = x^2 + 4 \mid x = -1 \implies u = 5 \\ du = 2x\,dx \mid x = 2 \implies u = 8 \end{array} \right\};$

$$\int_{-1}^{2} \dfrac{x}{x^2+4}\,dx = \dfrac{1}{2}\int_{5}^{8}\dfrac{1}{u}\,du = \dfrac{1}{2}\left[\ln |u|\right]_5^8 = \dfrac{1}{2}\ln \dfrac{8}{5}$$

37. $\left\{ \begin{array}{ll} u = \ln x \mid x = 1 \implies u = 0 \\ du = -\dfrac{dx}{x}\mid x = e \implies u = 1 \end{array} \right\};$

$$\int_1^e \dfrac{\ln x^3}{x}\,dx = \int_1^e \dfrac{3\ln x}{x}\,dx = 3\int_0^1 u\,du = 3\left[\dfrac{u^2}{2}\right]_0^1 = \dfrac{3}{2}$$

39. $\left\{ \begin{array}{l} u = \sin^{-1} x \\ du = \dfrac{dx}{\sqrt{1-x^2}} \end{array} \right\};\ \int \dfrac{\sin^{-1}x}{\sqrt{1-x^2}}\,dx = \int u\,du = \dfrac{1}{2}u^2 + C = \dfrac{1}{2}\left(\sin^{-1}x\right)^2 + C$

41. $\left\{ \begin{array}{l} u = e^{2x} \\ du = 2e^{2x}\,dx \end{array} \right\};\ \int \dfrac{e^{2x}}{\sqrt{1-e^{4x}}}\,dx = \dfrac{1}{2}\int \dfrac{1}{\sqrt{1-u^2}}\,du = \dfrac{1}{2}\sin^{-1}u + C = \dfrac{1}{2}\sin^{-1}\left(e^{2x}\right) + C$

43. $\left\{ \begin{array}{l} u = \ln x \\ du = \frac{1}{x}\,dx \end{array} \right\};\ \int \dfrac{1}{x\ln x}\,dx = \int \dfrac{1}{u}\,du = \ln |u| + C = \ln | \ln x| + C$

45. $\left\{ \begin{array}{ll} u = \ln x \mid x = 1 \implies u = 0 \\ du = -\dfrac{dx}{x}\mid x = 4 \implies u = \ln 4 \end{array} \right\};$

$$\int_1^4 \dfrac{\sqrt{\ln x}}{x}\,dx = \int_1^{\ln 4} \sqrt{u}\,du = \left[\dfrac{2}{3}u^{3/2}\right]_0^{\ln 4} = \dfrac{2}{3}(\ln 4)^{3/2}$$

47. $\left\{ \begin{array}{ll} u = \cos x \mid x = 0 \implies u = 1 \\ du = -\sin x\,dx \mid x = \pi/4 \implies u = \sqrt{2}/2 \end{array} \right\};$

$$\int_0^{\pi/4} \frac{1+\sin x}{\cos^2 x}\,dx = \int_0^{\pi/4} \sec^2 x\,dx + \int_0^{\pi/4} \frac{\sin x}{\cos^2 x}\,dx$$

$$= [\tan x]_0^{\pi/4} - \int_1^{\sqrt{2}/2} \frac{du}{u^2}$$

$$= 1 + \left[\frac{1}{u}\right]_1^{\sqrt{2}/2} = \sqrt{2}$$

49.
$$\int_0^{\pi} \sqrt{1+\cos x}\,dx = \int_0^{\pi} \sqrt{2\cos^2\left(\frac{x}{2}\right)}\,dx$$

$$= \sqrt{2}\int_0^{\pi} \cos\left(\frac{x}{2}\right)\,dx \qquad \left[\cos\left(\frac{x}{2}\right) \geq 0 \text{ on } [0,\pi]\right]$$

$$= 2\sqrt{2}\left[\sin\left(\frac{x}{2}\right)\right]_0^{\pi} = 2\sqrt{2}$$

51. (a) $\displaystyle\int_0^{\pi} \sin^2 nx\,dx = \int_0^{\pi}\left[\frac{1}{2} - \frac{\cos 2nx}{2}\right]dx = \left[\frac{x}{2} - \frac{\sin 2nx}{4n}\right]_0^{\pi} = \frac{\pi}{2}$

(b) $\displaystyle\int_0^{\pi} \sin nx \cos nx\,dx = \frac{1}{2}\int_0^{\pi} \sin 2nx\,dx = -\left[\frac{\cos 2nx}{4n}\right]_0^{\pi} = 0$

(c) $\displaystyle\int_0^{\pi/n} \sin nx \cos nx\,dx = \frac{1}{2}\int_0^{\pi/n} \sin 2nx\,dx = -\left[\frac{\cos 2nx}{4n}\right]_0^{\pi/n} = 0$

53. (a) $\displaystyle\int \tan^3 x\,dx = \int \tan^2 x \tan x\,dx = \int (\sec^2 x - 1)\tan x\,dx$

$$= \int \sec^2 x \tan x\,dx - \int \tan x\,dx$$

$$= \int u\,du - \int \tan x\,dx \quad (u = \tan x,\ du = \sec^2 x\,dx)$$

$$= \frac{1}{2}u^2 - \ln|\sec x| + C = \frac{1}{2}\tan^2 x - \ln|\sec x| + C$$

(b) $\displaystyle\int \tan^5 x\,dx = \int \tan^3 x \tan^2 x\,dx = \int \tan^3 x (\sec^2 x - 1)\,dx$

$$= \int \tan^3 x \sec^2 x\,dx - \int \tan^3 x\,dx$$

$$= \int u^3\,du - \int \tan^3 x\,dx \quad (u = \tan x\ du = \sec^2 x\,dx)$$

$$= \frac{1}{4}u^4 - \frac{1}{2}\tan^2 x + \ln|\sec x| + C$$

$$= \frac{1}{4}\tan^4 x - \frac{1}{2}\tan^2 x + \ln|\sec x| + C$$

(c)
$$\int \tan^7 x \, dx = \int \tan^5 x \sec^2 x \, dx - \int \tan^5 x \, dx$$

$$= \frac{1}{6} \tan^6 x - \frac{1}{4} \tan^4 x + \frac{1}{2} \tan^2 x - \ln |\sec x| + C$$

(d)
$$\int \tan^{2k+1} x \, dx = \int \tan^{2k-1} x \tan^2 x \, dx = \int \tan^{2k-1} x \sec^2 x \, dx - \int \tan^{2k-1} x \, dx$$

$$= \frac{1}{2k} \tan^{2k} - \int \tan^{2k-1} x \, dx$$

55. (a)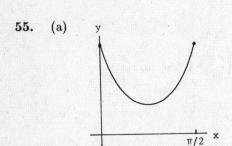

(b) $\sin x + \cos x = \sqrt{2} \left[\sin x \, \cos(\pi/4) + \cos x \, \sin(\pi/4) \right]$

$$= \sqrt{2} \sin\left(x + \frac{\pi}{4}\right);$$

$$A = \sqrt{2}, \quad B = \pi/4$$

(c) Area $= \displaystyle\int_0^{\pi/2} \frac{1}{\sin x + \cos x} \, dx = \frac{1}{\sqrt{2}} \int_0^{\pi/2} \frac{1}{\sin\left(x + \frac{\pi}{4}\right)} \, dx$

$$\left\{ \begin{array}{ll} u = x + \pi/4 | \; x = 0 & \implies \quad u = \pi/4 \\ du = dx | \; x = \pi/2 & \implies \quad u = 3\pi/4 \end{array} \right\};$$

$$\frac{1}{\sqrt{2}} \int_0^{\pi/2} \frac{1}{\sin\left(x + \frac{\pi}{4}\right)} \, dx = \frac{\sqrt{2}}{2} \int_{\pi/4}^{3\pi/4} \frac{1}{\sin u} \, du$$

$$= \frac{\sqrt{2}}{2} \int_{\pi/4}^{3\pi/4} \csc u \, du$$

$$= \frac{\sqrt{2}}{2} \left[\ln |\csc u - \cot u| \right]_{\pi/4}^{3\pi/4}$$

$$= \frac{\sqrt{2}}{2} \ln \left[\frac{\sqrt{2}+1}{\sqrt{2}-1} \right]$$

57. (a)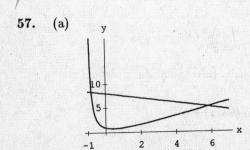

(b) $x_1 \cong -0.80, \quad x_2 \cong 5.80$

(c)
$$\text{Area} \cong \int_{-0.80}^{5.80} \left[8 - \frac{1}{2}x - \frac{x^2+1}{x+1} \right] dx = \int_{-0.80}^{5.80} \left[8 - \frac{1}{2}x - x + 1 - \frac{2}{x+1} \right] dx$$

$$= \int_{-0.80}^{5.80} \left[9 - \frac{3}{2}x - \frac{2}{x+1} \right] dx$$

$$= \left[9x - \frac{3}{4}x^2 - 2\ln|x+1| \right]_{-0.80}^{5.80} \cong 27.6$$

SECTION 8.2

1.
$$\boxed{\begin{array}{ll} u = x, & dv = e^{-x}\,dx \\ du = dx, & v = -e^{-x} \end{array}}$$
$$\int xe^{-x}\,dx = -xe^{-x} - \int -e^{-x}\,dx = -xe^{-x} - e^{-x} + C$$

3.
$$\int x^2 \ln x\,dx = \frac{1}{3}x^3 \ln x - \int \frac{1}{3}x^3 \left(\frac{1}{x}\right) dx$$

$$\boxed{\begin{array}{ll} u = \ln x, & dv = x^2\,dx \\ du = \dfrac{dx}{x}, & v = \dfrac{1}{3}x^3 \end{array}}$$

$$= \frac{1}{3}x^3 \ln x - \frac{1}{3}\int x^2\,dx$$

$$= \tfrac{1}{3}x^3 \ln x - \tfrac{1}{9}x^3 + C$$

$$\int_1^e x^2 \ln x\,dx = \left[\tfrac{1}{3}x^3 \ln x - \tfrac{1}{9}x^3 \right]_1^e = \tfrac{2}{9}e^3 + \tfrac{1}{9}$$

5.
$$\left\{ \begin{array}{l} t = -x^3 \\ dt = -3x^2\,dx \end{array} \right\}; \quad \int x^2 e^{-x^3}\,dx = -\frac{1}{3}\int e^t\,dt = -\frac{1}{3}e^t + C = -\frac{1}{3}e^{-x^3} + C$$

7.
$$\int x^2 e^{-x}\,dx = -x^2 e^{-x} - \int -2xe^{-x}\,dx = -x^2 e^{-x} + 2\int xe^{-x}\,dx$$

$$\boxed{\begin{array}{ll} u = x^2, & dv = e^{-x}\,dx \\ du = 2x\,dx, & v = -e^{-x} \end{array}}$$

$$= -x^2 e^{-x} + 2\left[-xe^{-x} - \int -e^{-x}\,dx \right]$$

$$\boxed{\begin{array}{ll} u = x, & dv = e^{-x}\,dx \\ du = dx, & v = -e^{-x} \end{array}}$$

$$= -x^2 e^{-x} + 2\left(-xe^{-x} - e^{-x} \right) + C$$

$$= -e^{-x}\left(x^2 + 2x + 2 \right) + C$$

$$\int_0^1 x^2 e^{-x}\,dx = \left[-e^{-x}\left(x^2 + 2x + 2 \right) \right]_0^1 = 2 - 5e^{-1}$$

9.
$$\int x^2(1-x)^{-1/2}\,dx = -2x^2(1-x)^{1/2} + 4\int x(1-x)^{1/2}\,dx$$

$$\begin{array}{ll} u = x^2, & dv = (1-x)^{-1/2}\,dx \\ du = 2x\,dx, & v = -2(1-x)^{1/2} \end{array}$$

$$= -2x^2(1-x)^{1/2} + 4\left[-\frac{2x}{3}(1-x)^{3/2} + \int \frac{2}{3}(1-x)^{3/2}\,dx\right]$$

$$\begin{array}{ll} u = x, & dv = (1-x)^{1/2}\,dx \\ du = dx, & v = -\frac{2}{3}(1-x)^{3/2} \end{array}$$

$$= -2x^2(1-x)^{1/2} - \frac{8x}{3}(1-x)^{3/2} - \frac{16}{15}(1-x)^{5/2} + C$$

Or, use the substitution $t = 1-x$ (no integration by parts needed) to obtain:

$$-2(1-x)^{1/2} + \tfrac{4}{3}(1-x)^{3/2} - \tfrac{2}{5}(1-x)^{5/2} + C.$$

11.
$$\int x\ln\sqrt{x}\,dx = \frac{1}{2}\int x\ln x\,dx = \frac{1}{2}\left[\frac{1}{2}x^2\ln x - \frac{1}{2}\int x\,dx\right]$$

$$\begin{array}{ll} u = \ln x, & dv = x\,dx \\ du = \dfrac{dx}{x}, & v = \dfrac{1}{2}x^2 \end{array}$$

$$= \tfrac{1}{4}x^2\ln x - \tfrac{1}{8}x^2 + C$$

$$\int_1^{e^2} x\ln\sqrt{x}\,dx = \left[\tfrac{1}{4}x^2\ln x - \tfrac{1}{8}x^2\right]_1^{e^2} = \tfrac{3}{8}e^4 + \tfrac{1}{8}$$

13.
$$\int \frac{\ln(x+1)}{\sqrt{x+1}}\,dx = 2\sqrt{x+1}\ln(x+1) - \int \frac{2\,dx}{\sqrt{x+1}}$$

$$\begin{array}{ll} u = \ln(x+1), & dv = \dfrac{dx}{\sqrt{x+1}} \\ du = \dfrac{dx}{x+1}, & v = 2\sqrt{x+1} \end{array}$$

$$= 2\sqrt{x+1}\ln(x+1) - 4\sqrt{x+1} + C$$

15.
$$\int (\ln x)^2\,dx = x(\ln x)^2 - 2\int \ln x\,dx$$

$$\begin{array}{ll} u = (\ln x)^2, & dv = dx \\ du = \dfrac{2\ln x}{x}\,dx, & v = x \end{array}$$

$$= x(\ln x)^2 - 2\left[x\ln x - \int dx\right]$$

$$\begin{array}{ll} u = \ln x, & dv = dx \\ du = \dfrac{dx}{x}, & v = x \end{array}$$

$$= x(\ln x)^2 - 2x\ln x + 2x + C$$

17.

$$\int x^3\, 3^x\, dx = \frac{x^3\, 3^x}{\ln 3} - \frac{3}{\ln 3}\int x^2\, 3^x\, dx$$

$$\boxed{\begin{array}{ll} u = x^3, & dv = 3^x\, dx \\ du = 3x^2\, dx, & v = \dfrac{3^x}{\ln 3} \end{array}}$$

$$= \frac{x^3\, 3^x}{\ln 3} - \frac{3}{\ln 3}\left[\frac{x^2\, 3^x}{\ln 3} - \frac{2}{\ln 3}\int x\, 3^x\, dx\right]$$

$$\boxed{\begin{array}{ll} u = x^2, & dv = 3^x\, dx \\ du = 2x\, dx & v = \dfrac{3^x}{\ln 3} \end{array}}$$

$$= \frac{x^3\, 3^x}{\ln 3} - \frac{3x^2\, 3^x}{(\ln 3)^2} + \frac{6}{(\ln 3)^2}\int x\, 3^x\, dx$$

$$= \frac{x^3\, 3^x}{\ln 3} - \frac{3x^2\, 3^x}{(\ln 3)^2} + \frac{6}{(\ln 3)^2}\left[\frac{x3^x}{\ln 3} - \frac{1}{\ln 3}\int 3^x\, dx\right]$$

$$\boxed{\begin{array}{ll} u = x, & dv = 3^x\, dx \\ du = dx, & v = \dfrac{3^x}{\ln 3} \end{array}}$$

$$= 3^x\left[\frac{x^3}{\ln 3} - \frac{3x^2}{(\ln 3)^2} + \frac{6x}{(\ln 3)^3} - \frac{6}{(\ln 3)^4}\right] + C$$

19.

$$\int x\,(x+5)^{14}\, dx = \frac{x}{15}\,(x+5)^{15} - \frac{1}{15}\int (x+5)^{15}\, dx$$

$$\boxed{\begin{array}{ll} u = x, & dv = (x+5)^{14}\, dx \\ du = dx, & v = \tfrac{1}{15}\,(x+5)^{15} \end{array}}$$

$$= \tfrac{1}{15}x\,(x+5)^{15} - \tfrac{1}{240}\,(x+5)^{16} + C$$

Or, use the substitution $t = x + 5$ (integration by parts not needed) to obtain:

$$\tfrac{1}{16}\,(x+5)^{16} - \tfrac{1}{3}\,(x+5)^{15} + C.$$

21.

$$\int x\cos \pi x\, dx = \frac{1}{\pi}x\sin x - \frac{1}{\pi}\int \sin \pi x\, dx$$

$$\boxed{\begin{array}{ll} u = x, & dv = \cos \pi x\, dx \\ du = dx, & v = \dfrac{1}{\pi}\sin \pi x \end{array}}$$

$$= \frac{1}{\pi}x\sin \pi x + \frac{1}{\pi^2}\cos \pi x + C$$

$$\int_0^{1/2} x\,\cos \pi x\, dx = \left[\frac{1}{\pi}x\sin \pi x + \frac{1}{\pi^2}\cos \pi x\right]_0^{1/2} = \frac{1}{2\pi} - \frac{1}{\pi^2}$$

23.
$$\int x^2 (x+1)^9 \, dx = \frac{x^2}{10} (x+1)^{10} - \frac{1}{5} \int x (x+1)^{10} \, dx$$

$$\boxed{\begin{array}{ll} u = x^2, & dv = (x+1)^9 \, dx \\ du = 2x \, dx, & v = \frac{1}{10} (x+1)^{10} \end{array}}$$

$$= \frac{x^2}{10} (x+1)^{10} - \frac{1}{5} \left[\frac{x}{11} (x+1)^{11} - \frac{1}{11} \int (x+1)^{11} \, dx \right]$$

$$\boxed{\begin{array}{ll} u = x, & dv = (x+1)^{10} \, dx \\ du = dx, & v = \frac{1}{11} (x+1)^{11} \end{array}}$$

$$= \frac{x^2}{10} (x+1)^{10} - \frac{x}{55} (x+1)^{11} + \frac{1}{660} (x+1)^{12} + C$$

25.
$$\int e^x \sin x \, dx = -e^x \cos x + \int e^x \cos x \, dx$$

$$\boxed{\begin{array}{ll} u = e^x, & dv = \sin x \, dx \\ du = e^x \, dx, & v = -\cos x \end{array}}$$

$$= -e^x \cos x + e^x \sin x - \int e^x \sin x \, dx$$

$$\boxed{\begin{array}{ll} u = e^x, & dv = \cos x \, dx \\ du = e^x \, dx, & v = \sin x \end{array}}$$

Adding $\int e^x \sin x \, dx$ to both sides, we get

$$2 \int e^x \sin x \, dx = -e^x \cos x + e^x \sin x$$

so that

$$\int e^x \sin x \, dx = \frac{1}{2} e^x (\sin x - \cos x) + C.$$

27
$$\int \ln (1+x^2) \, dx = x \ln (1+x^2) - 2 \int \frac{x^2}{1+x^2} \, dx$$

$$\boxed{\begin{array}{ll} u = \ln (1+x^2), & dv = dx \\ du = \frac{2x}{1+x^2} \, dx, & v = x \end{array}}$$

$$= x \ln (1+x^2) - 2 \int \frac{x^2 + 1 - 1}{1+x^2} \, dx$$

$$= x \ln (1+x^2) - 2 \int \left(1 - \frac{1}{1+x^2} \right) dx$$

$$= x \ln (1+x^2) - 2x + 2 \tan^{-1} x + C$$

$$\int_0^1 \ln(1+x^2) \, dx = \left[x \ln(1+x^2) - 2x + 2 \tan^{-1} x \right]_0^1 = \ln 2 - 2 + \frac{\pi}{2}$$

29.
$$\int x^n \ln x \, dx = \frac{x^{n+1} \ln x}{n+1} - \frac{1}{n+1} \int x^n \, dx$$

$$\boxed{\begin{array}{ll} u = \ln x, & dv = x^n \, dx \\ du = \frac{dx}{x}, & v = \frac{x^{n+1}}{n+1} \end{array}}$$

$$= \frac{x^{n+1} \ln x}{n+1} - \frac{x^{n+1}}{(n+1)^2} + C$$

31.

$$\{t = x^2, \quad dt = 2x\,dx\}; \qquad \int x^3 \sin x^2\,dx = \frac{1}{2}\int t \sin t\,dt$$

$$\boxed{\begin{aligned} u &= t, & dv &= \sin t\,dt \\ du &= dt, & v &= -\cos t \end{aligned}} \quad\longrightarrow\quad = \frac{1}{2}\left[-t\cos t + \int \cos t\,dt\right]$$

$$= \frac{1}{2}(-t\cos t + \sin t) + C$$

$$= -\tfrac{1}{2}x^2\cos x^2 + \tfrac{1}{2}\sin x^2 + C$$

33. To calculate $\int x^4 e^x\,dx$ we could integrate by parts four times, each time selecting $dv = e^x\,dx$. Instead we shall guess the antiderivative. We have seen that

$$\int xe^x\,dx = (x-1)\,e^x + C \quad \text{and} \quad \int x^2 e^x\,dx = (x^2 - 2x + 2)\,e^x + C.$$

Thus we guess that

$$\int x^4 e^x\,dx = P(x)e^x + C$$

with P a polynomial of degree 4:

$$P(x) = Ax^4 + Bx^3 + Dx^2 + Ex + F.$$

From $\dfrac{d}{dx}\left[P(x)e^x\right] = x^4 e^x$,

$$\left(Ax^4 + Bx^3 + Dx^2 + Ex + F\right)e^x + (4Ax^3 + 3Bx^2 + 2Dx + E)e^x = x^4 e^x$$

or

$$Ax^4 + (B + 4A)\,x^3 + (D + 3B)\,x^2 + (E + 2D)\,x + (F + E) + x^4.$$

Comparing coefficients, we get

$$A = 1, \quad B + 4A = 0 \implies B = -4, \quad D + 3B = 0 \implies D = 12,$$
$$E + 2D = 0 \implies E = -24, \quad F + E = 0 \implies F = 24.$$

Thus,

$$\int x^4 e^x\,dx = (x^4 - 4x^3 + 12x^2 - 24x + 24)\,e^x + C.$$

35.
$$\left\{\begin{array}{l|l} u = 2x & x = 0 \implies u = 0 \\ du = 2\,dx & x = 1/4 \implies u = 1/2 \end{array}\right\};$$

$$\int_0^{1/4} \sin^{-1} 2x\,dx = \frac{1}{2}\int_0^{1/2} \sin^{-1} u\,du$$

$$= \frac{1}{2}\left[u\sin^{-1} u + \sqrt{1 - u^2}\right]_0^{1/2} \qquad [\text{by}(8.2.5)]$$

$$= \frac{1}{2}\left[\frac{\pi}{12} + \frac{\sqrt{3}}{2} - 1\right] = \frac{\pi}{24} + \frac{\sqrt{3} - 2}{4}$$

37. Let $u = \sec^{-1} x$, $dv = dx$. Then $du = \dfrac{1}{x\sqrt{x^2 - 1}}\,dx$, $v = x$, and

$$\int \sec^{-1} x \, dx = x \sec^{-1} x - \int x \, \frac{1}{x\sqrt{x^2 - 1}} \, dx$$

$$= x \sec^{-1} x - \int \frac{1}{\sqrt{x^2 - 1}} \, dx$$

One way to evaluate the integral on the right-hand side uses (7.10.4):

$$\frac{d}{dx} \left(\cosh^{-1} x \right) = \frac{1}{\sqrt{x^2 - 1}} \quad (x > 1).$$

Thus,

$$\int \sec^{-1} x \, dx = x \sec^{-1} x - \cosh^{-1} x + C \quad (x \geq 1)$$

Now, by (10.7.2), $\cosh^{-1} x = \ln \left(x + \sqrt{x^2 - 1} \right)$, $x \geq 1$ and so

$$\int \sec^{-1} x \, dx = x \sec^{-1} x - \ln \left(x + \sqrt{x^2 - 1} \right) + C \quad (x \geq 1)$$

and

$$\int \sec^{-1} x \, dx = x \sec^{-1} x - \ln \left| x + \sqrt{x^2 - 1} \right| + C \quad (|x| \geq 1)$$

Another way to evaluate $\int \dfrac{1}{\sqrt{x^2 - 1}} \, dx$ uses the substitutions:

$$x = \sec u, \quad dx = \sec u \tan u \, du, \quad \sqrt{x^2 - 1} = \tan u$$

Then

$$\int \frac{1}{\sqrt{x^2 - 1}} \, dx = \int \frac{\sec u \tan u}{\tan u} \, du = \int \sec u \, du$$

$$= \ln |\sec u + \tan u| + C = \ln \left| x + \sqrt{x^2 - 1} \right| + C$$

Thus,

$$\int \sec^{-1} \, dx = x \sec^{-1} x - \ln \left| x + \sqrt{x^2 - 1} \right| + C$$

39.

$$\left\{ \begin{array}{l} u = x^2 \\ du = 2x \, dx \end{array} \; \middle| \; \begin{array}{lcl} x = 0 & \Longrightarrow & u = 0 \\ x = 1 & \Longrightarrow & u = 1 \end{array} \right\};$$

$$\int_0^1 x \tan^{-1} x \, dx = \frac{1}{2} \int_0^1 \tan^{-1} u \, du$$

$$= \frac{1}{2} \left[u \tan^{-1} u - \frac{1}{2} (1 + x^2) \right]_0^1 \quad \text{[by 8.2.6]}$$

$$= \frac{\pi}{8} - \frac{1}{2} \ln 2$$

41.

$$\int x^2 \cosh 2x\,dx = \tfrac{1}{2}x^2 \sinh 2x - \int x \sinh 2x\,dx$$

$u = x^2,$	$dv = \cosh 2x\,dx$
$du = 2x\,dx,$	$v = \tfrac{1}{2}\sinh 2x$

$$= \tfrac{1}{2}x^2 \sinh 2x - \tfrac{1}{2}x \cosh 2x + \tfrac{1}{2}\int \cosh 2x\,dx$$

$u = x,$	$dv = \sinh 2x\,dx$
$du = dx,$	$v = \tfrac{1}{2}\cosh 2x$

$$= \tfrac{1}{2}x^2 \sinh 2x - \tfrac{1}{2}x \cosh 2x + \tfrac{1}{4}\sinh 2x + C$$

43.

$$\int \sin(\ln x)\,dx = x \sin(\ln x) - \int \cos(\ln x)\,dx$$

$u = \sin(\ln x),$	$dv = dx$
$du = \cos(\ln x)\dfrac{1}{x}\,dx,$	$v = x$

$$= x \sin(\ln x) - x \cos(\ln x) - \int \sin(\ln x)\,dx$$

$u = \cos(\ln x),$	$dv = dx$
$du = -\sin(\ln x)\dfrac{1}{x}\,dx,$	$v = x$

Adding $\displaystyle\int \sin(\ln x)\,dx$ to both sides, we get

$$2\int \sin(\ln x)\,dx = x \sin(\ln x) - x \cos(\ln x)$$

so that

$$\int \sin(\ln x)\,dx = \tfrac{1}{2}\left[x \sin(\ln x) - x \cos(\ln x)\right] + C$$

45. Let $u = \ln x,\ \ du = \tfrac{1}{x}\,dx.$ Then

$$\int \frac{1}{x}\sin^{-1}(\ln x)\,dx = \int \sin^{-1} u\,du = u \sin^{-1} u + \sqrt{(1 - u^2)} + C$$

$$= (\ln x)\sin^{-1}(\ln x) + \sqrt{1 - (\ln x)^2} + C$$

47. $A = \displaystyle\int_0^{1/2} \sin^{-1}x\,dx = \left[x \sin^{-1} x + \sqrt{1 - x^2}\right]_0^{1/2} = \dfrac{\pi}{12} + \dfrac{\sqrt{3} - 2}{2}$

49. (a) $A = \displaystyle\int_1^e \ln x\,dx = [x \ln x - x]_1^e = 1$

(b) $\bar{x}A = \displaystyle\int_1^e x \ln x\,dx = \left[\tfrac{1}{2}x^2 \ln x - \tfrac{1}{4}x^2\right]_1^e = \tfrac{1}{4}\left(e^2 + 1\right),\quad \bar{x} = \tfrac{1}{4}\left(e^2 + 1\right)$

$\bar{y}A = \displaystyle\int_1^e \tfrac{1}{2}\left(\ln x\right)^2\,dx = \tfrac{1}{2}\left[x\left(\ln x\right)^2 - 2x \ln x + 2x\right]_1^e = \tfrac{1}{2}e - 1,\quad \bar{y} = \tfrac{1}{2}e - 1$

(c) $V_x = 2\pi \bar{y}A = \pi\left(e - 2\right),\quad V_y = 2\pi \bar{x}A = \tfrac{1}{2}\pi\left(e^2 + 1\right)$

51. $\bar{x} = \dfrac{1}{e - 1},\qquad \bar{y} = \dfrac{1}{4}\left(e + 1\right)$ **53.** $\bar{x} = \tfrac{1}{2}\pi,\qquad \bar{y} = \tfrac{1}{8}\pi$

55. (a) $M = \int_0^1 e^{kx}\, dx = \dfrac{1}{k}\left(e^k - 1\right)$

(b) $x_M M = \int_0^1 x e^{kx}\, dx = \dfrac{(k-1)\,e^k + 1}{k^2}, \qquad x_M = \dfrac{(k-1)\,e^k + 1}{k\,(e^k - 1)}$

57. $V_y = \int_0^1 2\pi x e^{\alpha x}\, dx = 2\pi\left[\dfrac{1}{\alpha^2}\left(\alpha x e^{\alpha x} - e^{\alpha x}\right)\right]_0^1 = \dfrac{2\pi}{\alpha^2}\left(\alpha e^\alpha - e^\alpha + 1\right)$

59. $V_y = \int_0^1 2\pi x \cos\tfrac{1}{2}\pi x\, dx = \left[4x \sin\tfrac{1}{2}\pi x + \dfrac{8}{\pi}\cos\tfrac{1}{2}\pi x\right]_0^1 = 4 - \dfrac{8}{\pi}$

61. $V_y = \int_0^1 2\pi x^2 e^x\, dx = 2\pi\,(e - 2)$ (see Example 6)

63.
$$V_x = \int_0^1 \pi e^{2x}\, dx = \pi\left[\tfrac{1}{2}e^{2x}\right]_0^1 = \tfrac{1}{2}\pi\left(e^2 - 1\right)$$

$$\bar{x} V_x = \int_0^1 \pi x e^{2x}\, dx = \pi\left[\tfrac{1}{2}x e^{2x} - \tfrac{1}{4}e^{2x}\right]_0^1 = \tfrac{1}{4}\pi\left(e^2 + 1\right)$$

$$\bar{x} = \dfrac{e^2 + 1}{2\left(e^2 - 1\right)}$$

65.
$$A = \int_0^1 \cosh x\, dx = \left[\sinh x\right]_0^1 = \sinh 1 = \dfrac{e - e^{-1}}{2} = \dfrac{e^2 - 1}{2e}$$

$$\bar{x} A = \int_0^1 x \cosh x\, dx = \left[x \sinh x - \cosh x\right]_0^1 = \sinh 1 - \cosh 1 + 1 = \dfrac{2\,(e - 1)}{2e}$$

$$\bar{y} A = \int_0^1 \tfrac{1}{2}\cosh^2 x\, dx = \tfrac{1}{4}\left[\sinh x \cosh x + x\right]_0^1 = \tfrac{1}{4}\left(\sinh 1 \cosh 1 + 1\right) = \dfrac{e^4 + 4e^2 - 1}{16e^2}$$

Therefore $\bar{x} = \dfrac{2}{e+1}$ and $\bar{y} = \dfrac{e^4 + 4e^2 - 1}{8e\,(e^2 - 1)}$.

67.
$$\boxed{\begin{array}{ll} u = x^2, & dv = e^{-x}\, dx \\[4pt] du = 2x\, dx, & v = -e^{-x} \end{array}}\ ; \qquad \int x^n e^{ax}\, dx = \dfrac{x^n e^{ax}}{a} - \dfrac{n}{a}\int x^{n-1} e^{ax}\, dx$$

69.
$$\int x^3 e^{2x}\, dx = \tfrac{1}{2}x^3 e^{2x} - \tfrac{3}{2}\int x^2 e^{2x}\, dx = \tfrac{1}{2}x^3 e^{2x} - \tfrac{3}{2}\left[\tfrac{1}{2}x^2 e^{2x} - \int x e^{2x}\, dx\right]$$

$$= \tfrac{1}{2}x^3 e^{2x} - \tfrac{3}{4}x^2 e^{2x} - \tfrac{3}{4}x e^{2x} - \tfrac{3}{4}\int e^{2x}\, dx$$

$$= \tfrac{1}{2}x^3 e^{2x} - \tfrac{3}{4}x^2 e^{2x} - \tfrac{3}{4}x e^{2x} - \tfrac{3}{8}e^{2x} + C$$

71.
$$\int (\ln x)^3 \, dx = x(\ln x)^3 - 3 \int (\ln x)^2 \, dx = x(\ln x)^3 - 3x(\ln x)^2 + 6 \int \ln x \, dx$$
$$= x(\ln x)^3 - 3x(\ln x)^2 + 6x \ln x - 6 \int dx$$
$$= x(\ln x)^3 - 3x(\ln x)^2 + 6x \ln x - 6x + C$$

73. Let $u = f(x)$, $dv = g''(x) \, dx$. Then $du = f'(x) \, dx$, $v = g'(x)$, and

$$\int_a^b f(x)g''(x) \, dx = [f(x)g'(x)]_a^b - \int_a^b f'(x)g'(x) \, dx = - \int_a^b f'(x)g'(x) \, dx \quad \text{since } f(a) = f(b) = 0$$

Now let $u = f'(x)$, $dv = g'(x) \, dx$. Then $du = f''(x) \, dx$, $v = g(x)$, and

$$- \int_a^b f'(x)g'(x) \, dx = [-f'(x)g(x)]_a^b + \int f''(x)g(x) \, dx = \int g(x)f''(x) \, dx \quad \text{since } g(a) = g(b) = 0$$

Therefore, if f and g have continuous second derivatives, and if $f(a) = g(a) = f(b) = g(b) = 0$, then

$$\int_a^b f(x)g''(x) \, dx = \int_a^b g(x)f''(x) \, dx$$

75. Let P be a regular partition of the interval $[0, n]$. The present value of R dollars continuously compounded at the rate r on the interval $[t_{i-1}, t_i]$ is approximately $Re^{-rt} \Delta t$. Therefore, it follows that the present value of the revenue stream over the time interval $[0, n]$ is given by the definite integral

$$P.V. = \int_0^n Re^{-rt} \, dt$$

77. (a) $\displaystyle P.V. = \int_0^2 (1000 + 60t) e^{-0.05t} \, dt$

$$= 1000 \int_0^2 e^{-t/20} \, dt + 60 \int_0^2 te^{-t/20} \, dt$$

$$= 1000 \left[-20e^{-t/20} \right]_0^2 + 60 \left[-20te^{-t/20} - 400e^{-t/20} \right]_0^2$$
$$\underbrace{\qquad}_{\text{(by parts)}}$$

$$= \left[-(44000 + 1200t) e^{-t/20} \right]_0^2 = 44000 - 46400 \, e^{-0.10} \cong \$2016$$

(b) $\displaystyle P.V. = \int_0^2 (1000 + 60t) e^{-0.1t} \, dt$

$$= 1000 \left[-10e^{-t/10} \right]_0^2 + 60 \left[-10te^{-t/10} - 100e^{-t/10} \right]_0^2$$
$$\underbrace{\qquad}_{\text{(by parts)}}$$

$$= \left[-(16000 + 600t) e^{-t/10} \right]_0^2 = 16000 - 17200e^{-0.20} \cong \$1918$$

SECTION 8.3

1. $\int \sin^3 x \, dx = \int (1 - \cos^2 x) \sin x \, dx = \frac{1}{3} \cos^3 x - \cos x + C$

3. $\int_0^{\pi/6} \sin^2 3x \, dx = \int_0^{\pi/6} \frac{1 - \cos 6x}{2} \, dx = \left[\frac{1}{2} x - \frac{1}{12} \sin 6x \right]_0^{\pi/6} = \frac{\pi}{12}$

5.
$$\int \cos^4 x \sin^3 x \, dx = \int \cos^4 x \, (1 - \cos^2 x) \sin x \, dx$$
$$= \int (\cos^4 x - \cos^6 x) \sin x \, dx$$
$$= -\tfrac{1}{5} \cos^5 x + \tfrac{1}{7} \cos^7 x + C$$

7.
$$\int \sin^3 x \cos^3 x \, dx = \int \sin^3 x \, (1 - \sin^2 x) \cos x \, dx$$
$$= \int (\sin^3 x - \sin^5 x) \cos x \, dx$$
$$= \tfrac{1}{4} \sin^4 x - \tfrac{1}{6} \sin^6 x + C$$

9
$$\int \sin^2 x \cos^3 x \, dx = \int \sin^2 x \, (1 - \sin^2 x) \cos x \, dx$$
$$= \int (\sin^2 x - \sin^4 x) \cos x \, dx$$
$$= \tfrac{1}{3} \sin^3 x - \tfrac{1}{5} \sin^5 x + C$$

11.
$$\int \sin^4 x \, dx = \int \left(\frac{1 - \cos 2x}{2} \right)^2 \, dx$$
$$= \frac{1}{4} \int \left(1 - 2 \cos 2x + \cos^2 2x \right) \, dx$$
$$= \frac{1}{4} \int \left(1 - 2 \cos 2x + \frac{1 + \cos 4x}{2} \right) \, dx$$
$$= \int \left(\frac{3}{8} - \frac{1}{2} \cos 2x + \frac{1}{8} \cos 4x \right) \, dx$$
$$= \tfrac{3}{8} x - \tfrac{1}{4} \sin 2x + \tfrac{1}{32} \sin 4x + C$$

$$\int_0^{\pi} \sin^4 x \, dx = \left[\tfrac{3}{8} x - \tfrac{1}{4} \sin 2x + \tfrac{1}{32} \sin 4x \right]_0^{\pi} = \tfrac{3}{8} \pi$$

13.

$$\int \sin 2x \cos 3x \, dx = \int \frac{1}{2} \left[\sin(-x) + \sin 5x \right] dx$$

$$= \int \frac{1}{2} (-\sin x + \sin 5x) \, dx$$

$$= \tfrac{1}{2} \cos x - \tfrac{1}{10} \cos 5x + C$$

15.

$$\int \sin^2 x \sin 2x \, dx = \int \sin^2 x \, (2 \sin x \cos x) \, dx$$

$$= 2 \int \sin^3 x \cos x \, dx$$

$$= \tfrac{1}{2} \sin^4 x + C$$

17.

$$\int \sin^4 x \cos^4 x \, dx = \int \left(\frac{\sin 2x}{2} \right)^4 dx = \frac{1}{16} \int \left(\frac{1 - \cos 4x}{2} \right)^2 dx$$

$$= \frac{1}{64} \int (1 - 2 \cos 4x + \cos^2 4x) \, dx$$

$$= \frac{1}{64} \int \left(1 - 2 \cos 4x + \frac{1 + \cos 8x}{2} \right) dx$$

$$= \frac{1}{64} \left(\frac{3x}{2} - \frac{\sin 4x}{2} + \frac{\sin 8x}{16} \right) + C$$

19.

$$\int \sin^6 x \, dx = \int \left(\frac{1 - \cos 2x}{2} \right)^3 dx$$

$$= \frac{1}{8} \int (1 - 3 \cos 2x + 3 \cos^2 2x - \cos^3 2x) \, dx$$

$$= \frac{1}{8} \int \left[1 - 3 \cos 2x + 3 \left(\frac{1 + \cos 4x}{2} \right) - \cos 2x \, (1 - \sin^2 2x) \right] dx$$

$$= \frac{1}{8} \int \left(\frac{5}{2} - 4 \cos 2x + \frac{3}{2} \cos 4x + \sin^2 2x \cos 2x \right) dx$$

$$= \tfrac{5}{16} x - \tfrac{1}{4} \sin 2x + \tfrac{3}{64} \sin 4x + \tfrac{1}{48} \sin^3 2x + C$$

21.

$$\int \cos^7 x \, dx = \int (1 - \sin^2 x)^3 \cos x \, dx$$

$$= \int (1 - 3 \sin^2 x + 3 \sin^4 x - \sin^6 x) \cos x \, dx$$

$$= \int (\cos x - 3 \sin^2 x \cos x + 3 \sin^4 x \cos x - \sin^6 x \cos x) \, dx$$

$$= \sin x - \sin^3 x + \tfrac{3}{5} \sin^5 x - \tfrac{1}{7} \sin^7 x + C$$

23.
$$\int_0^{\pi/2} \cos 3x \cos 2x \, dx = \int_0^{\pi/2} \frac{1}{2}(\cos x + \cos 5x) \, dx$$
$$= \left[\tfrac{1}{2}\sin x + \tfrac{1}{10}\sin 5x \right]_0^{\pi/2} = \tfrac{3}{5}$$

25.
$$\int \sin 5x \sin 2x \, dx = \int \frac{1}{2}(\cos 3x - \cos 7x) \, dx$$
$$= \tfrac{1}{6}\sin 3x - \tfrac{1}{14}\sin 7x + C$$

27.
$$\int_{-1/6}^{1/3} \sin^4 3\pi x \cos^3 3\pi x \, dx = \int_{-1/6}^{1/3} \sin^4 3\pi x \cos^2 3\pi x \cos 3\pi x \, dx$$
$$= int_{-1/6}^{1/3} \sin^4 3\pi x \, (1 - \sin^2 3\pi x \cos 3\pi x \, dx$$
$$= \int_{-1}^{0} u^4 (1 - u^2) \frac{1}{3\pi} \, du \quad [u = \sin 3\pi x, \ \ du = 3\pi \cos 3\pi x \, dx]$$
$$= \frac{1}{3\pi} \left[\tfrac{1}{5}u^5 - \tfrac{1}{7}u^7 \right]_{-1}^{0} = \frac{2}{105\pi}$$

29.
$$\int_0^{\pi/4} \cos 4x \sin 2x \, dx = \int_0^{\pi/4} \frac{1}{2}(\sin 6x - \sin 2x) \, dx$$
$$= \left[-\tfrac{1}{12}\cos 6x + \tfrac{1}{4}\cos 2x \right]_0^{\pi/4} = -\tfrac{1}{6}$$

31.
$$\int \sin(x/2) \cos 2x \, dx = \int \tfrac{1}{2}\left(\sin\left(\tfrac{5}{2}x\right) - \sin\left(\tfrac{3}{2}x\right)\right) dx$$
$$= \tfrac{1}{3}\cos\left(\tfrac{3}{2}x\right) - \tfrac{1}{5}\cos\left(\tfrac{5}{2}x\right) + C$$

33.
$$\left\{ \begin{array}{l} x = \tan u \\ dx = \sec^2 u \, du \end{array} \right\}; \qquad \int \frac{dx}{(x^2+1)^3} = \int \frac{\sec^2 u \, du}{(1 + \tan^2 u)^3} = \int \cos^4 u \, du$$
$$= \int \left(\frac{1 + \cos 2u}{2} \right)^2 du$$
$$= \frac{1}{4} \int (1 + 2\cos 2u + \cos^2 2u) \, du$$
$$= \frac{1}{4} \int \left(1 + 2\cos 2u + \frac{1 + \cos 4u}{2} \right) du$$
$$= \frac{1}{4} \left(u + \sin 2u + \frac{u}{2} + \frac{\sin 4u}{8} \right) + C$$
$$= \frac{3}{8}u + \frac{\sin u \cos u}{2} + \frac{\sin 2u \cos 2u}{16} + C$$

$$= \frac{3}{8}\tan^{-1}x + \frac{x}{2(x^2+1)} + \frac{x(1-x^2)}{8(x^2+1)^2} + C$$

$(\sin 2u = 2\sin u\cos u, \quad \cos 2u = \cos^2 u - \sin^2 u, \quad \text{and} \quad \sin u = x/\sqrt{1+x^2}, \quad \cos u = 1/\sqrt{1+x^2}\,)$

35. $\begin{Bmatrix} x+1 = \tan u \\ dx = \sec^2 u\, du \end{Bmatrix}$; $\quad \displaystyle\int \frac{dx}{[(x+1)^2+1]^2} = \int \frac{\sec^2 u\, du}{(1+\tan^2 u)^2}$

$$= \int \cos^2 u\, du$$

$$= \int \frac{1+\cos 2u}{2}\, du$$

$$= \frac{u}{2} + \frac{\sin 2u}{4} + C$$

$$= \tfrac{1}{2}(u + \sin u\cos u) + C$$

$$= \frac{1}{2}\left[\tan^{-1}(x+1) + \frac{x+1}{x^2+2x+2}\right] + C$$

37. $A = \displaystyle\int_0^\pi \sin^2 x\, dx = \int_0^\pi \tfrac{1}{2}(1-\cos 2x)\, dx = \tfrac{1}{2}\left[x - \tfrac{1}{2}\sin 2x\right]_0^\pi = \frac{\pi}{2}$

39. $V = \displaystyle\int_0^\pi \pi\left(\sin^2 x\right)^2 dx = \pi\int_0^\pi \sin^4 x\, dx = \pi\int_0^\pi \left(\tfrac{1}{2}(1-\cos 2x)\right)^2 dx$

$$= \frac{\pi}{4}\int_0^\pi \left(1 - 2\cos 2x + \cos^2 2x\right) dx$$

$$= \frac{\pi}{4}\left[x - \sin 2x\right]_0^\pi + \frac{\pi}{8}\int_0^\pi (1 + \cos 4x)\, dx$$

$$= \frac{\pi^2}{4} + \frac{\pi}{8}\left[x + \tfrac{1}{4}\sin 4x\right]_0^\pi = \frac{3\pi^2}{8}$$

41. Suppose $m \neq n$:

$$\int \sin mx \sin nx\, dx = \int \tfrac{1}{2}[\cos(m-n)x - \cos(m+n)x]\, dx$$

$$= \frac{\sin(m-n)x}{2(m-n)} - \frac{\sin(m+n)x}{2(m+n)} + C$$

Now suppose that $m = n$:

$$\int \sin mx \sin nx\, dx = \int \sin^2 mx\, dx = \int \tfrac{1}{2}(1-\cos 2mx)\, dx$$

$$= \frac{1}{2}\left(x - \frac{1}{2m}\sin 2mx\right) + C = \frac{x}{2} - \frac{\sin 2mx}{4m} + C$$

43.   Suppose $m \neq n$:

$$\int_{-\pi}^{\pi} \sin mx \sin nx \, dx = \left[\frac{\sin(m-n)x}{2(m-n)} - \frac{\sin(m+n)x}{2(m+n)} \right]_{-\pi}^{\pi} = 0$$

Suppose $m = n$:

$$\int_{-\pi}^{\pi} \sin^2 mx \, dx = \left[\frac{x}{2} - \frac{\sin 2mx}{4m} \right]_{-\pi}^{\pi} = \pi$$

45.   Let $u = \cos^{n-1} x$, $dv = \cos x \, dx$. Then $du = (n-1)\cos^{n-2} x(-\sin x) \, dx$, $v = \sin x$ and

$$\int \cos^n x \, dx = \cos^{n-1} x \sin x + (n-1) \int \cos^{n-2} x \sin^2 x \, dx$$

$$= \cos^{n-1} x \sin x + (n-1) \int \cos^{n-2} x (1 - \cos^2 x) \, dx$$

$$= \cos^{n-1} x \sin x + (n-1) \int \cos^{n-2} x \, dx - (n-1) \int \cos^n x \, dx$$

Adding $(n-1) \int \cos^n x \, dx$ to both sides, we get

$$n \int \cos^n x \, dx = \cos^{n-1} x \sin x + (n-1) \int \cos^{n-2} x \, dx$$

so that

$$\int \cos^n x \, dx = \frac{1}{n} \cos^{n-1} x \sin x + \frac{(n-1)}{n} \int \cos^{n-2} x \, dx$$

47. $\displaystyle \int_0^{\pi/2} \sin^7 x \, dx = \frac{6 \cdot 4 \cdot 2}{7 \cdot 5 \cdot 3} = \frac{16}{35}$

SECTION 8.4

1. $\displaystyle \int \tan^2 3x \, dx = \int (\sec^2 3x - 1) \, dx = \frac{1}{3} \tan 3x - x + C$

3. $\displaystyle \int \sec^2 \pi x \, dx = \frac{1}{\pi} \tan \pi x + C$

5. $\displaystyle \int \tan^3 x \, dx = \int (\sec^2 x - 1) \tan x \, dx$

$$= \int \tan x \sec^2 x \, dx - \int \tan x \, dx$$

$$= \frac{1}{2} \tan^2 x + \ln |\cos x| + C$$

7. $\displaystyle \int \tan^2 x \sec^2 x \, dx = \frac{1}{3} \tan^3 x + C$

9.
$$\int \csc^3 x \, dx = -\csc x \cot x - \int \csc x \cot^2 x \, dx$$

$$\boxed{\begin{aligned} u &= \csc x, & dv &= \csc^2 x \, dx \\ du &= -\csc x \cot x \, dx, & v &= -\cot x \end{aligned}}$$

$$= -\csc x \cot x - \int \csc x \left(\csc^2 x - 1\right) dx$$

Thus

$$2\int \csc^3 x \, dx = -\csc x \cot x + \int \csc x \, dx$$

so that

$$\int \csc^3 x \, dx = -\frac{1}{2}\csc x \cot x + \frac{1}{2}\ln|\csc x - \cot x| + C.$$

11.
$$\int \cot^4 x \, dx = \int \cot^2 x \left(\csc^2 x - 1\right) dx$$

$$= \int \left(\cot^2 x \csc^2 x - \cot^2 x\right) dx$$

$$= \int \left(\cot^2 x \csc^2 x - \csc^2 x + 1\right) dx$$

$$= -\tfrac{1}{3}\cot^3 x + \cot x + x + C$$

13.
$$\int \cot^3 x \csc^3 x \, dx = \int \left(\csc^2 x - 1\right) \csc^3 x \cot x \, dx$$

$$= \int \left(\csc^4 x - \csc^2 x\right) \csc x \cot x \, dx$$

$$= -\tfrac{1}{5}\csc^5 x + \tfrac{1}{3}\csc^3 x + C$$

15.
$$\int \csc^4 2x \, dx = \int \left(1 + \cot^2 2x\right) \csc^2 2x \, dx = -\frac{1}{2}\cot 2x - \frac{1}{6}\cot^3 2x + C$$

17.
$$\int \cot^2 x \csc x \, dx = -\cot x \csc x - \int \csc^3 x \, dx$$

$$\boxed{\begin{aligned} u &= \cot x, & dv &= \cot x \csc x \, dx \\ du &= -\csc x^2 \, dx, & v &= -\csc x \end{aligned}}$$

$$= -\cot x \csc x - \int \csc x \left(1 + \cot^2 x\right) dx.$$

Thus,

$$2\int \cot^2 x \csc x \, dx = -\cot x \csc x - \int \csc x \, dx$$

so that

$$\int \cot^2 x \csc x \, dx = -\frac{1}{2}\cot x \csc x - \frac{1}{2}\ln|\csc x - \cot x| + C.$$

19.
$$\int \tan^5 3x \, dx = \int \tan^3 3x \, (\sec^2 3x - 1) \, dx$$

$$= \int \tan^3 3x \sec^2 3x \, dx - \int \tan^3 3x \, dx$$

$$= \int \tan^3 3x \sec^2 3x \, dx - \int (\tan 3x \sec^2 3x - \tan 3x) \, dx$$

$$= \tfrac{1}{12} \tan^4 3x - \tfrac{1}{6} \tan^2 3x + \tfrac{1}{3} \ln|\sec 3x| + C$$

21.
$$\int \sec^5 x \, dx = \tan x \sec^3 x - 3 \int \sec^3 x \tan^2 x \, dx$$

$$\boxed{\begin{aligned} u &= \sec^3 x, & dv &= \sec^2 x \, dx \\ du &= 3\sec^3 x \tan x \, dx, & v &= \tan x \end{aligned}}$$

$$= \tan x \sec^3 x - 3 \int \sec^5 x \, dx + 3 \int \sec^3 x \, dx.$$

Rearranging terms, we get
$$4 \int \sec^5 x \, dx = \tan x \sec^3 x + 3 \int \sec^3 x \, dx.$$

We have already seen that
$$\int \sec^3 x \, dx = \frac{1}{2} \sec x \tan x + \frac{1}{2} \ln|\sec x + \tan x| + C.$$

Therefore
$$\int \sec^5 x \, dx = \frac{1}{4} \tan x \sec^3 x + \frac{3}{8} \sec x \tan x + \frac{3}{8} \ln|\sec x + \tan x| + C.$$

23.
$$\int \tan^4 x \sec^4 x \, dx = \int \tan^4 x \, (\tan^2 x + 1) \sec^2 x \, dx$$

$$= \int (\tan^6 x + \tan^4 x) \sec^2 x \, dx$$

$$= \tfrac{1}{7} \tan^7 x + \tfrac{1}{5} \tan^5 x + C$$

25.
$$\int e^{2x} \tan^2 \left(e^{2x}\right) \sec^2 \left(e^{2x}\right) \, dx = \frac{1}{2} \int \tan^2 u \sec^2 u \, du \quad [u = e^{2x}, \ du = 2e^{2x} \, dx]$$

$$= \frac{1}{2} \int v^2 \, dv \quad [v = \tan u, \ dv = \sec^2 u \, du]$$

$$= \tfrac{1}{6} v^3 + C = \tfrac{1}{6} \left[\tan \left(e^{2x}\right)\right]^3 + C$$

27.
$$\left\{ \begin{aligned} u &= \tan x \, | \, x = 0 & \Longrightarrow & \quad u = 0 \\ du &= \sec^2 x \, dx \, | \, x = \pi/4 & \Longrightarrow & \quad u = 1 \end{aligned} \right\};$$

$$\int_0^{\pi/4} \tan^3 x \sec^2 x \, dx = \int_0^1 u^3 \, du = \left[\tfrac{1}{4} u^4\right]_0^1 = \tfrac{1}{4}$$

29.
$$\int_0^{\pi/6} \tan^2 2x \, dx = \int_0^{\pi/6} (\sec^2 2x - 1) \, dx = \left[\tfrac{1}{2} \tan 2x - x\right]_0^{\pi/6} = \frac{\sqrt{3}}{2} - \frac{\pi}{6}$$

31. $\left\{\begin{array}{l} u = \csc x \mid x = \pi/6 \implies u = 2 \\ du = -\csc x \cot x\, dx \mid x = \pi/3 \implies u = 2/\sqrt{3} \end{array}\right\};$

$$\int_{\pi/6}^{\pi/3} \cot^3 x \csc^3 x\, dx = \int_{\pi/6}^{\pi/3} \left(\csc^2 x - 1\right) \csc^2 x \csc x \cot x\, dx$$

$$= -\int_{2}^{2/\sqrt{3}} \left(u^2 - 1\right) u^2\, du$$

$$= -\int_{2}^{2/\sqrt{3}} \left(u^4 - u^2\right)\, du$$

$$= -\left[\tfrac{1}{5} u^5 + \tfrac{1}{3} u^3\right]_{2}^{2/\sqrt{3}} = \frac{8\sqrt{3} + 504}{135}$$

33. $V = \displaystyle\int_0^{\pi/4} \pi \left[1^2 - \tan^2 x\right] dx = \pi \int_0^{\pi/4} \left[2 - \sec^2 x\right] dx = \pi \left[2x - \tan x\right]_0^{\pi/4} = \frac{\pi^2}{2} - \pi$

35. $V = \displaystyle\int_0^{\pi/4} \pi \left[(\tan x + 1)^2 - 1^2\right] dx = \pi \int_0^{\pi/4} \left[\tan^2 x + 2\tan x\right] dx$

$$= \pi \int_0^{\pi/4} \left(\sec^2 x + 2\tan x - 1\right) dx$$

$$= \pi \left[\tan x + 2\ln|\sec x| - x\right]_0^{\pi/4} = \pi \left[\ln 2 + 1 - \frac{\pi}{4}\right]$$

37.
$$\int \cot^n x\, dx = \int \cot^{n-2} x \cot^2 x\, dx$$

$$= \int \cot^{n-2} x \left(\csc^2 x - 1\right) dx$$

$$= -\frac{\cot^{n-1} x}{n-1} - \int \cot^{n-2} x\, dx$$

39. $\displaystyle\int \csc^n x\, dx = \int \csc^{n-2} x \csc^2 x\, dx$

Let $u = \csc^{n-2} x$, $dv = \csc^2 x\, dx$. Then $du = -(n-2)\csc^{n-2} x \cot x\, dx$, $v = -\cot x$ and

$$\int \csc^n x\, dx = -(n-2)\csc^{n-2} x \cot x - (n-2)\int \csc^{n-2} x \cot^2 x\, dx$$

$$= -\csc^{n-2} x \cot x - (n-2)\int \csc^{n-2} x \left(\csc^2 x - 1\right) dx$$

$$= -\csc^{n-2} x \cot x - (n-2)\int \left(\csc^n x - \csc^{n-2} x\right) dx$$

Adding $(n-2)\displaystyle\int \csc^n x\, dx$ to both sides, we get

$$(n-1)\int \csc^n x\, dx = -\csc^{n-2} x \cot x + (n-2)\int \csc^{n-2} x\, dx.$$

Thus,

$$\int \csc^n x\, dx = \frac{-\csc^{n-2} x \cot x}{n-1} + \frac{n-2}{n-1}\int \csc^{n-2} x\, dx.$$

SECTION 8.5

1. $\left\{\begin{array}{l} x = a\sin u \\ dx = a\cos u\,du \end{array}\right\};$ $\displaystyle\int \frac{dx}{\sqrt{a^2 - x^2}} = \int \frac{a\cos u\,du}{a\cos u}$

$$= \int du = u + C = \sin^{-1}\left(\frac{x}{a}\right) + C$$

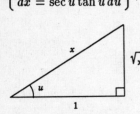

3. $\left\{\begin{array}{l} x = \sqrt{5}\sin u \\ dx = \sqrt{5}\cos u\,du \end{array}\right\};$ $\displaystyle\int \frac{dx}{(5 - x^2)^{3/2}} = \int \frac{\sqrt{5}\cos u\,du}{(5\cos^2 u)^{3/2}}$

$$= \frac{1}{5}\int \sec^2 u\,du$$

$$= \frac{1}{5}\tan u + C = \frac{x}{5\sqrt{5 - x^2}} + C$$

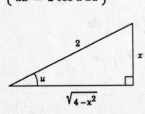

$$\int_0^1 \frac{dx}{(5 - x^2)^{3/2}} = \left[\frac{x}{5\sqrt{5 - x^2}}\right]_0^1 = \tfrac{1}{10}$$

5. $\left\{\begin{array}{l} x = \sec u \\ dx = \sec u\tan u\,du \end{array}\right\};$ $\displaystyle\int \sqrt{x^2 - 1}\,dx = \int \tan^2 u\sec u\,du$

$$= \int (\sec^3 u - \sec u)\,du$$

$$= \tfrac{1}{2}\sec u\tan u - \tfrac{1}{2}\ln|\sec u + \tan u| + C$$

Section 8.4

$$= \tfrac{1}{2}x\sqrt{x^2 - 1} - \tfrac{1}{2}\ln|x + \sqrt{x^2 - 1}| + C$$

7. $\left\{\begin{array}{l} x = 2\sin u \\ dx = 2\cos u\,du \end{array}\right\};$ $\displaystyle\int \frac{x^2}{\sqrt{4 - x^2}}\,dx = \int \frac{4\sin^2 u}{2\cos u}2\cos u\,du$

$$= 2\int (1 - \cos 2u)\,du$$

$$= 2u - \sin 2u + C$$

$$= 2u - 2\sin u\cos u + C$$

$$= 2\sin^{-1}\left(\frac{x}{2}\right) - \frac{1}{2}x\sqrt{4 - x^2} + C$$

9. $\left\{\begin{array}{l} u = 1 - x^2 \\ du = -2x\,dx \end{array}\right\};$ $\displaystyle\int \frac{x}{(1 - x^2)^{3/2}}\,dx = -\frac{1}{2}\int \frac{du}{u^{3/2}} = u^{-1/2} + C = \frac{1}{\sqrt{1 - x^2}} + C$

11. $\left\{\begin{array}{l} x = \sin u \\ dx = \cos u\, du \end{array}\right\};$ $\displaystyle\int \frac{x^2}{(1-x^2)^{3/2}}\, dx = \int \frac{\sin^2 u}{\cos^3 u} \cos u\, du = \int \tan^2 u\, du$

$$= \int (\sec^2 u - 1)\, du = \tan u - u + C$$

$$= \frac{x}{\sqrt{1-x^2}} - \sin^{-1} x + C$$

$$\int_0^{1/2} \frac{x^2}{(1-x^2)^{3/2}}\, dx = \left[\frac{x}{\sqrt{1-x^2}} - \sin^{-1} x\right]_0^{1/2} = \frac{2\sqrt{3} - \pi}{6}$$

13. $\left\{\begin{array}{l} u = 4 - x^2 \\ du = -2x\, dx \end{array}\right\};$ $\displaystyle\int x\sqrt{4-x^2}\, dx = -\frac{1}{2}\int u^{1/2}\, du = -\frac{1}{3}u^{3/2} + C$

$$= -\tfrac{1}{3}(4-x^2)^{3/2} + C$$

15. $\left\{\begin{array}{l} x = 5\sin u \,|\, x = 0 \implies u = 0 \\ dx = 5\cos u\, du \,|\, x = 5 \implies u = \pi/2 \end{array}\right\};$

$$\int_0^5 x^2 \sqrt{25-x^2}\, dx = \int_0^{\pi/2} (5\sin u)^2 (5\cos u)^2\, du$$

$$= 625 \int_0^{\pi/2} (\sin^2 u - \sin^4 u)\, du$$

$$= 625\left[\frac{1}{2}\cdot\frac{\pi}{2} - \frac{3\cdot 1}{4\cdot 2}\cdot\frac{\pi}{2}\right] = \frac{625\pi}{16} \qquad \text{[see Exercise 46, Section 8.3]}$$

17. $\left\{\begin{array}{l} x = \sqrt{8}\tan u \\ dx = \sqrt{8}\sec^2 u\, du \end{array}\right\};$ $\displaystyle\int \frac{x^2}{(x^2+8)^{3/2}}\, dx = \int \frac{8\tan^2 u}{(8\sec^2 u)^{3/2}}\sqrt{8}\sec^2 u\, du$

$$= \int \frac{\tan^2 u}{\sec u}\, du = \int \frac{\sec^2 u - 1}{\sec u}\, du$$

$$= \int (\sec u - \cos u)\, du$$

$$= \ln|\sec u + \tan u| - \sin u + C$$

$$= \ln\left(\frac{\sqrt{x^2+8} + x}{\sqrt{8}}\right) - \frac{x}{\sqrt{x^2+8}} + C$$

$$(\text{absorb} - \ln\sqrt{8} \text{ in } C) \longrightarrow$$

$$= \ln\left(\sqrt{x^2+8} + x\right) - \frac{x}{\sqrt{x^2+8}} + C$$

19. $\left\{\begin{array}{l} x = a\sin u \\ dx = a\cos u\, du \end{array}\right\};$ $\displaystyle\int \frac{dx}{x\sqrt{a^2-x^2}} = \int \frac{a\cos u\, du}{a\sin u\,(a\cos u)} = \frac{1}{a}\int \csc u\, du$

$$= \frac{1}{a}\ln|\csc u - \cot u| + C$$

$$= \frac{1}{a}\ln\left|\frac{a - \sqrt{a^2-x^2}}{x}\right| + C$$

21. $\left\{ \begin{array}{l} x = a \sec u \\ dx = a \sec u \tan u \, du \end{array} \right\}$; $\displaystyle\int \frac{dx}{\sqrt{x^2 - a^2}} = \int \frac{a \sec u \tan u \, du}{a \tan u} = \int \sec u \, du$

$$= \ln |\sec u + \tan u| + C$$

$$= \ln \left| \frac{x + \sqrt{x^2 - a^2}}{a} \right| + C$$

(absorb $- \ln a$ in C)

$$= \ln \left| x + \sqrt{x^2 - a^2} \right| + C$$

23. $\left\{ \begin{array}{l} e^x = \sec u \\ e^x dx = \sec u \tan u \, du \end{array} \right\}$; $\displaystyle\int e^x \sqrt{e^{2x} - 1} \, dx = \int \tan^2 u \sec u \, du$

$$= \int (\sec^3 u - \sec u) \, du$$

$$= \tfrac{1}{2} \sec u \tan u - \tfrac{1}{2} \ln |\sec u + \tan u| + C$$

$$= \tfrac{1}{2} e^x \sqrt{e^{2x} - 1} - \tfrac{1}{2} \ln \left(e^x + \sqrt{e^{2x} - 1} \right) + C$$

25. $\left\{ \begin{array}{l} x = 3 \tan u \\ dx = 3 \sec^2 u \, du \end{array} \right\}$; $\displaystyle\int \frac{x^3}{\sqrt{9 + x^2}} \, dx = \int \frac{27 \tan^3 u}{3 \sec u} \cdot 3 \sec^2 u \, du$

$$= 27 \int \tan^3 u \sec u \, du$$

$$= 27 \int (\sec^2 u - 1) \sec u \tan u \, du$$

$$= 27 \left[\tfrac{1}{3} \sec^3 u - \sec u \right] + C$$

$$= \tfrac{1}{3} (9 + x^2)^{3/2} - 9 (9 + x^2)^{1/2} + C$$

$$\int_0^3 \frac{x^3}{\sqrt{9 + x^2}} \, dx = \left[\tfrac{1}{3} (9 + x^2)^{3/2} - 9 (9 + x^2)^{1/2} \right]_0^3 = 18 - 9\sqrt{2}$$

27. $\left\{ \begin{array}{l} x = a \tan u \\ dx = a \sec^2 u \, du \end{array} \right\}$; $\displaystyle\int \frac{dx}{x^2 \sqrt{a^2 + x^2}} = \int \frac{a \sec^2 u \, du}{a^2 \tan^2 u \, (a \sec u)}$

$$= \frac{1}{a^2} \int \frac{\sec u}{\tan^2 u} \, du$$

$$= \frac{1}{a^2} \int \cot u \csc u \, du$$

$$= -\frac{1}{a^2} \cos u + C = -\frac{1}{a^2 x} \sqrt{a^2 + x^2} + C$$

29. $\left\{\begin{array}{l} x = a \sec u \\ dx = a \sec u \tan u \, du \end{array}\right\}$; $\displaystyle\int \frac{dx}{x^2\sqrt{x^2-a^2}} = \int \frac{a \sec u \tan u \, du}{a^2 \sec^2 u \, (a \tan u)}$

$$= \frac{1}{a^2} \int \cos u \, du$$

$$= \frac{1}{a^2} \sin u + C$$

$$= \frac{1}{a^2 x}\sqrt{x^2-a^2} + C$$

31. $\left\{\begin{array}{ll} x = 2 \sec u \, \big| \, x = 2 & \implies \quad u = 0 \\ dx = 2 \sec u \tan u \, du \, \big| \, x = 2\sqrt{2} & \implies \quad u = \pi/4 \end{array}\right\}$;

$$\int_2^{2\sqrt{2}} \frac{\sqrt{x^2-4}}{x} \, dx = \int_0^{\pi/4} \frac{2 \tan u}{2 \sec u} \cdot 2 \sec u \tan u \, du$$

$$= 2\int_0^{\pi/4} \tan^2 u \, du$$

$$= 2\int_0^{\pi/4} \left(\sec^2 u - 1\right) du$$

$$= 2\left[\tan u - u\right]_0^{\pi/4} = 2 - \frac{\pi}{2}$$

33. $\left\{\begin{array}{l} e^x = 3 \sec u \\ e^x dx = 3 \sec u \tan u \, du \end{array}\right\}$; $\displaystyle\int \frac{dx}{e^x\sqrt{e^{2x}-9}} = \int \frac{\tan u \, du}{3 \sec u \, (3 \tan u)}$

$$= \frac{1}{9} \int \cos u \, du$$

$$= \tfrac{1}{9} \sin u + C$$

$$= \tfrac{1}{9} e^{-x}\sqrt{e^{2x}-9} + C$$

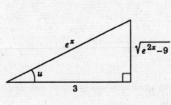

35. $(x^2 - 4x + 4)^{\frac{3}{2}} = \left\{\begin{array}{ll} (x-2)^3, & x > 2 \\ (2-x)^3, & x < 2 \end{array}\right.$

$$\int \frac{dx}{(x^2-4x+4)^{3/2}} = \left\{\begin{array}{ll} -\dfrac{1}{2(x-2)^2} + C, & x > 2 \\[2mm] \dfrac{1}{2(2-x)^2} + C, & x < 2 \end{array}\right.$$

37. $\left\{\begin{array}{l} x - 3 = \sin u \\ dx = \cos u \, du \end{array}\right\}$; $\displaystyle\int x\sqrt{6x-x^2-8} \, dx = \int x\sqrt{1-(x-3)^2} \, dx$

$$= \int (3 + \sin u)(\cos u)\cos u \, du$$

$$= \int (3\cos^2 u + \cos^2 u \sin u) \, du$$

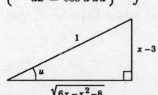

$$= \int \left[3\left(\frac{1 + \cos 2u}{2} \right) + \cos^2 u \sin u \right] du$$

$$= \frac{3u}{2} + \frac{3}{4} \sin 2u - \frac{1}{3} \cos^3 u + C$$

$$= \tfrac{3}{2} \sin^{-1}(x - 3) + \tfrac{3}{2}(x - 3)\sqrt{6x - x^2 - 8} - \tfrac{1}{3}(6x - x^2 - 8)^{3/2} + C$$

39. $\left\{ \begin{array}{l} x + 1 = 2\tan u \\ dx = 2\sec^2 u\, du \end{array} \right\}$; $\displaystyle\int \frac{x}{(x^2 + 2x + 5)^2}\, dx = \int \frac{x}{[(x+1)^2 + 4]^2}\, dx$

$$= \int \frac{2\tan u - 1}{(4\sec^2 u)^2} 2\sec^2 u\, du$$

$$= \frac{1}{8} \int \frac{2\tan u - 1}{\sec^2 u}\, du$$

$$= \frac{1}{8} \int (2\sin u \cos u - \cos^2 u)\, du$$

$$= \frac{1}{8} \int \left(2\sin u \cos u - \frac{1 + \cos 2u}{2} \right) du$$

$$= \frac{1}{8} \left(\sin^2 u - \frac{u}{2} - \frac{\sin 2u}{4} \right) + C$$

$$= \frac{1}{8} \left[\left(\frac{x+1}{\sqrt{x^2+2x+5}} \right)^2 - \frac{1}{2}\tan^{-1}\left(\frac{x+1}{2} \right) - \frac{1}{4} \overbrace{(2)\left(\frac{x+1}{\sqrt{x^2+2x+5}} \right)\left(\frac{2}{\sqrt{x^2+2x+5}} \right)}^{\sin 2u = 2\sin u \cos u} \right] + C$$

$$= \frac{x^2 + x}{8(x^2 + 2x + 5)} - \frac{1}{16}\tan^{-1}\left(\frac{x+1}{2} \right) + C$$

41. $\left\{ \begin{array}{l} x + 2 = 3\tan u \\ dx = 3\sec^2 u\, du \end{array} \right\}$; $\displaystyle\int \frac{x+3}{\sqrt{x^2 + 4x + 13}}\, dx = \int \frac{x+3}{\sqrt{(x+2)^2 + 9}}\, dx$

$$= \int \frac{3\tan u + 1}{3\sec u} 3\sec^2 u\, du$$

$$= \int (3\sec u \tan u + \sec u)\, du$$

$$= 3\sec u + \ln|\sec u + \tan u| + C$$

$$= \sqrt{x^2 + 4x + 13} + \ln\left| \frac{\sqrt{x^2 + 4x + 13}}{3} + \frac{x+2}{3} \right| + C$$

(absorb $-\ln 3$ in C)

$$= \sqrt{x^2 + 4x + 13} + \ln\left| x + 2 + \sqrt{x^2 + 4x + 13} \right| + C$$

43. $\left\{ \begin{array}{l} x - 3 = \sin u \\ dx = \cos u\, du \end{array} \right\};$ $\int \sqrt{6x - x^2 - 8}\, dx = \int \sqrt{1 - (x - 3)^2}\, dx$

$$= \int \cos^2 u\, du$$

$$= \int \frac{1 + \cos 2u}{2}\, du$$

$$= \frac{u}{2} + \frac{\sin 2u}{4} + C$$

$$= \tfrac{1}{2} \left[\sin^{-1}(x - 3) + (x - 3)\sqrt{6x - x^2 - 8} \right] + C$$

45. Let $x = a \tan u$. Then $dx = a \sec^2 u\, du$, $\sqrt{x^2 + a^2} = a \sec u$, and

$$\int \frac{dx}{(x^2 + a^2)^n} = \int \frac{a \sec^2 u}{a^{2n} \sec^{2n} u}\, du = \frac{1}{a^{2n-1}} \int \cos^{2n-2} u\, du$$

47. $\int \frac{1}{(x^2 + 1)^3}\, dx = \int \cos^4 u\, du$

$$= \tfrac{1}{4} \cos^3 u \sin u + \tfrac{3}{8} \cos u \sin u + \tfrac{3}{8} u + C \quad \text{[by the reduction formula (8.3.2)]}$$

$$= \tfrac{3}{8} \tan^{-1} x + \frac{3x}{8(x^2 + 1)} + \frac{x}{4(x^2 + 1)^2} + C$$

49. (a) Let $x = a \tan u$, $dx = a \sec^2 u\, du$ and $\sqrt{x^2 + a^2} = a \sec u$.

$$\int \frac{1}{\sqrt{x^2 + a^2}}\, dx = \int \frac{a \sec^2 u}{a \sec u}\, du = \int \sec u\, du$$

$$= \ln|\sec u + \tan u| + C = \ln\left(\frac{x}{a} + \frac{\sqrt{x^2 + a^2}}{a} \right) + C$$

$$= \ln\left(x + \sqrt{x^2 + a^2} \right) - \ln a + C = \ln\left(x + \sqrt{x^2 + a^2} \right) + C \quad \text{[absorb } -\ln a \text{ in } C]$$

(b) Let $x = a \sinh u$, $dx = a \cosh u\, du$ and $\sqrt{x^2 + a^2} = a \cosh u$.

$$\int \frac{1}{\sqrt{x^2 + a^2}}\, dx = \int \frac{a \cosh u}{a \cosh u}\, du = \int du = u + C = \sinh^{-1}\left(\frac{x}{a} \right) + C$$

(c) See Section 7.10.

51. $$V = \int_0^1 \pi \left(\frac{1}{1 + x^2} \right)^2 dx = \pi \int_0^1 \frac{1}{(1 + x^2)^2}\, dx$$

$$= \pi \int_0^{\pi/4} \cos^2 u\, du \quad [x = \tan u, \text{ see Ex.45}]$$

$$= \frac{\pi}{2} \int_0^{\pi/4} (1 + \cos 2u)\, du$$

$$= \frac{\pi}{2} \left[u + \tfrac{1}{2} \sin 2u \right]_0^{\pi/4} = \frac{\pi^2}{8} + \frac{\pi}{4}$$

53.   We need only consider angles θ between 0 and π. Assume first that $0 \leq \theta \leq \frac{\pi}{2}$.

The area of the triangle is

$\frac{1}{2} r^2 \sin \theta \cos \theta.$

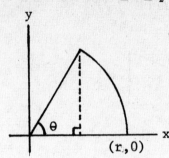

The area of the other region is given by:

$$\int_{r \cos \theta}^{r} \sqrt{r^2 - x^2}\, dx = \left[\frac{x}{2}\sqrt{r^2 - x^2} + \frac{r^2}{2}\sin^{-1}\frac{x}{r}\right]_{r \cos \theta}^{r}$$

$$= \frac{\pi r^2}{4} - \frac{r^2}{2}\sin \theta \cos \theta - \frac{r^2}{2}\sin^{-1}(\cos \theta)$$

$$= \frac{r^2 \theta}{2} - \frac{r^2}{2}\sin \theta \cos \theta$$

Thus, the area of the sector is $A = \frac{1}{2} r^2 \theta.$

Now suppose that $\frac{\pi}{2} < \theta \leq \pi.$

Then $A = \frac{1}{2}\pi r^2 - \frac{1}{2} r^2 (\pi - \theta) = \frac{1}{2} r^2 \theta.$

55. $M = \int_{0}^{a} \frac{dx}{\sqrt{x^2 + a^2}} = \left[\ln\left(x + \sqrt{x^2 + a^2}\right)\right]_{0}^{a} = \ln\left(1 + \sqrt{2}\right)$

$x_M M = \int_{0}^{a} \frac{x}{\sqrt{x^2 + a^2}}\, dx = \left[\sqrt{x^2 + a^2}\right]_{0}^{a} = (\sqrt{2} - 1)a \qquad x_M = \frac{(\sqrt{2} - 1)a}{\ln\left(1 + \sqrt{2}\right)}$

57.
$$A = \int_{a}^{\sqrt{2}a} \sqrt{x^2 - a^2}\, dx = \left[\frac{1}{2}x\sqrt{x^2 - a^2} - \frac{1}{2}a^2 \ln\left|x + \sqrt{x^2 - a^2}\right|\right]_{a}^{\sqrt{2}a}$$

$$= \frac{1}{2}a^2\left[\sqrt{2} - \ln\left(\sqrt{2} + 1\right)\right]$$

$$\overline{x}A = \int_{a}^{\sqrt{2}a} x\sqrt{x^2 - a^2}\, dx = \frac{1}{3}a^3, \qquad \overline{y}A = \int_{a}^{\sqrt{2}a} \left[\frac{1}{2}(x^2 - a^2)\right] dx = \frac{1}{6}a^3(2 - \sqrt{2})$$

$$\overline{x} = \frac{2a}{3[\sqrt{2} - \ln\left(\sqrt{2} + 1\right)]}, \qquad \overline{y} = \frac{(2 - \sqrt{2})a}{3[\sqrt{2} - \ln\left(\sqrt{2} + 1\right)]}$$

59. $V_y = 2\pi \overline{R} A = \frac{2}{3}\pi a^3$ $\overline{y} V_y = \int_a^{\sqrt{2}a} \pi x(x^2 - a^2)\, dx = \frac{1}{4}\pi a^4, \quad \overline{y} = \frac{3}{8}a$

$$(6.4.6)$$

61. (a)

(b) $A = \int_3^6 \frac{\sqrt{x^2 - 9}}{x^2}\, dx$

$\quad = \int_0^{\pi/3} \frac{3\tan u}{9\sec^2 u} \cdot 3\sec u \tan u\, du \quad [x = 3\sec u]$

$\quad = \int_0^{\pi/3} (\sec u - \cos u)\, du$

$\quad = \left[\ln|\sec u + \tan u| - \sin u\right]_0^{\pi/3}$

$\quad = \ln(2 + \sqrt{3}) - \dfrac{\sqrt{3}}{2}$

(c) $\overline{x} A = \int_3^6 x \cdot \frac{\sqrt{x^2 - 9}}{x^2}\, dx = \int_0^{\pi/3} (3\sec^2 u - 3)\, du = [3\tan u - 3u]_0^{\pi/3} = 3\sqrt{3} - \pi$

Thus, $\overline{x} = \dfrac{2\left(3\sqrt{3} - \pi\right)}{2\ln\left(2 + \sqrt{3}\right) - \sqrt{3}}.$

$\overline{y} A = \int \frac{1}{2}\left(\frac{\sqrt{x^2 - 9}}{x^2}\right)^2 dx = \frac{1}{2}\int_3^6 \left(\frac{1}{x^2} - \frac{9}{x^4}\right)dx = \frac{1}{2}\left[-\frac{1}{x} + \frac{3}{x^3}\right]_3^6 = \frac{5}{144}$

Thus, $\overline{y} = \dfrac{5}{72\left[2\ln\left(2 + \sqrt{3}\right) - \sqrt{3}\right]}.$

SECTION 8.6

1. $\dfrac{1}{x^2 + 7x + 6} = \dfrac{1}{(x + 1)(x + 6)} = \dfrac{A}{x + 1} + \dfrac{B}{x + 6}$

$1 = A(x + 6) + B(x + 1)$

$x = -6: \quad 1 = -5B \implies B = -1/5$

$x = -1: \quad 1 = 5A \implies A = 1/5$

$\dfrac{1}{x^2 + 7x + 6} = \dfrac{1/5}{x + 1} - \dfrac{1/5}{x + 6}$

3. $\dfrac{x}{x^4 - 1} = \dfrac{x}{(x^2 + 1)(x + 1)(x - 1)} = \dfrac{Ax + B}{x^2 + 1} + \dfrac{C}{x + 1} + \dfrac{D}{x - 1}$

$x = (Ax + B)(x^2 - 1) + C(x - 1)(x^2 + 1) + D(x + 1)(x^2 + 1)$

$x = 1:$ $1 = 4D$ $\implies$ $D = 1/4$

$x = -1:$ $-1 = -4C$ $\implies$ $C = 1/4$

$x = 0:$ $-B - C + D = 0 \implies B = 0$

$x = 2:$ $6A + 5C + 15D = 2 \implies A = -1/2$

$$\frac{x}{x^4 - 1} = \frac{1/4}{x - 1} + \frac{1/4}{x + 1} - \frac{x/2}{x^2 + 1}$$

5. $\dfrac{x^2 - 3x - 1}{x^3 + x^2 - 2x} = \dfrac{x^2 - 3x - 1}{x(x + 2)(x + 1)} = \dfrac{A}{x} + \dfrac{B}{x + 2} + \dfrac{C}{x - 1}$

$x^2 - 3x - 1 = A(x + 2)(x - 1) + Bx(x - 1) + Cx(x + 2)$

$x = 0:$ $-1 = -2A$ $\implies$ $A = 1/2$

$x = -2:$ $9 = 6B$ $\implies$ $B = 3/2$

$x = 1:$ $-3 = 3C \implies C = -1$

$$\frac{x^2 - 3x - 1}{x^3 + x^2 - 2x} = \frac{1/2}{x} + \frac{3/2}{x + 2} - \frac{1}{x - 1}$$

7. $\dfrac{2x^2 + 1}{x^3 - 6x^2 + 11x - 6} = \dfrac{2x^2 + 1}{(x - 1)(x - 2)(x - 3)} = \dfrac{A}{x - 1} + \dfrac{B}{x - 2} + \dfrac{C}{x - 3}$

$2x^2 + 1 = A(x - 2)(x - 3) + B(x - 1)(x - 3) + C(x - 1)(x - 2)$

$x = 1:$ $3 = 2A$ $\implies$ $A = 3/2$

$x = 2:$ $9 = -B$ $\implies$ $B = -9$

$x = 3:$ $19 = 2C \implies C = 19/2$

$$\frac{2x^2 + 1}{x^3 - 6x^2 + 11x - 6} = \frac{3/2}{x - 1} - \frac{9}{x - 2} + \frac{19/2}{x - 3}$$

9. $\dfrac{7}{(x - 2)(x + 5)} = \dfrac{A}{x - 2} + \dfrac{B}{x + 5}$

$7 = A(x + 5) + B(x - 2)$

$x = -5:$ $7 = -7B$ $\implies$ $B = -1$

$x = 2:$ $7 = 7A$ $\implies$ $A = 1$

$$\int \frac{7}{(x - 2)(x + 5)}\, dx = \int \left(\frac{1}{x - 2} - \frac{1}{x + 5} \right) = \ln|x - 2| - \ln|x + 5| + C = \ln\left| \frac{x - 2}{x + 5} \right| + C$$

11. $\dfrac{2x^2 + 3}{x^2(x - 1)} = \dfrac{A}{x} + \dfrac{B}{x^2} + \dfrac{C}{x - 1}$

$2x^2 + 3 = Ax(x - 1) + B(x - 1) + Cx^2$

$x = 0:$ $3 = -B$ $\implies$ $B = -3$

$x = 1:$ $5 = C$ $\implies$ $C = 5$

$x = -1:$ $5 = 2A - 2B + C$ $\implies$ $A = -3$

$$\int \frac{2x^2 + 3}{x^2(x - 1)}\, dx = \int \left(-\frac{3}{x} - \frac{3}{x^2} + \frac{5}{x - 1} \right) dx = -3\ln|x| + \frac{3}{x} + 5\ln|x - 1| + C$$

13. We carry out the division until the numerator has degree smaller than the denominator:

$$\frac{x^5}{(x-2)^2} = \frac{x^5}{x^2 - 4x + 4} = x^3 + 4x^2 + 12x + 32 + \frac{80x - 128}{(x-2)^2}.$$

Then,

$$\frac{80x - 128}{(x-2)^2} = \frac{80x - 160 + 32}{(x-2)^2} = \frac{80}{x-2} + \frac{32}{(x-2)^2}$$

$$\int \frac{x^5}{(x-2)^2}\, dx = \int \left(x^3 + 4x^2 + 12x + 32 + \frac{80}{x-2} + \frac{32}{(x-2)^2} \right) dx$$

$$= \frac{1}{4}x^4 + \frac{4}{3}x^3 + 6x^2 + 32x + 80\ln|x-2| - \frac{32}{x-2} + C.$$

15.
$$\frac{x+3}{x^2 - 3x + 2} = \frac{A}{x-1} + \frac{B}{x-2}$$

$$x + 3 = A(x-2) + B(x-1)$$

$x = 1$: $\quad 4 = -A \quad \Longrightarrow \quad A = -4$

$x = 2$: $\quad 5 = B \quad \Longrightarrow \quad B = 5$

$$\int \frac{x+3}{x^2 - 3x + 2}\, dx = \int \left(\frac{-4}{x-1} + \frac{5}{x-2} \right) dx = -4\ln|x-1| + 5\ln|x-2| + C$$

17. $\displaystyle \int \frac{dx}{(x-1)^3} = \int (x-1)^{-3}\, dx = -\frac{1}{2}(x-1)^{-2} + C = -\frac{1}{2(x-1)^2} + C$

19.
$$\frac{x^2}{(x-1)^2(x+1)} = \frac{A}{x-1} + \frac{B}{(x-1)^2} + \frac{C}{x+1}$$

$$x^2 = A(x-1)(x+1) + B(x+1) + C(x-1)^2$$

$x = 1$: $\quad 1 = 2B \quad \Longrightarrow \quad B = 1/2$

$x = -1$: $\quad 1 = 4C \quad \Longrightarrow \quad C = 1/4$

$x = 0$: $\quad 0 = -A + B + C \quad \Longrightarrow \quad A = 3/4$

$$\int \frac{x^2}{(x-1)^2(x+1)}\, dx = \int \left(\frac{3/4}{x-1} + \frac{1/2}{(x-1)^2} + \frac{1/4}{x+1} \right) dx$$

$$= \frac{3}{4}\ln|x-1| - \frac{1}{2(x-1)} + \frac{1}{4}\ln|x+1| + C$$

21.
$$x^4 - 16 = (x^2 - 4)(x^2 + 4) = (x-2)(x+2)(x^2 + 4)$$

$$\frac{1}{x^4 - 16} = \frac{A}{x-2} + \frac{B}{x+2} + \frac{Cx + D}{x^2 + 4}$$

$$1 = A(x+2)(x^2+4) + B(x-2)(x^2+4) + (Cx+D)(x^2-4)$$

$x = 2:\quad 1 = 32A \implies A = 1/32$

$x = -2:\quad 1 = -32B \implies B = -1/32$

$x = 0:\quad 1 = 8A - 8B - 4D \implies D = -1/8$

$x = 1:\quad 1 = 15A - 5B - 3C - 3D \implies C = 0$

$$\int \frac{dx}{x^4 - 16} = \int \left(\frac{1/32}{x-2} - \frac{1/32}{x+2} - \frac{1/8}{x^2+4} \right) dx$$

$$= \frac{1}{32} \ln|x-2| - \frac{1}{32} \ln|x+2| - \frac{1}{8}\left(\frac{1}{2} \tan^{-1} \frac{x}{2} \right) + C$$

$$= \frac{1}{32} \ln \left| \frac{x-2}{x+2} \right| - \frac{1}{16} \tan^{-1} \frac{x}{2} + C$$

23. $\dfrac{x^3 + 4x^2 - 4x - 1}{(x^2+1)^2} = \dfrac{Ax+B}{x^2+1} + \dfrac{Cx+D}{(x^2+1)^2}$

$$x^3 + 4x^2 - 4x - 1 = (Ax+B)(x^2+1) + (Cx+D)$$

$x = 0:\quad -1 = B + D \qquad\qquad \implies D = -B - 1$

$\qquad\qquad\qquad\qquad\qquad\qquad\qquad\qquad\qquad \implies B = 4,\ \ D = -5$

$x = 1:\quad 0 = 2A + 2B + C + D \implies 6 = 4B + 2D$

$x = -1:\quad 6 = -2A + 2B - C + D \qquad 6 = -2A + 8 - C - 5$

$\qquad\qquad\qquad\qquad\qquad\qquad\qquad\qquad\qquad \implies A = 1,$
$\qquad\qquad\qquad\qquad\qquad\qquad\qquad\qquad\qquad C = -5$

$x = 2:\quad 15 = 10A + 5B + 2C + D \qquad 15 = 10A + 20 + 2C - 5$

$$\int \frac{x^3 + 4x^2 - 4x - 1}{(x^2+1)^2}\, dx = \int \left(\frac{x}{x^2+1} + \frac{4}{x^2+1} - \frac{5x}{(x^2+1)^2} - \frac{5}{(x^2+1)^2} \right) dx$$

$(*)$
$$= \frac{1}{2} \ln(x^2+1) + 4\tan^{-1} x + \frac{5}{2(x^2+1)} - 5 \int \frac{dx}{(x^2+1)^2}$$

For this last integral we set

$$\left\{ \begin{array}{l} x = \tan u \\ dx = \sec^2 u\, du \end{array} \right\};\quad \int \frac{dx}{(x^2+1)^2} = \int \frac{\sec^2 u\, du}{(1 + \tan^2 u)^2} = \int \cos^2 u\, du$$

$$= \frac{1}{2} \int (1 + \cos 2u)\, du$$

$$= \tfrac{1}{2}\left(u + \tfrac{1}{2} \sin 2u\right) + C = \tfrac{1}{2}(u + \sin u \cos u) + C$$

$$= \frac{1}{2}\left(\tan^{-1} x + \frac{x}{1+x^2} \right) + C.$$

Substituting this result in (∗) and rearranging the terms, we get

$$\int \frac{x^3 + 4x^2 - 4x + 1}{(x^2 + 1)^2}\, dx = \frac{1}{2}\ln(x^2 + 1) + \frac{3}{2}\tan^{-1} x + \frac{5(1 - x)}{2(1 + x^2)} + C.$$

25.
$$\frac{1}{x^4 + 4} = \frac{Ax + B}{x^2 + 2x + 2} + \frac{Cx + D}{x^2 - 2x + 2} \qquad \text{(using the hint)}$$

$$1 = (Ax + B)(x^2 - 2x + 2) + (Cx + D)(x^2 + 2x + 2)$$

$$\left.\begin{array}{l} x = 0: \quad 1 = 2B + 2D \\ x = 1: \quad 1 = A + B + 5C + 5D \\ x = -1: \quad 1 = -5A + 5B - C + D \\ x = 2: \quad 1 = 4A + 2B + 20C + 10D \end{array}\right\} \implies \begin{array}{l} A = 1/8 \\ B = 1/4 \\ C = -1/8 \\ D = 1/4 \end{array}$$

$$\int \frac{dx}{x^4 + 4} = \frac{1}{8}\int \frac{x + 2}{x^2 + 2x + 2}\, dx - \frac{1}{8}\int \frac{x - 2}{x^2 - 2x + 2}\, dx$$

$$= \frac{1}{8}\int \frac{x + 1}{x^2 + 2x + 2}\, dx + \frac{1}{8}\int \frac{dx}{(x + 1)^2 + 1} - \frac{1}{8}\int \frac{x - 1}{x^2 - 2x + 2}\, dx + \frac{1}{8}\int \frac{dx}{(x - 1)^2 + 1}$$

$$= \frac{1}{16}\ln(x^2 + 2x + 2) + \frac{1}{8}\tan^{-1}(x + 1) - \frac{1}{16}\ln(x^2 - 2x + 2) + \frac{1}{8}\tan^{-1}(x - 1) + C$$

$$= \frac{1}{16}\ln\left(\frac{x^2 + 2x + 2}{x^2 - 2x + 2}\right) + \frac{1}{8}\tan^{-1}(x + 1) + \frac{1}{8}\tan^{-1}(x - 1) + C$$

27.
$$\frac{x - 3}{x^3 + x^2} = \frac{x - 3}{x^2(x + 1)} = \frac{A}{x} + \frac{B}{x^2} + \frac{C}{x + 1}$$

$$x - 3 = Ax(x + 1) + B(x + 1) + Cx^2$$

$$x = 0: \quad -3 = B$$

$$x = -1: \quad -4 = C$$

$$x = 1: \quad -2 = 2A + 2B + C \implies A = 4$$

$$\int \frac{x - 3}{x^3 + x^2}\, dx = 4\int \frac{1}{x}\, dx - 3\int \frac{1}{x^2}\, dx - 4\int \frac{1}{x + 1}\, dx$$

$$= 4\ln|x| + \frac{3}{x} - 4\ln|x + 1| + C = \frac{3}{x} + 4\ln\left|\frac{x}{x + 1}\right| + C$$

29.
$$\frac{x + 1}{x^3 + x^2 - 6x} = \frac{x + 1}{x(x - 2)(x + 3)} = \frac{A}{x} + \frac{B}{x - 2} + \frac{C}{x + 3}$$

$$x + 1 = A(x - 2)(x + 3) + Bx(x + 3) + Cx(x - 2)$$

$$x = 0: \quad 1 = -6A \implies A = -1/6$$

$$x = 2: \quad 3 = 10B \implies B = 3/10$$

$$x = -3: \quad -2 = 15C \implies C = -2/15$$

$$\int \frac{x+1}{x^3+x^2-6x}\,dx = -\tfrac{1}{6}\int \tfrac{1}{x}\,dx + \tfrac{3}{10}\int \tfrac{1}{x-2}\,dx - \tfrac{2}{15}\int \tfrac{1}{x+3}\,dx$$

$$= -\tfrac{1}{6}\ln|x| + \tfrac{3}{10}\ln|x-2| - \tfrac{2}{15}\ln|x+3| + C$$

31. $$\int_0^2 \frac{x}{x^2+5x+6}\,dx = \int_0^2 \frac{x}{(x+2)(x+3)}\,dx = \int_0^2 \left(\frac{3}{x+3} - \frac{2}{x+2}\right)dx$$

$$= \left[3\ln|x+3| - 2\ln|x+2|\right]_0^2 = \ln\left(\frac{125}{108}\right)$$

33. $$\int_3^6 \frac{2x}{x^3-2x^2-4x+8}\,dx = \int_3^6 \frac{2x}{(x+2)(x-2)^2}\,dx = \int_3^6 \left[\frac{1/4}{x-2} + \frac{1}{(x-2)^2} - \frac{1/4}{x+2}\right]dx$$

$$= \left[\tfrac{1}{4}\ln|x-2| - \tfrac{1}{x-2} - \tfrac{1}{4}\ln|x+2|\right]_3^6 = \tfrac{1}{4}\ln\left(\frac{5}{2}\right) + \frac{3}{4}$$

35. $$\int_1^3 \frac{x^2-4x+3}{x^3+2x^2+x}\,dx = \int_1^3 \frac{x^2-4x+3}{x(x+1)^2}\,dx = \int_1^3 \left(\frac{3}{x} - \frac{2}{x+1} - \frac{8}{(x+1)^2}\right)dx$$

$$= \left[3\ln|x| - 2\ln|x+1| + \frac{8}{x+1}\right]_1^3 = \ln\left(\frac{27}{4}\right) - 2$$

37. Note that $$y = \frac{1}{x^2-1} = \frac{1}{2}\left[\frac{1}{x-1} - \frac{1}{x+1}\right]$$

and thus $$\frac{d^0 y}{dx^0} = \left(\frac{1}{2}\right)(-1)^0\,0!\left[\frac{1}{(x-1)^{0+1}} - \frac{1}{(x+1)^{0+1}}\right].$$

The rest is a routine induction.

39. $$A = \int_0^1 \frac{dx}{x^2+1} = \left[\tan^{-1} x\right]_0^1 = \frac{1}{4}\pi$$

$$\bar{x}A = \int_0^1 \frac{x}{x^2+1}\,dx = \frac{1}{2}\left[\ln(x^2+1)\right]_0^1 = \frac{1}{2}\ln 2$$

$$\bar{y}A = \int_0^1 \frac{dx}{2(x^2+1)^2} = \frac{1}{2}\left[\tan^{-1} x + \frac{x}{x^2+1}\right]_0^1 = \frac{1}{8}(\pi+2)$$

$$\bar{x} = \frac{2\ln 2}{\pi}; \quad \bar{y} = \frac{\pi+2}{2\pi}$$

41. (a)

(b) $$A = \int_0^4 \frac{x}{x^2+5x+6}\,dx$$

$$= \int_0^4 \frac{x}{(x+2)(x+3)}\,dx$$

$$= \int_0^4 \left(\frac{3}{x+3} - \frac{2}{x+2}\right)dx$$

$$= \left[3\ln|x+3| - 2\ln|x+2|\right]_0^4 = 3\ln 7 - 5\ln 3$$

43. (a)

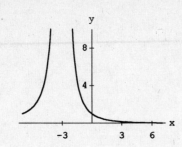

(b) $A = \int_{-2}^{9} \frac{9-x}{(x+3)^2}\, dx$

$$= \int_{-2}^{9} \left(\frac{12}{(x+3)^2} - \frac{1}{x+3} \right) dx$$

$$= \left[-\ln|x+3| - \frac{12}{x+3} \right]_{-2}^{9} = 11 - \ln 12$$

45. We assume that $C = 0$ at time $t = 0$.   (a) Let $A_0 = B_0$. Then

$$\frac{dC}{dt} = k(A_0 - C)^2 \quad \text{and} \quad \frac{dC}{(A_0 - C)^2} = k\, dt \qquad \text{see Section 7.6.}$$

Integrating, we get

$$\int \frac{1}{(A_0 - C)^2}\, dC = \int k\, dt$$

$$\frac{1}{A_0 - C} = kt + M \qquad M \text{ an arbitrary constant.}$$

Since $C(0) = 0$, $M = \dfrac{1}{A_0}$ and

$$\frac{1}{A_0 - C} = kt + \frac{1}{A_0}.$$

Solving this equation for C gives

$$C(t) = \frac{kA_0^2 t}{1 + kA_0 t}.$$

(b) Suppose that $A_0 \neq B_0$. Then

$$\frac{dC}{dt} = k(A_0 - C)(B_0 - C) \quad \text{and} \quad \frac{dC}{(A_0 - C)(B_0 - C)} = k\, dt.$$

Integrating, we get

$$\int \frac{1}{(A_0 - C)(B_0 - C)}\, dC = \int k\, dt$$

$$\frac{1}{B_0 - A_0} \int \left(\frac{1}{A_0 - C} - \frac{1}{B_0 - C} \right) dC = \int k\, dt$$

$$\frac{1}{B_0 - A_0} \left[-\ln(A_0 - C) + \ln(B_0 - C) \right] = kt + M$$

$$\frac{1}{B_0 - A_0} \ln\left(\frac{B_0 - C}{A_0 - C} \right) = kt + M \qquad M \text{ an arbitrary constant}$$

Since $C(0) = 0$, $M = \frac{1}{B_0 - A_0} \ln\left(\frac{B_0}{A_0} \right)$ and

$$\frac{1}{B_0 - A_0} \ln\left(\frac{B_0 - C}{A_0 - C} \right) = kt + \frac{1}{B_0 - A_0}.$$

Solving this equation for C, gives

$$C(t) = \frac{A_0 B_0 \left(e^{kA_0 t} - e^{kB_0 t}\right)}{A_0 e^{kA_0 t} - B_0 e^{kB_0 t}}.$$

47. (a)
$$m\frac{dv}{dt} = -\alpha v - \beta v^2$$

$$\frac{dv}{v(\alpha + \beta v)} = -\frac{1}{m}\,dt$$

$$\int \frac{1}{v(\alpha + \beta v)}\,dv = -\int \frac{1}{m}\,dt$$

$$\frac{1}{\alpha}\int \frac{1}{v}\,dv - \frac{\beta}{\alpha}\int \frac{1}{\alpha + \beta v}\,dv = -\int \frac{1}{m}\,dt$$

$$\frac{1}{\alpha}\ln v - \frac{1}{\alpha}\ln(\alpha + \beta v) = -\frac{1}{m}t + M, \quad M \text{ an arbitrary constant}$$

$$\ln\left(\frac{v}{\alpha + \beta v}\right) = -\frac{\alpha}{m}t + M$$

$$\frac{v}{\alpha + \beta v} = Ke^{-\alpha t/m} \quad \left[K = e^M\right]$$

Solving this equation for v we get $\quad v(t) = \dfrac{\alpha K}{e^{\alpha t/m} - \beta K} = \dfrac{\alpha}{Ce^{\alpha t/m} - \beta} \quad [C = 1/K].$

(b) Setting $v(0) = v_0$, we get   (c) $\lim\limits_{t\to\infty} v(t) = 0$

$$C = \frac{\alpha + \beta v_0}{v_0} \quad \text{and}$$

$$v(t) = \frac{\alpha v_0}{(\alpha + \beta v_0)e^{\alpha t/m} - \beta v_0}$$

SECTION 8.7

1. $\left\{\begin{array}{l} x = u^2 \\ dx = 2u\,du \end{array}\right\};\quad \displaystyle\int \frac{dx}{1 - \sqrt{x}} = \int \frac{2u\,du}{1 - u} = 2\int \left(\frac{1}{1-u} - 1\right)du$

$$= -2(\ln|1 - u| + u) + C = -2(\sqrt{x} + \ln|1 - \sqrt{x}|) + C$$

3. $\left\{\begin{array}{l} u^2 = 1 + e^x \\ 2u\,du = e^x\,dx \end{array}\right\};\quad \displaystyle\int \sqrt{1 + e^x}\,dx = \int u\cdot\frac{2u\,du}{u^2 - 1} = 2\int \left(1 + \frac{1}{u^2 - 1}\right)du$

$$= 2\int \left(1 + \frac{1}{2}\left[\frac{1}{u-1} - \frac{1}{u+1}\right]\right)du$$

$$= 2u + \ln|u - 1| - \ln|u + 1| + C$$

$$= 2\sqrt{1 + e^x} + \ln\left[\frac{\sqrt{1 + e^x} - 1}{\sqrt{1 + e^x} + 1}\right] + C$$

$$= 2\sqrt{1 + e^x} + \ln\left[\frac{\left(\sqrt{1 + e^x} - 1\right)^2}{e^x}\right] + C$$

$$= 2\sqrt{1 + e^x} + 2\ln\left(\sqrt{1 + e^x} - 1\right) - x + C$$

5   (a) $\left\{\begin{array}{l} u^2 = 1 + x \\ 2u\,du = dx \end{array}\right\};$ $\displaystyle\int x\sqrt{1+x}\,dx = \int (u^2 - 1)(u)2u\,du$

$$= \int (2u^4 - 2u^2)\,du$$

$$= \tfrac{2}{5}u^5 - \tfrac{2}{3}u^3 + C$$

$$= \tfrac{2}{5}(x+1)^{5/2} - \tfrac{2}{3}(x+1)^{3/2} + C$$

(b) $\left\{\begin{array}{l} u = 1 + x \\ du = dx \end{array}\right\};$ $\displaystyle\int x\sqrt{1+x}\,dx = \int (u-1)\sqrt{u}\,du = \int \left(u^{3/2} - u^{1/2}\right)du$

$$= \tfrac{2}{5}u^{5/2} - \tfrac{2}{3}u^{3/2} + C$$

$$= \tfrac{2}{5}(1+x)^{5/2} - \tfrac{2}{3}(1+x)^{3/2} + C$$

7. $\left\{\begin{array}{l} u^2 = x - 1 \\ 2u\,du = dx \end{array}\right\};$ $\displaystyle\int (x+2)\sqrt{x-1}\,dx = \int (u+3)(u)2u\,du$

$$= \int (2u^4 + 6u^2)\,du$$

$$= \tfrac{2}{5}u^5 + 2u^3 + C$$

$$= \tfrac{2}{5}(x-1)^{5/2} + 2(x-1)^{3/2} + C$$

9. $\left\{\begin{array}{l} u^2 = 1 + x^2 \\ 2u\,du = 2x\,dx \end{array}\right\};$ $\displaystyle\int \frac{x^3}{(1+x^2)^3}\,dx = \int \frac{x^2}{(1+x^2)^3}x\,dx = \int \frac{u^2 - 1}{u^6}u\,du$

$$= \int (u^{-3} - u^{-5})\,du = \frac{1}{2}u^{-2} + \frac{1}{4}u^{-4} + C$$

$$= \frac{1}{4(1+x^2)^2} - \frac{1}{2(1+x^2)} + C$$

$$= -\frac{1+2x^2}{4(1+x^2)^2} + C$$

11. $\left\{\begin{array}{l} u^2 = x \\ 2u\,du = dx \end{array}\right\};$ $\displaystyle\int \frac{\sqrt{x}}{\sqrt{x}-1}\,dx = \int \left(\frac{u}{u-1}\right)2u\,du = 2\int \left(u+1+\frac{1}{u-1}\right)du$

$$= u^2 + 2u + 2\ln|u-1| + C$$

$$= x + 2\sqrt{x} + 2\ln|\sqrt{x}-1| + C$$

13. $\left\{ \begin{array}{l} u^2 = x - 1 \\ 2u\,du = dx \end{array} \right\};$ $\displaystyle\int \frac{\sqrt{x-1}+1}{\sqrt{x-1}-1}\,dx = \int \frac{u+1}{u-1}2u\,du = \int \left(2u + 4 + \frac{4}{u-1}\right) du$

$$= u^2 + 4u + 4\ln|u-1| + C$$

$$= x - 1 + 4\sqrt{x-1} + 4\ln|\sqrt{x-1}-1| + C$$

(absorb -1 in C)

$$= x + 4\sqrt{x-1} + 4\ln|\sqrt{x-1}-1| + C$$

15. $\left\{ \begin{array}{l} u^2 = 1 + e^x \\ 2u\,du = e^x\,dx \end{array} \right\};$ $\displaystyle\int \frac{dx}{\sqrt{1+e^x}} = \int \left(\frac{1}{u}\right) \frac{2u\,du}{u^2 - 1} = \int \left[\frac{1}{u-1} - \frac{1}{u+1}\right] du$

$$= \ln|u-1| - \ln|u+1| + C$$

$$= \ln\left[\frac{\sqrt{1+e^x}-1}{\sqrt{1+e^x}+1}\right] + C$$

$$= \ln\left[\frac{(\sqrt{1+e^x}-1)^2}{e^x}\right] + C$$

$$= 2\ln(\sqrt{1+e^x}-1) - x + C$$

17. $\left\{ \begin{array}{l} u^2 = x + 4 \\ 2u\,du = dx \end{array} \right\};$ $\displaystyle\int \frac{x}{\sqrt{x+4}}\,dx = \int \frac{u^2-4}{u}2u\,du = \int (2u^2 - 8)\,du$

$$= \tfrac{2}{3}u^3 - 8u + C$$

$$= \tfrac{2}{3}(x+4)^{3/2} - 8(x+4)^{1/2} + C$$

$$= \tfrac{2}{3}(x-8)\sqrt{x+4} + C$$

19. $\left\{ \begin{array}{l} u^2 = 4x + 1 \\ 2u\,du = 4\,dx \end{array} \right\};$ $\displaystyle\int 2x^2(4x+1)^{-5/2}\,dx = \int 2\left(\frac{u^2-1}{4}\right)^2 (u^{-5})\frac{u}{2}\,du$

$$= \frac{1}{16}\int (1 - 2u^{-2} + u^{-4})\,du = \frac{1}{16}u + \frac{1}{8}u^{-1} - \frac{1}{48}u^{-3} + C$$

$$= \tfrac{1}{16}(4x+1)^{1/2} + \tfrac{1}{8}(4x+1)^{-1/2} - \tfrac{1}{48}(4x+1)^{-3/2} + C$$

21. $\left\{ \begin{array}{l} u^2 = ax + b \\ 2u\,du = a\,dx \end{array} \right\};$ $\displaystyle\int \frac{x}{(ax+b)^{3/2}}\,dx = \int \frac{\dfrac{u^2-b}{a}}{u^3}\frac{2u}{a}\,du$

$$= \frac{2}{a^2}\int (1 - bu^{-2})\,du$$

$$= \frac{2}{a^2}(u + bu^{-1}) + C = \frac{2u^2 + 2b}{a^2 u} + C$$

$$= \frac{4b + 2ax}{a^2\sqrt{ax+b}} + C$$

23.
$$\left\{ \begin{array}{ll} u = \tan(x/2), & dx = \dfrac{2}{1+u^2}\,du \\[3mm] \sin x = \dfrac{2u}{1+u^2}, & \cos x = \dfrac{1-u^2}{1+u^2} \end{array} \right\};$$

$$\int \frac{1}{1+\cos x - \sin x}\,dx = \int \frac{1}{1 + \dfrac{1-u^2}{1+u^2} - \dfrac{2u}{1+u^2}} \cdot \frac{2}{1+u^2}\,du$$

$$= \int \frac{1}{1-u}\,du$$

$$= -\ln|1-u| + C = -\ln\left|1 - \tan\left(\frac{x}{2}\right)\right| + C$$

25.
$$\left\{ \begin{array}{ll} u = \tan(x/2), & dx = \dfrac{2}{1+u^2}\,du \\[3mm] \sin x = \dfrac{2u}{1+u^2} & \end{array} \right\};$$

$$\int \frac{1}{2+\sin x}\,dx = \int \frac{1}{2 + \dfrac{2u}{1+u^2}} \cdot \frac{2}{1+u^2}\,du$$

$$= \int \frac{1}{u^2 + u + 1}\,du = \int \frac{1}{\left(u+\frac{1}{2}\right)^2 + \left(\frac{\sqrt{3}}{2}\right)^2}\,du$$

$$= \frac{2}{\sqrt{3}} \tan^{-1}\left(\frac{u+\frac{1}{2}}{\frac{\sqrt{3}}{2}}\right) + C$$

$$= \frac{2}{\sqrt{3}} \tan^{-1}\left[\frac{1}{\sqrt{3}}\left(2\tan(x/2) + 1\right)\right] + C$$

27.
$$\left\{ \begin{array}{ll} u = \tan(x/2), & dx = \dfrac{2}{1+u^2}\,du \\[3mm] \sin x = \dfrac{2u}{1+u^2}, & \tan x = \dfrac{2u}{1-u^2} \end{array} \right\};$$

$$\int \frac{1}{\sin x + \tan x}\,dx = \int \frac{1}{\dfrac{2u}{1+u^2} + \dfrac{2u}{1-u^2}} \cdot \frac{2}{1+u^2}\,du$$

$$= \int \frac{1-u^2}{2u}\,du = \frac{1}{2}\int \left(\frac{1}{u} - u\right)\,du$$

$$= \frac{1}{2}\left(\ln|u| - \frac{1}{2}u^2\right) + C = \frac{1}{2}\ln\|\tan(x/2)\| - \frac{1}{4}[\tan(x/2)]^2 + C$$

29.
$$\left\{ \begin{array}{ll} u = \tan(x/2), & dx = \dfrac{2}{1+u^2}\,du \\[3mm] \sin x = \dfrac{2u}{1+u^2}, & \cos x = \dfrac{1-u^2}{1+u^2} \end{array} \right\};$$

$$\int \frac{1 - \cos x}{1 + \sin x}\, dx = \int \frac{1 - \dfrac{1 - u^2}{1 + u^2}}{1 + \dfrac{2u}{1 + u^2}} \cdot \frac{2}{1 + u^2}\, du$$

$$= \int \frac{4u^2}{(1 + u^2)(u + 1)^2}\, du$$

$$= \int \left[\frac{2u}{1 + u^2} - \frac{2}{u + 1} + \frac{2}{(u + 1)^2} \right] du$$

$$= \ln(u^2 + 1) - 2\ln|u + 1| - \frac{2}{u + 1} + C$$

$$= \ln\left[\frac{u^2 + 1}{(u + 1)^2} - \frac{2}{u + 1} \right] + C$$

$$= \ln\left[\frac{\tan^2(x/2) + 1}{(\tan(x/2) + 1)^2} \right] - \frac{2}{\tan(x/2) + 1} + C = \ln\left| \frac{1}{1 + \sin x} \right| - \frac{2}{\tan(x/2) + 1} + C$$

31. $\left\{ \begin{array}{l} u^2 = x \\ 2u\,du = dx \end{array} \right\};$ $\displaystyle \int \frac{x^{3/2}}{x + 1}\, dx = \int \frac{u^3}{u^2 + 1}\, 2u\,du$

$$= \int \frac{2u^4}{u^2 + 1}\, du = \int \left[2u^2 - 2 + \frac{2}{u^2 + 1} \right] du$$

$$= \tfrac{2}{3}u^3 - 2u + 2\tan^{-1} u + C = \tfrac{2}{3}x^{3/2} - 2x^{1/2} + 2\tan^{-1} x^{1/2} + C$$

$$\int_0^4 \frac{x^{3/2}}{x + 1}\, dx = \left[\tfrac{2}{3}x^{3/2} - 2x^{1/2} + 2\tan^{-1} x^{1/2} \right]_0^4 = \tfrac{4}{3} + 2\tan^{-1} 2$$

33.
$$\int_0^{\pi/2} \frac{\sin 2x}{2 + \cos x}\, dx = \int_0^{\pi/2} \frac{2\sin x \cos x}{2 + \cos x}\, dx$$

$$= \int_0^1 \frac{2u}{2 + u}\, du \qquad [u = \cos x, \quad du = -\sin x\,dx]$$

$$= \int_0^1 \left(2 - \frac{4}{2 + u} \right) du$$

$$= [2u - 4\ln|2 + u|]_0^1 = 2 + 4\ln\left(\tfrac{2}{3}\right)$$

35.
$$\left\{ \begin{array}{ll} u = \tan(x/2), & dx = \dfrac{2}{1 + u^2}\, du \\[2mm] \sin x = \dfrac{2u}{1 + u^2}, & \cos x = \dfrac{1 - u^2}{1 + u^2} \end{array} \right\};$$

$$\int \frac{1}{\sin x - \cos x - 1}\, dx = \int \frac{1}{\dfrac{2u}{1 + u^2} - \dfrac{1 - u^2}{1 + u^2} - 1} \cdot \frac{2}{1 + u^2}\, du$$

$$= \int \frac{1}{u - 1}\, du$$

$$= \ln|u - 1| + C = \ln|\tan(x/2) - 1| + C$$

$$\int_0^{\pi/3} \frac{1}{\sin x - \cos x - 1}\, dx = [\ln|\tan(x/2) - 1|]_0^{\pi/3} = \ln\left(\frac{\sqrt{3} - 1}{\sqrt{3}} \right)$$

37. $\begin{cases} u = \tan(x2), \quad dx = \dfrac{2}{1+u^2}\, du \\[2mm] \sin x = \dfrac{2u}{1+u^2}, \quad \cos x = \dfrac{1-u^2}{1+u^2} \end{cases}$;

$$\int \sec x\, dx = \int \frac{1}{\cos x}\, dx = \int \frac{1}{\dfrac{1-u^2}{1+u^2}} \cdot \frac{2}{1+u^2}\, du$$

$$= 2\int \frac{1}{1-u^2}\, du = 2\int \left[\frac{1/2}{1-u} + \frac{1/2}{1+u}\right] du$$

$$= \int \left[\frac{1}{1-u} + \frac{1}{1+u}\right] du = -\ln|1-u| + \ln|1+u| + C$$

$$= \ln\left|\frac{1+\tan(x/2)}{1-\tan(x/2)}\right| + C$$

39. $\displaystyle \int \csc x\, dx = \int \frac{\sin x}{\sin^2 x}\, dx = \int \frac{\sin x}{1-\cos^2 x}\, dx$

$$= -\int \frac{1}{1-u^2}\, du \qquad [u = \cos x, \quad du = -\sin x\, dx]$$

$$= \frac{1}{2}\int \left[\frac{1}{u-1} - \frac{1}{u+1}\right] du$$

$$= \frac{1}{2}\left[\ln|u-1| - \ln|u+1|\right] + C = \ln\sqrt{\frac{1-\cos x}{1+\cos x}} + C$$

41. $\begin{cases} u = \tanh(x/2), \quad dx = \dfrac{2}{1-u^2}\, du \\[2mm] \cosh x = \dfrac{1+u^2}{1-u^2}, \quad \operatorname{sech} x = \dfrac{1-u^2}{1+u^2} \end{cases}$;

$$\int \operatorname{sech} x\, dx = \int \frac{1-u^2}{1+u^2} \cdot \frac{2}{1-u^2}\, du$$

$$= \int \frac{2}{1+u^2}\, du = 2\tan^{-1} u + C = 2\tan^{-1}(\tanh(x/2)) + C$$

43. $\begin{cases} u = \tanh(x/2), \quad dx = \dfrac{2}{1-u^2}\, du \\[2mm] \sinh x = \dfrac{2u}{1-u^2}, \quad \cosh x = \dfrac{1+u^2}{1-u^2} \end{cases}$;

$$\int \frac{1}{\sinh x + \cosh x}\, dx = \int \frac{1}{\dfrac{2u}{1-u^2} + \dfrac{1+u^2}{1-u^2}} \cdot \frac{2}{1-u^2}\, du$$

$$= \int \frac{2}{(1+u)^2}\, du = \frac{-2}{u+1} + C = \frac{-2}{\tanh(x/2)+1} + C$$

SECTION 8.8

1. (a) $L_{12} = \frac{12}{12}[0 + 1 + 4 + 9 + 16 + 25 + 36 + 49 + 64 + 81 + 100 + 121] = 506$

   (b) $R_{12} = \frac{12}{12}[1 + 4 + 9 + 16 + 25 + 36 + 49 + 64 + 81 + 100 + 121 + 144] = 650$

   (c) $M_6 = \frac{12}{6}[1 + 9 + 25 + 49 + 81 + 121] = 572$

   (d) $T_{12} = \frac{12}{24}[0 + 2(1 + 4 + 9 + 16 + 25 + 36 + 49 + 64 + 81 + 100 + 121) + 144] = 578$

   (e) $S_6 = \frac{12}{36}[0 + 144 + 2(4 + 16 + 36 + 64 + 100) + 4(1 + 9 + 25 + 49 + 81 + 121)] = 576$

   $$\int_0^{12} x^2 \, dx = \left[\frac{1}{3}x^3\right]_0^{12} = 576$$

3. (a) $L_6 = \frac{3}{6}\left[\frac{1}{1+0} + \frac{1}{1+1/8} + \frac{1}{1+1} + \frac{1}{1+27/8} + \frac{1}{1+8} + \frac{1}{1+125/8}\right]$

   $= \frac{1}{2}\left[1 + \frac{8}{9} + \frac{1}{2} + \frac{8}{35} + \frac{1}{9} + \frac{8}{133}\right] \cong 1.394$

   (b) $R_6 = \frac{3}{6}\left[\frac{1}{1+1/8} + \frac{1}{1+1} + \frac{1}{1+27/8} + \frac{1}{1+8} + \frac{1}{1+125/8} + \frac{1}{1+27}\right]$

   $= \frac{1}{2}\left[\frac{8}{9} + \frac{1}{2} + \frac{8}{35} + \frac{1}{9} + \frac{8}{133} + \frac{1}{28}\right] \cong 0.9122$

   (c) $M_3 = \frac{3}{3}\left[\frac{1}{1+1/8} + \frac{1}{1+27/8} + \frac{1}{1+125/8}\right] = \frac{8}{9} + \frac{8}{35} + \frac{8}{133} \cong 1.1852$

   (d) $T_6 = \frac{3}{12}\left[1 + 2\left(\frac{8}{9} + \frac{1}{2} + \frac{8}{35} + \frac{1}{9} + \frac{8}{133}\right) + \frac{1}{28}\right] \cong 1.1533$

   (e) $S_3 = \frac{3}{18}\left\{1 + \frac{1}{28} + 2\left[\frac{1}{2} + \frac{1}{9}\right] + 4\left[\frac{8}{9} + \frac{8}{35} + \frac{8}{133}\right]\right\} \cong 1.1614$

5. (a) $\frac{1}{4}\pi \cong T_4 = \frac{1}{8}\left[1 + 2\left(\frac{1}{1+1/16} + \frac{1}{1+1/4} + \frac{1}{1+9/16}\right) + \frac{1}{1+1}\right]$

   $= \frac{1}{8}\left[1 + 2\left(\frac{16}{17} + \frac{4}{5} + \frac{16}{25}\right) + \frac{1}{2}\right] \cong 0.7828$

   (b) $\frac{1}{4}\pi \cong S_4 = \frac{1}{24}\left[1 + \frac{1}{2} + 2\left(\frac{16}{17} + \frac{4}{5} + \frac{16}{25}\right) + 4\left(\frac{64}{65} + \frac{64}{73} + \frac{64}{89} + \frac{64}{113}\right)\right] \cong 0.7854$

7. (a) $M_4 = \frac{2}{4}\left[\cos\left(\frac{-3}{4}\right)^2 + \cos\left(\frac{-1}{4}\right)^2 + \cos\left(\frac{1}{4}\right)^2 + \cos\left(\frac{3}{4}\right)^2\right] \cong 1.8440$

   (b) $T_8 = \frac{2}{16}\left[\cos(-1)^2 + 2\cos\left(\frac{-3}{4}\right)^2 + 2\cos\left(\frac{-1}{2}\right)^2 + 2\cos\left(\frac{-1}{4}\right)^2 + \right.$

   $\left. 2\cos(0)^2 + 2\cos\left(\frac{1}{4}\right)^2 + 2\cos\left(\frac{1}{2}\right)^2 + 2\cos\left(\frac{3}{4}\right)^2 + \cos(1)^2\right] \cong 1.7915$

(c) $S_4 = \dfrac{2}{24} \left\{ \cos(-1)^2 + \cos(1)^2 + 2\left[\cos\left(\tfrac{-1}{2}\right)^2 + \cos(0)^2 + \cos\left(\tfrac{1}{2}\right)^2\right] + \right.$

$$\left. 4\left[\cos\left(\tfrac{-3}{4}\right)^2 + \cos\left(\tfrac{-1}{4}\right)^2 + \cos\left(\tfrac{1}{4}\right)^2 + \cos\left(\tfrac{3}{4}\right)^2\right] \right\} \cong 1.8090$$

9. (a) $T_{10} = \dfrac{2}{20}\left[e^{-0^2} + 2e^{-(1/5)^2} + 2e^{-(2/5)^2} + 2e^{-(3/5)^2} + 2e^{-(4/5)^2} + 2e^{-1^2} + 2e^{-(6/5)^2} + \right.$

$$\left. 2e^{-(7/5)^2} + 2e^{-(8/5)^2} + 2e - (9/5)^2 + e - 2^2 \right] \cong 0.8818$$

(b) $S_5 = \dfrac{2}{30}\left\{ e^{-0^2} + e^{-2^2} + 2\left[e^{-(2/5)^2} + e^{-(4/5)^2} + e^{-(6/5)^2} + e^{-(8/5)^2}\right] + \right.$

$$\left. 4\left[e^{-(1/5)^2} + e^{-(3/5)^2} + e^{-1^2} + e^{-(7/5)^2} + e^{-(9/5)^2}\right]\right\} \cong 0.8821$$

11. Such a curve passes through the three points

$$(a_1, b_1), \quad (a_2, b_2), \quad (a_3, b_3)$$

iff

$$b_1 = a_1{}^2 A + a_1 B + C, \quad b_2 = a_2{}^2 A + a_2 B + C, \quad b_3 = a_3{}^2 A + a_3 B + C,$$

which happens iff

$$A = \frac{b_1(a_2 - a_3) - b_2(a_1 - a_3) + b_3(a_1 - a_2)}{(a_1 - a_3)(a_1 - a_2)(a_2 - a_3)},$$

$$B = -\frac{b_1(a_2{}^2 - a_3{}^2) - b_2(a_1{}^2 - a_3{}^2) + b_3(a_1{}^2 - a_2{}^2)}{(a_1 - a_3)(a_1 - a_2)(a_2 - a_3)},$$

$$C = \frac{a_1{}^2(a_2 b_3 - a_3 b_2) - a_2{}^2(a_1 b_3 - a_3 b_1) + a_3{}^2(a_1 b_2 - a_2 b_1)}{(a_1 - a_3)(a_1 - a_2)(a_2 - a_3)}.$$

13. (a) $\left| \dfrac{(b-a)^3}{12n^2} f''(c) \right| = \dfrac{27}{12n^2}\dfrac{1}{4c^{3/2}} \le \dfrac{9}{16n^2} < 0.01 \implies n^2 > \left(\dfrac{15}{2}\right)^2 \implies n \ge 8$

(b) $\left| \dfrac{(b-a)^5}{180n^4} f^{(4)}(c) \right| = \dfrac{243}{180n^4}\dfrac{15}{16c^{7/2}} \le \dfrac{81}{64n^4} < 0.01 \implies n > \dfrac{3}{2}\sqrt{5} \implies n \ge 4$

15. (a) $\left| \dfrac{(b-a)^3}{12n^2} f''(c) \right| = \dfrac{27}{12n^2}\dfrac{1}{4c^{3/2}} \le \dfrac{9}{16n^2} < 0.00001 \implies n > 75\sqrt{10} \implies n \ge 238$

(b) $\left| \dfrac{(b-a)^5}{180n^4} f^{(4)}(c) \right| = \dfrac{243}{180n^4}\dfrac{15}{16c^{7/2}} \le \dfrac{81}{64n^4} < 0.00001 \implies n > 15\left(\dfrac{5}{2}\right)^{1/4} \implies n \ge 19$

17. (a) $\left| \dfrac{(b-a)^3}{12n^2} f''(c) \right| = \dfrac{\pi^3}{12n^2}\sin c \le \dfrac{\pi^3}{12n^2} < 0.001 \implies n > 5\pi\sqrt{\dfrac{10\pi}{3}} \implies n \ge 51$

(b) $\left| \dfrac{(b-a)^5}{180n^4} f^{(4)}(c) \right| = \dfrac{\pi^5}{180n^4}\sin c \le \dfrac{\pi^5}{180n^4} < 0.001 \implies n > \pi\left(\dfrac{50\pi}{9}\right)^{1/4} \implies n \ge 7$

19. (a) $\left| \frac{(b-a)^3}{12n^2} f''(c) \right| = \frac{8}{12n^2} e^c \leq \frac{8}{12n^2} e^3 < 0.01 \implies n > 10e\sqrt{\frac{2e}{3}} \implies n \geq 37$

(b) $\left| \frac{(b-a)^5}{180n^4} f^{(4)}(c) \right| = \frac{32}{180n^4} e^c \leq \frac{8}{45n^4} e^3 < 0.01 \implies n > 2\left(\frac{10e^3}{9}\right)^{1/4} \implies n \geq 5$

21. (a) $\left| \frac{(b-a)^3}{12n^2} f''(c) \right| = \left| \frac{8}{12n^2} 2e^{-c^2}(2c^2 - 1) \right| \leq \frac{8}{3n^2} e^{-3/2} < 0.0001$

$\implies n > 100\sqrt{\frac{8}{3} e^{-3/2}} \implies n \geq 78$

(b) $\left| \frac{(b-a)^5}{2880n^4} f^{(4)}(c) \right| = \left| \frac{32}{2880n^4} 4e^{-c^2}\left(4c^4 - 12c^2 + 3\right) \right| \leq \frac{32}{2880n^4} 12 < 0.0001$

$\implies n > 10\left[\frac{32 \cdot 12}{2880}\right]^{1/4} \implies n \geq 7$

23. $f^{(4)}(x) = 0$ for all x;   therefore by (8.8.3) the theoretical error is zero

25. (a) $\left| T_2 - \int_0^1 x^2\,dx \right| = \frac{3}{8} - \frac{1}{3} = \frac{1}{24} = E_2^T$

(b) $\left| S_1 - \int_0^1 x^4\,dx \right| = \frac{5}{24} - \frac{1}{5} = \frac{1}{120} = E_1^S$

27. Let f be twice differentiable on $[a,b]$ with $f(x) > 0$ and $f''(x) > 0$, and let $P = \{x0, x1, x_2, \ldots, x_n\}$ be a regular partition of $[a, b]$. Figure A shows a typical subinterval with the approximating trapezoid ABCD. Since the area under the curve is less than the area of the trapezoid, we can conclude that

$$\int_a^b f(x)\,dx \leq T_n.$$

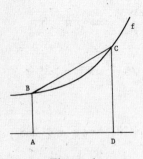

Figure A

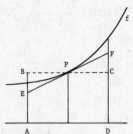

Figure B

Now consider Figure B. Since the triangles EBP and PFC are congruent, the area of the rectangle ABCD equals the area of the trapezoid AEFD, and since the area under the curve is greater than the area of AEFD it follows that

$$M_n \leq \int_a^b f(x)\,dx.$$

PROJECTS AND EXPLORATIONS

8.1.   (a) Since $f(x) = e^{-x^2}$ is continuous for all x, B is differentiable on $(-\infty, \infty)$

by Theorem 5.3.5. Since differentiability implies continuity, B is continuous on $(-\infty, \infty)$.

(b) By Theorem 5.3.5, $B'(x) = e^{-x^2}$. Since $B'(x) > 0$, B is increasing on $(-\infty, \infty)$.

(c) $B''(x) = -2xe^{-x^2}$;   the graph of B has a point of inflection at $(0, 1)$.

(d) Since $B(-x) = -B(x)$, B is an odd function. Use a graphing utility to show that

$$\lim_{x \to \infty} B(x) = 0.8862. \quad \text{By symmetry,} \quad \lim_{x \to -\infty} B(x) = -0.8862.$$

(e) Since B is increasing on $(-\infty, \infty)$, B has an inverse; $B^{-1}(0.5) \cong 0.5510$.

(f)

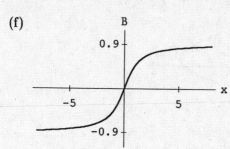
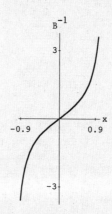

8.3.   (a) $F'(x) = \cos(\sin x)$ and $F''(x) = -\sin(\sin x)\cos x$;

The graph of F has a point of inflection at $(\pi/2, F(\pi/2))$; The graph of F is concave

down on $(1, \pi/2)$ and concave up on $(\pi/2, 3)$.

(b) On $[1, \pi/2]$ with $n = 10$:

trapezoidal rule: 0.59443   midpoint rule: 0.59448

On $[\pi/2, 3]$ with $n = 10$:

trapezoidal rule: 2.38756   midpoint rule: 2.38641

(c) Upperbound: 2.982;   lowerbound: 2.981

CHAPTER 9

SECTION 9.1

1. $y^2 = 8x$

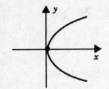

3. $(x + 1)^2 = -12(y - 3)$

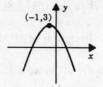

5. $(x - 1)^2 = 4y$

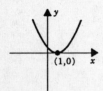

7. $(y - 1)^2 = -2(x - \frac{3}{2})$

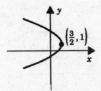

9. $y^2 = 2x$

vertex $(0, 0)$

focus $(\frac{1}{2}, 0)$

axis $y = 0$

directrix $x = -\frac{1}{2}$

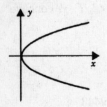

11. $x^2 = \frac{1}{2}(y + \frac{1}{2})$

vertex $(0, -\frac{1}{2})$

focus $(0, -\frac{3}{8})$

axis $x = 0$

directrix $y = -\frac{5}{8}$

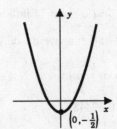

13. $(x + 2)^2 = -8(y - \frac{3}{2})$

vertex $(-2, \frac{3}{2})$

focus $(-2, -\frac{1}{2})$

axis $x = -2$

directrix $y = \frac{7}{2}$

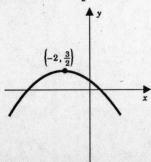

15. $(y + \frac{1}{2})^2 = x - \frac{3}{4}$

vertex $(\frac{3}{4}, -\frac{1}{2})$

focus $(1, -\frac{1}{2})$

axis $y = -\frac{1}{2}$

directrix $x = \frac{1}{2}$

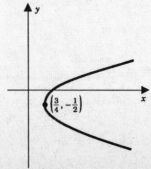

17. $\dfrac{x^2}{9} + \dfrac{y^2}{4} = 1$

center $(0,0)$

foci $(\pm\sqrt{5},\, 0)$

length of major axis 6

length of minor axis 4

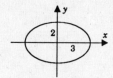

19. $\dfrac{x^2}{4} + \dfrac{y^2}{6} = 1$

center $(0,0)$

foci $(0,\, \pm\sqrt{2})$

length of major axis $2\sqrt{6}$

length of minor axis 4

21. $\dfrac{x^2}{9} + \dfrac{(y-1)^2}{4} = 1$

center $(0,1)$

foci $(\pm\sqrt{5},\, 1)$

length of major axis 6

length of minor axis 4

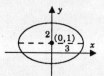

23. $\dfrac{(x-1)^2}{16} + \dfrac{y^2}{64} = 1$

center $(1,0)$

foci $(1,\, \pm 4\sqrt{3})$

length of major axis 16

length of minor axis 8

25. Foci $(-1,0),(1,0)$ $\implies$ center $(0,0)$, $c = 1$, and major axis parallel to x-axis.

Major axis 6 $\implies$ $a = 3$. Thus, $b = \sqrt{8}$.

Equation: $\dfrac{x^2}{9} + \dfrac{y^2}{8} = 1$.

27. Foci at $(1,3)$ and $(1,9)$ $\implies$ center $(1,6)$, $c = 3$, and major axis parallel to y-axis.

Minor axis 8 $\implies$ $b = 4$. Thus, $a = 5$.

Equation: $\dfrac{(x-1)^2}{16} + \dfrac{(y-6)^2}{25} = 1$.

29. Focus $(1,1)$ and center $(1,3)$ $\implies$ $c = 2$ and major axis parallel to y-axis.

Major axis 10 $\implies$ $a = 5$. Thus, $b = \sqrt{21}$.

Equation: $\dfrac{(x-1)^2}{21} + \dfrac{(y-3)^2}{25} = 1$.

31. Major axis 10 $\implies$ $a = 5$. Vertices at $(3,2)$ and $(3,-4)$ are then on minor axis parallel to y-axis.

Then, $b = 3$ and center is $(3, -1)$.

Equation: $\dfrac{(x-3)^2}{25} + \dfrac{(y+1)^2}{9} = 1$.

33. Foci $(-5, 0)$ and $(5, 0)$ $\implies$ $c = 5$ and center $(0, 0)$.

Transverse axis 6 $\implies$ $a = 3$. Thus, $b = 4$.

Equation: $\dfrac{x^2}{9} - \dfrac{y^2}{16} = 1$.

35. Foci $(0, -13)$ and $(0, 13)$ $\implies$ $c = 13$ and center $(0, 0)$.

Transverse axis 10 $\implies$ $a = 5$. Thus, $b = 12$.

Equation: $\dfrac{y^2}{25} - \dfrac{x^2}{144} = 1$.

37. Foci $(-5, 1)$ and $(5, 1)$ $\implies$ $c = 5$ and center $(0, 1)$.

Transverse axis 6 $\implies$ $a = 3$. Thus, $b = 4$.

Equation: $\dfrac{x^2}{9} - \dfrac{(y-1)^2}{16} = 1$.

39. Foci $(-1, -1)$ and $(-1, 1)$ $\implies$ $c = 1$ and center $(-1, 0)$.

Transverse axis $\frac{1}{2}$ $\implies$ $a = \frac{1}{4}$. Thus, $b = \frac{1}{4}\sqrt{15}$.

Equation: $\dfrac{y^2}{1/16} - \dfrac{(x+1)^2}{15/16} = 1$.

41. $x^2 - y^2 = 1$

center $(0, 0)$
transverse axis 2
vertices $(\pm 1, 0)$
foci $(\pm\sqrt{2}, 0)$
asymptotes $y = \pm x$

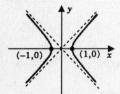

43. $\dfrac{x^2}{9} - \dfrac{y^2}{16} = 1$

center $(0, 0)$
transverse axis 6
vertices $(\pm 3, 0)$
foci $(\pm 5, 0)$
asymptotes $y = \pm\frac{4}{3}x$

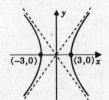

45. $\dfrac{y^2}{16} - \dfrac{x^2}{9} = 1$

center $(0,0)$
transverse axis 8
vertices $(0, \pm 4)$
foci $(0, \pm 5)$
asymptotes $y = \pm \frac{4}{3}x$

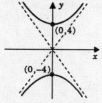

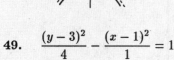

47. $\dfrac{(x-1)^2}{9} - \dfrac{(y-3)^2}{16} = 1$

center $(1, 3)$
transverse axis 6
vertices $(4, 3)$ and $(-2, 3)$
foci $(6, 3)$ and $(-4, 3)$
asymptotes $y - 3 = \pm \frac{4}{3}(x - 1)$

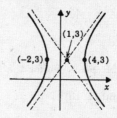

49. $\dfrac{(y-3)^2}{4} - \dfrac{(x-1)^2}{1} = 1$

center $(1, 3)$
transverse axis 4
vertices $(1, 5)$ and $(1, 1)$
foci $(1, 3 \pm \sqrt{5})$
asymptotes $y - 3 = \pm 2(x - 1)$

51. $\sqrt{(x-1)^2 + (y-2)^2} = \dfrac{|x + y + 1|}{\sqrt{2}}$ simplifies to $(x - y)^2 = 6x + 10y - 9$

53. Directrix has equation $x - y - 6 = 0$ since it has

slope 1 and passes through the point $(4, -2)$.

$$\sqrt{x^2 + (y-2)^2} = \frac{|x - y - 6|}{\sqrt{2}}$$

This simplifies to $(x + y)^2 = -12x + 20y + 28$.

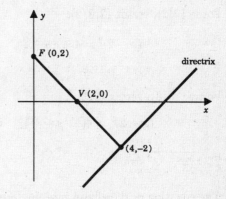

55. $P(x, y)$ is on the parabola with directrix l: $Ax + By + C = 0$ and focus $F(a, b)$ iff $d(P, l) = d(P, F)$

which happens iff $\dfrac{|Ax + By + C|}{\sqrt{A^2 + B^2}} = \sqrt{(x - a)^2 + (y - b)^2}.$

Squaring both sides of this equation and simplifying, we obtain

$$(Ay - Bx)^2 = (2aS + 2AC)x + (2bS + 2BC)y + c^2 - (a^2 + b^2)S$$

with $S = A^2 + B^2 \neq 0$.

57. We can choose the coordinate system so that the parabola has an equation of the form $y = \alpha x^2, \alpha > 0$. One of the points of intersection is then the origin and the other is of the form $(c, \alpha c^2)$. We will assume that $c > 0$.

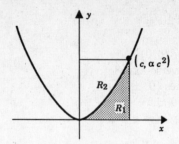

$$\text{area of } R_1 = \int_0^c \alpha x^2 dx = \frac{1}{3}\alpha c^3 = \frac{1}{3}A,$$
$$\text{area of } R_2 = A - \frac{1}{3}A = \frac{2}{3}A.$$

59. There are two possible positions for the focus:

$$(2, 2) \text{ and } (2, 10).$$

[The point $(5, 6)$ is equidistant from the focus and the directrix. This distance is 5. The points on the line $x = 2$ which are 5 units from $(5, 6)$ are $(2, 2)$ and $(2, 10)$.] These in turn give rise to two parabolas.

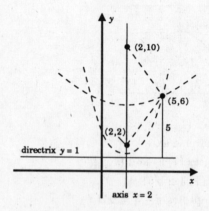

Focus $(2, 2)$, vertex $(2, 3/2)$:

$$(x - 2)^2 = 4(\tfrac{1}{2})(y - \tfrac{3}{2}), \quad \text{which simplifies to} \quad x^2 - 4x + 7 = 2y.$$

Focus $(2, 10)$, vertex $(2, 11/2)$:

$$(x - 2)^2 = 4(\tfrac{9}{2})(y - \tfrac{11}{2}), \quad \text{which simplifies to} \quad x^2 - 4x + 103 = 18y.$$

61. $2\sqrt{\pi^2 a^4 - A^2}/\pi a$

63. The equation of the ellipse is of the form

$$\frac{(x - 5)^2}{25} + \frac{y^2}{25 - c^2} = 1.$$

Substitute $x = 3$ and $y = 4$ in that equation and you find that $c = \pm \frac{5}{21}\sqrt{5}$. The foci are at $\left(5 \pm \frac{5}{21}\sqrt{5}, 0\right)$.

65. $e = \frac{3}{5}$ **67.** $e = \frac{4}{5}$

69. E_1 is fatter than E_2, more like a circle.

71. The ellipse tends to a line segment of length $2a$.

73. $x^2/9 + y^2 = 1$

75.   By the hint, $xy = X^2 - Y^2 = 1$. In the XY-system $a = 1$, $b = 1$, $c = \sqrt{2}$. We have center $(0,0)$, vertices $(\pm 1, 0)$, foci $(\pm\sqrt{2}, 0)$ and asymptotes $Y = \pm X$. Using

$$x = X + Y \quad \text{and} \quad y = X - Y$$

to convert to the xy-system, we find center $(0, 0)$, vertices $(1, 1)$ and $(-1, -1)$, foci $(\sqrt{2}, \sqrt{2})$ and $(-\sqrt{2}, -\sqrt{2})$, asymptotes $y = 0$ and $x = 0$, tranverse axis $2\sqrt{2}$.

77.
$$A = \frac{2b}{a}\int_a^{2a} \sqrt{x^2 - a^2}\, dx = \frac{2b}{a}\left[\frac{x}{2}\sqrt{x^2 - a^2} - \frac{a^2}{2}\ln\left(x + \sqrt{x^2 - a^2}\right)\right]_a^{2a}$$
$$= [2\sqrt{3} - \ln(2 + \sqrt{3})]ab$$

79. $e = \frac{5}{3}$ | **81.** $e = \sqrt{2}$

83.   The branches of H_1 open up less quickly than the branches of H_2.

85.   The hyperbola tends to a pair of parallel lines separated by the tranverse axis.

87.   In this case the length of the latus rectum is the width of the parabola at height $y = c$. With $y = c$, $4c^2 = x^2$,   and $x = \pm 2c$. The length of the latus rectum is thus $4c$.

89.
$$A = \int_{-2c}^{2c}\left(c - \frac{x^2}{4c}\right)dx = 2\int_0^{2c}\left(c - \frac{x^2}{4c}\right)dx = 2\left[cx - \frac{x^3}{12c}\right]_0^{2c} = \frac{8}{3}c^2$$
$$\overline{x} = 0 \quad \text{by symmetry}$$
$$\overline{y}A = \int_{-2c}^{2c}\frac{1}{2}\left(c^2 - \frac{x^4}{16c^2}\right)dx = \int_0^{2c}\left(c^2 - \frac{x^4}{16c^2}\right)dx = \left[c^2 x - \frac{x^5}{80c^2}\right]_0^{2c} = \frac{8}{5}c^3$$
$$\overline{y} = \left(\frac{8}{5}c^3\right)/\left(\frac{8}{3}c^2\right) = \frac{3}{5}c$$

91. $\dfrac{kx}{p(0)} = \tan\theta = \dfrac{dy}{dx}, \quad y = \dfrac{k}{2\,p(0)}x^2 + C$

In our figure $C = y(0) = 0$. Thus the equation of the cable is $y = kx^2/2p(0)$, the equation of a parabola.

93.   Start with any two parabolas γ_1, γ_2. By moving them we can see to it that they have equations of the following form:

$$\gamma_1: x^2 = 4c_1 y, \quad c_1 > 0; \qquad \gamma_2: x^2 = 4c_2 y, \quad c_2 > 0.$$

Now we change the scale for γ_2 so that the equation for γ_2 will look exactly like the equation for γ_1. Set $X = (c_1/c_2)\,x$, $\quad Y = (c_1/c_2)\,y$. Then

$$x^2 = 4c_2 y \quad \Longrightarrow \quad (c_2/c_1)^2 X^2 = 4c_2\,(c_2/c_1)\,Y \quad \Longrightarrow \quad X^2 = 4c_1 Y.$$

Now γ_2 has exactly the same equation as γ_1; only the scale, the units by which we measure distance, has changed.

95. Measure distances in miles and time in seconds. Place the origin at A and let $P(x,y)$ be the site of the crash. Then

$$d(P,B) - d(P,A) = (4)(0.20) = 0.80.$$

This places P on the right branch of the hyperbola
$$\frac{(x+1)^2}{(0.4)^2} - \frac{y^2}{1 - (0.4)^2} = 1.$$

Also

$$d(P,C) - d(P,A) = 6(0.20) = 1.20.$$

This places P on the left branch of the hyperbola
$$\frac{(x-1)^2}{(0.6)^2} - \frac{y^2}{1 - (0.6)^2} = 1.$$

Solve the two equations simultaneously keeping in mind the conditions of the problem and you will find that $x \cong -0.248$ and $y \cong 1.459$. The impact takes place about a quarter of a mile west of A and one and a half miles north.

SECTION 9.2

1–7.

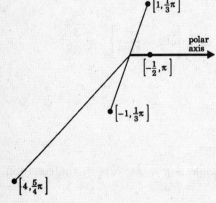

$$[1, \tfrac{1}{3}\pi]$$
polar axis
$$[-\tfrac{1}{2}, \pi]$$
$$[-1, \tfrac{1}{3}\pi]$$
$$[4, \tfrac{5}{4}\pi]$$

9. $x = 3\cos\tfrac{1}{2}\pi = 0$

$\quad y = 3\sin\tfrac{1}{2}\pi = 3$

11. $x = -\cos(-\pi) = 1$

$\quad y = -\sin(-\pi) = 0$

13. $x = -3\cos\left(-\tfrac{1}{3}\pi\right) = -\tfrac{3}{2}$

$\quad y = -3\sin\left(-\tfrac{1}{3}\pi\right) = \tfrac{3}{2}\sqrt{3}$

15. $x = 3\cos\left(-\tfrac{1}{2}\pi\right) = 0$

$\quad y = 3\sin\left(-\tfrac{1}{2}\pi\right) = -3$

17. $\left. \begin{array}{l} r^2 = 0^2 + 1^2, \quad r = \pm 1 \\ r = 1: \quad \cos\theta = 0 \text{ and } \sin\theta = 1 \\ \theta = \frac{1}{2}\pi \end{array} \right\}$ $\left[1, \frac{1}{2}\pi + 2n\pi\right], \quad \left[-1, \frac{3}{2}\pi + 2n\pi\right]$

19. $\left. \begin{array}{l} r^2 = (-3)^2 + 0^2 = 9, \quad r = \pm 3 \\ r = 3: \quad \cos\theta = -1 \text{ and } \sin\theta = 0 \\ \theta = \pi \end{array} \right\}$ $[3, \pi + 2n\pi], \quad [3, 2n\pi]$

21. $\left. \begin{array}{l} r^2 = 2^2 + (-2)^2 = 8, \quad r = \pm 2\sqrt{2} \\ r + 2\sqrt{2}: \quad \cos\theta = \frac{1}{2}\sqrt{2}, \quad \sin\theta = -\frac{1}{2}\sqrt{2} \\ \theta = \frac{7}{4}\pi \end{array} \right\}$ $\left[2\sqrt{2}, \frac{7}{4}\pi + 2n\pi\right], \quad \left[-2\sqrt{2}, \frac{3}{4}\pi + 2n\pi\right]$

23. $\left. \begin{array}{l} r^2 = \left(4\sqrt{3}\right)^2 + 4^2 = 64, \quad r \pm 8 \\ r = 8: \quad \cos\theta = \frac{1}{2}\sqrt{3}, \quad \sin\theta = \frac{1}{2} \\ r = \frac{1}{6}\pi \end{array} \right\}$ $\left[8, \frac{1}{6}\pi + 2n\pi\right], \quad \left[-8, \frac{7}{6}\pi + 2n\pi\right]$

25.
$$d^2 = (x_1 - x_2)^2 + (y_1 - y_2)^2 = (r_1\cos\theta_1 - r_2\cos\theta_2)^2 + (r_1\sin\theta_1 - r_2\sin\theta_2)^2$$

$$= r_1{}^2\cos^2\theta_1 - 2r_1 r_2\cos\theta_1\cos\theta_2 + r_2{}^2\cos^2\theta_2$$

$$+ r_1{}^2\sin^2\theta_1 - 2r_1 r_2\sin\theta_1\sin\theta_2 + r_2{}^2\sin^2\theta_2$$

$$= r_1{}^2 + r_2{}^2 - 2r_1 r_2\left(\cos\theta_1\cos\theta_2 + \sin\theta_1\sin\theta_2\right)$$

$$= r_1{}^2 + r_2{}^2 - 2r_1 r_2\cos\left(\theta_1 - \theta_2\right)$$

$$d = \sqrt{r_1{}^2 + r_2{}^2 - 2r_1 r_2\cos\left(\theta_1 - \theta_2\right)}$$

27.   (a) $\left[\frac{1}{2}, \frac{11}{6}\pi\right]$ (b) $\left[\frac{1}{2}, \frac{5}{6}\pi\right]$ (c) $\left[\frac{1}{2}, \frac{7}{6}\pi\right]$

29.   (a) $\left[2, \frac{2}{3}\pi\right]$ (b) $\left[2, \frac{5}{3}\pi\right]$ (c) $\left[2, \frac{1}{3}\pi\right]$

31.   about the x-axis?: $r = 2 + \cos(-\theta) \implies r = 2 + \cos\theta, \quad$ yes.

about the y-axis?: $r = 2 + \cos(\pi - \theta) \implies r = 2 - \cos\theta, \quad$ no.

about the origin?: $r = 2 + \cos(\pi + \theta) \implies r = 2 - \cos\theta, \quad$ no.

33.   about the x-axis?: $r\left(\sin(-\theta) + \cos(-\theta)\right) = 1 \implies r\left(-\sin\theta + \cos\theta\right) = 1, \quad$ no.

about the y-axis?: $r\left(\sin(\pi - \theta) + \cos(\pi - \theta)\right) = 1 \implies r\left(\sin\theta - \cos\theta\right) = 1, \quad$ no.

about the origin?: $r\left(\sin(\pi + \theta) + \cos(\pi + \theta)\right) = 1 \implies r\left(-\sin\theta - \cos\theta\right) = 1, \quad$ no.

35.   about the x-axis?: $r^2\sin(-2\theta) = 1 \implies -r^2\sin 2\theta = 1, \quad$ no.

about the y-axis?: $r^2\sin(2(\pi - \theta)) = 1 \implies -r^2\sin 2\theta = 1, \quad$ no.

about the origin?: $r^2\sin(2(\pi + \theta)) = 1 \implies r^2\sin 2\theta = 1, \quad$ yes.

37.
$$x = 2$$
$$r \cos \theta = 2$$

39.
$$2xy = 1$$
$$2 (r \cos \theta)(r \sin \theta) = 1$$
$$r^2 \sin 2\theta = 1$$

41. $x^2 + (y-2)^2 = 4$

$x^2 + y^2 - 4y = 0$

$r^2 - 4r \sin \theta = 0$

$$r = 4 \sin \theta$$

[note: division by r okay
since $[0, 0,]$ on curve]

43.
$$y = x$$
$$r \sin \theta = r \cos \theta$$
$$\tan \theta = 1$$
$$\theta = \pi/4$$

45. $x^2 + y^2 + x = \sqrt{x^2 + y^2}$

$r^2 + r \cos \theta = r$

$$r = 1 - \cos \theta$$

47.
$$(x^2 + y^2)^2 = 2xy$$
$$r^4 = 2(r \cos \theta)(r \sin \theta)$$
$$r^2 = \sin 2\theta$$

49. the horizontal line $y = 4$

51. the line $y = \sqrt{3}x$

53.
$$r = 2 (1 - \cos \theta)^{-1}$$
$$r - r \cos \theta = 2$$
$$\sqrt{x^2 + y^2} - x = 2$$
$$x^2 + y^2 = (x + 2)^2$$
$$y^2 = 4(x + 1)$$

a parabola

55.
$$r = 3 \cos \theta$$
$$r^2 = 3r \cos \theta$$
$$x^2 + y^2 = 3x$$

a circle

57. the line $y = 2x$

59. the vertical line $x = 0$

59.
$$r = \frac{4}{2 - \cos \theta}$$
$$2r - r \cos \theta = 4$$
$$2\sqrt{x^2 + y^2} - x = 4$$
$$4(x^2 + y^2) = (x + 4)^2$$
$$3x^2 + 4y^2 - 8x = 16$$

an ellipse

61.
$$r = \frac{4}{1 - \cos \theta}$$
$$r - r \cos \theta = 4$$
$$\sqrt{x^2 + y^2} - x = 4$$
$$x^2 + y^2 = (x + 4)^2$$
$$y^2 = 8x + 16$$

a parabola

63.
$$r = a \sin\theta + b \cos\theta$$
$$r^2 = a r \sin\theta + b r \cos\theta$$
$$x^2 + y^2 = ay + bz$$
$$\left(x - \frac{b}{2}\right)^2 + \left(y - \frac{a}{2}\right)^2 = \frac{a^2 + b^2}{4}$$

center: $(b/2, a/2)$;   radius: $\frac{1}{2}\sqrt{a^2 + b^2}$

65.
$$\frac{1}{2}(r\cos\theta + d) = r$$
$$r = \frac{d}{2 - \cos\theta}$$

SECTION 9.3

1.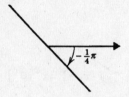

3.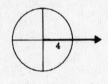

5.

7.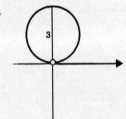

9.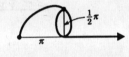

11.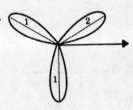

13.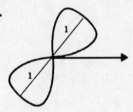

15.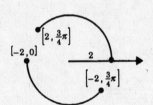

17.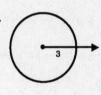

19.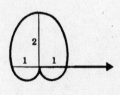

21.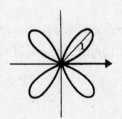

23.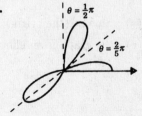

25.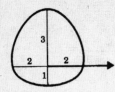

27.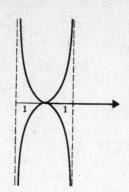

29.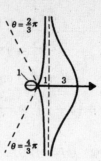

31.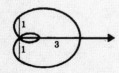

33.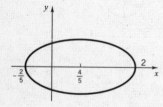

35.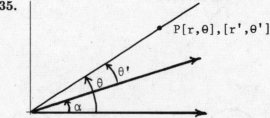

37. (a) $r = \dfrac{ce}{1 - e\,\cos\theta}$

$r - r\cos\theta = ec$

$\sqrt{x^2 + y^2} - e\,x = ec$

$x^2 + y^2 = e^2(x + c)^2$

$(1 - e^2)x^2 - 2e^2cx + y^2 = e^2c^2$

$\left(x - \dfrac{e^2c}{1 - e^2}\right)^2 + \dfrac{y^2}{1 - e^2} = \dfrac{e^2c^2}{(1 - e^2)^2}$

$\dfrac{\left(x - e^2c/(1 - e^2)\right)^2}{\left(ec/(1 - e^2)\right)^2} + \dfrac{y^2}{\left(ec/\sqrt{1 - e^2}\right)^2} = 1$

(b) center: $\left(\dfrac{e^2c}{1 - e^2}, 0\right)$;

length of major axis: $\dfrac{2ec}{1 - e^2}$;

length of minor axis: $\dfrac{2ec}{\sqrt{1 - e^2}}$

39. (a) a point (the origin);   (b) the graph "approaches" a parabola

(c) the graph is an ellipse;   center at $\left(\tfrac{2}{3}d, 0\right)$, length of major axis $\tfrac{8}{3}d$.

41.   "Butterfly" curves. The graph for the case $k = 2$ is:

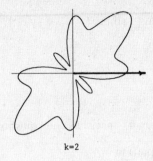

k=2

SECTION * 9.4

1.   (a) $\frac{1}{2}$ unit to the right of the pole   (c)

     (b) 2 units

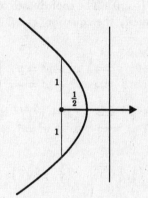

3.   the parabola of Exercise 1 rotated by π radians

5.   (a) $e = \frac{2}{3}$ (b) 2 units to the left of the   (g)

     pole and $\frac{2}{5}$ units to the right of the pole

     (c) $\frac{4}{5}$ units to the left of the pole

     (d) $\frac{8}{5}$ units to the left of the pole

     (e) $\frac{4}{5}\sqrt{5}$ units (about 1.79 units)

     (f) $\frac{4}{3}$ units

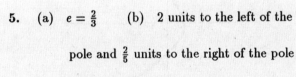

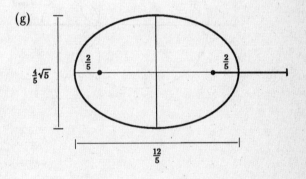

7.   the ellipse of Exercise 5 rotated by $\frac{1}{2}\pi$ radians

9.   (a) $e = 2$ (b) 2 units to the right of the pole   (f)

     and 6 units to the right of the pole

     (c) 4 units to the right of the pole

     (d) 8 units to the right of the pole   (e) 12 units

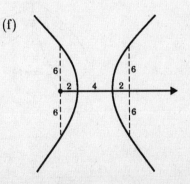

11.   the hyperbola of Exercise 9 rotated by $\frac{3}{2}\pi$ radians

13.   ellipse: $\dfrac{x^2}{48} + \dfrac{(y+4)^2}{64} = 1$ **15.**   ellipse: $x^2/9 + (y-4)^2/25 = 1$

SECTION 9.5

1.   yes; $[1, \pi] = [-1, 0]$ and the pair $r = -1,\ \theta = 0$ satisfies the equation

3.   yes;   the pair $r = \frac{1}{2},\ \theta = \frac{1}{2}\pi$ satisfies the equation

5. $[2, \pi] = [-2, 0]$. The coordinates of $[-2, 0]$ satisfy the equation $r^2 = 4\cos\theta$, and the coordinates of $[2, \pi]$ satisfy the equation $r = 3 + \cos\theta$.

7. $\begin{bmatrix} r = \sin\theta & \Longrightarrow & x^2 + y^2 = y \\ r = -\cos\theta & \Longrightarrow & x^2 + y^2 = -x \end{bmatrix} \Longrightarrow \left\{ x + y = 0, 2x^2 = -x \right\} \Longrightarrow x = 0, -\frac{1}{2};\quad (0,0), \left(-\frac{1}{2}, \frac{1}{2}\right)$

9. $\begin{bmatrix} r = \cos^2\theta & \Longrightarrow & (x^2 + y^2)^{3/2} = x^2 \\ r = -1 & \Longrightarrow & x^2 + y^2 = 1 \end{bmatrix} \Longrightarrow x = \pm 1,\ y = 0;\quad (1,0), (-1,0)$

11. $(0,0),\quad \left(\frac{1}{4}, \frac{1}{4}\sqrt{3}\right),\quad \left(\frac{1}{4}, -\frac{1}{4}\sqrt{3}\right)$ **13.** $\left(\frac{3}{2}, 2\right)$

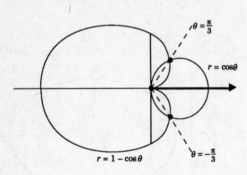

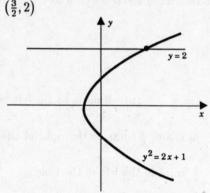

15. $(0,0),\quad (1,0)$

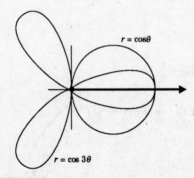

17.   **(a)**

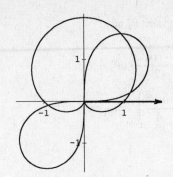

(b)   The curves intersect at the pole and at:

$[1.172, 0.173], \quad [1.86, 1.036], \quad [0.90, 3.245]$

19.   **(a)**

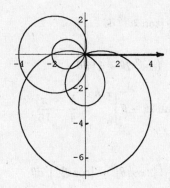

(b)   The curves intersect at the pole and at:

$r = 1 - 3\sin\theta$	$r = 2 - 5\sin\theta$
$[-2, 0]$	$[2, \pi]$
$[3.800, 3.510]$	$[3.800, 3.510]$
$[2.412, 4.223]$	$[-2.412, 1.081]$
$[-1.267, 0.713]$	$[-1.267, 0.713]$

SECTION 9.6

1.

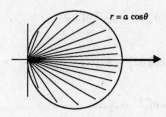

$r = a\cos\theta$

$$A = \int_{-\pi/2}^{\pi/2} \frac{1}{2} \left[a\cos\theta \right]^2 d\theta$$

$$= a^2 \int_0^{\pi/2} \frac{1 + \cos 2\theta}{2} \, d\theta$$

$$= a^2 \left[\frac{\theta}{2} + \frac{\sin 2\theta}{4} \right]_0^{\pi/2} = \frac{1}{4}\pi a^2$$

3.

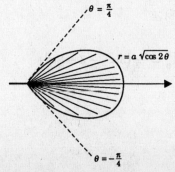

$\theta = \frac{\pi}{4}$

$r = a\sqrt{\cos 2\theta}$

$\theta = -\frac{\pi}{4}$

$$A = \int_{-\pi/4}^{\pi/4} \frac{1}{2} \left[a\sqrt{\cos 2\theta} \right]^2 d\theta$$

$$= a^2 \int_0^{\pi/4} \cos 2\theta \, d\theta$$

$$= a^2 \left[\frac{\sin 2\theta}{2} \right]_0^{\pi/4} = \frac{1}{2}a^2$$

5.

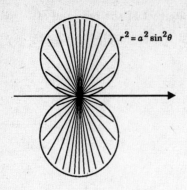

$$A = 2 \int_0^\pi \frac{1}{2} \left(a^2 \sin^2 \theta \right) d\theta$$

$$= a^2 \int_0^\pi \frac{1 - \cos 2\theta}{2} d\theta$$

$$= a^2 \left[\frac{\theta}{2} - \frac{\sin 2\theta}{4} \right]_0^\pi = \frac{1}{2} \pi a^2$$

7.

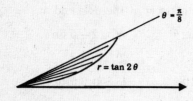

$$A = \int_0^{\pi/8} \frac{1}{2} \left[\tan 2\theta \right]^2 d\theta$$

$$= \frac{1}{2} \int_0^{\pi/8} \left(\sec^2 2\theta - 1 \right) d\theta$$

$$= \frac{1}{2} \left[\frac{1}{2} \tan 2\theta - \theta \right]_0^{\pi/8} = \frac{1}{4} - \frac{\pi}{16}$$

9.

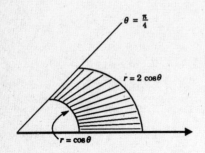

$$A = \int_0^{\pi/4} \frac{1}{2} \left([2 \cos \theta]^2 - [\cos \theta]^2 \right) d\theta$$

$$= \frac{3}{2} \int_0^{\pi/4} \frac{1 + \cos 2\theta}{2} d\theta$$

$$= \frac{3}{2} \left[\frac{\theta}{2} + \frac{\sin 2\theta}{4} \right]_0^{\pi/4} = \frac{3}{16} \pi + \frac{3}{8}$$

11.

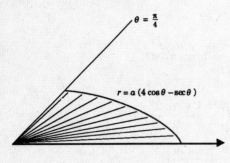

$$A = \int_0^{\pi/4} \frac{1}{2} \left[a \left(4 \cos \theta - \sec \theta \right) \right]^2 d\theta$$

$$= \frac{a^2}{2} \int_0^{\pi/4} \left[16 \cos^2 \theta - 8 + \sec^2 \theta \right] d\theta$$

$$= \frac{a^2}{2} \int_0^{\pi/4} \left[8 \left(1 + \cos 2\theta \right) - 8 + \sec^2 \theta \right] d\theta$$

$$= \frac{a^2}{2} \left[4 \sin 2\theta + \tan \theta \right]_0^{\pi/4} = \frac{5}{2} a^2$$

13.

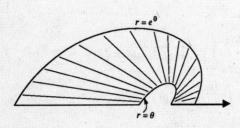

$$A = \int_0^\pi \frac{1}{2} \left([e^\theta]^2 - [\theta]^2 \right) d\theta$$

$$= \frac{1}{2} \int_0^\pi \left(e^{2\theta} - \theta^2 \right) d\theta$$

$$= \frac{1}{2} \left[\frac{1}{2} e^{2\theta} - \frac{1}{3} \theta^3 \right]_0^\pi = \frac{1}{12} \left(3 e^{2\pi} - 3 - 2\pi^3 \right)$$

15.

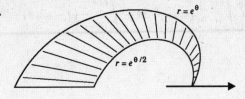

$$A = \int_0^\pi \frac{1}{2} \left(\left[e^\theta \right]^2 - \left[e^{\theta/2} \right]^2 \right) d\theta$$

$$= \frac{1}{2} \int_0^\pi \left(e^{2\theta} - e^\theta \right) d\theta$$

$$= \frac{1}{2} \left[\frac{1}{2} e^{2\theta} - e^\theta \right]_0^\pi = \frac{1}{4} \left(e^{2\pi} + 1 - 2e^\pi \right)$$

17.

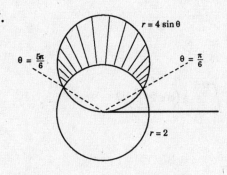

$$A = \int_{\pi/6}^{5\pi/6} \frac{1}{2} \left([4\sin\theta]^2 - [2]^2 \right) d\theta$$

19.

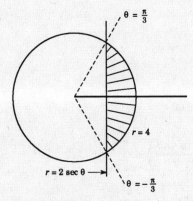

$$A = \int_{-\pi/3}^{\pi/3} \frac{1}{2} \left([4]^2 - [2\sec\theta]^2 \right) d\theta$$

21.

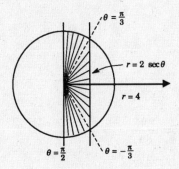

$$A = 2 \left\{ \int_0^{\pi/3} \frac{1}{2} (2\sec\theta)^2 \, d\theta + \int_{\pi/3}^{\pi/2} \frac{1}{2} (4)^2 \, d\theta \right\}$$

23.

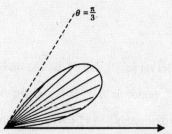

$$A = \int_0^{\pi/3} \frac{1}{2} (2\sin 3\theta)^2 \, d\theta$$

25.

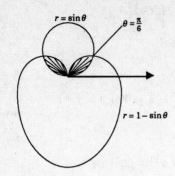

$$A = 2\left\{\int_0^{\pi/6} \frac{1}{2}(\sin\theta)^2\,d\theta + \int_{\pi/6}^{\pi/2} \frac{1}{2}(1-\sin\theta)^2\,d\theta\right\}$$

27.

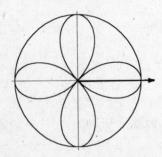

$$A = \pi - 8\int_0^{\pi/4} \frac{1}{2}(\cos 2\theta)^2\,d\theta$$

29. The area of one petal of the curve $r = a\cos 2n\theta$ is given by:

$$2\int_0^{\pi/4n} \frac{1}{2}(a\cos 2n\theta)^2\,d\theta = a^2\int_0^{\pi/4n} \cos^2 2n\theta\,d\theta$$

$$= a^2\int_0^{\pi/4n}\left(\frac{1}{2} + \frac{\cos 4n\theta}{2}\right)d\theta$$

$$= a^2\left[\frac{1}{2}\theta + \frac{\sin 4n\theta}{8n}\right]_0^{\pi/4n} = \frac{\pi a^2}{8n}$$

The total area enclosed by $r = a\cos 2n\theta$ is $\dfrac{\pi a^2}{2}$.

The area of one petal of the curve $r = a\sin 2n\theta$ is given by:

$$A = 2\int_0^{\pi/4n} \frac{1}{2}(a\sin 2n\theta)^2\,d\theta = a^2\int_0^{\pi/4n}\left(\frac{1}{2} + \frac{\cos 4n\theta}{2}\right)d\theta = \frac{\pi a^2}{8n}$$

and the total area enclosed by the cruve is $\dfrac{\pi a^2}{2}$.

31. Let $P = \{\alpha = \theta_0, \theta_1, \theta_2, \ldots, theta_n = \beta\}$ be a partition of the interval $[\alpha, \beta]$. Let θ_i^* be the midpoint of $[\theta_{i-1}, \theta_i]$ and let $r_i^* = f(\theta_i^*)$. The area of the ith "triangular" region is $\frac{1}{2}(r_i^*)\Delta\theta_i$, where $\Delta\theta_i = \theta_i - \theta_{i-1}$, and the rectangular coordinates of its centroid are(approximately) $\left(\frac{2}{3}r_i^*\cos\theta_i^*, \frac{2}{3}r_i^*\sin\theta_i^*\right)$.

The centroid $(\overline{x}_p, \overline{y}_p)$ of the union of the triangular regions satisfies the following equations

$$\overline{x}_p A_p = \frac{1}{3}\left(r_1^*\right)^3 \cos\theta_1 \Delta\theta_1 + \frac{1}{3}\left(r_2^*\right)^3 \cos\theta_2 \Delta\theta_2 + \cdots + \frac{1}{3}\left(r_n^*\right)^3 \cos\theta_n \Delta\theta_n$$

$$\overline{y}_p A_p = \frac{1}{3}\left(r_1^*\right)^3 \sin\theta_1 \Delta\theta_1 + \frac{1}{3}\left(r_2^*\right)^3 \sin\theta_2 \Delta\theta_2 + \cdots + \frac{1}{3}\left(r_n^*\right)^3 \sin\theta_n \Delta\theta_n$$

As $\|P\| \to 0$, the union of the triangular regions tends to the region Ω and the equations above tend to

$$\overline{x}\, A = \int_\alpha^\beta \frac{1}{3} r^3 \cos\theta \, d\theta$$

$$\overline{x}\, A = \int_\alpha^\beta \frac{1}{3} r^3 \cos\theta \, d\theta$$

The reslut follows from the fact that $A = \int_\alpha^\beta \frac{1}{2} r^2 \cos\theta \, d\theta$.

33. Since the region enclosed by the cardiod $r = 1 + \cos\theta$ is symmetric with respect to the x-axis, $\overline{y} = 0$.

To find $\overline{x}$:

$$A = \int_0^{2\pi} r^2 \, d\theta = \int_0^{2\pi} (1 + \cos\theta)^2 \, d\theta$$

$$= \int_0^{2\pi} (1 + 2\cos\theta + \cos^2\theta)\, d\theta$$

$$= \int_0^{2\pi} \left(\frac{3}{2} + 2\cos\theta \frac{1}{2}\cos 2\theta\right) d\theta$$

$$= \left[\frac{3}{2} + 2\sin\theta + \frac{1}{4}\sin 2\theta\right]_0^{2\pi} = 3\pi$$

and

$$\frac{2}{3}\int_0^{2\pi} r^3 \cos\theta \, d\theta = \frac{2}{3}\int_0^{2\pi} (1 + \cos\theta)^3 \cos\theta \, d\theta$$

$$= \frac{2}{3}\int_0^{2\pi} \left(\cos\theta + 3\cos^2\theta + 3\cos^3\theta + \cos^4\theta\right) d\theta$$

$$= \frac{2}{3}\int_0^{2\pi} \left(\frac{15}{8} + 4\cos\theta + 2\cos 2\theta + \frac{1}{8}\cos 4\theta - 3\sin^2\theta \cos\theta\right) d\theta$$

$$= \frac{2}{3}\left[\frac{15}{8}\theta + 4\sin\theta + \sin 2\theta + \frac{1}{32}\sin 4\theta - \sin^3\theta\right]_0^{2\pi} = \frac{5}{2}\pi$$

Thus $\quad \overline{x} = \dfrac{5\pi/2}{3\pi} = \dfrac{5}{6}.$

35.   (a) $y^2 = x^2 \left(\dfrac{a-x}{a+x} \right)$

(b) Let $a = 2$

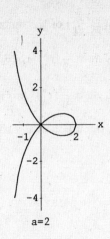

$a = 2$

$$r^2 \sin^2 \theta = r^2 \cos^2 \theta \left(\frac{a - r\cos\theta}{a + r\cos\theta} \right)$$

$$\sin^2 \theta (a + r\cos\theta) = \cos^2 \theta (a - r\cos\theta)$$

$$r\cos\theta = a\cos 2\theta$$

$$r = a\cos 2\theta \sec\theta$$

(c) $A = \displaystyle\int_{3\pi/4}^{5\pi/4} \frac{1}{2} a^2 \cos^2 2\theta \, \sec^2 \theta \, d\theta$

$\displaystyle = 2\int_{3\pi/4}^{5\pi/4} \cos^2 2\theta \, \sec^2 \theta \, d\theta \qquad (a=2)$

$\displaystyle = 2\int_{3\pi/4}^{5\pi/4} \frac{\left(2\cos^2\theta - 1\right)^2}{\cos^2\theta} \, d\theta$

$\displaystyle = 2\int_{3\pi/4}^{5\pi/4} \left(4\cos^2\theta - 4 + \sec^2\theta\right) d\theta$

$\displaystyle = 2\int_{3\pi/4}^{5\pi/4} \left(-2 + 2\cos 2\theta + \sec^2\theta\right) d\theta$

$\displaystyle = 2\left[-2\theta + \sin 2\theta + \tan\theta\right]_{3\pi/4}^{5\pi/4} = 8 - 2\pi$

SECTION 9.7

1. $4x = (y - 1)^2$

3. $y = 4x^2 + 1, \quad x \geq 0$

5. $9x^2 + 4y^2 = 36$

7. $1 + x^2 = y^2$

9. $y = 2 - x^2, \quad -1 \leq x \leq 1$

11. $2y - 6 = x, \quad -4 \leq x \leq 4$

13. $y = x - 1$

15. $xy = 1$

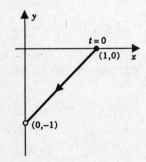

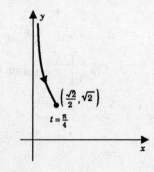

17. $2x + y = 11$

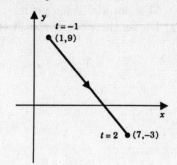

19. $x = \sin \frac{1}{2}\pi y$

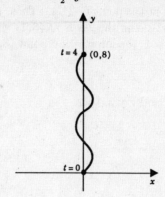

21. $1 + x^2 = y^2$

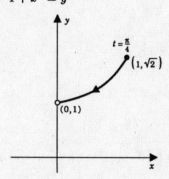

23. (a) $x(t) = -\sin 2\pi t, \quad y(t) = \cos 2\pi t$

(b) $x(t) = \sin 4\pi t, \quad y(t) = \cos 4\pi t$

(c) $x(t) = \cos \frac{1}{2}\pi t, \quad y(t) = \sin \frac{1}{2}\pi t$

(d) $x(t) = \cos \frac{3}{2}\pi t, \quad y(t) = -\sin \frac{3}{2}\pi t$

25. $x(t) = \tan \frac{1}{2}\pi t, \quad y(t) = 2$

27. $x(t) = 3 + 5t, \quad y(t) = 7 - 2t$

29. $x(t) = \sin^2 \pi t, \quad y(t) = -\cos \pi t$

31. $x(t) = (2 - t)^2, \quad y(t) = (2 - t)^3$

33. $x(t) = t(b - a) + a, \quad y(t) = f(t(b - a) + a)$

35.

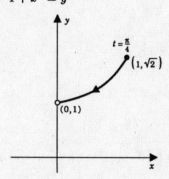

$$A = \int_0^{2\pi} x(t) \, y'(t) \, dt$$

$$= r^2 \int_0^{2\pi} (1 - \cos t) \, dt$$

$$= r^2 \left[t - \sin t\right]_0^{2\pi} = 2\pi r^2$$

37. (a) $V_x = 2\pi \bar{y} A = 2\pi \left(\frac{3}{4} r\right) \left(2\pi r^2\right) = 3\pi^2 r^3$

(b) $V_y = 2\pi \bar{x} A = 2\pi (\pi r) \, 2\pi r^2 = 4\pi^3 r^3$

39. $x(t) = -a \cos t, \quad y(t) = b \sin t \qquad t \in [0, \pi]$

41. (a) Equation for the ray: $y + 2x = 17, \quad x \geq 6$.

Equation for the circle: $(x - 3)^2 + (y - 1)^2 = 25$.

Simultaneous solution of these equations gives the points of intersection: $(6, 5)$ and $(8, 1)$.

(b)   The particle on the ray is at $(6, 5)$ when $t = 0$.   However, when $t = 0$ the particle on the circle is at the point $(-2, 1)$. Thus, the intersection point $(6, 5)$ is not a collision point. The particle on the ray is at $(8, 1)$ when $t = 1$. Since the particle on the circle is also at $(8, 1)$ when $t = 1$, the intersection point $(8, 1)$ is a collision point.

43.   If $x(r) = x(s)$ and $r \neq s$, then

$$r^2 - 2r = s^2 - 2s$$

$$r^2 - s^2 = 2r - 2s$$

(1) $$r + s = 2.$$

If $y(r) = y(s)$ and $r \neq s$, then

$$r^3 - 3r^2 + 2r = s^3 - 3s^2 + 2s$$

$$\left(r^3 - s^3\right) - 3\left(r^2 - s^2\right) + 2\left(r - s\right) = 0$$

(2) $$\left(r^2 + rs + s^2\right) - 3\left(r + s\right) + 2 = 0.$$

Simultaneous solution of (1) and (2) gives $r = 0$ and $r = 2$. Since $(x(0), y(0)) = (0, 0) = (x(2), y(2))$, the curve intersects itself at the origin.

45.   Suppose that $r, s \in [0, 4]$ and $r \neq s$.

$$x(r) = x(s)   \implies   \sin 2\pi r = \sin 2\pi s.$$

$$y(r) = y(s)   \implies   2r - r^2 = 2s - s^2   \implies   2(r - s) = r^2 - s^2   \implies   2 = r + s.$$

Now we solve the equations simultaneously:

$$\sin 2\pi r = \sin\left[2\pi\left(2 - r\right)\right] = -\sin 2\pi r$$

$$2\sin 2\pi r = 0$$

$$\sin 2\pi r = 0.$$

Since $r \in [0, 4]$, $r = 0, \frac{1}{2}, 1, \frac{3}{2}, 2, \frac{5}{2}, 3, \frac{7}{2}, 4$.

Since $s \in [0, 4]$ and $r \neq s$ and $r + s = 2$,   we are left with $r = 0, \frac{1}{2}, \frac{3}{2}, 2$. Note that

$$\left(x(0), y(0)\right) = (0, 0) = (x(2), y(2))   \text{and}   \left(x\left(\tfrac{1}{2}\right), y\left(\tfrac{1}{2}\right)\right) = \left(0, \tfrac{3}{4}\right) = \left(x\left(\tfrac{3}{2}\right), y\left(\tfrac{3}{2}\right)\right).$$

The curve intersects itself at $(0, 0)$ and $\left(0, \frac{3}{4}\right)$.

47.   (a)   The coefficient a affects the amplitude and the period.

(b) $\dfrac{dy}{dx} = \dfrac{dy/d\theta}{dx/d\theta}$

$$= \frac{a \sin \theta}{a(1 - \cos \theta)} = \frac{\sin \theta}{1 - \cos \theta}$$

You can verify that $\dfrac{dy}{dx} \to -\infty$ as $\theta \to 2\pi^-$; $\dfrac{dy}{dx} \to \infty$ as $\theta \to 2\pi^+$.

(c)   The curve has a vertical cusp at $\theta = 2\pi$.

49. See the answer section in the text.

51. Equation (9.7.8) with $x_0 = 0$, $y_0 = 0$, $g = 32$: $y = -\dfrac{16}{v_0{}^2}(\sec^2\theta)x^2 + (\tan\theta)x$.

53. $y = 0$ (and $x \neq 0$) $\implies$ $\frac{1}{16}v_0{}^2\cos\theta\sin\theta$

55. The range $\frac{1}{16}v_0{}^2\sin\theta\cos\theta = \frac{1}{32}v_0{}^2\sin 2\theta$ is clearly maximal when $\theta = \frac{1}{4}\pi$ for then $\sin 2\theta = 1$.

SECTION 9.8

1. $x'(1) = 1$, $y'(1) = 3$,   slope 3,   point $(1,0)$;   tangent $y = 3(x - 1)$

3. $x'(0) = 2$, $y'(0) = 0$,   slope 0,   point $(0,1)$;   tangent $y = 1$

5. $x'(1/2) = 1$, $y'(1/2) = -3$,   slope -3,   point $\left(\frac{1}{4}, \frac{9}{4}\right)$;   tangent $y - \frac{9}{4} = -3\left(x - \frac{1}{4}\right)$

7. $x'\left(\dfrac{\pi}{4}\right) = -\dfrac{3}{4}\sqrt{2}$, $y'\left(\dfrac{\pi}{4}\right) = \dfrac{3}{4}\sqrt{2}$,   slope -1,   point $\left(\dfrac{1}{4}\sqrt{2}, \dfrac{1}{4}\sqrt{2}\right)$;

tangent $y - \frac{1}{4}\sqrt{2} = -\left(x - \frac{1}{4}\sqrt{2}\right)$

9. $x(\theta) = \cos\theta\,(4 - 2\sin\theta)$, $y(\theta) = \sin\theta\,(4 - 2\sin\theta)$,   point $(4,0)$

$x'(\theta) = -4\sin\theta - 2\left(\cos^2\theta - \sin^2\theta\right)$, $y'(\theta) = 4\cos\theta - 4\sin\theta\cos\theta$

$x'(0) = -2$, $y'(0) = 4$,   slope -2,   tangent $y = -2\,(x - 4)$

11. $x(\theta) = \dfrac{4\cos\theta}{5 - \cos\theta}$, $y(\theta) = \dfrac{4\sin\theta}{5 - \cos\theta}$,   point $\left(0, \dfrac{4}{5}\right)$

$x'(\theta) = \dfrac{-20\sin\theta}{(5 - \cos\theta)^2}$, $y'(\theta) = \dfrac{4\,(5\cos\theta - 1)}{(5 - \cos\theta)^2}$

$x'\left(\dfrac{\pi}{2}\right) = -\dfrac{4}{5}$, $y'\left(\dfrac{\pi}{2}\right) = -\dfrac{4}{25}$,   slope $\dfrac{1}{5}$,   tangent $y - \dfrac{4}{5} = \dfrac{1}{5}x$

13. $x(\theta) = \dfrac{\cos\theta\,(\sin\theta - \cos\theta)}{\sin\theta + \cos\theta}$, $y(\theta) = \dfrac{\sin\theta\,(\sin\theta - \cos\theta)}{\sin\theta + \cos\theta}$,   point $(-1,0)$

$x'(\theta) = \dfrac{\sin\theta\,\cos 2\theta + 2\cos\theta}{(\sin\theta + \cos\theta)^2}$, $y'(\theta) = \dfrac{2\sin\theta - \cos\theta\,\cos 2\theta}{(\sin\theta + \cos\theta)^2}$

$x'(0) = 2$, $y'(0) = -1$,   slope $-\frac{1}{2}$,   tangent $y = -\frac{1}{2}(x + 1)$

15. $x(t) = t, \quad y(t) = t^3$

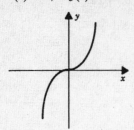

$x'(0) = 1, \quad y'(0) = 0, \quad$ slope 0

tangent $y = 0$

17. $x(t) = t^{5/3}, \quad y(t) = t$

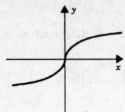

$x'(0) = 0, \quad y'(0) = 1, \quad$ slope undefined

tangent $x = 0$

19. $x'(t) = 3 - 3t^2, \quad y'(t) = 1$

$x'(t) = 0 \implies t = \pm 1; \quad y'(t) \neq 0$

(a)   none

(b)   at $(2, 2)$ and $(-2, 0)$

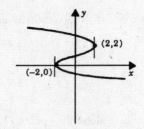

21.   curve traced once completely with $t \in [0, 2\pi)$

$x'(t) = -4\cos t, \quad y'(t) = -3\sin t$

$x'(t) = 0 \implies t = \dfrac{\pi}{2}, \dfrac{3\pi}{2};$

$y'(t) = 0 \implies t = 0, \pi$

(a)   at $(3, 7)$ and $(3, 1)$

(b)   at $(-1, 4)$ and $(7, 4)$

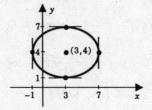

23. $x'(t) = 2t - 2, \quad y'(t) = 3t^2 - 6t + 2$

$x'(t) = 0 \implies t = 1$

$y'(t) = 0 \implies t = 1 \pm \tfrac{1}{3}\sqrt{3}$

(a)   at $\left(-\tfrac{2}{3}, \pm\tfrac{2}{9}\sqrt{3}\right)$

(b)   at $(-1, 0)$

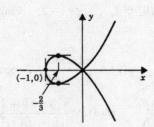

25.   curve traced completely with $t \in [0, 2\pi)$

$x'(t) = -\sin t, \quad y'(t) = 2\cos 2t$

$x'(t) = 0 \implies t = 0, \pi$

$y'(t) = 0 \implies t = \dfrac{\pi}{4}, \dfrac{3\pi}{4}, \dfrac{5\pi}{4}, \dfrac{7\pi}{4}$

(a)   at $\left(\pm\tfrac{1}{2}\sqrt{2}, \pm 1\right)$

(b)   at $(\pm 1, 0)$

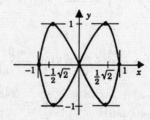

27. First, we find the values of t when the curve passes through $(2, 0)$.

$$y(t) = 0 \implies t^4 - 4t^2 = 0 \implies t = 0, \pm 2.$$

$$x(-2) = 2, \quad x(0) = 2, \quad x(2) = -2.$$

The curve passes through $(2, 0)$ at $t = -2$ and $t = 0$.

$$x'(t) = -1 - \frac{\pi}{2} \sin \frac{\pi t}{4}, \quad y'(t) = 4t^3 - 8t.$$

At $t = -2$, $\quad x'(-2) = \frac{\pi}{2} - 1$, $\quad y'(t) = -16$, $\quad$ tangent $y = \frac{32}{2 - \pi}(x - 2)$.

At $t = 0$, $\quad x'(0) = -1$, $\quad y'(0) = 0$, $\quad$ tangent $y = 0$.

29. The slope of $\overline{OP}$ is $\tan \theta_1$. The curve $r = f(\theta)$ can be parametrized by setting

$$x(\theta) = f(\theta) \cos \theta, \qquad y(\theta) = f(\theta) \sin \theta.$$

Differentiation gives

$$x'(\theta) = -f(\theta) \sin \theta + f'(\theta) \cos \theta, \qquad y'(\theta) = f(\theta) \cos \theta + f'(\theta) \sin \theta.$$

If $\quad f'(\theta_1) = 0$, $\quad$ then

$$x'(\theta_1) = -f(\theta_1) \sin \theta_1, \qquad y'(\theta_1) = f(\theta_1) \cos \theta_1.$$

Since $\quad f(\theta_1) \neq 0$, $\quad$ we have

$$m = \frac{y'(\theta_1)}{x'(\theta_1)} = -\cot \theta_1 = -\frac{1}{\text{slope of } \overline{OP}}.$$

31. $\quad x'(t) = 3t^2, \quad y'(t) = 2t$ **33.** $\quad x'(t) = 5t^4, \quad y'(t) = 3t^2$

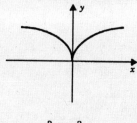   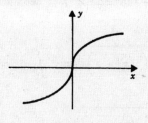

$x^2 = y^3$ $x^3 = y^5$

35. $\quad x'(t) = 2t, \quad y'(t) = 2t$

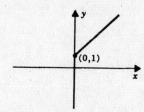

ray: $y = x + 1$, $\quad x \geq 0$

37. By (9.8.5), $\quad \dfrac{d^2 y}{dx^2} = \dfrac{(-\sin t)(-\sin t) - (\cos t)(-\cos t)}{(-\sin t)^3} = \dfrac{-1}{\sin^3 t}$. At $t = \dfrac{\pi}{6}$, $\dfrac{d^2 y}{dx^2} = -8$.

39. By (9.8.5), $\quad \dfrac{d^2 y}{dx^2} = \dfrac{(e^t)(e^{-t}) - (-e^{-t})(e^t)}{(e^t)^3} = 2e^{-3t}$. At $t = 0$, $\dfrac{d^2 y}{dx^2} = 2$.

SECTION 9.9

1. $L = \int_0^1 \sqrt{1 + 2^2}\,dx = \sqrt{5}$

3. $L = \int_1^4 \sqrt{1 + \left[\frac{3}{2}\left(x - \frac{4}{9}\right)^{1/2}\right]^2}\,dx = \int_1^4 \frac{3}{2}\sqrt{x}\,dx = \left[x^{3/2}\right]_1^4 = 7$

5. $L = \int_0^3 \sqrt{1 + \left(\frac{1}{2}\sqrt{x} - \frac{1}{2\sqrt{x}}\right)^2}\,dx = \int_0^3 \left(\frac{1}{2}\sqrt{x} + \frac{1}{2\sqrt{x}}\right)\,dx = \left[\frac{1}{3}x^{3/2} + x^{1/2}\right]_0^3 = 2\sqrt{3}$

7. $L = \int_0^1 \sqrt{1 + \left[x\left(x^2 + 2\right)^{1/2}\right]^2}\,dx = \int_0^1 \left(x^2 + 1\right)\,dx = \left[\frac{1}{3}x^3 + x\right]_0^1 = \frac{4}{3}$

9. $L = \int_1^5 \sqrt{1 + \left[\frac{1}{2}\left(x - \frac{1}{x}\right)\right]^2}\,dx = \int_1^5 \frac{1}{2}\left(x + \frac{1}{x}\right)\,dx = \left[\frac{1}{2}\left(\frac{1}{2}x^2 + \ln x\right)\right]_1^5 = 6 + \frac{1}{2}\ln 5 \cong 6.80$

11. $L = \int_1^8 \sqrt{1 + \left[\frac{1}{2}\left(x^{1/3} - x^{-1/3}\right)\right]^2}\,dx = \int_1^8 \frac{1}{2}\left(x^{1/3} + x^{-1/3}\right)\,dx = \frac{1}{2}\left[\frac{3}{4}x^{4/3} + \frac{3}{2}x^{2/3}\right]_1^8 = \frac{63}{8}$

13. $L = \int_0^{\pi/4} \sqrt{1 + \tan^2 x}\,dx = \int_0^{\pi/4} \sec x\,dx = \left[\ln|\sec x + \tan x|\right]_0^{\pi/4} = \ln\left(1 + \sqrt{2}\right) \cong 0.88$

15. $L = \int_1^2 \sqrt{1 + \left(\sqrt{x^2 - 1}\right)^2}\,dx = \int_1^2 x\,dx = \left[\frac{1}{2}x^2\right]_1^2 = \frac{3}{2}$

17. $L = \int_0^1 \sqrt{1 + \left[\sqrt{3 - x^2}\right]^2}\,dx = \int_0^1 \sqrt{4 - x^2}\,dx = \int_0^{\pi/6} \sqrt{4\cos^2 u}\,du$

$\qquad\qquad\qquad\qquad (x = 2\sin u)$

$\qquad = 2\int_0^{\pi/6} \cos u\,du = 2\left[\sin u\right]_0^{\pi/6} = 1$

19. $v(t) = \sqrt{(2t)^2 + 2^2} = 2\sqrt{t^2 + 1}$

initial speed $= v(0) = 2,$ terminal speed $= v\left(\sqrt{3}\right) = 4$

$$s = \int_0^{\sqrt{3}} 2\sqrt{t^2 + 1}\,dt = 2\int_0^{\pi/3} \sec^3 u\,du = 2\left[\frac{1}{2}\sec u\tan u + \frac{1}{2}\ln|\sec u + \tan u|\right]_0^{\pi/3}$$

$\qquad\qquad (t = \tan u) \qquad\qquad\qquad\quad (\text{by parts})$

$$= 2\sqrt{3} + \ln\left(2 + \sqrt{3}\right) \cong 4.78$$

21. $v(t) = \sqrt{(2t)^2 + (3t^2)^2}\, dt = t\,(4 + 9t^2)^{1/2}$

initial speed $= v(0) = 0,$ terminal speed $= v(1) = \sqrt{13}$

$$s = \int_0^1 t\,(4 + 9t^2)^{1/2}\, dt = \left[\frac{1}{27}\,(4 + 9t^2)^{3/2}\right]_0^1 = \frac{1}{27}\left(13\sqrt{13} - 8\right)$$

23. $v(t) = \sqrt{[e^t \cos t + e^t \sin t]^2 + [e^t \cos t - e^t \sin t]^2} = \sqrt{2}\,e^t$

initial speed $= v(0) = \sqrt{2},$ terminal speed $= \sqrt{2}\,e^\pi$

$$s = \int_0^\pi \sqrt{2}\,e^t\, dt = \left[\sqrt{2}\,e^t\right]_0^\pi = \sqrt{2}\,(e^\pi - 1)$$

25. $L = \displaystyle\int_0^{2\pi} \sqrt{[x'(\theta)]^2 + [y'(\theta)]^2}\, d\theta = \int_0^{2\pi} \sqrt{a^2(1 - \cos\theta)^2 + a^2 \sin^2\theta}\, d\theta$

$$= a\int_0^{2\pi} \sqrt{2(1 - \cos\theta)}\, d\theta = 2a\int_0^{2\pi} \sin\frac{\theta}{2}\, d\theta = -4a\left[\cos\frac{\theta}{2}\right]_0^{2\pi} = 8a$$

27. (a) $L = \displaystyle\int_0^{2\pi} \sqrt{(-3a\sin\theta - 3a\sin 3\theta)^2 + (3a\cos\theta - 3a\cos 3\theta)^2}\, d\theta$

$$= 3a\int_0^{2\pi} \sqrt{\sin^2\theta + 2\sin\theta\sin 3\theta + \sin^2 3\theta + \cos^2\theta - 2\cos\theta\cos 3\theta + \cos^2 3\theta}\, d\theta$$

$$= 3a\int_0^{2\pi} \sqrt{2(1 - \cos 4\theta)}\, d\theta = 6a\int_0^{2\pi} |\sin 2\theta|\, d\theta$$

$$= 24a\int_0^{\pi/2} \sin 2\theta\, d\theta = -12a\,[\cos 2\theta]_0^{\pi/2} = 24a$$

(b) The result follows from the identities: $\cos 3\theta = 4\cos^3\theta - 3\cos\theta;$ $\sin 3\theta = 3\sin\theta - 4\sin^3\theta$

29. $L = $ circumference of circle of radius $1 = 2\pi$

31. $L = \displaystyle\int_0^{4\pi} \sqrt{[e^\theta]^2 + [e^\theta]^2}\, d\theta = \int_0^{4\pi} \sqrt{2}\,e^\theta\, d\theta = \left[\sqrt{2}\,e^\theta\right]_0^{4\pi} = \sqrt{2}\,(e^{4\pi} - 1)$

33. $L = \displaystyle\int_0^{2\pi} \sqrt{[e^{2\theta}]^2 + [2e^{2\theta}]^2}\, d\theta = \int_0^{2\pi} \sqrt{5}\,e^{2\theta}\, d\theta = \left[\frac{1}{2}\sqrt{5}\,e^{2\theta}\right]_0^{2\pi} = \frac{1}{2}\sqrt{5}\,(e^{4\pi} - 1)$

35. $L = \displaystyle\int_0^{\pi/2} \sqrt{(1 - \cos\theta)^2 + \sin^2\theta}\, d\theta = \int_0^{\pi/2} \sqrt{2 - 2\cos\theta}\, d\theta$

$$= \int_0^{\pi/2} \left(2\sin\frac{1}{2}\theta\right)\, d\theta = \left[-4\cos\frac{1}{2}\theta\right]_0^{\pi/2} = 4 - 2\sqrt{2}$$

37. $s = \int_0^1 \sqrt{\left[\dfrac{1}{1+t^2}\right]^2 + \left[\dfrac{-t}{1+t^2}\right]^2}\, dt = \int_0^1 \dfrac{dt}{\sqrt{1+t^2}}$

$\qquad = \int_0^{\pi/4} \sec u\, du = [\ln |\sec u + \tan u|]_0^{\pi/4} = \ln\left(1+\sqrt{2}\right)$

$\qquad \underline{\quad}(t = \tan u)$

initial speed $= v(0) = 1,$ terminal speed $= v(1) = \tfrac{1}{2}\sqrt{2}$

39. $c = 1;$ the curve $y = e^x$ is the curve $y = \ln x$ reflected in the line $y = x$

41. $L = \int_a^b \sqrt{1+\sinh^2 x}\, dx = \int_a^b \sqrt{\cosh^2 x}\, dx = \int_a^b \cosh x\, dx = A$

43. (a) Express GPE + KE as a function of t and verify that the derivative with respect to t is zero.

(b) From (a) we learn that throughout the motion

$$32my + \tfrac{1}{2}mv^2 = C.$$

At the time of firing $y = 0$ and $v = |v_0| = v_0.$ Therefore

$$32my + \tfrac{1}{2}mv^2 = \tfrac{1}{2}mv_0{}^2.$$

At impact $y = 0,$ $\tfrac{1}{2}mv^2 = \tfrac{1}{2}mv_0{}^2,$ and $v = v_0.$

45. $\sqrt{1 + [f(x)]^2} = \sqrt{1 + \tan^2 [\alpha(x)]} = |\sec [\alpha(x)]|$

47. (a)   (b) $L = \int_{-1}^1 \sqrt{9t^4 - 2t^2 + 1}\, dt \cong 2.7156$

49. (a) $L = \int_0^{2\pi} \sqrt{a^2 \sin^2 t + b^2 \cos^2 t}\, dt = 4\int_0^{\pi/2} \sqrt{a^2(1 - \cos^2 t) + b^2 \cos^2 t}\, dt$

$\qquad = 4a \int_0^{\pi/2} \sqrt{1 - e^2 \cos^2 t}\, dt,$ where $e = \dfrac{\sqrt{a^2 - b^2}}{a}$

(b) $L = 4 \int_0^{\pi/2} \sqrt{25 - 9\cos^2 t}\, dt \cong 28.3617$

SECTION 9.10

1. $L = $ length of the line segment $= 1$

$(\overline{x}, \overline{y}) = (\frac{1}{2}, 4)$ (the midpoint of the line segment)

$A_x = $ lateral surface area of cylinder of radius 4 and side $1 = 16\pi$.

3. $L = \int_0^3 \sqrt{1 + \left(\frac{4}{3}\right)^2}\, dx = \left(\frac{5}{3}\right) 3 = 5$

$\overline{x}L = \int_0^3 x\sqrt{1 + \left(\frac{4}{3}\right)^2}\, dx = \frac{5}{3}\left[\frac{1}{2}x^2\right]_0^3 = \frac{15}{2}, \quad \overline{x} = \frac{3}{2}$

$\overline{y}L = \int_0^3 \frac{4}{3}x\sqrt{1 + \left(\frac{4}{3}\right)^2}\, dx = \left(\frac{4}{3}\right)\left(\frac{15}{2}\right) = 10, \quad \overline{y} = 2$

$A_x = 2\pi\overline{y}L = 2\pi(2)(5) = 20\pi$

5. $L = \int_0^2 \sqrt{(3)^2 + (4)^2}\, dt = (2)(5) = 10$

$\overline{x}L = \int_0^2 3t\sqrt{(3)^2 + (4)^2}\, dt = 15\left[\frac{1}{2}t^2\right]_0^2 = 30, \quad \overline{x} = 3$

$\overline{y}L = \int_0^2 4t\sqrt{(3)^2 + (4)^2}\, dt = 20\left[\frac{1}{2}t^2\right]_0^2 = 40, \quad \overline{y} = 4$

$A_x = 2\pi\overline{y}L = 2\pi(4)(10) = 80\pi$

7. $L = \int_0^{\pi/6} \sqrt{4\sin^2 t + 4\cos^2 t}\, dt = 2\left(\frac{\pi}{6}\right) = \frac{1}{3}\pi$

$\overline{x}L = \int_0^{\pi/6} 2\cos t\sqrt{4\sin^2 t + 4\cos^2 t}\, dt = 4\left[\sin t\right]_0^{\pi/6} = 2, \quad \overline{x} = \frac{6}{\pi}$

$\overline{y}L = \int_0^{\pi/6} 2\sin t\sqrt{4\sin^2 t + 4\cos^2 t}\, dt = 4\left[-\cos t\right]_0^{\pi/6} = 4 - 2\sqrt{3}, \quad \overline{y} = 6\left(2 - \sqrt{3}\right)/\pi$

$A_x = 2\pi\overline{y}L = 2\pi(6(2 - \sqrt{3})/\pi)\frac{1}{3}\pi = 4\pi(2 - \sqrt{3})$

9. $x(t) = a\cos t, \quad y = a\sin t; \quad t \in [\frac{1}{3}\pi, \frac{2}{3}\pi]$

$$L = \int_{\pi/3}^{2\pi/3} \sqrt{a^2 \sin^2 t + a^2 \cos^2 t} \, dt = \frac{1}{3}\pi a$$

by symmetry $\bar{x} = 0$

$$\bar{y}L = \int_{\pi/3}^{2\pi/3} a \sin t \sqrt{a^2 \sin^2 t + a^2 \cos^2 t} \, dt = a^2 \int_{\pi/3}^{2\pi/3} \sin t \, dt$$

$$= a^2 \left[-\cos t\right]_{\pi/3}^{2\pi/3} = a^2, \quad \bar{y} = 3a/\pi$$

$$A_x = 2\pi \bar{y} L = 2\pi a^2$$

11. $A_x = \int_0^2 \frac{2}{3}\pi x^3 \sqrt{1+x^4} \, dx = \frac{1}{9}\pi \left[(1+x^4)^{3/2}\right]_0^2 = \frac{1}{9}(17\sqrt{17}-1) \cong 7.68$

13. $A_x = \int_0^1 \frac{1}{2}\pi x^3 \sqrt{1+\frac{9}{16}x^4} \, dx = \frac{4}{27}\pi \left[\left(1+\frac{9}{16}x^4\right)\right]_0^1 = \frac{61}{432}\pi$

15. $A_x = \int_0^{\pi/2} 2\pi \cos x \sqrt{1+\sin^2 x} \, dx = \int_0^1 2\pi \sqrt{1+u^2} \, du$

$$u = \sin x$$

$$= 2\pi \left[\tfrac{1}{2}u\sqrt{1+u^2} + \tfrac{1}{2}\ln\left(u+\sqrt{1+u^2}\right)\right]_0^1 = \pi \left[\sqrt{2}+\ln\left(1+\sqrt{2}\right)\right]$$

(8.5.1)

17. $A_x = \int_0^{\pi/2} 2\pi(e^\theta \sin\theta)\sqrt{[e^\theta \cos\theta - e^\theta \sin\theta]^2 + [e^\theta \sin\theta + e^\theta \cos\theta]^2} \, d\theta$

$$= 2\pi\sqrt{2} \int_0^{\pi/2} e^{2\theta} \sin\theta \, d\theta$$

$$= 2\pi\sqrt{2} \left[\tfrac{1}{5}\left(2e^{2\theta}\sin\theta - e^{2\theta}\cos\theta\right)\right]_0^{\pi/2} = \frac{2}{5}\sqrt{2}\,\pi\left(2e^\pi + 1\right)$$

└─(by parts twice)

19. (a) $A = \int_0^{2\pi} y(\theta)x'(\theta) \, d\theta$ [see (9.7.4)]

$$= \int_0^{2\pi} a^2(1-\cos\theta)^2 \, d\theta$$

$$= a^2 \int_0^{2\pi} (1 - 2\cos\theta + \cos^2\theta) \, d\theta$$

$$= a^2 \int_0^{2\pi} \left(\frac{3}{2} - 2 \cos\theta + \frac{1}{2} \cos 2\theta \right) d\theta$$

$$= a^2 \left[\frac{3}{2}\theta - 2 \sin\theta + \frac{1}{4} \sin 2\theta \right]_0^{2\pi}$$

$$= 3\pi a^2$$

(b) $A = \int_0^{2\pi} 2\pi \, y(\theta) \sqrt{[x'(\theta)]^2 + [y'(\theta)]^2} \, d\theta$ (9.10.2)

$$= \int_0^{2\pi} 2\pi \, a(1 - \cos\theta) \sqrt{a^2(1 - \cos\theta)^2 + a^2 \sin^2\theta} \, d\theta$$

$$= 2\pi a^2 \int_0^{2\pi} (1 - \cos\theta) \sqrt{2 - 2\cos\theta} \, d\theta$$

$$= 4\pi a^2 \int_0^{2\pi} (1 - \cos\theta) \sin\frac{\theta}{2} \, d\theta$$

$$= 4\pi a^2 \int_0^{2\pi} \left(2 \sin\frac{\theta}{2} - 2 \cos^2\frac{\theta}{2} \sin\frac{\theta}{2} \right) d\theta$$

$$= 4\pi a^2 \left[-4 \cos\frac{\theta}{2} \right]_0^{2\pi} + \frac{16\pi a^2}{3} \left[\cos^3(\theta/2) \right]_0^{2\pi}$$

$$= \frac{64\pi a^2}{3}$$

21. $A = \frac{1}{2}\theta s_2^{\ 2} - \frac{1}{2}\theta s_1^{\ 2}$

$$= \frac{1}{2}(\theta s_2 + \theta s_1)(s_2 - s_1)$$

$$= \frac{1}{2}(2\pi R + 2\pi r)s = \pi(R + r)s$$

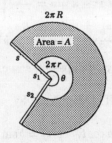

23. (a) The centroids of the 3, 4, 5 sides are the midpoints $\left(\frac{3}{2}, 0 \right)$, $(3, 2)$, $\left(\frac{3}{2}, 2 \right)$.

(b) $\overline{x}(3 + 4 + 5) = \frac{3}{2}(3) + 3(4) + \frac{3}{2}(5)$, $12\overline{x} = 24$, $\overline{x} = 2$

$\overline{y}(3 + 4 + 5) = 0(3) + 2(4) + 2(5)$, $12\overline{y} = 18$, $\overline{y} = \frac{3}{2}$

(c) $A = \frac{1}{2}(3)(4) = 6$

$$\overline{x}A = \int_0^3 x\left(\frac{4}{3}x \right) dx = \int_0^3 \frac{4}{3}x^2 dx = \frac{4}{9}\left[x^3 \right]_0^3 = 12, \quad \overline{x} = 2$$

$$\overline{y}A = \int_0^3 \frac{1}{2}\left(\frac{4}{3}x \right)^2 dx = \int_0^3 \frac{8}{9}x^2 dx = \frac{8}{27}\left[x^3 \right]_0^3 = 8, \quad \overline{x} = \frac{4}{3}$$

(d) $\overline{x}(4 + 5) = 3(4) + \frac{3}{2}(5)$, $9\overline{x} = \frac{39}{2}$, $\overline{x} = \frac{13}{6}$

$\overline{y}(4 + 5) = 2(4) + 2(5)$, $9\overline{y} = 18$, $\overline{y} = 2$

(e) $A_x = 2\pi(2)(5) = 20\pi$ (f) $A_x = 2\pi(2)(4 + 5) = 36\pi$

25. $A_x = 2\pi \bar{y} L = 2\pi(b)(2\pi a) = 4\pi^2 ab$

27. The band can be obtained by revolving about the x-axis the graph of the function

$$f(x) = \sqrt{r^2 - x^2}, \qquad x \in [a, b].$$

A straightforward calculation shows that the surface area of the band is $2\pi r(b - a)$.

29. (a) Parametrize the upper half of the ellipse by

$$x(t) = a\cos t, \quad y(t) = b\sin t; \qquad t \in [0, \pi].$$

Here

$$\sqrt{[x'(t)]^2 + [y'(t)]^2} = \sqrt{a^2 \sin^2 t + b^2 \cos^2 t} = \sqrt{a^2 - (a^2 - b^2)\cos^2 t},$$

which, with $c = \sqrt{a^2 - b^2}$, can be written $\sqrt{a^2 - c^2 \cos^2 t}$.   Therefore,

$$A = \int_0^\pi 2\pi b \sin t \sqrt{a^2 - c^2 \cos^2 t}\, dt = 4\pi b \int_0^{\pi/2} \sin t \sqrt{a^2 - c^2 \cos^2 t}\, dt.$$

Setting $u = c\cos t$, we have $du = -c\sin t$ and

$$A = -\frac{4\pi b}{c} \int_c^0 \sqrt{a^2 - u^2}\, du = \frac{4\pi b}{c}\left[\frac{u}{2}\sqrt{a^2 - u^2} + \frac{a^2}{2}\sin^{-1}\left(\frac{u}{a}\right)\right]_0^c$$

$$= 2\pi b^2 + \frac{2\pi a^2 b}{c}\sin^{-1}\left(\frac{c}{a}\right) = 2\pi b^2 + \frac{2\pi ab}{e}\sin^{-1} e$$

where e is the eccentricity of ellipse: $e = c/a$.

(b) Parametrize the right half of the ellipse by

$$x(t) = a\cos t, \quad y(t) = b\sin t; \qquad t \in [-\tfrac{1}{2}\pi, \tfrac{1}{2}\pi].$$

Again $\sqrt{[x'(t)]^2 + [y'(t)]^2} = \sqrt{a^2 - c^2 \cos^2 t}$ where $c = \sqrt{a^2 - b^2}$.

Therefore

$$A = \int_{-\pi/2}^{\pi/2} 2\pi a\cos t \sqrt{a^2 - c^2 \cos^2 t}\, dt.$$

Set $u = c\sin t$.   Then $du = c\cos t\, dt$ and

$$A = \frac{2\pi a}{c} \int_{-c}^c \sqrt{b^2 + u^2}\, du = \frac{2\pi a}{c}\left[\frac{u}{2}\sqrt{b^2 + u^2} + \frac{b^2}{2}\ln\left|u + \sqrt{b^2 + u^2}\right|\right]_{-c}^c$$

Routine calculation gives

$$A = 2\pi a^2 + \frac{\pi b^2}{e}\ln\left|\frac{1 + e}{1 - e}\right|.$$

31. Such a hemisphere can be obtained by revolving about the x-axis the curve

$$x(t) = r\cos t, \quad y(t) = r\sin t; \quad t \in [0, \tfrac{1}{2}\pi].$$

Therefore,
$$\overline{x}A = \int_0^{\pi/2} 2\pi (r\cos t)(r\sin t)\sqrt{r^2\sin^2 t + r^2\cos^2 t}\, dt$$

$$= \int_0^{\pi/2} 2\pi r^3 \sin t \cos t\, dt = \pi r^3 \left[\sin^2 t\right]_0^{\pi/2} = \pi r^3.$$

$$A = 2\pi r^2; \quad \overline{x} = \overline{x}A/A = \tfrac{1}{2}r.$$

The centroid lies on the midpoint of the axis of the hemisphere.

33.   Such a surface can be obtained by revolving about the x-axis the graph of the function

$$f(x) = \left(\frac{R-r}{h}\right)x + r, \quad x \in [0, h].$$

Formula (9.10.8) gives

$$\overline{x}A = \int_0^h 2\pi x f(x)\sqrt{1 + [f'(x)]^2}\, dx$$

$$= \frac{2\pi}{h}\sqrt{h^2 + (R-h)^2} \int_0^h \left[\left(\frac{R-r}{h}\right)x^2 + rx\right] dx$$

$$= \frac{\pi}{3}\sqrt{h^2 + (R-r)^2}\,(2R+r)h$$

$$A = \pi(R+r)s = \pi(R+r)\sqrt{h^2 + (R-r)^2} \quad \text{and} \quad \overline{x} = \frac{\overline{x}A}{A} = \left(\frac{2R+r}{R+r}\right)\frac{h}{3}.$$

The centroid of the surface lies on the axis of the cone $\left(\dfrac{2R+r}{R+r}\right)\dfrac{h}{3}$ units from the base of radius r.

SECTION * 9.11

1.   Referring to Figure 9.11.1 we have

$$x(\theta) = \overline{OB} - \overline{AB} = R\theta - R\sin\theta = R(\theta - \sin\theta)$$

$$y(\theta) = \overline{BQ} - \overline{QC} = R - R\cos\theta = R(1 - \cos\theta).$$

3.   (a) The slope at P is

$$m = \frac{y'(\theta)}{x'(\theta)} = \frac{\sin\theta}{1 - \cos\theta}.$$

The line tangent to the cycloid at P has equation

$$y - R(1 - \cos\theta) = \frac{\sin\theta}{1 - \cos\theta}[x - R(\theta - \sin\theta)].$$

The top of the circle is the point $(R\theta, 2R)$. Its coordinates satisfy the equation for the tangent:

$$2R - R(1 - \cos\theta) \overset{?}{=} \frac{\sin\theta}{1 - \cos\theta}[R\theta - R(\theta - \sin\theta)]$$

$$R(1 + \cos\theta) \overset{?}{=} \frac{R\sin^2\theta}{1 - \cos\theta}$$

$$1 - \cos^2\theta \overset{\checkmark}{=} \sin^2\theta.$$

(b)   In view of the symmetry and repetitiveness of the curve we can assume that $\theta \in (0, \pi)$. Then

$$\tan\alpha = \frac{\sin\theta}{1 - \cos\theta} = \frac{\sin\theta}{2\sin^2\frac{1}{2}\theta} = \frac{\sin\frac{1}{2}\theta\cos\frac{1}{2}\theta}{\sin^2\frac{1}{2}\theta} = \cot\frac{1}{2}\theta$$

and

$$\alpha = \tfrac{1}{2}\pi - \tfrac{1}{2}\theta = \tfrac{1}{2}(\pi - \theta).$$

5. $L = \int_0^{2\pi} \sqrt{[R(1 - \cos\theta)]^2 + [R\sin\theta]^2}\, d\theta$

$$= R\int_0^{2\pi} \sqrt{2 - 2\cos\theta}\, d\theta$$

$$= R\int_0^{2\pi} \sqrt{4\sin^2\left(\frac{\theta}{2}\right)}\, d\theta = 2R\int_0^{2\pi}\sin\frac{\theta}{2}\, d\theta = 4R\left[-\cos\frac{\theta}{2}\right]_0^{2\pi} = 8R$$

7. $\bar{x} = \pi R$ by symmetry

$$\bar{y}A = \int_0^{2\pi} \frac{1}{2}[y(\theta)]^2 x'(\theta)\, d\theta$$

$$= \int_0^{2\pi} \frac{1}{2}R^2(1 - \cos\theta)^2[R(1 - \cos\theta)]\, d\theta$$

$$= \frac{1}{2}R^3\int_0^{2\pi} (1 - 3\cos\theta + 3\cos^2\theta - \cos^3\theta)\, d\theta = \frac{5}{2}\pi R^3$$

$A = 3\pi R^2$ (by Exercise 6) $\bar{y} = \left(\frac{5}{2}\pi R^3\right)/\left(3\pi R^2\right) = \frac{5}{6}R$

9. $V_y = 2\pi\bar{x}A = 2\pi(\pi R)(3\pi R^2) = 6\pi^3 R^3$

11. $A_x = 2\pi\bar{y}L = 2\pi\left(\frac{4}{3}R\right)(8R) = \frac{64}{3}\pi R^2$ ($\bar{y} = \frac{4}{3}R$ by Exercise 10)

13. $\dfrac{dy}{dx} = \dfrac{y'(\phi)}{x'(\phi)} = \dfrac{R\sin\phi}{R(1 + \cos\phi)} = \dfrac{2\sin\frac{1}{2}\phi\cos\frac{1}{2}\phi}{2\cos^2\frac{1}{2}\phi} = \tan\frac{1}{2}\phi;\quad \alpha = \frac{1}{2}\phi$

15.   (a)   Already shown more generally in Example 6 of Section 9.9.

(b)   Combining $d^2s/dt^2 = -g\sin\alpha$ with $s = 4R\sin\alpha$, we have

$$\frac{d^2s}{dt^2} = -\frac{g}{4R}s.$$

This is simple harmonic motion with angular frequency $\omega = \frac{1}{2}\sqrt{g/R}$ (see Section 18.7)

and period $T = 2\pi/\omega = 4\pi\sqrt{R/g}$.

PROJECTS AND EXPLORATIONS

9.1.   (a)   The focii are at $\left(\pm\dfrac{3\sqrt{3}}{2}, 0\right)$. By the law of cosines:

$$c = \frac{(x + 3\sqrt{3}/2)^2 + y^2 + (x - 3\sqrt{3}/2)^2 + y^2 - (3\sqrt{3})^2}{2\sqrt{(x + 3\sqrt{3}/2)^2 + y^2}\sqrt{(x - 3\sqrt{3}/2)^2 + y^2}}$$

$$= \frac{x^2 - 6}{\sqrt{x^4 - 24x^2 + 144}} \quad (\text{set } y^2 = \frac{9}{4} - \frac{1}{4}x^2)$$

$$= \frac{x^2 - 6}{\sqrt{(x^2 - 12)^2}}$$

Since $-3 \le x \le 3$ it follows that $\quad c(x) = \dfrac{x^2 - 6}{12 - x^2}$

(b)   Clearly c is an even function. The extreme points of c (if any) must be symmetric with respect to the y-axis.

(c)   The graph indicates that c has a minimum value at $x = 0$. This minimum represents a maximum angle; sit opposite the center of the stage.

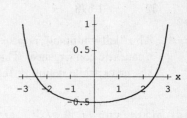

(d)   Assume that $a > b$. The focii are at $(\pm\sqrt{a^2 - b^2}, 0)$ and

$$c(x) = \frac{(a^2 - b^2)x^2 + a^2(2b^2 - a^2)}{\sqrt{(a^2 - b^2)^2 x^4 - 2a^4(a^2 - b^2)x^2 + a^8}}$$

c has a minimum at $x = 0$.

(e) $c(-a) = c(a) = 1$ and $c(0) = \dfrac{2b^2 - a^2}{a^2}$

Thus, by the intermediate-value theorem (Theorem 2.6.1), $c(x) = 0$ has a solution if $2b^2 - a^2 < 0$.

9.3.   (a) $f(x) = x^5$ on $[0,1]$;   arc length: $L = \int_0^1 \sqrt{1 + 5x^4}\, dx \cong 1.6405660$

n	Summation	error
1	1.41421	2.26×10^{-1}
2	1.59115	4.49×10^{-2}
4	1.62842	1.21×10^{-2}
8	1.63754	3.02×10^{-3}
16	1.63980	7.56×10^{-4}
32	1.64037	1.89×10^{-4}

Doubling the intervals appears to quarter the error.

(b)

Arc length is always greater than the length of the straight line segment connecting the endpoints of the arc.

(c)
n	Left Endpoint	error	Mid-Point	error
1	1.000	0.641	1.048	0.593
2	1.024	0.617	1.436	0.205
4	1.230	0.411	1.590	0.051
8	1.410	0.231	1.628	0.013
16	1.519	0.122	1.637	0.004
32	1.578	0.063	1.640	0.001

The "left endpoint" approximation has linear convergence; the "midpoint" approximation has quadratic convergence. That is, using left endpoints, doubling the number of intervals "halves" the error while doubling the number of intervals using midpoints quarters the error.

(d) $g(x) = \sin x$ on $[0, \pi/2]$;   arc length: $\int_0^{\pi/2} \sqrt{1 + \cos^2 x}\, dx \cong 1.910099.$

n	Sum	error	Left endpt.	error	Midpt.	error
1	1.862	0.048	2.221	-0.311	1.924	-0.014
2	1.895	0.015	2.073	-0.163	1.910	-1.2×10^{-4}
4	1.906	0.004	1.991	-0.081	1.910	-3.6×10^{-8}
8	1.909	0.001	1.951	-0.041	1.910	7.8×10^{-12}
16	1.910		1.930	-0.020	1.910	7.8×10^{-12}
32	1.91003		1.920	-0.010	1.910	7.8×10^{-12}

CHAPTER 10

SECTION 10.1

1. lub $= 2$; glb $= 0$

3. no lub; glb $= 0$

5. lub $= 2$; glb $= -2$

7. no lub; glb $= 2$

9. lub $= 2\frac{1}{2}$; glb $= 2$

11. lub $= 1$; glb $= 0.9$

13. lub $= e$; glb $= 0$

15. lub $= \frac{1}{2}(-1 + \sqrt{5})$; glb $= \frac{1}{2}(-1 - \sqrt{5})$

17. no lub; no glb

19. no lub; no glb

21. glb $S = 0$, $0 \leq \left(\frac{1}{11}\right)^3 < 0 + 0.001$

23. glb $S = 0$, $0 \leq \left(\frac{1}{10^{2n-1}}\right) < 0 + \left(\frac{1}{10^k}\right)$ $\left(n > \frac{k+1}{2}\right)$

25. Let $\epsilon > 0$. The condition $m \leq s$ is satisfied by all numbers s in S. All we have to show therefore is that there is some number s in S such that

$$s < m + \epsilon.$$

Suppose on the contrary that there is no such number in S. We then have

$$m + \epsilon \leq x \quad \text{for all} \quad x \in S.$$

This makes $m + \epsilon$ a lower bound for S. But this cannot be, for then $m + \epsilon$ is a lower bound for S that is *greater* than m, and by assumption, m is the *greatest* lower bound.

27. Let $c = $ lub S. Since $b \in S$, $b \leq c$. Since b is an upper bound for S, $c \leq b$. Thus, $b = c$.

29. (a) Suppose that K is an upper bound for S and k is a lower bound. Let t be any element of T. Then $t \in S$ which implies that $k \leq t \leq K$. Thus K is an upper bound for T and k is a lower bound, and T is bounded.

(b) Let $a = $ glb S. Then $a \leq t$ for all $t \in T$. Therefore, $a \leq $ glb T. Similarly, if $b = $ lub S, then $t \leq b$ for all $t \in T$, so lub $T \leq b$. It now follows that glb $S \leq $ glb $T \leq $ lub $T \leq $ lub S.

31. Let c be a positive number and let $S = \{c, 2c, 3c, \ldots\}$. Choose any positive number M and consider the positive number M/c. Since the set of positive integers is not bounded above, there exists a positive integer k such that $k \geq M/c$. This implies that $kc \geq M$. Since $kc \in S$, it follows that S is not bounded above.

33. (a)

a_1	a_2	a_3	a_4	a_5
1.4142	1.6818	1.8340	1.9152	1.9571

a_6	a_7	a_8	a_9	a_{10}
1.9785	1.9892	1.9946	1.9973	1.9986

(b)   Let S be the set of positive integers for which $a_n < 2$. Then $1 \in S$ since

$$a_1 = \sqrt{2} \cong 1.4142 < 2.$$

Assume that $k \in S$. Since $a_{k+1}^2 = 2a_k < 4$, it follows that $a_{k+1} < 2$. Thus $k+1 \in S$ and S is the set of positive integers.

(c)   Yes, $a_n \to 2$ as $n \to \infty$.

(d)   Let c be a positive number. Then c is the least upper bound of the set

$$S = \left\{ \sqrt{c}, \ \sqrt{c\sqrt{c}}, \ \sqrt{c\sqrt{c\sqrt{c}}}, \ldots \right\}.$$

SECTION 10.2

1. $a_n = 2 + 3(n-1), \quad n = 1, 2, 3, \ldots$ 3. $a_n = \dfrac{(-1)^{n-1}}{2n-1}, \quad n = 1, 2, 3, \ldots$

5. $a_n = \dfrac{n^2 + 1}{n}, \quad n = 1, 2, 3, \ldots$

7. $a_n = \begin{cases} n, & \text{if } n = 2k-1 \\ 1/n, & \text{if } n = 2k, \end{cases}$ where $k = 1, 2, 3, \ldots$

9.   decreasing;   bounded below by 0 and above by 2

11.   increasing;   bounded below by 1 but not bounded above

13. $\dfrac{n + (-1)^n}{n} = 1 + (-1)^n \dfrac{1}{n}$:   not monotonic;   bounded below by 0 and above by $\dfrac{3}{2}$

15.   decreasing;   bounded below by 0 and above by 0.9

17. $\dfrac{n^2}{n+1} = n - 1 + \dfrac{1}{n+1}$:   increasing;   bounded below by $\dfrac{1}{2}$ but not bounded above

19. $\dfrac{4n}{\sqrt{4n^2+1}} = \dfrac{2}{\sqrt{1 + 1/4n^2}}$ and $\dfrac{1}{4n^2}$ decreases to 0:   increasing;

    bounded below by $\dfrac{4}{5}\sqrt{5}$ and above by 2

21.   increasing;   bounded below by $\frac{2}{51}$ but not bounded above

23.   decreasing;   bounded below by 0 and above by $\frac{1}{2}(10^{10})$

25. $\dfrac{2n}{n+1} = 2 - \dfrac{2}{n+1}$ increases toward 2:   increasing;   bounded below by 0 and above by $\ln 2$

27.   decreasing;   bounded below by 1 and above by 4

29.   increasing;   bounded below by $\sqrt{3}$ and above by 2

31. $(-1)^{2n+1}\sqrt{n} = -\sqrt{n}$:   decreasing;   bounded above by -1 but not bounded below

33. $\dfrac{2^n - 1}{2^n} = 1 - \dfrac{1}{2^n}$:   increasing;   bounded below by $\dfrac{1}{2}$ and above by 1

35.   consider $\sin x$ as $x \to 0^+$:   decreasing;   bounded below by 0 and above by 1

37.   decreasing;   bounded below by 0 and above by $\frac{5}{6}$

39. $\dfrac{1}{n} - \dfrac{1}{n+1} = \dfrac{1}{n(n+1)}$:   decreasing;   bounded below by 0 and above by $\dfrac{1}{2}$

41.   Set $f(x) = \dfrac{\ln x}{x}$.   Then, $f'(x) = \dfrac{1 - \ln x}{x^2} < 0$ for $x > e$: decreasing;

bounded below by 0 and above by $\frac{1}{3}\ln 3$.

43.   Set $a_n = \dfrac{3^n}{(n+1)^2}$.   Then, $\dfrac{a_{n+1}}{a_n} = 3\left(\dfrac{n+1}{n+2}\right)^2 > 1$:   increasing;

bounded below by $\frac{3}{4}$ but not bounded above.

45.   For $n \geq 5$

$$\frac{a_{n+1}}{a_n} = \frac{5^{n+1}}{(n+1)!} \cdot \frac{n!}{5^n} = \frac{5}{n+1} < 1 \quad \text{and thus} \quad a_{n+1} < a_n.$$

Sequence is not nonincreasing: $a_1 = 5 < \frac{25}{2} = a_2$.

47.   boundedness: $0 < (c^n + d^n)^{1/n} < (2d^n)^{1/n} = 2^{1/n}d \leq 2d$

monotonicity : $a_{n+1}^{n+1} = c^{n+1} + d^{n+1} = cc^n + dd^n$

$$< (c^n + d^n)^{1/n}c^n + (c^n + d^n)^{1/n}d^n$$

$$= (c^n + d^n)^{1+1/n}$$

$$= (c^n + d^n)^{(n+1)/n}$$

$$= a_n^{n+1}$$

Taking the $(n+1)$st root of each side we have $a_{n+1} < a_n$.   The sequence is monotonic decreasing.

49. $a_1 = 1$, $a_2 = \frac{1}{2}$, $a_3 = \frac{1}{6}$, $a_4 = \frac{1}{24}$, $a_5 = \frac{1}{120}$, $a_6 = \frac{1}{720}$; $a_n = 1/n!$

51. $a_1 = a_2 = a_3 = a_4 = a_5 = a_6 = 1$; $a_n = 1$

53. $a_1 = 1$, $a_2 = 3$, $a_3 = 5$, $a_4 = 7$, $a_5 = 9$, $a_6 = 11$; $a_n = 2n - 1$

55. $a_1 = 1$, $a_2 = 4$, $a_3 = 13$, $a_4 = 40$, $a_5 = 121$, $a_6 = 364$; $a_n = \frac{1}{2}(3^n - 1)$

57. $a_1 = 1$, $a_2 = 4$, $a_3 = 9$, $a_4 = 16$, $a_5 = 25$, $a_6 = 36$; $a_n = n^2$

59. $a_1 = 1$, $a_2 = 3$, $a_3 = 4$, $a_4 = 8$, $a_5 = 16$, $a_6 = 32$; $a_n = 2^{n-1}$ $(n \geq 3)$

61. $a_1 = 2$, $a_2 = 1$, $a_3 = 2$, $a_4 = 1$, $a_5 = 2$, $a_6 = 1$; $a_n = \frac{1}{2}[3 - (-1)^n]$

63. $a_1 = 1$, $a_2 = 3$, $a_3 = 5$, $a_4 = 7$, $a_5 = 9$, $a_6 = 11$; $a_n = 2n - 1$

65.   First $a_1 = 2^1 - 1 = 1$.   Next suppose $a_k = 2^k - 1$ for some $k \geq 1$.   Then

$$a_{k+1} = 2a_k + 1 = 2\left(2^k - 1\right) + 1 = 2^{k+1} - 1.$$

67. First $a_1 = \dfrac{1}{2^0} = 1$.   Next suppose $a_k = \dfrac{k}{2^{k-1}}$ for some $k \geq 1$.   Then

$$a_{k+1} = \frac{k+1}{2k} a_k = \frac{k+1}{2k} \frac{k}{2^{k-1}} = \frac{k+1}{2^k}.$$

69.   (a)   If $r = 1$ then $S_n = n$ for $n = 1, 2, 3, \dots$

(b)

$$S_n = 1 + r + r^2 + \cdots + r^{n-1}$$
$$rS_n = r + r^2 + \cdots + r^n$$
$$S_n - rS_n = 1 - r^n$$
$$S_n = \frac{1 - r^n}{1 - r}, \qquad r \neq 1.$$

71.   (a)   Let S_n denote the distance traveled between the nth and $(n+1)$st bounce. Then

$$S_1 = 75 + 75 = 150, \qquad S_2 = \tfrac{3}{4}(75) + \tfrac{3}{4}(75) = 150\left(\frac{3}{4}\right), \dots, \qquad S_n = 150\left(\frac{3}{4}\right)^{n-1}$$

(b)   An object dropped from rest from a height h above the ground will hit the ground in $\frac{1}{4}\sqrt{h}$ seconds. Therefore it follows that the ball will be in the air

$$T_n = 2\left(\tfrac{1}{4}\right)\sqrt{\frac{S_n}{2}} = \frac{5\sqrt{3}}{2}\left(\frac{3}{4}\right)^{(n-1)/2} \qquad \text{seconds.}$$

73.   (a)   Let S be the set of positive integers for which $a_{n+1} > a_n$. Since $a_2 = 1 + \sqrt{a_1} = 2 > 1$, $1 \in S$. Assume that $a_k = 1 + \sqrt{a_{k-1}} > a_{k-1}$. Then

$$a_{k+1} = 1 + \sqrt{a_k} > 1 + \sqrt{a_{k-1}} = a_k.$$

Thus, $k \in S$ implies $k + 1 \in S$. It now follows that $\{a_n\}$ is an increasing sequence.

(b)   Since $\{a_n\}$ is an increasing sequence,

$$a_n = 1 + \sqrt{a_{n-1}} < 1 + \sqrt{a_n}, \quad \text{or} \quad a_n - \sqrt{a_n} - 1 < 0.$$

Rewriting the second inequality as

$$\left(\sqrt{a_n}\right)^2 - \sqrt{a_n} - 1 < 0$$

and solving for $\sqrt{a_n}$ it follows that $\sqrt{a_n} < \frac{1}{2}(1 + \sqrt{5})$. Hence, $a_n < \frac{1}{2}(3 + \sqrt{5})$ for all n.

(c) $a_2 = 2$, $a_3 \cong 2.4142$, $a_4 \cong 2.5538$, $a_5 \cong 2.6118, \dots$, $a_9 \cong 2.6179$, $\dots$, $a_{15} \cong 2.6180$; $\text{lub}\{a_n\} = \frac{1}{2}(3 + \sqrt{5}) \cong 2.6180$

1. diverges   **3.** converges to 0

5. converges to 1: $\dfrac{n-1}{n} = 1 - \dfrac{1}{n} \to 1$

7. converges to 0: $\dfrac{n+1}{n^2} = \dfrac{1}{n} + \dfrac{1}{n^2} \to 0$

9. converges to 0: $0 < \dfrac{2^n}{4^n+1} < \dfrac{2^n}{4^n} = \dfrac{1}{2^n} \to 0$

11. diverges   **13.** converges to 0

15. converges to 1: $\dfrac{n\pi}{4n+1} \to \dfrac{\pi}{4}$ so $\tan \dfrac{n\pi}{4n+1} \to \tan \dfrac{\pi}{4} = 1$

17. converges to $\dfrac{4}{9}$: $\dfrac{(2n+1)^2}{(3n-1)^2} = \dfrac{4+4/n+1/n^2}{9-6/n+1/n^2} \to \dfrac{4}{9}$

19. converges to $\dfrac{1}{2}\sqrt{2}$: $\dfrac{n^2}{\sqrt{2n^4+1}} = \dfrac{1}{\sqrt{2+1/n^4}} \to \dfrac{1}{\sqrt{2}}$

21. diverges: $\cos n\pi = (-1)^n$

23. converges to 1: $\dfrac{1}{\sqrt{n}} \to 0$ so $e^{1/\sqrt{n}} \to e^0 = 1$

25. diverges

27. converges to 0 : $\ln n - \ln(n+1) = \ln\left(\dfrac{n}{n+1}\right) \to \ln 1 = 0$

29. converges to $\dfrac{1}{2}$: $\dfrac{\sqrt{n+1}}{2\sqrt{n}} = \dfrac{1}{2}\sqrt{1+\dfrac{1}{n}} \to \dfrac{1}{2}$

31. converges to e^2: $\left(1+\dfrac{1}{n}\right)^{2n} = \left[\left(1+\dfrac{1}{n}\right)^n\right]^2 \to e^2$

33. diverges; since $2^n > n^3$ for $n \geq 10$, $\dfrac{2^n}{n^2} > \dfrac{n^3}{n^2} = n$

35. converges to 0: $\left|\dfrac{\sqrt{n}\sin(e^n\pi)}{n+1}\right| = \dfrac{|\sin(e^n\pi)|}{\sqrt{n}+\dfrac{1}{\sqrt{n}}} \leq \dfrac{1}{\sqrt{n}+\dfrac{1}{\sqrt{n}}} \to 0$

37. Set $\epsilon > 0$. Since $a_n \to L$, there exists N_1 such that

$$\text{if}\quad n \leq N_1, \quad \text{then}\quad |a_n - L| < \epsilon/2.$$

Since $b_n \to M$, there exists N_2 such that

$$\text{if} \quad n \geq N_2, \quad \text{then} \quad |b_n - M| < \epsilon/2.$$

Now set $N = \max\{N_1, N_2\}$. Then, for $n \geq N$,

$$|(a_n + b_n) - (L + M)| \leq |a_n - L| + |b_n - M| < \frac{\epsilon}{2} + \frac{\epsilon}{2} = \epsilon.$$

39. Since $\left(1 + \dfrac{1}{n}\right) \to 1$ and $\left(1 + \dfrac{1}{n}\right)^n \to e$,

$$\left(1 + \frac{1}{n}\right)^{n+1} = \left(1 + \frac{1}{n}\right)^n \left(1 + \frac{1}{n}\right) \to (e)(1) = e.$$

41. Suppose that $\{a_n\}$ is bounded and non-increasing. If L is the greatest lower bound of the range of this sequence, then $a_n \geq L$ for all n. Set $\epsilon > 0$. By Theorem 10.1.5 there exists a_k such that $a_k < L + \epsilon$. Since the sequence is non-increasing, $a_n \leq a_k$ for all $n \geq k$. Thus,

$$L \leq a_n < L + \epsilon \quad \text{or} \quad |a_n - L| < \epsilon \quad \text{for all} \quad n \geq k$$

and $a_n \to L$.

43. Let $\epsilon > 0$. Choose k so that, for $n \geq k$,

$$L - \epsilon < a_n < L + \epsilon, \quad L - \epsilon < c_n < L + \epsilon \quad \text{and} \quad a_n \leq b_n \leq c_n.$$

For such n,

$$L - \epsilon < b_n < L + \epsilon.$$

45. Let $\epsilon > 0$. Since $a_n \to L$, there exists a positive integer N such that $L - \epsilon < a_n < L + \epsilon$ for all $n \geq N$. Now $a_n \leq M$ for all n, so $L - \epsilon < M$, or $L < M + \epsilon$. Since ϵ is arbitrary, $L \leq M$.

47. By the continuity of f, $f(L) = f\left(\lim_{n \to \infty} a_n\right) = \lim_{n \to \infty} f(a_n) = \lim_{n \to \infty} a_{n+1} = L$.

49. Set $f(x) = x^{1/p}$. Since $\dfrac{1}{n} \to 0$ and f is continuous at 0, it follows by Theorem 10.3.12 that

$$\left(\frac{1}{n}\right)^{1/p} \to 0.$$

51. $a_n = e^{1-n} \to 0$

53. $a_n = \dfrac{1}{n!} \to 0$

55. $a_n = \dfrac{1}{2}[1 - (-1)^n] \quad$ diverges

57. $a_n = \dfrac{2^n - 1}{2^{n-1}} \to 2$

59. $L = 0, \quad n = 32$

61. $L = 0, \quad n = 4$

63. $L = 0, \quad n = 7$

65. $L = 0, \quad n = 65$

67. (a) $a_n = 1 + \sqrt{a_{n-1}}$ Suppose that $a_n \to L$ as $n \to \infty$. Then $a_{n-1} \to L$ as $n \to \infty$. Therefore $L = 1 + \sqrt{L}$ which implies that $L = \frac{1}{2}(3 + \sqrt{5})$.

(b) $a_n = \sqrt{3a_{n-1}}$ Suppose that $a_n \to L$ as $n \to \infty$. Then $a_{n-1} \to L$ as $n \to \infty$. Therefore

$L = \sqrt{3L}$ which implies that $L = 3$.

69. (a)
$$\begin{array}{ccccc} a_2 & a_3 & a_4 & a_5 & a_6 \\ 0.540302 & 0.857553 & 0.654290 & 0.793480 & 0.701369 \end{array}$$

$$\begin{array}{cccc} a_7 & a_8 & a_9 & a_{10} \\ 0.763960 & 0.722102 & 0.750418 & 0.73140 \end{array}$$

(b) L is a fixed point of $f(x) = \cos x$, that is, $\cos L = L$; $L \cong 0.739085$.

71. (a)
$$\begin{array}{ccccccc} a_2 & a_3 & a_4 & a_5 & a_6 & a_7 & a_8 \\ 2.000000 & 1.750000 & 1.732143 & 1.732051 & 1.732051 & 1.732051 & 1.732051 \end{array}$$

(b) $L = \dfrac{1}{2}\left(L + \dfrac{3}{L}\right)$ which implies $L^2 = 3$ or $L = \sqrt{3}$.

(c) Newton's method applied to the function $f(x) = x^2 - R$ gives

$$a_n = a_{n-1} - \frac{f(a_{n-1})}{f'(a_{n-1})} = a_{n-1} - \frac{a_{n-1}^2 - R}{2a_{n-1}}$$

$$= \frac{1}{2}\,a_{n-1} + \frac{1}{2}\,\frac{R}{a_{n-1}} = \frac{1}{2}\left(a_{n-1} + \frac{R}{a_{n-1}}\right), \quad n = 2, 3, \ldots .$$

SECTION 10.4

1. converges to 1: $2^{2/n} = (2^{1/n})^2 \to 1^2 = 1$

3. converges to 0: for $n > 3$, $0 < \left(\dfrac{2}{n}\right)^n < \left(\dfrac{2}{3}\right)^n \to 0$

5. converges to 0: $\dfrac{\ln(n+1)}{n} = \left[\dfrac{\ln(n+1)}{n+1}\right]\left(\dfrac{n+1}{n}\right) \to (0)(1) = 0$

7. converges to 0: $\dfrac{x^{100n}}{n!} = \dfrac{(x^{100})^n}{n!} \to 0$

9. converges to 1: $n^{\alpha/n} = (n^{1/n})^\alpha \to 1^\alpha = 1$

11. converges to 0: $\dfrac{3^{n+1}}{4^{n-1}} = 12\left(\dfrac{3^n}{4^n}\right) = 12\left(\dfrac{3}{4}\right)^n \to 12(0) = 0$

13. converges to 1: $(n+2)^{1/n} = e^{\frac{1}{n}\ln(n+2)}$ and, since

$$\frac{1}{n}\ln(n+2) = \left[\frac{\ln(n+2)}{n+2}\right]\left(\frac{n+2}{n}\right) \to (0)(1) = 0,$$

it follows that

$$(n+2)^{1/n} \to e^0 = 1.$$

15. converges to 1: $\displaystyle\int_0^n e^{-x}\,dx = 1 - \frac{1}{e^n} \to 1$

17. converges to π: integral $= 2\displaystyle\int_0^n \frac{dx}{1+x^2} = 2\tan^{-1} n \to 2\left(\frac{\pi}{2}\right) = \pi$

19. converges to 1:   recall (10.4.6)

21. converges to 0: $\displaystyle\frac{\ln(n^2)}{n} = 2\frac{\ln n}{n} \to 2(0) = 0$

23. diverges: since $\displaystyle\lim_{x\to 0}\frac{\sin x}{x} = 1,$

$$\frac{n}{\pi}\sin\frac{\pi}{n} = \frac{\sin(\pi/n)}{\pi/n} \to 1$$

and, for n sufficiently large,

$$n^2\sin\frac{\pi}{n} = n\pi\left(\frac{n}{\pi}\sin\frac{\pi}{n}\right) > n\pi\left(\frac{1}{2}\right) = \frac{n\pi}{2}$$

25. converges to 0: $\displaystyle\frac{5^{n+1}}{4^{2n-1}} = 20\left(\frac{5}{16}\right)^n \to 0$

27. converges to e^{-1}: $\displaystyle\left(\frac{n+1}{n+2}\right)^n = \left(1 - \frac{1}{n+2}\right)^n = \frac{\left(1 + \frac{(-1)}{n+2}\right)^{n+2}}{\left(1 + \frac{(-1)}{n+2}\right)^2} \to \frac{e^{-1}}{1} = e^{-1}$

29. converges to 0: $\displaystyle 0 < \int_n^{n+1} e^{-x^2}\,dx \le e^{-n^2}[(n+1)-n] = e^{-n^2} \to 0$

31. converges to 0: $\displaystyle\frac{n^n}{2^{n^2}} = \left(\frac{n}{2^n}\right)^n \to 0$ since $\displaystyle\frac{n}{2^n} \to 0$

33. converges to e^x: use (10.4.7)

35. converges to 0: $\displaystyle\left|\int_{-1/n}^{1/n}\sin x^2\,dx\right| \le \int_{-1/n}^{1/n}|\sin x^2|\,dx \le \int_{-1/n}^{1/n} 1\,dx = \frac{2}{n} \to 0$

37. $\displaystyle\sqrt{n+1}-\sqrt{n} = \frac{\sqrt{n+1}-\sqrt{n}}{\sqrt{n+1}+\sqrt{n}}\left(\sqrt{n+1}+\sqrt{n}\right) = \frac{1}{\sqrt{n+1}+\sqrt{n}} \to 0$

39.   (a) The length of each side of the polygon is $2r\sin(\pi/n)$. Therefore the perimeter, p_n, of the polygon is given by: $p_n = 2rn\sin(\pi/n)$.

  (b) $2rn\sin(\pi/n) \to 2\pi r$ as $n \to \infty$: The number $2rn\sin(\pi/n)$ is the perimeter of a regular polygon of n sides inscribed in a circle of radius r. As n tends to ∞, the perimeter of the polygon tends to the circumference of the circle.

41. By the hint, $\lim\limits_{n\to\infty} \dfrac{1+2+\cdots+n}{n^2} = \lim\limits_{n\to\infty} \dfrac{n(n+1)}{2n^2} = \lim\limits_{n\to\infty} \dfrac{1+1/n}{2} = \dfrac{1}{2}.$

43. By the hint, $\lim\limits_{n\to\infty} \dfrac{1^3+2^3+\cdots+n^3}{2n^4+n-1} = \lim\limits_{n\to\infty} \dfrac{n^2(n+1)^2}{4(2n^4+n-1)} = \lim\limits_{n\to\infty} \dfrac{1+2/n+1/n^2}{8+4/n^3-4/n^4} = \dfrac{1}{8}.$

45. (a)
$$m_{n+1} - m_n = \frac{1}{n+1}(a_1 + \cdots + a_n + a_{n+1}) - \frac{1}{n}(a_1 + \cdots + a_n)$$

$$= \frac{1}{n(n+1)}\left[na_{n+1} - (\overbrace{a_1 + \cdots + a_n}^{n}) \right]$$

$$> 0 \quad \text{since } \{a_n\} \text{ is increasing.}$$

(b) We begin with the hint

$$m_n < \frac{|a_1 + \cdots + a_j|}{n} + \frac{\epsilon}{2}\left(\frac{n-j}{n}\right).$$

Since j is fixed,

$$\frac{|a_1 + \cdots + a_j|}{n} \to 0$$

and therefore for n sufficiently large

$$\frac{|a_1 + \cdots + a_j|}{n} < \frac{\epsilon}{2}.$$

Since

$$\frac{\epsilon}{2}\left(\frac{n-j}{n}\right) < \frac{\epsilon}{2},$$

we see that, for n sufficiently large, $|m_n| < \epsilon$. This shows that $m_n \to 0$.

47. (a) Let S be the set of positive integers n $(n \geq 2)$ for which the inequalities hold. Since

$$\left(\sqrt{b}\right)^2 - 2\sqrt{ab} + \left(\sqrt{a}\right)^2 > 0 = \left(\sqrt{b} - \sqrt{a}\right)^2 > 0,$$

it follows that $\dfrac{a+b}{2} > \sqrt{ab}$ and so $a_1 > b_1$. Now,

$$a_2 = \frac{a_1 + b_1}{2} < a_1 \quad \text{and} \quad b_2 = \sqrt{a_1 b_1} > b_1.$$

Also, by the argument above,

$$a_2 = \frac{a_1 + b_1}{2} > \sqrt{a_1 b_1} = b_2,$$

and so $a_1 > a_2 > b_2 > b_1$. Thus $2 \in S$. Assume that $k \in S$. Then

$$a_{k+1} = \frac{a_k + b_k}{2} < \frac{a_k + a_k}{2} = a_k, \quad b_{k+1} = \sqrt{a_k b_k} > \sqrt{b_k^2} = b_k,$$

and

$$a_{k+1} = \frac{a_k + b_k}{2} > \sqrt{a_k b_k} = b_{k+1}.$$

Thus $k + 1 \in S$. Therefore, the inequalities hold for all $n \geq 2$.

(b) $\{a_n\}$ is a decreasing sequence which is bounded below.

$\{b_n\}$ is an increasing sequence which is bounded above.

Let $L_a = \lim\limits_{n\to\infty} a_n$, $L_b = \lim\limits_{n\to\infty} b_n$. Then

$$a_n = \frac{a_{n-1} + b_{n-1}}{2} \quad \text{implies} \quad L_a = \frac{L_a + L_b}{2} \quad \text{and} \quad L_a = L_b.$$

49. The numerical work suggests $L \cong 1$. Justification: Set $f(x) = \sin x - x^2$. Note that $f(0) = 0$ and $f'(x) = \cos x - 2x > 0$ for x close to 0. Therefore $\sin x - x^2 > 0$ for x close to 0 and $\sin 1/n - 1/n^2 > 0$ for n large. Thus, for n large,

$$\frac{1}{n^2} < \sin\frac{1}{n} < \frac{1}{n}$$

$$|\sin x| \le |x| \quad \text{for all } x$$

$$\left(\frac{1}{n^2}\right)^{1/n} < \left(\sin\frac{1}{n}\right)^{1/n} < \left(\frac{1}{n}\right)^{1/n}$$

$$\left(\frac{1}{n^{1/n}}\right)^2 < \left(\sin\frac{1}{n}\right)^{1/n} < \frac{1}{n^{1/n}}.$$

As $n \to \infty$ both bounds tend to 1 and therefore the middle term also tends to 1.

51. (a)

a_3	a_4	a_5	a_6	a_7	a_8	a_9	a_{10}
2	3	5	8	13	21	34	55

(b)

r_1	r_2	r_3	r_4	r_5	r_6
1	2	1.2	1.667	1.600	1.625

(c) Following the hint,

$$1 + \frac{1}{r_{n-1}} = 1 + \frac{1}{\frac{a_n}{a_{n-1}}} = 1 + \frac{a_{n-1}}{a_n} = \frac{a_n + a_{n-1}}{a_n} = \frac{a_{n+1}}{a_n} = r_n.$$

Now, if $r_n \to L$, then $r_{n-1} \to L$ and

$$1 + \frac{1}{L} = L \quad \text{which implies} \quad L = \frac{1 + \sqrt{5}}{2} \cong 1.618034.$$

SECTION 10.5

(We'll use $\star$ to indicate differentiation of numerator and denominator.)

1. $\lim\limits_{x\to 0+} \frac{\sin x}{\sqrt{x}} \overset{\star}{=} \lim\limits_{x\to 0+} 2\sqrt{x}\cos x = 0$

3. $\lim\limits_{x\to 0} \frac{e^x - 1}{\ln(1+x)} \overset{\star}{=} \lim\limits_{x\to 0}(1+x)e^x = 1$

5. $\lim\limits_{x\to \pi/2} \frac{\cos x}{\sin 2x} \overset{\star}{=} \lim\limits_{x\to \pi/2} \frac{-\sin x}{2\cos 2x} = \frac{1}{2}$

7. $\lim\limits_{x\to 0} \frac{2^x - 1}{x} \overset{\star}{=} \lim\limits_{x\to 0} 2^x \ln 2 = \ln 2$

9. $\displaystyle\lim_{x\to 1}\frac{x^{1/2}-x^{1/4}}{x-1} \overset{\star}{=} \lim_{x\to 1}\left(\frac{1}{2}x^{-1/2}-\frac{1}{4}x^{-3/4}\right)=\frac{1}{4}$

11. $\displaystyle\lim_{x\to 0}\frac{e^x-e^{-x}}{\sin x} \overset{\star}{=} \lim_{x\to 0}\frac{e^x+e^{-x}}{\cos x}=2$ **13.** $\displaystyle\lim_{x\to 0}\frac{x+\sin\pi x}{x-\sin\pi x} \overset{\star}{=} \lim_{x\to 0}\frac{1+\pi\cos\pi x}{1-\pi\cos\pi x}=\frac{1+\pi}{1-\pi}$

15. $\displaystyle\lim_{x\to 0}\frac{e^x+e^{-x}-2}{1-\cos 2x} \overset{\star}{=} \lim_{x\to 0}\frac{e^x-e^{-x}}{2\sin 2x} \overset{\star}{=} \lim_{x\to 0}\frac{e^x+e^{-x}}{4\cos 2x}=\frac{1}{2}$

17. $\displaystyle\lim_{x\to 0}\frac{\tan\pi x}{e^x-1} \overset{\star}{=} \lim_{x\to 0}\frac{\pi\sec^2\pi x}{e^x}=\pi$

19. $\displaystyle\lim_{x\to 0}\frac{1+x-e^x}{x(e^x-1)} \overset{\star}{=} \lim_{x\to 0}\frac{1-e^x}{xe^x+e^x-1} \overset{\star}{=} \lim_{x\to 0}\frac{-e^x}{xe^x+2e^x}=-\frac{1}{2}$

21. $\displaystyle\lim_{x\to 0}\frac{x-\tan x}{x-\sin x} \overset{\star}{=} \lim_{x\to 0}\frac{1-\sec^2 x}{1-\cos x} \overset{\star}{=} \lim_{x\to 0}\frac{-2\sec^2 x\tan x}{\sin x} = \lim_{x\to 0}\frac{-2\sec^2 x}{\cos x}=-2$

23. $\displaystyle\lim_{x\to 1^-}\frac{\sqrt{1-x^2}}{\sqrt{1-x^3}} = \lim_{x\to 1^-}\sqrt{\frac{1-x^2}{1-x^3}} = \sqrt{\frac{2}{3}}=\frac{1}{3}\sqrt{6}$ since $\displaystyle\lim_{x\to 1^-}\frac{1-x^2}{1-x^3} \overset{\star}{=} \lim_{x\to 1^-}\frac{2x}{3x^2}=\frac{2}{3}$

25. $\displaystyle\lim_{x\to \pi/2}\frac{\ln(\sin x)}{(\pi-2x)^2} \overset{\star}{=} \lim_{x\to \pi/2}\frac{-\cot x}{4(\pi-2x)} \overset{\star}{=} \lim_{x\to \pi/2}\frac{\csc^2 x}{-8}=-\frac{1}{8}$

27. $\displaystyle\lim_{x\to 0}\frac{\cos x-\cos 3x}{\sin(x^2)} \overset{\star}{=} \lim_{x\to 0}\frac{-\sin x+3\sin 3x}{2x\cos(x^2)} \overset{\star}{=} \lim_{x\to 0}\frac{-\cos x+9\cos 3x}{2\cos(x^2)-4x^2\sin(x^2)}=4$

29. $\displaystyle\lim_{x\to \pi/4}\frac{\sec^2 x-2\tan x}{1+\cos 4x} \overset{\star}{=} \lim_{x\to \pi/4}\frac{2\sec^2 x\,\tan x-2\sec^2 x}{-4\sin 4x}$

$\displaystyle\overset{\star}{=} \lim_{x\to \pi/4}\frac{2\sec^4 x+4\sec^2 x\tan^2 x-4\sec^2 x\tan x}{-16\cos 4x}=\frac{1}{2}$

31. $\displaystyle\lim_{x\to 0}\frac{\tan^{-1}x}{\tan^{-1}2x} \overset{\star}{=} \lim_{x\to 0}\frac{1}{1+x^2}\frac{1+4x^2}{2}=\frac{1}{2}$

33. $1;\quad \displaystyle\lim_{x\to \infty}\frac{\pi/2-\tan^{-1}x}{1/x} \overset{\star}{=} \lim_{x\to \infty}\frac{x^2}{1+x^2}=1$

35. $1;\quad \displaystyle\lim_{x\to \infty}\frac{1}{x[\ln(x+1)-\ln x]} = \lim_{x\to \infty}\frac{1/x}{\ln(1+1/x)} = \lim_{t\to 0^+}\frac{t}{\ln(1+t)} \overset{\star}{=} \lim_{t\to 0^+}(1+t)=1$

37. $\displaystyle\lim_{x\to 0}(2+x+\sin x)\ne 0,\quad \lim_{x\to 0}(x^3+x-\cos x)\ne 0$

39. The limit does not exist if $b\ne 1$. Therefore, $b=1$.

$$\lim_{x\to 0} \frac{\cos ax - 1}{2x^2} \overset{*}{=} \lim_{x\to 0} \frac{-a\sin ax}{4x} \overset{*}{=} \lim_{x\to 0} \frac{-a^2\cos ax}{4} = -\frac{a^2}{4}$$

Now, $-\dfrac{a^2}{4} = -4$ implies $a = \pm 4$.

41. $\displaystyle\lim_{x\to 0} \frac{1}{x} \int_0^x f(t)\, dt \overset{*}{=} \lim_{x\to 0} \frac{f(x)}{1} = f(0)$

43. $A(b) = 2\displaystyle\int_0^{\sqrt{b}} (b - x^2)\, dx = 2\left[bx - x^2 \right]_0^{\sqrt{b}} = \frac{4}{3} b\sqrt{b}$ and $T(b) = \dfrac{1}{2}\left(2\sqrt{b}\right) b = b\sqrt{b}.$

Thus, $\displaystyle\lim_{b\to 0} \frac{T(b)}{A(b)} = \frac{b\sqrt{b}}{\frac{4}{3} b\sqrt{b}} = \frac{3}{4}.$

45. (a)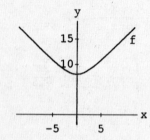

$f(x) \to \infty$ as $x \to \pm\infty$

(b) $f(x) \to 10$ as $x \to 4$

Confirmation: $\displaystyle\lim_{x\to 4} \frac{x^2 - 16}{\sqrt{x^2 + 9} - 5} \overset{*}{=} \lim_{x\to 4} \frac{2x}{x\left(x^2 + 9\right)^{-1/2}} = \lim_{x\to 4} 2\sqrt{x^2 + 9} = 10$

47. (a)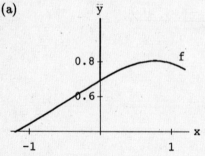

$f(x) \to 0.7$ as $x \to 0$

(b) Confirmation: $\displaystyle\lim_{x\to 0} \frac{2^{\sin x} - 1}{x} \overset{*}{=} \lim_{x\to 0} \frac{\ln(2)\, 2^{\sin x} \cos x}{1} = \ln 2 \cong 0.6931$

SECTION 10.6

(We'll use $\star$ to indicate differentiation of numerator and denominator.)

1. $\lim\limits_{x\to-\infty} \dfrac{x^2+1}{1-x} \overset{\star}{=} \lim\limits_{x\to-\infty} \dfrac{2x}{-1} = \infty$

3. $\lim\limits_{x\to\infty} \dfrac{x^3}{1-x^3} = \lim\limits_{x\to\infty} \dfrac{1}{1/x^3-1} = -1$

5. $\lim\limits_{x\to\infty} x^2 \sin\dfrac{1}{x} = \lim\limits_{h\to 0^+}\left[\left(\dfrac{1}{h}\right)\left(\dfrac{\sin h}{h}\right)\right] = \infty$

7. $\lim\limits_{x\to\frac{\pi}{2}^-} \dfrac{\tan 5x}{\tan x} = \lim\limits_{x\to\frac{\pi}{2}^-}\left[\left(\dfrac{\sin 5x}{\sin x}\right)\left(\dfrac{\cos x}{\cos 5x}\right)\right] = \dfrac{1}{5}$ since

$$\lim\limits_{x\to\frac{\pi}{2}^-} \dfrac{\sin 5x}{\sin x} = 1 \quad\text{and}\quad \lim\limits_{x\to\frac{\pi}{2}^-} \dfrac{\cos x}{\cos 5x} \overset{\star}{=} \lim\limits_{x\to\frac{\pi}{2}^-} \dfrac{\sin x}{5\sin 5x} = \dfrac{1}{5}$$

9. $\lim\limits_{x\to 0^+} x^{2x} = \lim\limits_{x\to 0^+} (x^x)^2 = 1^2 = 1$ [see (10.6.4)]

11. $\lim\limits_{x\to 0} x(\ln|x|)^2 = \lim\limits_{x\to 0} \dfrac{(\ln|x|)^2}{1/x} \overset{\star}{=} \lim\limits_{x\to 0} \dfrac{2\ln|x|}{-1/x} \overset{\star}{=} \lim\limits_{x\to 0} \dfrac{2}{1/x} = 0$

13. $\lim\limits_{x\to\infty} \dfrac{1}{x}\displaystyle\int_0^x e^{t^2}\,dt \overset{\star}{=} \lim\limits_{x\to\infty} \dfrac{e^{x^2}}{1} = \infty$

15. $\lim\limits_{x\to 0}\left[\dfrac{1}{\sin^2 x} - \dfrac{1}{x^2}\right] = \lim\limits_{x\to 0} \dfrac{x^2 - \sin^2 x}{x^2 \sin^2 x} \overset{\star}{=} \lim\limits_{x\to 0} \dfrac{2x - 2\sin x\cos x}{2x^2\sin x\cos x + 2x\sin^2 x}$

$$= \lim\limits_{x\to 0} \dfrac{2x - \sin 2x}{x^2\sin 2x + 2x\sin^2 x}$$

$$\overset{\star}{=} \lim\limits_{x\to 0} \dfrac{2 - 2\cos 2x}{2x^2\cos 2x + 4x\sin 2x + 2\sin^2 x}$$

$$\overset{\star}{=} \lim\limits_{x\to 0} \dfrac{4\sin 2x}{-4x^2\sin 2x + 12x\cos 2x + 6\sin 2x}$$

$$\overset{\star}{=} \lim\limits_{x\to 0} \dfrac{8\cos 2x}{-8x^2\cos 2x - 32x\sin 2x + 24\cos 2x} = \dfrac{1}{3}$$

17. $\lim\limits_{x\to 1} x^{1/(x-1)} = e$ since $\lim\limits_{x\to 1} \ln\left[x^{1/(x-1)}\right] = \lim\limits_{x\to 1} \dfrac{\ln x}{x-1} \overset{\star}{=} \lim\limits_{x\to 1} \dfrac{1}{x} = 1$

19. $\lim\limits_{x\to\infty}\left(\cos\dfrac{1}{x}\right)^x = 1$ since $\lim\limits_{x\to\infty} \ln\left[\left(\cos\dfrac{1}{x}\right)^x\right] = \lim\limits_{x\to\infty} \dfrac{\ln\left(\cos\dfrac{1}{x}\right)}{(1/x)}$

$$\overset{\star}{=} \lim\limits_{x\to\infty}\left(-\dfrac{\sin(1/x)}{\cos(1/x)}\right) = 0$$

21.
$$\lim_{x \to 0} \left[\frac{1}{\ln(1+x)} - \frac{1}{x} \right] = \lim_{x \to 0} \frac{x - \ln(1+x)}{x \ln(1+x)} \overset{\star}{=} \lim_{x \to 0} \frac{x}{x + (1+x)\ln(1+x)}$$
$$\overset{\star}{=} \lim_{x \to 0} \frac{1}{1 + 1 + \ln(1+x)} = \frac{1}{2}$$

23.
$$\lim_{x \to 0} \left[\frac{1}{x} - \cot x \right] = \lim_{x \to 0} \frac{\sin x - x \cos x}{x \sin x} \overset{\star}{=} \lim_{x \to 0} \frac{x \sin x}{\sin x + x \cos x}$$
$$\overset{\star}{=} \lim_{x \to 0} \frac{\sin x + x \cos x}{2 \cos x - x \sin x} = 0$$

25.
$$\lim_{x \to \infty} \left(\sqrt{x^2 + 2x} - x \right) = \lim_{x \to \infty} \left[\left(\sqrt{x^2 + 2x} - x \right) \left(\frac{\sqrt{x^2 + 2x} + x}{\sqrt{x^2 + 2x} + x} \right) \right]$$
$$= \lim_{x \to \infty} \frac{2x}{\sqrt{x^2 + 2x} + x} = \lim_{x \to \infty} \frac{2}{\sqrt{1 + 2/x} + 1} = 1$$

27. $\lim_{x \to \infty} \left(x^3 + 1 \right)^{1/\ln x} = e^3$ since

$$\lim_{x \to \infty} \ln \left[\left(x^3 + 1 \right)^{1/\ln x} \right] = \lim_{x \to \infty} \frac{\ln \left(x^3 + 1 \right)}{\ln x} \overset{\star}{=} \lim_{x \to \infty} \frac{\left(\dfrac{3x^2}{x^3 + 1} \right)}{1/x} = \lim_{x \to \infty} \frac{3}{1 + 1/x^3} = 3.$$

29. $\lim_{x \to \infty} (\cosh x)^{1/x} = e$ since

$$\lim_{x \to \infty} \ln \left[(\cosh x)^{1/x} \right] = \lim_{x \to \infty} \frac{\ln(\cosh x)}{x} \overset{\star}{=} \lim_{x \to \infty} \frac{\sinh x}{\cosh x} = 1.$$

31. $\lim_{x \to 0} (e^x + x)^{1/x} = e^2$ since

$$\lim_{x \to 0} \ln (e^x + x)^{1/x} = \lim_{x \to 0} \frac{\ln(e^x + x)}{x} \overset{\star}{=} \lim_{x \to 0} \frac{(e^x + 1)}{(e^x + x)} = 2.$$

33.
$$\lim_{x \to 0} \left(\frac{1}{\sin x} - \frac{1}{x} \right) = \lim_{x \to 0} \frac{x - \sin x}{x \sin x} \overset{\star}{=} \lim_{x \to 0} \frac{1 - \cos x}{\sin x + x \cos x}$$
$$\overset{\star}{=} \lim_{x \to 0} \frac{\sin x}{2 \cos x - x \sin x} = 0$$

35.
$$\lim_{x \to 1} \left(\frac{1}{\ln x} - \frac{x}{x-1} \right) = \lim_{x \to 1} \frac{x - 1 \ln x}{(x-1)\ln x} \overset{\star}{=} \lim_{x \to 1} \frac{-\ln x}{(x-1)(1/x) + \ln x}$$
$$= \lim_{x \to 1} \frac{-x \ln x}{x - 1 + x \ln x} \overset{\star}{=} \lim_{x \to 1} \frac{-\ln x - 1}{2 + \ln x} = -\frac{1}{2}$$

37. $0;\quad \dfrac{1}{n}\ln\dfrac{1}{n} = -\dfrac{\ln n}{n} \to 0$ **39.** $1;\quad \ln\left[(\ln n)^{1/n}\right] = \dfrac{1}{n}\ln(\ln n) \to 0$

41. $1;\quad \ln\left[(n^2+n)^{1/n}\right] = \dfrac{1}{n}\ln[n(n+1)] = \dfrac{\ln n}{n} + \dfrac{\ln(n+1)}{n} \to 0$

43. $0;\quad 0 \le \dfrac{n^2\ln n}{e^n} < \dfrac{n^3}{e^n},\quad \lim\limits_{x\to\infty}\dfrac{x^3}{e^x} = 0$

45.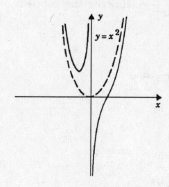

47.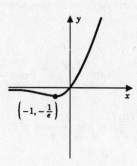

vertical asymptote y-axis

horizontal asymptote x-axis

49.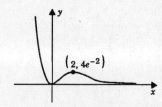

horizontal asymptote x-axis

51. $\dfrac{b}{a}\sqrt{x^2-a^2} - \dfrac{b}{a}x = \dfrac{\sqrt{x^2-a^2}+x}{\sqrt{x^2-a^2}+x}\left(\dfrac{b}{a}\right)\left(\sqrt{x^2-a^2}-x\right) = \dfrac{-ab}{\sqrt{x^2-a^2}+x} \to 0 \quad \text{as} \quad x \to \infty$

53. for instance, $\quad f(x) = x^2 + \dfrac{(x-1)(x-2)}{x^3}$

55. $\lim\limits_{x\to 0^-} -\dfrac{2x}{\cos x} \ne \lim\limits_{x\to 0^-}\dfrac{2}{-\sin x}.$ L'Hospital's rule does not apply here since $\lim\limits_{x\to 0^-}\cos x = 1.$

57. (a) Let S be the set of positive integers for which the statement is true. Since $\lim\limits_{x\to\infty}\dfrac{\ln x}{x} = 0,\quad 1\in S.$ Assume that $k\in S$. By L'Hospital's rule,

$$\lim\limits_{x\to\infty}\dfrac{(\ln x)^{k+1}}{x} \overset{*}{=} \lim\limits_{x\to\infty}\dfrac{(k+1)(\ln x)^k}{x} = 0 \quad \text{(since } k\in S\text{).}$$

Thus $k+1\in S$, and S is the set of positive integers.

(b) Choose any positive number α. Let $k - 1$ and k be positive integers such that $k - 1 \leq \alpha \leq k$. Then, for $x > e$,

$$\frac{(\ln x)^{k-1}}{x} \leq \frac{(\ln x)^{\alpha}}{x} \leq \frac{(\ln x)^{k}}{x}$$

and the result follows by the pinching theorem.

59. (a) L'Hospital's rule applied to the given limit results in $\displaystyle\lim_{x \to 0} \frac{2\,e^{-1/x^2}}{x^3}$. Rewrite the quotient as $\dfrac{1/x}{e^{1/x^2}}$. Then

$$\lim_{x \to 0} \frac{1/x}{e^{1/x^2}} \overset{\star}{=} \lim_{x \to \infty} \frac{-1/x^2}{(-2/x^3)\,e^{1/x^2}} = \lim_{x \to 0} \frac{x}{2\,e^{1/x^2}} = 0$$

(b) $\displaystyle\lim_{x \to 0} \frac{f(x) - f(0)}{x} = \lim_{x \to 0} \frac{e^{-1/x^2}}{x} = 0.$ Therefore, $f'(0) = 0$.

61. (a)

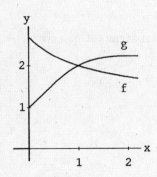

$$\lim_{x \to 0+} \left(1 + x^2\right)^{1/x} = 1.$$

(b) $\displaystyle\lim_{x \to 0+} \left(1 + x^2\right)^{1/x} = 1.$ since

$$\lim_{x \to 0+} \ln\left[\left(1 + x^2\right)^{1/x}\right] = \lim_{x \to 0+} \frac{\ln\left(1 + x^2\right)}{x} \overset{\star}{=} \lim_{x \to 0+} \frac{2x}{\left(1 + x^2\right)} = 0$$

63. (a)

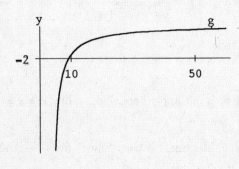

$$\lim_{x \to \infty} g(x) \cong -1.7.$$

(b) $\qquad \lim_{x \to \infty} g(x) = \lim_{x \to \infty} \left[\sqrt[3]{x^3 - 5x^2 + 2x + 1} - x \right]$

$$= \lim_{x \to \infty} \frac{-5x^2 + 2x + 1}{\left(\sqrt[3]{x^3 - 5x^2 + 2x + 1} \right)^2 + x \sqrt[3]{x^3 - 5x^2 + 2x + 1} + x^2}$$

$$= -\frac{5}{3} \cong -1.667$$

SECTION 10.7

1. 1; $\displaystyle\int_1^\infty \frac{dx}{x^2} = \lim_{b \to \infty} \int_1^b \frac{dx}{x^2} = \lim_{b \to \infty} \left[-\frac{1}{x} \right]_1^b = \lim_{b \to \infty} \left[1 - \frac{1}{b} \right] = 1$

3. $\dfrac{\pi}{4}$; $\displaystyle\int_0^\infty \frac{dx}{4 + x^2} = \lim_{b \to \infty} \int_0^b \frac{dx}{4 + x^2} = \lim_{b \to \infty} \left[\frac{1}{2} \tan^{-1} \frac{x}{2} \right]_0^b = \lim_{b \to \infty} \frac{1}{2} \tan^{-1} \left(\frac{b}{2} \right) = \frac{\pi}{4}$

5. diverges; $\displaystyle\int_0^\infty e^{px} \, dx = \lim_{b \to \infty} \int_0^b e^{px} \, dx = \lim_{b \to \infty} \left[\frac{1}{p} e^{px} \right]_0^b = \lim_{b \to \infty} \frac{1}{p} \left(e^{pb} - 1 \right) = \infty$

7. 6; $\displaystyle\int_0^8 \frac{dx}{x^{2/3}} = \lim_{a \to 0^+} \int_a^8 x^{-2/3} \, dx = \lim_{a \to 0^+} \left[3x^{1/3} \right]_a^8 = \lim_{a \to 0^+} \left[6 - 3a^{1/3} \right] = 6$

9. $\dfrac{\pi}{2}$; $\displaystyle\int_0^1 \frac{dx}{\sqrt{1 - x^2}} = \lim_{b \to 1^-} \int_0^b \frac{dx}{\sqrt{1 - x^2}} = \lim_{b \to 1^-} \sin^{-1} b = \frac{\pi}{2}$

11. 2; $\displaystyle\int_0^2 \frac{x}{\sqrt{4 - x^2}} \, dx = \lim_{b \to 2^-} \int_0^b x \left(4 - x^2 \right)^{-1/2} dx = \lim_{b \to 2^-} \left[-\left(4 - x^2 \right)^{1/2} \right]_0^b$

$$= \lim_{b \to 2^-} \left(2 - \sqrt{4 - b^2} \right) = 2$$

13. diverges; $\displaystyle\int_e^\infty \frac{\ln x}{x} \, dx = \lim_{b \to \infty} \int_e^b \frac{\ln x}{x} \, dx = \lim_{b \to \infty} \left[\frac{1}{2} \left(\ln x \right)^2 \right]_e^b$

$$= \lim_{b \to \infty} \left[\frac{1}{2} \left(\ln b \right)^2 - \frac{1}{2} \right] = \infty$$

15. $-\dfrac{1}{4}$;

$$\int_0^1 x \ln x \, dx = \lim_{a \to 0^+} \int_a^1 x \ln x \, dx = \lim_{a \to 0^+} \left[\frac{1}{2} x^2 \ln x - \frac{1}{4} x^2 \right]_a^1$$

(by parts)

$$= \lim_{a \to 0^+} \left[\frac{1}{4} a^2 - \frac{1}{2} a^2 \ln a - \frac{1}{4} \right] = -\frac{1}{4}$$

Note: $\displaystyle\lim_{t\to 0+} t^2 \ln t = \lim_{t\to 0+} \frac{\ln t}{1/t^2} \overset{*}{=} \lim_{t\to 0+} \frac{1/t}{-2/t^3} = -\frac{1}{2}\lim_{t\to 0+} t^2 = 0.$

17. π;
$$\int_{-\infty}^{\infty} \frac{dx}{1+x^2} = \lim_{a\to -\infty}\int_a^0 \frac{dx}{1+x^2} + \lim_{b\to\infty}\int_0^b \frac{dx}{1+x^2}$$

$$= \lim_{a\to -\infty}\left[\tan^{-1}x\right]_a^0 + \lim_{b\to\infty}\left[\tan^{-1}x\right]_0^b = -\left(-\frac{\pi}{2}\right) + \frac{\pi}{2} = \pi$$

19. diverges;
$$\int_{-\infty}^{\infty} \frac{dx}{x^2} = \lim_{a\to -\infty}\int_a^{-1}\frac{dx}{x^2} + \lim_{b\to 0-}\int_{-1}^{b}\frac{dx}{x^2} + \lim_{c\to 0+}\int_c^1\frac{dx}{x^2} + \lim_{d\to\infty}\int_1^d\frac{dx}{x^2};$$

and, $\displaystyle\lim_{c\to 0+}\int_c^1 \frac{dx}{x^2} = \lim_{c\to 0+}\left[-\frac{1}{x}\right]_c^1 = \lim_{c\to 0+}\left[\frac{1}{c}-1\right] = \infty$

21. $\ln 2$;
$$\int_1^{\infty}\frac{dx}{x(x+1)} = \lim_{b\to\infty}\int_1^b\left[\frac{1}{x}-\frac{1}{x+1}\right]dx$$

$$= \lim_{b\to\infty}\left[\ln\left(\frac{x}{x+1}\right)\right]_1^b = \lim_{b\to\infty}\left[\ln\left(\frac{b}{b+1}\right)-\ln\left(\frac{1}{2}\right)\right]$$

$$= 0 - \ln\tfrac{1}{2} = \ln 2$$

23. 4;
$$\int_3^5 \frac{x}{\sqrt{x^2-9}}\,dx = \lim_{a\to 3-}\int_a^5 x\left(x^2-9\right)^{-1/2}dx$$

$$= \lim_{a\to 3-}\left[\left(x^2-9\right)^{1/2}\right]_a^5 = \lim_{a\to 3-}\left[4 - \left(a^2-9\right)^{1/2}\right] = 4$$

25. $\displaystyle\int_{-3}^3 \frac{dx}{x(x+1)}$ diverges since $\displaystyle\int_0^3 \frac{dx}{x(x+1)}$ diverges:

$$\int_0^3 \frac{dx}{x(x+1)} = \int_0^3\left(\frac{1}{x}-\frac{1}{x+1}\right)dx = \lim_{a\to 0+}\left[\ln|x| - \ln|x+1|\right]_a^3$$

$$= \lim_{a\to 0+}\left[\ln 3 - \ln 4 - \ln a + \ln(a+1)\right] = \infty.$$

27. $\int_{-3}^{1} \dfrac{dx}{x^2 - 4}$ diverges since $\int_{-2}^{1} \dfrac{dx}{x^2 - 4}$ diverges:

$$\int_{-2}^{1} \frac{dx}{x^2 - 4} = \int_{-2}^{1} \frac{1}{4}\left[\frac{1}{x-2} - \frac{1}{x+2}\right] dx$$

$$= \lim_{a \to -2+} \left[\frac{1}{4}\left(\ln|x-2| - \ln|x+2|\right)\right]_{a}^{1}$$

$$= \lim_{a \to -2+} \frac{1}{4}\left[-\ln 3 - \ln|a-2| + \ln|a+2|\right] = -\infty.$$

29. diverges: $\displaystyle\int_{0}^{\infty} \cosh x\, dx = \lim_{b \to \infty} \int_{0}^{b} \cosh x\, dx = \lim_{b \to \infty} \left[\sinh x\right]_{0}^{b} = \infty$

31. $\dfrac{1}{2}$; $\displaystyle\int_{0}^{\infty} e^{-x} \sin x\, dx = \lim_{b \to \infty} \int_{0}^{b} e^{-x} \sin x\, dx = \lim_{b \to \infty} -\frac{1}{2}\left[e^{-x}\cos x + e^{-x}\sin x\right]_{0}^{b}$

(by parts)

$$= \lim_{b \to \infty} \frac{1}{2}\left[1 - e^{-b}\cos b - e^{-b}\sin b\right] = \frac{1}{2}$$

33. $2e - 2$; $\displaystyle\int_{0}^{1} \frac{e^{\sqrt{x}}}{\sqrt{x}}\, dx = \lim_{a \to 0+} \int_{a}^{1} \frac{e^{\sqrt{x}}}{\sqrt{x}}\, dx = \lim_{a \to 0+} \left[2\, e^{\sqrt{x}}\right]_{a}^{1} = 2(e - 1)$

35. $\displaystyle\int_{0}^{1} \sin^{-1} x\, dx = \left[x \sin^{-1} x\right]_{0}^{1} - \int_{0}^{1} \frac{x}{\sqrt{1-x^2}}\, dx = \frac{\pi}{2} - \lim_{a \to 1-} \int_{0}^{a} \frac{x}{\sqrt{1-x^2}}\, dx$

(by parts)

Now, $\displaystyle\int_{0}^{a} \frac{x}{\sqrt{1-x^2}}\, dx = -\frac{1}{2}\int_{1}^{1-a^2} \frac{1}{\sqrt{u}}\, du = \left[-\sqrt{u}\right]_{1}^{1-a^2} = 1 - \sqrt{1-a^2}$

$u = 1 - x^2$

Thus, $\displaystyle\int_{0}^{1} \sin^{-1} x\, dx = \frac{\pi}{2} - \lim_{a \to 1-}\left(1 - \sqrt{1-a^2}\right) = \frac{\pi}{2} - 1.$

37. $\displaystyle\int_{0}^{\infty} \frac{1}{\sqrt{x}\,(1+x)}\, dx = \int_{0}^{1} \frac{1}{\sqrt{x}\,(1+x)}\, dx + \int_{1}^{\infty} \frac{1}{\sqrt{x}\,(1+x)}\, dx$

$$= \lim_{a \to 0+} \int_{a}^{1} \frac{1}{\sqrt{x}\,(1+x)}\, dx + \lim_{b \to \infty} \int_{1}^{b} \frac{1}{\sqrt{x}\,(1+x)}\, dx$$

Now, $\displaystyle\int \frac{1}{\sqrt{x}\,(1+x)}\,dx = \int \frac{2}{1+u^2}\,du = 2\tan^{-1} u + C = 2\tan^{-1}\sqrt{x} + C.$

$$u = \sqrt{x}$$

Therefore, $\displaystyle\lim_{a\to 0^+}\int_a^1 \frac{1}{\sqrt{x}\,(1+x)}\,dx = \lim_{a\to 0^+}\left[2\tan^{-1}\sqrt{x}\right]_a^1 = \lim_{a\to 0^+} 2\left[\pi/4 - \tan^{-1} a\right] = \frac{\pi}{2}$

and $\displaystyle\lim_{b\to\infty}\int_1^b \frac{1}{\sqrt{x}\,(1+x)}\,dx = \lim_{b\to\infty}\left[2\tan^{-1}\sqrt{x}\right]_1^b = \lim_{b\to\infty} 2\left[\tan^{-1} b - \pi/4\right] = \frac{\pi}{2}.$

Thus, $\displaystyle\int_0^\infty \frac{1}{\sqrt{x}\,(1+x)}\,dx = \pi.$

39. (a)

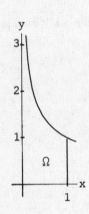

(b) $\displaystyle A = \int_0^1 \frac{1}{\sqrt{x}}\,dx = \lim_{a\to 0^+}\int_a^1 \frac{1}{\sqrt{x}}\,dx$

$$= \lim_{a\to 0^+}\left[2\sqrt{x}\right]_a^1 = 2$$

(c) $\displaystyle V = \int_0^1 \pi\left(\frac{1}{\sqrt{x}}\right)^2 dx = \pi\int_0^1 \frac{1}{x}\,dx = \pi\lim_{a\to 0^+}\int_a^1 \frac{1}{x}\,dx = \pi\lim_{a\to 0^+}\left[\ln x\right]_a^1$ diverges

41. (a)

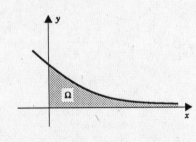

(b) $\displaystyle A = \int_0^\infty e^{-x}\,dx = 1$

(c) $\displaystyle V_x = \int_0^\infty \pi e^{-2x}\,dx = \pi/2$

(d) $\displaystyle V_y = \int_0^\infty 2\pi x e^{-x}\,dx = \lim_{b\to\infty}\int_0^b 2\pi x e^{-x}\,dx = \lim_{b\to\infty}\left[2\pi(-x-1)e^{-x}\right]_0^b$

(by parts) ⟶

$$= 2\pi\left(1 - \lim_{b\to\infty}\frac{b+1}{e^b}\right) = 2\pi(1-0) = 2\pi$$

(e) $A = \int_0^\infty 2\pi e^{-x} \sqrt{1 + e^{-2x}}\, dx = \lim_{b \to \infty} \int_0^b 2\pi e^{-x} \sqrt{1 + e^{-2x}}\, dx$

$$\int_0^b 2\pi e^{-x} \sqrt{1 + e^{-2x}}\, dx = -2\pi \int_1^{e^{-b}} \sqrt{1 + u^2}\, du$$
$$u = e^{-x}$$
$$= -\pi \left[u\sqrt{1 + u^2} + \ln\left(1 + \sqrt{1 + u^2}\right) \right]_1^{e^{-b}}$$
$$= \pi \left[\sqrt{2} + \ln\left(1 + \sqrt{2}\right) - e^{-b}\sqrt{1 + e^{-2b}} - \ln\left(1 + \sqrt{1 + e^{-2b}}\right) \right]$$

Taking the limit of this last expression as $b \to \infty$, we have

$$A = \pi \left[\sqrt{2} + \ln\left(1 + \sqrt{2}\right) \right].$$

43. (a) The interval $[0, 1]$ causes no problem. For $x \ge 1$, $e^{-x^2} \le e^{-x}$ and $\int_1^\infty e^{-x}\, dx$ is finite.

(b) $V_y = \int_0^\infty 2\pi x e^{-x^2}\, dx = \lim_{b \to \infty} \int_0^b 2\pi x e^{-x^2}\, dx = \lim_{b \to \infty} \pi \left[-e^{-x^2} \right]_0^b = \lim_{b \to \infty} \pi \left(1 - e^{-b^2} \right) = \pi$

45. (a)

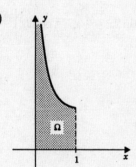

(b) $A = \lim_{a \to 0^+} \int_a^1 x^{-1/4}\, dx = \lim_{a \to 0^+} \left[\frac{4}{3} x^{3/4} \right]_a^1 = \frac{4}{3}$

(c) $V_x = \lim_{a \to 0^+} \int_a^1 \pi x^{-1/2}\, dx = \lim_{a \to 0^+} \left[2\pi x^{1/2} \right]_a^1 = 2\pi$

(d) $V_y = \lim_{a \to 0^+} \int_a^1 2\pi x^{3/4}\, dx = \lim_{a \to 0^+} \left[\frac{8\pi}{7} x^{7/4} \right]_a^1 = \frac{8}{7}\pi$

47. converges by comparison with $\int_1^\infty \dfrac{dx}{x^{3/2}}$

49. diverges since for x large the integrand is greater than $\dfrac{1}{x}$ and $\int_1^\infty \dfrac{dx}{x}$ diverges

51. converges by comparison with $\int_1^\infty \dfrac{dx}{x^{3/2}}$

53. $r(\theta) = ae^{c\theta},\quad r'(\theta) = ace^{c\theta}$

$$L = \int_{-\infty}^{\theta_1} \sqrt{a^2 e^{2c\theta} + a^2 c^2 e^{2c\theta}} \, d\theta$$

(9.9.3)

$$= \left(a\sqrt{1+c^2}\right) \left(\lim_{b \to -\infty} \int_b^{\theta_1} e^{c\theta} \, d\theta\right)$$

$$= \left(a\sqrt{1+c^2}\right) \left(\lim_{b \to -\infty} \left[\frac{e^{c\theta}}{c}\right]_b^{\theta_1}\right)$$

$$= \left(\frac{a\sqrt{1+c^2}}{c}\right) \left(\lim_{b \to -\infty} \left[e^{c\theta_1} - e^{cb}\right]\right) = \left(\frac{a\sqrt{1+c^2}}{c}\right) e^{c\theta_1}$$

55. $F(s) = \int_0^\infty e^{-sx} \cdot 1 \, dx = \lim_{b \to \infty} \int_0^b e^{-sx} \, dx = \lim_{b \to \infty} \left[-\frac{1}{s} e^{-sx}\right]_0^b = \frac{1}{s}$ provided $s > 0$.

Thus, $F(s) = \dfrac{1}{s}$; $\operatorname{dom}(F) = (0, \infty)$.

57. $F(s) = \int_0^\infty e^{-sx} \cos 2x \, dx = \lim_{b \to \infty} \int_0^b e^{-sx} \cos 2x \, dx$

Using integration by parts $\int e^{-sx} \cos 2x \, dx = \dfrac{4}{s^2 + 4} \left[\dfrac{1}{2} e^{-sx} \sin 2x - \dfrac{s}{4} e^{-sx} \cos 2x\right] + C.$

Therefore,

$$F(s) = \lim_{b \to \infty} \frac{4}{s^2 + 4} \left[\frac{1}{2} e^{-sx} \sin 2x - \frac{s}{4} e^{-sx} \cos 2x\right]_0^b$$

$$= \frac{4}{s^2 + 4} \lim_{b \to \infty} \left[\frac{1}{2} e^{-sb} \sin 2b - \frac{s}{4} e^{-sb} \cos 2b + \frac{s}{4}\right] = \frac{4}{s^2 + 4} \cdot \frac{s}{4} = \frac{s}{s^2 + 4}$$ provided $s > 0$.

Thus, $F(s) = \dfrac{s}{s^2 + 4}$; $\operatorname{dom}(F) = (0, \infty)$.

59.   The function f is nonnegative on $(-\infty, \infty)$ and

$$\int_{-\infty}^\infty f(x) \, dx = \int_{-\infty}^0 0 \, dx + \int_0^\infty \frac{6x}{(1+3x^2)^2} \, dx = \int_0^\infty \frac{6x}{(1+3x^2)^2} \, dx$$

Now, $\int \dfrac{6x}{(1+3x^2)^2} \, dx = -\dfrac{1}{1+3x^2} + C.$

Therefore,

$$\int_{-\infty}^{\infty} f(x)\, dx = \lim_{b\to\infty} \left[-\frac{1}{1+3x^2} \right]_0^b = \lim_{b\to\infty} \left(1 - \frac{1}{1+3b^2} \right) = 1.$$

61.
$$\mu = \int_{-\infty}^{\infty} x f(x)\, dx = \int_{-\infty}^{0} 0\, dx + \int_{0}^{\infty} kx e^{-kx}\, dx = \lim_{b\to\infty} \int_0^b kx e^{-kx}\, dx$$

Using integration by parts, $\quad \int kx e^{-kx}\, dx = -x e^{-kx} - \frac{1}{k} e^{-kx} + C.$

Therefore,

$$\mu = \int_{-\infty}^{\infty} x f(x)\, dx = \lim_{b\to\infty} \left[-x e^{-kx} - \frac{1}{k} e^{-kx} \right]_0^b = \lim_{b\to\infty} \left[-b e^{-kb} - \frac{1}{k} e^{-kb} + \frac{1}{k} \right] = \frac{1}{k}$$

63. Observe that

$$F(t) = \int_1^t f(x)\, dx$$

is continuous and increasing, that

$$a_n = \int_1^n f(x)\, dx$$

is increasing, and that

(∗)
$$a_n \le \int_1^t f(x)\, dx \le a_{n+1} \quad \text{for} \quad t \in [n, n+1].$$

If

$$\int_1^{\infty} f(x)\, dx$$

converges, then F, being continuous, is bounded and, by (∗), $\{a_n\}$ is bounded and therefore convergent. If $\{a_n\}$ converges, then $\{a_n\}$ is bounded and, by (∗), F is bounded. Being increasing, F is also convergent; i.e., $\int_1^{\infty} f(x)\, dx$ converges.

PROJECTS AND EXPLORATIONS

10.1. (a) $\displaystyle\int_1^\infty \frac{1}{x^3+1}\,dx \le \int_1^\infty \frac{1}{x^3}\,dx = \lim_{b\to\infty} \int_1^b \frac{1}{x^3}\,dx = \lim_{b\to\infty} \left[-\frac{1}{2x^2}\right]_0^b = \frac{1}{2}$

(b) $\displaystyle\frac{1}{x^3+1} = \frac{1/3}{x+1} + \frac{-x/3+2/3}{x^2-x+1} = \frac{1}{3}\frac{1}{x+1} - \frac{1}{6}\frac{2x-1}{x^2-x+1} + \frac{1}{2}\frac{1}{(x-1/2)^2+3/4}$

$$\int_1^\infty \frac{1}{x^3+1}\,dx = \lim_{b\to\infty} \int_1^b \frac{1}{x^3+1}\,dx$$

$$= \lim_{b\to\infty} \left[\frac{1}{3}\ln|x+1| - \frac{1}{6}\ln|x^2-x+1| + \frac{1}{\sqrt{3}}\tan^{-1}\left(\frac{2x-1}{\sqrt{3}}\right)\right]_0^b$$

$$= \lim_{b\to\infty} \left[\frac{1}{6}\ln\left|\frac{x^2+2x+1}{x^2-x+1}\right| + \frac{1}{\sqrt{3}}\tan^{-1}\left(\frac{2x-1}{\sqrt{3}}\right)\right]_0^b$$

$$= \frac{1}{\pi}3\sqrt{3} - \frac{\ln 4}{6} \cong 0.373551$$

(c) $F'(x) = \dfrac{1}{x^3+1} > 0$ on $[1,\infty)$.

(d) A TI-85 was used to calculate the entries in the following table.

x	$F(x)$
10	0.36855278
100	0.37350072
1000	0.37355023
10,000	0.37355072
20,000	0.37355073
25,000	0.37355073
26,000	$1.01393399 \times 10^{-5}$
30,000	$7.62158383 \times 10^{-6}$
100,000	$6.88311604 \times 10^{-7}$

10.3. (a) $f_{0.1}(4) = \left(1 + \dfrac{0.1}{4}\right)^4 \cong 1.10381289$; quarterly compounding

$f_{0.1}(12) = \left(1 + \dfrac{0.1}{12}\right)^{12} \cong 1.10471307$; monthly compounding

$f_{0.1}(365) = \left(1 + \dfrac{0.1}{365}\right)^{365} \cong 1.10515578$; daily compounding

$f_{0.1}(8760) = \left(1 + \dfrac{0.1}{8760}\right)^{8760} \cong 1.10517029$; hourly compounding

$$f_{0.1}(525,600) = \left(1 + \frac{0.1}{525,600}\right)^{525,600} \cong 1.10517094; \quad \text{interest compounded every minute}$$

$$f_{0.1}(31,536,600) = \left(1 + \frac{0.1}{31,536,600}\right)^{31,536,600525,600} \cong 1.10517026; \quad \text{interest compounded}$$

every second

(b)

	$r = 0.01$	$r = 0.05$	$r = 0.75$	$r = 0.12$
$n = 4$	1.01003756	1.05094534	1.07713587	1.12550881
$n = 12$	1.01004596	1.05116190	1.07763260	1.12582503
$n = 365$	1.01005003	1.05126750	1.07787585	1.12747462
$n = 8760$	1.01005016	1.05127095	1.07788380	1.12749592
$n = 525,600$	1.01005018	1.05127111	1.07788417	1.12749683
$n = 31,536,600$	1.01005043	1.05127244	1.07788452	1.12749676

(c)
$$\lim_{x \to \infty} f_r(x) = \lim_{x \to \infty} \left(1 + \frac{r}{x}\right)^x = \lim_{x \to \infty} \left[\left(1 + \frac{r}{x}\right)^{x/r}\right]^r$$

$$= \lim_{t \to \infty} \left[\left(1 + \frac{1}{t}\right)^t\right]^r \quad (t = x/r)$$

$$= e^r$$

(d)

	$x = 0.1$	$x = 0.01$	$x = 0.001$	$x = 0.0001$
$n = 4$	1.10381289	1.01003756	1.00100038	1.00010001
$n = 12$	1.10471306	1.01004596	1.00100046	1.00010000
$n = 365$	1.10515578	1.01005003	1.00100050	1.00010000
$n = 8760$	1.10517028	1.01005016	1.00100050	1.00010000
$n = 525,600$	1.10517090	1.01005017	1.00100050	1.00010000
$n = 31,536,600$	1.10517092	1.01005017	1.00100050	1.00010000

As the interest rate goes to 0 you earn no interest no matter how often you compound.

(e) The derivatives of f appear to be small and positive, while the derivatives of g appear to be greater than 1 and roughly constant for different values of n.

(f) For fixed x, f is clearly an increasing function of r (increasing the interest rate increases the earnings, exponentially). For fixed r, f is an increasing function of x, although this may be difficult to see just by looking at a graph; try adjusting the scales.

CHAPTER 11

SECTION 11.1

1. $1 + 4 + 7 = 12$

3. $1 + 2 + 4 + 8 = 15$

5. $2 - 4 + 8 - 16 = -10$

7. $\frac{1}{2} + \frac{1}{4} + \frac{1}{8} + \frac{1}{16} = \frac{15}{16}$

9. $-\frac{1}{6} + \frac{1}{24} - \frac{1}{120} = -\frac{2}{15}$

11. $1 + \frac{1}{4} + \frac{1}{16} + \frac{1}{64} = \frac{85}{64}$

13. $\displaystyle\sum_{n=1}^{11}(2n - 1)$

15. $\displaystyle\sum_{n=1}^{25} 2n$

17. $\displaystyle\sum_{n=1}^{81}(-1)^{n-1}\sqrt{n}$

19. $\displaystyle\sum_{k=1}^{n} m_k \, \Delta x_k$

21. $\displaystyle\sum_{k=1}^{n} f(x_k^*) \, \Delta x_k$

23. $\displaystyle\sum_{k=0}^{5}(-1)^k a^{5-k} b^k$

25. $\displaystyle\sum_{k=0}^{4} a_k x^{4-k}$

27. $\displaystyle\sum_{k=0}^{4}(-1)^k(k + 1)x^k$

29. $\displaystyle\sum_{k=3}^{10}\frac{1}{2^k}, \quad \sum_{i=0}^{7}\frac{1}{2^{i+3}}$

31. $\displaystyle\sum_{k=3}^{10}(-1)^{k+1}\frac{k}{k+1}, \quad \sum_{i=0}^{7}(-1)^i\frac{i+3}{i+4}$

33. Set $k = n + 3$. Then $n = -1$ when $k = 2$ and $n = 7$ when $k = 10$.

$$\sum_{k=2}^{10}\frac{k}{k^2 + 1} = \sum_{n=-1}^{7}\frac{n+3}{(n+3)^2 + 1} = \sum_{n=-1}^{7}\frac{n+3}{n^2 + 6n + 10}$$

35. Set $k = n - 3$. Then $n = 7$ when $k = 4$ and $n = 28$ when $k = 25$.

$$\sum_{k=4}^{25}\frac{1}{k^2 - 9} = \sum_{n=7}^{28}\frac{1}{(n-3)^2 - 9} = \sum_{n=7}^{28}\frac{1}{n^2 - 6n}$$

37. (a) $\displaystyle(1 - x)\sum_{k=0}^{n} x^k = \sum_{k=0}^{n}(x^k - x^{k+1}) = (1 - x) + (x - x^2) + \cdots + (x^n - x^{n+1}) = 1 - x^{n+1}$

(b) $a_n = \dfrac{1 - (\frac{1}{3})^{n+1}}{1 - \frac{1}{3}} = \dfrac{3}{2} - \dfrac{3}{2}\left(\dfrac{1}{3}\right)^{n+1} \rightarrow \dfrac{3}{2}$

39. $|a_n - L| < \epsilon$ for $n \geq k$ iff $|a_{n-p} - L| < \epsilon$ for $n \geq k + p$

41. True for $n = 1$: $\displaystyle\sum_{k=1}^{1} k = 1 = \frac{1}{2}(1)(2)$.

Suppose true for $n = p$. Then

$$\sum_{k=1}^{p+1} k = \sum_{k=1}^{p} k + (p + 1) = \frac{1}{2}(p)(p + 1) + (p + 1) = \frac{1}{2}[p(p + 1) + 2(p + 1)]$$

$$= \frac{1}{2}(p + 1)(p + 2) = \frac{1}{2}(p + 1)[(p + 1) + 1]$$

and thus true for $n = p + 1$.

43. True for $n = 1$: $\displaystyle\sum_{k=1}^{1} k^2 = 1 = \frac{1}{6}(1)(2)(3)$.

Suppose true for $n = p$. Then

$$\sum_{k=1}^{p+1} k^2 = \sum_{k=1}^{p} k^2 + (p+1)^2 = \frac{1}{6}(p)(p+1)(2p+1) + (p+1)^2 = \frac{1}{6}(p+1)[p(2p+1) + 6(p+1)]$$

$$= \frac{1}{6}(p+1)[2p^2 + 7p + 6] = \frac{1}{6}(p+1)[(p+2)(2p+3)]$$

$$= \frac{1}{6}(p+1)[(p+1) + 1][2(p+1) + 1]$$

and thus true for $n = p + 1$.

45. $\displaystyle\sum_{k=1}^{10}(2k+3) = 2\sum_{k=1}^{10} k + \sum_{k=1}^{10} 3 = 2 \cdot \frac{(10)(11)}{2} + 3 \cdot 10 = 140$

47. $\displaystyle\sum_{k=1}^{8}(2k-1)^2 = \sum_{k=1}^{8}(4k^2 - 4k + 1) = 4\sum_{k=1}^{8} k^2 - 4\sum_{k=1}^{8} k + \sum_{k=1}^{8} 1$

$$= 4 \cdot \frac{(8)(9)(17)}{6} - 4 \cdot \frac{(8)(9)}{2} + 8 = 680$$

SECTION 11.2

1. $\dfrac{1}{4}$; $\quad s_n = \dfrac{1}{4 \cdot 5} + \dfrac{1}{5 \cdot 6} + \cdots + \dfrac{1}{(n+1)(n+2)}$

$$= \left(\frac{1}{4} - \frac{1}{5}\right) + \left(\frac{1}{5} - \frac{1}{6}\right) + \cdots + \left(\frac{1}{n+1} - \frac{1}{n+2}\right) = \frac{1}{4} - \frac{1}{n+2} \to \frac{1}{4}$$

3. $\dfrac{1}{2}$; $\quad s_n = \dfrac{1}{2}\left[\dfrac{1}{1 \cdot 2} + \dfrac{1}{2 \cdot 3} + \cdots + \dfrac{1}{(n)(n+1)}\right]$

$$= \frac{1}{2}\left[\left(1 - \frac{1}{2}\right) + \left(\frac{1}{2} - \frac{1}{3}\right) + \cdots + \left(\frac{1}{n} - \frac{1}{n+1}\right)\right] = \frac{1}{2}\left[1 - \frac{1}{n+1}\right] \to \frac{1}{2}$$

5. $\dfrac{11}{18}$; $\quad s_n = \dfrac{1}{1 \cdot 4} + \dfrac{1}{2 \cdot 5} + \cdots + \dfrac{1}{n(n+3)}$

$$= \frac{1}{3}\left[\left(1 - \frac{1}{4}\right) + \left(\frac{1}{2} - \frac{1}{5}\right) + \cdots + \left(\frac{1}{n} - \frac{1}{n+3}\right)\right]$$

$$= \frac{1}{3}\left[1 + \frac{1}{2} + \frac{1}{3} - \frac{1}{n+1} - \frac{1}{n+2} - \frac{1}{n+3}\right] \to \frac{1}{3}\left(1 + \frac{1}{2} + \frac{1}{3}\right) = \frac{11}{18}$$

7. $\dfrac{10}{3}$; $\displaystyle\sum_{k=0}^{\infty}\dfrac{3}{10^k}=3\sum_{k=0}^{\infty}\left(\dfrac{1}{10}\right)^k=3\left(\dfrac{1}{1-1/10}\right)=\dfrac{30}{9}=\dfrac{10}{3}$

9. $\dfrac{67000}{999}$; $\displaystyle\sum_{k=0}^{\infty}\dfrac{67}{1000^k}=67\sum_{k=0}^{\infty}\left(\dfrac{1}{1000}\right)^k=67\left(\dfrac{1}{1-1/1000}\right)=\dfrac{67000}{999}$

11. 4; $\displaystyle\sum_{k=0}^{\infty}\left(\dfrac{3}{4}\right)^k=\dfrac{1}{1-3/4}=4$

13. $-\dfrac{3}{2}$; $\displaystyle\sum_{k=0}^{\infty}\dfrac{1-2^k}{3^k}=\sum_{k=0}^{\infty}\left(\dfrac{1}{3}\right)^k-\sum_{k=0}^{\infty}\left(\dfrac{2}{3}\right)^k=\dfrac{1}{1-1/3}-\dfrac{1}{1-2/3}=\dfrac{3}{2}-3=-\dfrac{3}{2}$

15. $\dfrac{1}{2}$; geometric series with $a=\dfrac{1}{4}$ and $r=\dfrac{1}{2}$, sum $=\dfrac{a}{1-r}=\dfrac{1}{2}$

17. 24; geometric series with $a=8$ and $r=\dfrac{2}{3}$, sum $=\dfrac{a}{1-r}=24$

19. $\displaystyle\sum_{k=1}^{\infty}\dfrac{7}{10^k}=\dfrac{7/10}{1-1/10}=\dfrac{7}{9}$

21. $\displaystyle\sum_{k=1}^{\infty}\dfrac{24}{100^k}=\dfrac{24/100}{1-1/100}=\dfrac{8}{33}$

23. $\displaystyle\sum_{k=1}^{\infty}\dfrac{112}{1000^k}=\dfrac{112/1000}{1-1/1000}=\dfrac{112}{999}$

25. $\dfrac{62}{100}+\dfrac{1}{100}\displaystyle\sum_{k=1}^{\infty}\dfrac{45}{100^k}=\dfrac{62}{100}+\dfrac{1}{100}\left(\dfrac{45/100}{1-1/100}\right)=\dfrac{687}{1100}$

27. Let $x=0.\overbrace{a_1a_2\cdots a_n}\overbrace{a_1a_2\cdots a_n}\cdots$.   Then

$$x=\sum_{k=1}^{\infty}\dfrac{a_1a_2\cdots a_n}{(10^n)^k}=a_1a_2\cdots a_n\sum_{k=1}^{\infty}\left(\dfrac{1}{10^n}\right)^k$$

$$=a_1a_2\cdots a_n\left[\dfrac{1}{1-1/10^n}-1\right]=\dfrac{a_1a_2\cdots a_n}{10^n-1}.$$

29. $\dfrac{1}{1+x}=\dfrac{1}{1-(-x)}=\displaystyle\sum_{k=0}^{\infty}(-x)^k=\sum_{k=0}^{\infty}(-1)^k x^k$

31. $\dfrac{x}{1-x}=x\left(\dfrac{1}{1-x}\right)=x\displaystyle\sum_{k=0}^{\infty}(x^k)=\sum_{k=0}^{\infty}x^{k+1}$

33. $\dfrac{x}{1+x^2}=x\left[\dfrac{1}{1-(-x^2)}\right]=x\displaystyle\sum_{k=0}^{\infty}(-x^2)^k=\sum_{k=0}^{\infty}(-1)^k x^{2k+1}$

35. $\dfrac{1}{1+4x^2} = \dfrac{1}{1-(-4x^2)} = \displaystyle\sum_{k=0}^{\infty}(-4x^2)^k = \sum_{k=0}^{\infty}(-1)^k(2x)^{2k}$

37. $1 + \dfrac{3}{2} + \dfrac{9}{4} + \dfrac{27}{8} + \dfrac{81}{16} + \cdots = \displaystyle\sum_{k=0}^{\infty}\left(\dfrac{3}{2}\right)^k$

This is a geometric series with $x = \frac{3}{2} > 1$. Therefore the series diverges.

39. $\displaystyle\lim_{k\to\infty}\left(\dfrac{k+1}{k}\right)^k = e \neq 0$

41. $4 + \dfrac{1}{3}\displaystyle\sum_{k=0}^{\infty}\left(\dfrac{1}{12}\right)^k = 4 + \dfrac{4}{11}$ o'clock

43. In the general case of a ball which rebounds σh feet if dropped from a height of h feet ($0 < \sigma < 1$), we find for a ball dropped initially from h_0 feet that the total distance traveled is given by

$$d = h_0 + 2\sigma h_0 + 2\sigma^2 h_0 + 2\sigma^3 h_0 + \cdots$$

$$= h_0 + 2h_0\sigma[1 + \sigma + \sigma^2 + \cdots]$$

$$= h_0 + 2h_0\sigma\sum_{k=0}^{\infty}\sigma^k = h_0 + \dfrac{2h_0\sigma}{1-\sigma}.$$

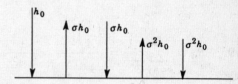

Here, $h_0 = 6$ ft and $\sigma = \frac{1}{3}$ so that distance = 12 ft.

45. A principal x deposited now at $r\%$ interest compounded annually will grow in k years to

$$x\left(1 + \dfrac{r}{100}\right)^k.$$

This means that in order to be able to withdraw n_k dollars after k years one must place

$$n_k\left(1 + \dfrac{r}{100}\right)^{-k}$$

dollars on deposit today. To extend this process in perpetuity as described in the text, the total deposit must be

$$\sum_{k=1}^{\infty} n_k\left(1 + \dfrac{r}{100}\right)^{-k}.$$

47. $\displaystyle\sum_{n=1}^{\infty}\left(\dfrac{9}{10}\right)^n = \dfrac{\frac{9}{10}}{1 - \frac{9}{10}} = 9$ or $9

49. $A = 4^2 + (2\sqrt{2})^2 + 2^2 + (\sqrt{2})^2 + 1^2 + \cdots + \left[4\left(\dfrac{1}{\sqrt{2}}\right)^n\right]^2 + \cdots$

$$= \sum_{n=0}^{\infty}\left[4\left(\dfrac{1}{\sqrt{2}}\right)^n\right]^2 = 16\sum_{n=0}^{\infty}\left(\dfrac{1}{2}\right)^n = 16 \cdot \dfrac{1}{1 - \frac{1}{2}} = 32$$

51. Let $L = \displaystyle\sum_{k=0}^{\infty} a_k$. Then

$$L = \sum_{k=0}^{\infty} a_k = \sum_{k=0}^{n} a_k + \sum_{k=n+1}^{\infty} a_k = s_n + R_n.$$

Therefore, $R_n = L - s_n$ and since $s_n \to L$ as $n \to \infty$, it follows that $R_n \to 0$ as $n \to \infty$.

53. $s_0 = 1,\ s_1 = 0,\ s_2 = 1,\ \dots,;\quad s_n = \dfrac{1 + (-1)^n}{2},\ n = 0, 1, 2, \dots .$

55. $s_n = \displaystyle\sum_{k=1}^{n} \ln\left(\dfrac{k+1}{k}\right) = [\ln(n+1) - \ln(n)] + [\ln n - \ln(n-1)] + \cdots + [\ln 2 - \ln 1] = \ln(n+1) \to \infty$

57. (a) $s_n = \displaystyle\sum_{k=1}^{n}(d_k - d_{k+1}) = d_1 - d_{n+1} \to d_1$

(b) We use part (a).

(i) $\displaystyle\sum_{k=1}^{\infty} \dfrac{\sqrt{k+1} - \sqrt{k}}{\sqrt{k(k+1)}} = \sum_{k=1}^{\infty}\left[\dfrac{1}{\sqrt{k}} - \dfrac{1}{\sqrt{k+1}}\right] = 1$

(ii) $\displaystyle\sum_{k=1}^{\infty} \dfrac{2k+1}{2k^2(k+1)^2} = \sum_{k=1}^{\infty} \dfrac{1}{2}\left[\dfrac{1}{k^2} - \dfrac{1}{(k+1)^2}\right] = \dfrac{1}{2}$

59. $R_n = \displaystyle\sum_{k=n+1}^{\infty} \dfrac{1}{4^k} = \dfrac{\left(\frac{1}{4}\right)^{n+1}}{1 - \frac{1}{4}} = \dfrac{1}{3 \cdot 4^n};$

$\dfrac{1}{3 \cdot 4^n} < 0.0001 \implies 4^n > 3333.33 \implies n > \dfrac{\ln 3333.33}{\ln 4} \cong 5.85$

Take $N = 6$.

61. $R_n = \displaystyle\sum_{k=n+1}^{\infty} \dfrac{1}{k(k+2)} = \dfrac{1}{2}\sum_{k=n+1}^{\infty}\left(\dfrac{1}{k} - \dfrac{1}{k+2}\right) = \dfrac{1}{2}\left(\dfrac{1}{n+1} + \dfrac{1}{n+2}\right);$

$\dfrac{1}{2}\left(\dfrac{1}{n+1} + \dfrac{1}{n+2}\right) < 0.0001 \implies n \geq 9999.$ Take $N = 9999$.

63. $|R_n| = \left|\displaystyle\sum_{k=n+1}^{\infty} x^k\right| = \left|\dfrac{x^{n+1}}{1-x}\right| = \dfrac{|x|^{n+1}}{1-x};$

$$\dfrac{|x|^{n+1}}{1-x} < \epsilon$$

$$|x|^{n+1} < \epsilon(1-x)$$

$$(n+1)\ln|x| < \ln\epsilon(1-x)$$

$$n+1 > \dfrac{\ln\epsilon(1-x)}{\ln|x|} \qquad [\text{recall}\ \ln|x| < 0]$$

$$n > \dfrac{\ln\epsilon(1-x)}{\ln|x|} - 1$$

Take N to be smallest integer which is greater than $\dfrac{\ln\epsilon(1-x)}{\ln|x|}$.

SECTION 11.3

1. converges; basic comparison with $\sum \dfrac{1}{k^2}$ 3. converges; basic comparison with $\sum \dfrac{1}{k^2}$

5. diverges; basic comparison with $\sum \dfrac{1}{k+1}$ 7. diverges; limit comparison with $\sum \dfrac{1}{k}$

9. converges; integral test, $\displaystyle\int_1^\infty \dfrac{\tan^{-1} x}{1+x^2}\, dx = \lim_{b\to\infty}\left[\dfrac{1}{2}(\tan^{-1} x)^2\right]_1^b = \dfrac{3\pi^2}{32}$

11. diverges; p-series with $p = \dfrac{2}{3} \le 1$ 13. diverges; divergence test, $\left(\dfrac{3}{4}\right)^{-k} \not\to 0$

15. diverges; basic comparison with $\sum \dfrac{1}{k}$

17. diverges; divergence test, $\dfrac{1}{2+3^{-k}} \to \dfrac{1}{2} \ne 0$

19. converges; limit comparison with $\sum \dfrac{1}{k^2}$

21. diverges; integral test, $\displaystyle\int_2^\infty \dfrac{dx}{x\ln x} = \lim_{b\to\infty}\left[\ln(\ln x)\right]_2^b = \infty$

23. converges; limit comparison with $\sum \dfrac{1}{k^2}$ 25. diverges; limit comparison with $\sum \dfrac{1}{k}$

27. converges; limit comparison with $\sum \dfrac{1}{k^{3/2}}$

29. converges; integral test, $\displaystyle\int_1^\infty x e^{-x^2}\, dx = \lim_{b\to\infty}\left[-\dfrac{1}{2}e^{-x^2}\right]_1^b = \dfrac{1}{2e}$

31. Converges; basic comparison with $\sum \dfrac{3}{k^2}$, $2 + \sin k \le 3$ for all k.

33. Recall that $1 + 2 + 3 + \cdots + k = \dfrac{k(k+1)}{2}$. Therefore

$$\sum \frac{1}{1+2+3+\cdots+k} = \sum \frac{2}{k(k+1)}.$$ This series converges; direct comparison with $\sum \dfrac{2}{k^2}$

35. Use the integral test:

Let $u = \ln x,\ du = \dfrac{1}{x}\, dx$: $\displaystyle\int \frac{1}{x(\ln x)^p}\, dx = \int u^{-p}\, du = \frac{u^{1-p}}{1-p} + C.$

$$\int_1^\infty \frac{1}{x(\ln x)^p}\, dx = \lim_{b\to\infty}\int_1^b \frac{1}{x(\ln x)^p}\, dx = \lim_{b\to\infty}\frac{1}{1-p}(\ln a)^{1-p}$$

The series converges for $p > 1$.

37. (a) Use the integral test: $\displaystyle\int_0^\infty e^{-\alpha x}\, dx = \lim_{b\to\infty}\left[-\frac{1}{\alpha}e^{-\alpha x}\right]_0^b = \frac{1}{\alpha}$ converges.

(b) Use the integral test: $\displaystyle\int_0^\infty xe^{-\alpha x}\,dx = \lim_{b\to\infty}\left[-\frac{1}{\alpha xe^{-\alpha x}} - \frac{1}{\alpha^2}e^{-\alpha x}\right]_0^b = \frac{1}{\alpha^2}$ converges.

(c) The proof follows by induction using parts (a) and (b) and the reduction formula

$$\int x^n e^{ax}\,dx = \frac{x^n e^{ax}}{a} - \frac{n}{a}\int x^{n-1}e^{ax}\,dx \quad \text{[see Exercise 67, Section 8.2]}$$

39. (a) $\displaystyle\sum_{k=1}^4 \frac{1}{k^3} \cong 1.1777$ (b) $\displaystyle\frac{1}{2\cdot 5^2} < R_4 < \frac{1}{2\cdot 4^2}$

$$0.02 < R_4 < 0.0313$$

(c) $\displaystyle 1.1777 + 0.02 = 1.1977 < \sum_{k=1}^\infty \frac{1}{k^3} < 1.1777 + 0.0313 = 1.2090$

41. (a) Put $p = 2$ and $n = 100$ in the estimates in Exercise 38. The result is: $\displaystyle\frac{1}{101} < R_{100} < \frac{1}{100}$.

(b) $\displaystyle R_n < \frac{1}{(2-1)n^{2-1}} < 0.0001 \implies n > 10,000$ Take $n = 10,001$.

43. (a) $\displaystyle R_n < \frac{1}{(4-1)n^{4-1}} < 0.0001 \implies n^3 > 3333 \implies n > 14.94 :$ Take $n = 15$.

(b) $\displaystyle R_n < \frac{1}{(4-1)n^{4-1}} < 0.0005 \implies n^3 > 666.67 \implies n > 8.74 :$ Take $n = 9$.

$$\sum_{k=1}^\infty \frac{1}{k^4} \cong \sum_{k=1}^9 \frac{1}{k^4} \cong 1.082$$

45. (a) If $a_k/b_k \to 0$, then $a_k/b_k < 1$ for all $k \geq K$ for some K. But then $a_k < b_k$ for all $k \geq K$

and, since $\sum b_k$ converges, $\sum a_k$ converges. [The Basic Comparison Theorem 11.3.5.]

(b) Similar to (a) except that this time we appeal to part (ii) of Theorem 11.3.5.

(c) $\displaystyle\sum a_k = \sum \frac{1}{k^2}$ converges, $\displaystyle\sum b_k = \sum \frac{1}{k^{3/2}}$ converges, $\displaystyle\frac{1/k^2}{1/k^{3/2}} = \frac{1}{\sqrt{k}} \to 0$

$\displaystyle\sum a_k = \sum \frac{1}{k^2}$ converges, $\displaystyle\sum b_k = \sum \frac{1}{\sqrt{k}}$ diverges, $\displaystyle\frac{1/k^2}{1/\sqrt{k}} = \frac{1}{k^{3/2}} \to 0$

(d) $\displaystyle\sum b_k = \sum \frac{1}{\sqrt{k}}$ diverges, $\displaystyle\sum a_k = \sum \frac{1}{k^2}$ converges, $\displaystyle\frac{1/k^2}{1/\sqrt{k}} = 1/k^{3/2} \to 0$

$\displaystyle\sum b_k = \sum \frac{1}{\sqrt{k}}$ diverges, $\displaystyle\sum a_k = \sum \frac{1}{k}$ diverges, $\displaystyle\frac{1/k}{1/\sqrt{k}} = \frac{1}{\sqrt{k}} \to 0$

47. (a) Since $\sum a_k$ converges, $a_k \to 0$. Therefore there exists a positive integer N such that $0 < a_k < 1$ for $k \geq N$. Thus, for $k \geq N$, $a_k^2 < a_k$ and so $\sum a_k^2$ converges by the comparison test.

(b) $\sum a_k$ may either converge or diverge: $\sum 1/k^4$ and $\sum 1/k^2$ both converge; $\sum 1/k^2$ converges and $\sum 1/k$ diverges.

49. $0 < L - \sum_{k=1}^{n} f(k) = L - s_n = \sum_{k=n+1}^{\infty} f(k) < \int_{n}^{\infty} f(x)\,dx$ [see the proof of the integral test]

51.
$$L - s_n < \int_{n}^{\infty} xe^{-x^2}\,dx = \lim_{b\to\infty} \int_{n}^{b} xe^{-x^2}\,dx$$

$$= \lim_{b\to\infty} \left[-\frac{1}{2}e^{-x^2} \right]_{n}^{b} = \frac{1}{2}e^{-n^2}$$

$\frac{1}{2}e^{-n^2} < 0.001 \implies e^{n^2} > 500 \implies n > 2.49;$ take n=3.

53. (a) Set $f(x) = x^{1/4} - \ln x$. Then

$$f'(x) = \frac{1}{4}x^{-3/4} - \frac{1}{x} = \frac{1}{4x}(x^{1/4} - 4).$$

Since $f(e^{12}) = e^3 - 12 > 0$ and $f'(x) > 0$ for $x > e^{12}$, we have that

$$n^{1/4} > \ln n \quad \text{and therefore} \quad \frac{1}{n^{5/4}} > \frac{\ln n}{n^{3/2}}$$

for sufficiently large n. Since $\sum \frac{1}{n^{5/4}}$ is a convergent p-series, $\sum \frac{\ln n}{n^{3/2}}$ converges

by the basic comparison test.

(b) By L'Hospital's rule

$$\lim_{\to\infty} \frac{(\ln x)/x^{3/2}}{1/x^{5/4}} = \lim_{x\to\infty} \frac{\ln x}{x^{1/4}} \overset{\star}{=} \lim_{x\to\infty} \frac{1/x}{\frac{1}{4}x^{-3/4}} = \lim_{x\to\infty} \frac{4}{x^{1/4}} = 0.$$

Thus, the limit comparison test does not apply.

1. converges; ratio test: $\frac{a_{k+1}}{a_k} = \frac{10}{k+1} \to 0$ **3.** converges; root test: $(a_k)^{1/k} = \frac{1}{k} \to 0$

5. diverges; divergence test: $\frac{k!}{100^k} \to \infty$ **7.** diverges; limit comparison with $\sum \frac{1}{k}$

9. converges; root test: $(a_k)^{1/k} = \frac{2}{3}k^{1/k} \to \frac{2}{3}$ **11.** diverges; limit comparison with $\sum \frac{1}{\sqrt{k}}$

13. diverges; ratio test: $\frac{a_{k+1}}{a_k} = \frac{k+1}{10^4} \to \infty$ **15.** converges; basic comparison with $\sum \frac{1}{k^{3/2}}$

17. converges; basic comparison with $\sum \frac{1}{k^2}$

19. diverges; integral test: $\int_2^\infty \frac{1}{x}(\ln x)^{-1/2} dx = \lim_{b \to \infty} \left[2(\ln x)^{1/2}\right]_2^b = \infty$

21. diverges; divergence test: $\left(\frac{k}{k+100}\right)^k = \left(1 + \frac{100}{k}\right)^{-k} \to e^{-100} \neq 0$

23. diverges; limit comparison with $\sum \frac{1}{k}$ **25.** converges; ratio test: $\frac{a_{k+1}}{a_k} = \frac{\ln(k+1)}{e \ln k} \to \frac{1}{e}$

27. converges; basic comparison with $\sum \frac{1}{k^{3/2}}$

29. converges; ratio test: $\frac{a_{k+1}}{a_k} = \frac{2(k+1)}{(2k+1)(2k+2)} \to 0$

31. converges; ratio test: $\frac{a_{k+1}}{a_k} = \frac{(k+1)(2k+1)(2k+2)}{(3k+1)(3k+2)(3k+3)} \to \frac{4}{27}$

33. converges; ratio test: $\frac{a_{k+1}}{a_k} = \frac{1}{(k+1)^{1/2}}\left(\frac{k+1}{k}\right)^{k/2} \to 0 \cdot \sqrt{e} = 0$

35. converges; root test: $(a_k)^{1/k} = \frac{k}{3^k} \to 0$

37. $\frac{1}{2} + \frac{2}{3^2} + \frac{4}{4^3} + \frac{8}{5^4} + \cdots = \sum_{k=0}^\infty \frac{2^k}{(k+2)^{k+1}}$

converges; root test: $(a_k)^{1/k} = \frac{2}{(k+2)^{1+1/k}} \to 0$

39. $\frac{1}{4} + \frac{1 \cdot 3}{4 \cdot 7} + \frac{1 \cot 3 \cdot 5}{4 \cdot 7 \cdot 10} + \cdots = \sum_{k=0}^\infty \frac{1 \cdot 3 \cdots (1+2k)}{4 \cdot 7 \cdots (4+3k)}$

converges; ratio test: $\frac{a_{k+1}}{a_k} = \frac{3+2k}{7+3k} \to \frac{2}{3}$

41. By the hint

$$\sum_{k=1}^\infty k\left(\frac{1}{10}\right)^k = \frac{1}{10}\sum_{k=1}^\infty k\left(\frac{1}{10}\right)^{k-1} = \frac{1}{10}\left[\frac{1}{1-1/10}\right]^2 = \frac{10}{81}.$$

43. The series $\sum_{k=0}^\infty \frac{k!}{k^k}$ converges (see Exercise 26). Therefore, $\lim_{k \to \infty} \frac{k!}{k^k} = 0$ by Theorem 11.2.6.

45. Use the ratio test:

$$\frac{a_{k+1}}{a_k} = \frac{\frac{[(k+1)!]^2}{[p(k+1)]!}}{\frac{(k!)^2}{(pk)!}} = (k+1)^2 \frac{(pk)!}{(pk)!(pk+1)\cdots(pk+p)} = \frac{(k+1)^2}{(pk+1)\cdots(pk+p)}$$

Thus

$$\frac{a_{k+1}}{a_k} \to \begin{cases} \dfrac{1}{4}, & \text{if } p = 2 \\[2mm] 0, & \text{if } p > 2 \end{cases}$$

The series converges for all $p \geq 2$.

47. Fix x and use the ratio test:

$$\frac{a_{k+1}}{a_k} = \frac{\dfrac{|x|^{k+1}}{k+1}}{\dfrac{|x|^k}{k}} = \frac{k}{k+1}|x| \to |x| \quad \text{as } k \to \infty$$

The series converges for $|x| < 1$.

49. Fix x and use the ratio test:

$$\frac{a_{k+1}}{a_k} = \frac{\dfrac{2^{k+1}|x|^{k+1}}{(k+1)!}}{\dfrac{2^k|x|^k}{k!}} = \frac{2}{k+1}|x| \to 0 \quad \text{as } k \to \infty$$

The series converges for all x.

51. Set $b_k = a_k r^k$. If $(a_k)^{1/k} \to \rho$ and $\rho < \dfrac{1}{r}$, then

$$(b_k)^{1/k} = (a_k r^k)^{1/k} = (a_k)^{1/k} r \to \rho r < 1$$

and thus, by the root test, $\Sigma b_k = \Sigma a_k r^k$ converges.

SECTION 11.5

1. diverges; $a_k \nrightarrow 0$ 3.   diverges; $\dfrac{k}{k+1} \to 1 \neq 0$

5. (a) does not converge absolutely;   integral test,

$$\int_1^\infty \frac{\ln x}{x}\,dx = \lim_{b \to \infty}\left[\frac{1}{2}(\ln x)^2\right]_1^b = \infty$$

   (b) converges conditionally;   Theorem 11.5.4

7. diverges;   limit comparison with $\sum \dfrac{1}{k}$

another approach: $\sum\left(\dfrac{1}{k} - \dfrac{1}{k!}\right) = \sum \dfrac{1}{k} - \sum \dfrac{1}{k!}$ diverges since $\sum \dfrac{1}{k}$ diverges and

$\sum \dfrac{1}{k!}$ converges

9. (a) does not converge absolutely;   limit comparison with $\sum \frac{1}{k}$

    (b) converges conditionally;   Theorem 11.5.4

11. diverges; $a_k \not\to 0$

13. (a) does not converge absolutely;

$$(\sqrt{k+1} - \sqrt{k}) \cdot \frac{(\sqrt{k+1} + \sqrt{k})}{(\sqrt{k+1} + \sqrt{k})} = \frac{1}{\sqrt{k+1} + \sqrt{k}}$$

and

$$\sum \frac{1}{\sqrt{k} + \sqrt{k+1}} > \sum \frac{1}{2\sqrt{k+1}} = 2 \sum \frac{1}{\sqrt{k+1}} \qquad (\text{a } p\text{-series with } p < 1)$$

    (b) converges conditionally;   Theorem 11.5.4

15. converges absolutely (terms already positive); basic comparison,

$$\sum \sin\left(\frac{\pi}{4k^2}\right) \leq \sum \frac{\pi}{4k^2} = \frac{\pi}{4} \sum \frac{1}{k^2} \qquad (|\sin x| \leq |x|)$$

17. converges absolutely;   ratio test, $\dfrac{a_{k+1}}{a_k} = \dfrac{k+1}{2k} \to \dfrac{1}{2}$

19. (a) does not converge absolutely;   limit comparison with $\sum \frac{1}{k}$

    (b) converges conditionally;   Theorem 11.5.4

21. diverges; $a_k = \dfrac{4^{k-2}}{e^k} = \dfrac{1}{16}\left(\dfrac{4}{e}\right)^k \not\to 0$

23. diverges; $a_k = k \sin(1/k) = \dfrac{\sin(1/k)}{1/k} \to 1 \neq 0$

25. converges absolutely;   ratio test, $\dfrac{a_{k+1}}{a_k} = \dfrac{(k+1)e^{-(k+1)}}{k\,e^{-k}} = \dfrac{k+1}{k}\dfrac{1}{e} \to \dfrac{1}{e}$

27. diverges; $\sum (-1)^k \dfrac{\cos \pi k}{k} = \sum (-1)^k \dfrac{(-1)^k}{k} = \sum \dfrac{1}{k}$

29. converges absolutely;   basic comparison

$$\sum \left| \frac{\sin(\pi k/4)}{k^2} \right| \leq \sum \frac{1}{k^2}$$

31. diverges; $a_k \not\to 0$

33. Use (11.5.5); $|s - s_{80}| < a_{81} = \dfrac{1}{\sqrt{82}} \cong 0.1104$

35. Use (11.5.5); $|s - s_9| < a_{10} = \dfrac{1}{10^3} = 0.001$

37. $\dfrac{10}{11}$;   geometric series with $a = 1$ and $r = -\dfrac{1}{10}$,   sum $= \dfrac{a}{1-r} = \dfrac{10}{11}$

39. Use (11.5.5); $\quad |s - s_n| < a_{n+1} = \dfrac{1}{\sqrt{n+2}} < 0.005 \quad \Longrightarrow \quad n \geq 39,998$

41. Use (11.5.5).

(a) $n = 4$; $\quad \dfrac{1}{(n+1)!} < 0.01 \quad \Longrightarrow \quad 100 < (n+1)!$

(b) $n = 6$; $\quad \dfrac{1}{(n+1)!} < 0.001 \quad \Longrightarrow \quad 1000 < (n+1)!$

43. No.   For instance, set $a_{2k} = 2/k$ and $a_{2k+1} = 1/k$.

45. (a) Since $\sum |a_k|$ converges, $\sum |a_k|^2 = \sum a_k^2$ converges (Exercise 47, Section 11.3).

(b) $\sum \dfrac{1}{k^2}$ converges, $\sum (-1)^k \dfrac{1}{k}$ is not absolutely convergent.

47. See the proof of Theorem 11.8.2.

49. (a) $\displaystyle\sum_{k=1}^{\infty} \dfrac{(-1)^{k-1}(a+b) + (a-b)}{2k} = \sum_{k=1}^{\infty} \dfrac{(-1)^{k-1}(a+b)}{2k} + \sum_{k=1}^{\infty} \dfrac{a-b}{2k}$

(b) The series is absolutely convergent if $a = b = 0$; conditionally convergent if $a = b \neq 0$;

divergent if $a \neq b$.

SECTION 11.6

1. $-1 + x + \frac{1}{2}x^2 - \frac{1}{24}x^4$

3. $-\frac{1}{2}x^2 - \frac{1}{12}x^4$

5. $1 - x + x^2 - x^3 + x^4 - x^5$

7. $x + \frac{1}{3}x^3 + \frac{2}{15}x^5$

9. $P_0(x) = 1$, $\quad P_1(x) = 1 - x$, $\quad P_2(x) = 1 - x + 3x^2$, $\quad P_3(x) = 1 - x + 3x^2 + 5x^3$

11. $\displaystyle\sum_{k=0}^{n} (-1)^k \dfrac{x^k}{k!}$

13. $\displaystyle\sum_{k=0}^{m} \dfrac{x^{2k}}{(2k)!}$ where $m = \dfrac{n}{2}$ and n is even

15. $f^{(k)}(x) = r^k e^{rx}$ and $f^{(k)}(0) = r^k$, $k = 0, 1, 2, \ldots$. Thus, $P_n(x) = \displaystyle\sum_{k=0}^{n} \dfrac{r^k}{k!} x^k$

17. The Taylor polynomial

$$P_n(0.5) = 1 + (0.5) + \dfrac{(0.5)^2}{2!} + \cdots + \dfrac{(0.5)^n}{n!}$$

estimates $e^{0.5}$ within

$$|R_{n+1}(0.5)| \leq e^{0.5} \dfrac{|0.5|^{n+1}}{(n+1)!} < 2 \dfrac{(0.5)^{n+1}}{(n+1)!}.$$

Since

$$2 \dfrac{(0.5)^4}{4!} = \dfrac{1}{8(24)} < 0.01,$$

we can take $n = 3$ and be sure that

$$P_3(0.5) = 1 + (0.5) + \frac{(0.5)^2}{2} + \frac{(0.5)^3}{6} = \frac{79}{48}$$

differs from $\sqrt{e}$ by less than 0.01. Our calculator gives

$$\frac{79}{48} \cong 1.645833 \quad \text{and} \quad \sqrt{e} \cong 1.6487213.$$

19.   At $x = 1$, the sine series gives

$$\sin 1 = 1 - \frac{1}{3!} + \frac{1}{5!} - \frac{1}{7!} + \cdots.$$

This is a convergent alternating series with decreasing terms. The first term of magnitude less than 0.01 is $1/5! = 1/120$. Thus

$$1 - \frac{1}{3!} = 1 - \frac{1}{6} = \frac{5}{6}$$

differs from $\sin 1$ by less than 0.01. Our calculator gives

$$\frac{5}{6} \cong 0.8333333 \quad \text{and} \quad \sin 1 \cong 0.84114709.$$

The estimate

$$1 - \frac{1}{3!} + \frac{1}{5!} = \frac{101}{120}$$

is much more accurate:

$$\frac{101}{120} \cong 0.8416666.$$

21.   At $x = 1$, the cosine series gives

$$\cos 1 = 1 - \frac{1}{2!} + \frac{1}{4!} - \frac{1}{6!} + \frac{1}{8!} + \cdots.$$

This is a convergent alternating series with decreasing terms. The first term of magnitude less than 0.01 is $1/6! = 1/720$. Thus

$$1 - \frac{1}{2!} + \frac{1}{4!} = 1 - \frac{1}{2} + \frac{1}{24} = \frac{13}{24}$$

differs from $\cos 1$ by less than 0.01. Our calculator gives

$$\frac{13}{24} \cong 0.5416666 \quad \text{and} \quad \cos 1 \cong 0.5403023.$$

23.   First convert $10°$ to radians: $10° = \frac{10}{180}\pi \cong 0.1745$ radians

At $x = 0.1745$, the sine series gives

$$\sin 0.1745 = 0.1745 - \frac{(0.1745)^3}{3!} + \frac{(0.1745)^5}{5!} - \cdots.$$

This is a convergent alternating series with decreasing terms. The first term of magnitude less than 0.01 is $(0.1745)^3/3! \cong 0.00089$. Thus 0.1745 differs from $\sin 10°$ by less than 0.01. Our calculator gives $\sin 10° \cong 0.1736$

25. $f(x) = e^{2x}; \quad f^{(5)}(x) = 2^5 e^{2x}; \quad R_5(x) = \frac{2^5 e^{2c}}{5!} x^5 = \frac{4}{15} e^{2c} x^5,$ where c is between 0 and x.

27. $f(x) = \cos 2x;$ $f^{(5)}(x) = -2^5 \sin 2x$
$$R_5(x) = \frac{-2^5 \sin 2c}{5!} x^5 = -\frac{4}{15} \sin(2c) \, x^5,$$

where c is between 0 and x.

29. $f(x) = \tan x;$ $f'''(x) = 6 \sec^4 x - 4 \sec^2 x$
$$R_3(x) = \frac{6 \sec^4 c - 4 \sec^2 c}{3!} x^3 = \frac{3 \sec^4 c - 2 \sec^2 c}{3} x^3,$$

where c is between 0 and x.

31. $f(x) = \tan^{-1} x;$ $f'''(x) = \dfrac{6x^2 - 2}{(1 + x^2)^3}$
$$R_3(x) = \frac{6c^2 - 2}{3! \, (1 + c^2)^3} x^3 = \frac{3c^2 - 1}{3 \, (1 + c^2)^3} x^3,$$

where c is between 0 and x.

33. $f(x) = e^{-x};$ $f^{(k)}(x) = (-1)^k e^{-x},$ $k = 0, 1, 2, \dots$
$$R_{n+1}(x) = \frac{(-1)^{n+1} e^{-c}}{(n + 1)!} x^{n+1},$$

where c is between 0 and x.

35. $f(x) = \dfrac{1}{1 - x};$ $f^{(k)}(x) = \dfrac{k!}{(1 - x)^{k+1}},$ $k = 0, 1, 2, \dots$
$$R_{n+1}(x) = \frac{(n + 1)!}{(1 - c)^{n+2}(n + 1)!} x^{n+1} = \frac{1}{(1 - c)^{n+2}} x^{n+1},$$

where c is between 0 and x.

37. By (11.6.8)
$$P_n(x) = x - \frac{x^2}{2} + \frac{x^3}{3} - \frac{x^4}{4} + \dots + (-1)^{n+1} \frac{x^n}{n}.$$

For $0 \le x \le 1$ we know from (11.5.5) that
$$|P_n(x) - \ln(1 + x)| < \frac{x^{n+1}}{n + 1}.$$

(a) $n = 4;$ $\dfrac{(0.5)^{n+1}}{n + 1} \le 0.01$ $\implies$ $100 \le (n + 1)2^{n+1}$ $\implies$ $n \ge 4$

(b) $n = 2;$ $\dfrac{(0.3)^{n+1}}{n + 1} \le 0.01$ $\implies$ $100 \le (n + 1)\left(\dfrac{10}{3}\right)^{n+1}$ $\implies$ $n \ge 2$

(c) $n = 999;$ $\dfrac{(1)^{n+1}}{n + 1} \le 0.001$ $\implies$ $1000 \le n + 1$ $\implies$ $n \ge 999$

39. $f(x) = e^x;$ $f^{(n)}(x) = e^x;$ $R_{n+1}(x) = \dfrac{e^c}{(n + 1)!} x^{n+1},$ $|c| < |x|$

(a) We want $|R_{n+1}(1/2)| < .00005$: for $0 < c < \frac{1}{2}$, we have
$$|R_{n+1}(1/2)| = \frac{e^c}{(n + 1)!} \left(\frac{1}{2}\right)^{n+1} < \frac{e^{1/2}}{(n + 1)!} \left(\frac{1}{2}\right)^{n+1} < \frac{2}{2^{n+1}(n + 1)!} < 0.00005$$

You can verify that this inequality is satisfied if $n \ge 5$.

$$P_5(x) = 1 + x + \frac{x^2}{2!} + \frac{x^3}{3!} + \frac{x^4}{4!} + \frac{x^5}{5!}$$

$$P_5(1/2) = 1 + \frac{1}{2} + \frac{1}{8} + \frac{1}{48} + \frac{1}{320} + \frac{1}{3840} \cong 1.6492$$

(b) We want $|R_{n+1}(-1)| < .0005$: for $-1 < c < 0$, we have

$$|R_{n+1}(-1)| = \frac{e^c}{(n+1)!} |(-1)^{n+1}| < \frac{1}{(n+1)!} < 0.0005$$

You can verify that this inequality is satisfied if $n \geq 7$.

$$P_7(x) = \sum_{k=0}^{7} \frac{x^k}{k!}; \quad P_7(-1) = \sum_{k=0}^{7} \frac{(-1)^k}{k!} \cong 0.368$$

41.   The result follows from the fact that $\quad P^{(k)}(0) = \left\{ \begin{array}{ll} k! a_k, & 0 \leq k \leq n \\ 0, & n < k \end{array} \right]$.

43.

$$\frac{d^k}{dx^k}(\sinh x) = \left\{ \begin{array}{ll} \sinh x, & \text{if } k \text{ is odd} \\ \cosh x, & \text{if } k \text{ is even} \end{array} \right.$$

Thus

$$\frac{d^k}{dx^k}(\sinh x)\Big|_{x=0} = \left\{ \begin{array}{ll} 0, & \text{if } k \text{ is odd} \\ 1, & \text{if } k \text{ is even} \end{array} \right.$$

and

$$\sinh x = x + \frac{x^3}{3!} + \frac{x^5}{5!} + \cdots = \sum_{k=0}^{\infty} \frac{x^{(2k+1)}}{(2k+1)!}$$

45.   Set $t = ax$.   Then, $\quad e^{ax} = e^t = \sum_{k=0}^{\infty} \frac{t^k}{k!} = \sum_{k=0}^{\infty} a^k \frac{x^k}{k!}, \quad (-\infty, \infty)$.

47.   Set $t = ax$.   Then, $\quad \cos ax = \cos t = \sum_{k=0}^{\infty} \frac{(-1)^k}{(2k)!} t^{2k} = \sum_{k=0}^{\infty} \frac{(-1)^k a^{2k}}{(2k)!} x^{2k}, \quad (-\infty, \infty)$.

49.   By the hint

$$\ln(a + x) = \ln\left[a\left(1 + \frac{x}{a}\right)\right] = \ln a + \ln\left(1 + \frac{x}{a}\right) = \ln a + \sum_{k=1}^{\infty} \frac{(-1)^{k+1}}{ka^k} x^k.$$

By (11.6.8) the series converges for $\quad -1 < \frac{x}{a} \leq 1; \quad$ that is, $\quad -a < x \leq a$.

51. $\ln 2 = \ln\left(\frac{1 + 1/3}{1 - 1/3}\right) \cong 2\left[\frac{1}{3} + \frac{1}{3}\left(\frac{1}{3}\right)^3 + \frac{1}{5}\left(\frac{1}{3}\right)^5\right] = \frac{842}{1215}$.

Our calculator gives $\quad \frac{842}{1215} \cong 0.6930041 \quad$ and $\quad \ln 2 \cong 0.6931471$.

53.   Set $\quad u = (x - t)^k, \qquad dv = f^{(k+1)}(t) \, dt$
$\qquad du = -k(x - t)^{k-1} \, dt, \quad v = f^{(k)}(t)$.

Then,
$$-\frac{1}{k!}\int_0^x f^{(k+1)}(t)(x-t)^k\,dt$$
$$=-\frac{1}{k!}\left[(x-t)^k f^{(k)}(t)\right]_0^x-\frac{1}{k!}\int_0^x k(x-t)^{k-1}f^{(k)}(t)\,dt$$
$$=\frac{f^{(k)}(0)}{k!}x^k-\frac{1}{(k-1)!}\int_0^x f^{(k)}(t)(x-t)^{k-1}\,dt.$$

The given identity follows.

55. (a)

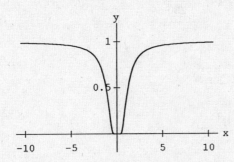

(b) Let $g(x)=\dfrac{x^{-n}}{e^{1/x^2}}$. Then $\lim\limits_{x\to 0} g(x)$ has the form ∞/∞. Successive applications of L'Hospital's rule will finally produce a quotient of the form $\dfrac{cx^k}{e^{1/x^2}}$, where k is a nonnegative integer and c is a constant. It follows that $\lim\limits_{x\to 0} g(x)=0$.

(c) $f'(0)=\lim\limits_{x\to 0}\dfrac{e^{-1/x^2}-0}{x}=0$ by part (b). Assume that $f^{(k)}(0)=0$. Then
$$f^{(k+1)}(0)=\lim_{x\to 0}\frac{f^{(k)}(x)-0}{x}=\lim_{x\to 0}\frac{f^{(k)}(x)}{x}.$$

Now, $f^{(k)}(x)/x$ is a sum of terms of the form $ce^{-1/x^2}/x^n$, where n is a positive integer and c is a constant.

Again by part (b), $f^{(k+1)}(0)=0$. Therefore, $f^{(n)}(0)=0$ for all n.

(d) 0   (e) x=0

57.

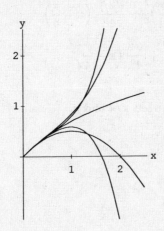

SECTION 11.7

1. $f(x) = \sqrt{x} = x^{1/2};$ $\qquad\qquad\qquad$ $f(4) = 2$

$\quad f'(x) = \frac{1}{2}x^{-1/2};$ $\qquad\qquad\qquad$ $f'(4) = \frac{1}{4}$

$\quad f''(x) = -\frac{1}{4}x^{-3/2};$ $\qquad\qquad\qquad$ $f''(4) = -\frac{1}{32}$

$\quad f'''(x) = \frac{3}{8}x^{-5/2};$ $\qquad\qquad\qquad$ $f'''(4) = \frac{3}{256}$

$\quad f^{(4)}(x) = -\frac{15}{16}x^{-7/2}$

$\quad P_3(x) = 2 + \frac{1}{4}(x-4) - \frac{1/32}{2!}(x-4)^2 + \frac{3/256}{3!}(x-4)^3$

$\qquad\quad = 2 + \frac{1}{4}(x-4) - \frac{1}{64}(x-4)^2 + \frac{1}{512}(x-4)^3$

$\quad R_4(x) = \frac{f^{(4)}(c)}{4!}(x-4)^4 = -\frac{15}{16}\cdot\frac{1}{4!}c^{-7/2}(x-4)^4 = -\frac{5}{128c^{7/2}}(x-4)^4,$ where c is between 4 and x.

3. $f(x) = \sin x;$ $\qquad\qquad\qquad$ $f(\pi/4) = \frac{\sqrt{2}}{2}$

$\quad f'(x) = \cos x;$ $\qquad\qquad\qquad$ $f'(\pi/4) = \frac{\sqrt{2}}{2}$

$\quad f''(x) = -\sin x;$ $\qquad\qquad\qquad$ $f''(\pi/4) = -\frac{\sqrt{2}}{2}$

$\quad f'''(x) = -\cos x;$ $\qquad\qquad\qquad$ $f'''(\pi/4) = -\frac{\sqrt{2}}{2}$

$\quad f^{(4)}(x) = \sin x;$ $\qquad\qquad\qquad$ $f^{(4)}(\pi/4) = \frac{\sqrt{2}}{2}$

$\quad f^{(5)}(x) = \cos x$

$\quad P_4(x) = \frac{\sqrt{2}}{2} + \frac{\sqrt{2}}{2}\left(x-\frac{\pi}{4}\right) - \frac{\sqrt{2}/2}{2!}\left(x-\frac{\pi}{4}\right)^2 - \frac{\sqrt{2}/2}{3!}\left(x-\frac{\pi}{4}\right)^3 + \frac{\sqrt{2}/2}{4!}\left(x-\frac{\pi}{4}\right)^4$

$\qquad\quad = \frac{\sqrt{2}}{2} + \frac{\sqrt{2}}{2}\left(x-\frac{\pi}{4}\right) - \frac{\sqrt{2}}{4}\left(x-\frac{\pi}{4}\right)^2 - \frac{\sqrt{2}}{12}\left(x-\frac{\pi}{4}\right)^3 + \frac{\sqrt{2}}{48}\left(x-\frac{\pi}{4}\right)^4$

$\quad R_5(x) = \frac{f^{(5)}(c)}{5!}\left(x-\frac{\pi}{4}\right)^5 = \frac{\cos c}{120}\left(x-\frac{\pi}{4}\right)^5,$ where c is between $\pi/4$ and x.

5. $f(x) = \tan^{-1}(x)$ $\qquad\qquad\qquad$ $f(1) = \frac{\pi}{4}$

$\quad f'(x) = \frac{1}{1+x^2}$ $\qquad\qquad\qquad$ $f'(1) = \frac{1}{2}$

$\quad f''(x) = \frac{-2x}{(1+x^2)^2}$ $\qquad\qquad\qquad$ $f''(1) = -\frac{1}{2}$

$$f'''(x) = \frac{6x^2 - 2}{(1 + x^2)^2} \qquad\qquad f'''(1) = \tfrac{1}{2}$$

$$f^{(4)}(x) = \frac{24(x - x^3)}{(1 + x^2)^3}$$

$$P_3(x) = \frac{\pi}{4} + \frac{1}{2}(x - 1) - \frac{1/2}{2!}(x - 1)^2 + \frac{1/2}{3!}(x - 1)^3 = \frac{\pi}{4} + \frac{1}{2}(x - 1) - \frac{1}{4}(x - 1)^2 + \frac{1}{12}(x - 1)^3$$

$$R_4(x) = \frac{f^{(4)}(c)}{4!}(x - 1)^4 = \frac{24(c - c^3)}{(1 + c^2)^3} \cdot \frac{1}{4!}(x - 1)^4 = \frac{c - c^3}{(1 + c^2)^3}(x - 1)^4, \quad \text{where } c \text{ is between 1 and } x.$$

7. $g(x) = 6 + 9(x - 1) + 7(x - 1)^2 + 3(x - 1)^3, \quad (-\infty, \infty)$

9. $g(x) = -3 + 5(x + 1) - 19(x + 1)^2 + 20(x + 1)^3 - 10(x + 1)^4 + 2(x + 1)^5, \quad (-\infty, \infty)$

11. $g(x) = \dfrac{1}{1 + x} = \dfrac{1}{2 + (x - 1)} = \dfrac{1}{2}\left[\dfrac{1}{1 + \left(\dfrac{x - 1}{2}\right)}\right] = \dfrac{1}{2}\displaystyle\sum_{k=0}^{\infty}(-1)^k\left(\dfrac{x - 1}{2}\right)^k$

(geometric series)

$$= \sum_{k=0}^{\infty}(-1)^k\frac{(x - 1)^k}{2^{k+1}} \quad \text{for} \quad \left|\frac{x - 1}{2}\right| < 1 \quad \text{and thus for} \quad -1 < x < 3$$

13. $g(x) = \dfrac{1}{1 - 2x} = \dfrac{1}{5 - 2(x + 2)} = \dfrac{1}{5}\left[\dfrac{1}{1 - \frac{2}{5}(x + 2)}\right] = \dfrac{1}{5}\displaystyle\sum_{k=0}^{\infty}\left[\dfrac{2}{5}(x + 2)\right]^k$

(geometric series)

$$= \sum_{k=0}^{\infty}\frac{2^k}{5^{k+1}}(x + 2)^k \quad \text{for} \quad \left|\frac{2}{5}(x + 2)\right| < 1 \quad \text{and thus for} \quad -\frac{9}{2} < x < \frac{1}{2}$$

15. $g(x) = \sin x = \sin\left[(x - \pi) + \pi\right] = \sin(x - \pi)\cos\pi + \cos(x - \pi)\sin\pi$

$$= -\sin(x - \pi) = -\sum_{k=0}^{\infty}(-1)^k\frac{(x - \pi)^{2k+1}}{(2k + 1)!}$$

(11.6.6)

$$= \sum_{k=0}^{\infty}(-1)^{k+1}\frac{(x - \pi)^{2k+1}}{(2k + 1)!}, \quad (-\infty, \infty)$$

17. $g(x) = \cos x = \cos\left[(x - \pi) + \pi\right] = \cos(x - \pi)\cos\pi - \sin(x - \pi)\sin\pi$

$$= -\cos(x - \pi) = -\sum_{k=0}^{\infty}(-1)^k\frac{(x - \pi)^{2k}}{(2k)!} = \sum_{k=0}^{\infty}(-1)^{k+1}\frac{(x - \pi)^{2k}}{(2k)!}, \quad (-\infty, \infty)$$

(11.6.7)

19. $g(x) = \sin \frac{1}{2}\pi x = \sin\left[\frac{\pi}{2}(x-1) + \frac{\pi}{2}\right]$

$\qquad = \sin\left[\frac{\pi}{2}(x-1)\right]\cos\frac{\pi}{2} + \cos\left[\frac{\pi}{2}(x-1)\right]\sin\frac{\pi}{2}$

$\qquad = \cos\left[\frac{\pi}{2}(x-1)\right] = \sum_{k=0}^{\infty}(-1)^k\left(\frac{\pi}{2}\right)^{2k}\frac{(x-1)^{2k}}{(2k)!}, \quad (-\infty, \infty)$

$\qquad$ (11.6.7) ⌐

21. $g(x) = \ln(1+2x) = \ln[3 + 2(x-1)] = \ln\left[3\left(1 + \frac{2}{3}(x-1)\right)\right]$

$\qquad = \ln 3 + \ln\left[1 + \frac{2}{3}(x-1)\right] = \ln 3 + \sum_{k=1}^{\infty}\frac{(-1)^{k+1}}{k}\left[\frac{2}{3}(x-1)\right]^k$

$\qquad$ (11.6.8) ⌐

$\qquad = \ln 3 + \sum_{k=1}^{\infty}\frac{(-1)^{k+1}}{k}\left(\frac{2}{3}\right)^k(x-1)^k.$

This result holds if $\quad -1 < \frac{2}{3}(x-1) \le 1, \quad$ which is to say, if $\quad -\frac{1}{2} < x \le \frac{5}{2}.$

23.
$$\begin{aligned} g(x) &= x\ln x \\ g'(x) &= 1 + \ln x \\ g''(x) &= x^{-1} \\ g'''(x) &= -x^{-2} \\ g^{(iv)}(x) &= 2x^{-3} \\ &\vdots \\ g^{(k)}(x) &= (-1)^k(k-2)!x^{1-k}, \quad k \ge 2. \end{aligned}$$

Then, $\quad g(2) = 2\ln 2, \; g'(2) = 1 + \ln 2, \quad$ and $\quad g^{(k)}(2) = \dfrac{(-1)^k(k-2)!}{2^{k-1}}, \quad k \ge 2.$

Thus, $\quad g(x) = 2\ln 2 + (1 + \ln 2)(x-2) + \sum_{k=2}^{\infty}\dfrac{(-1)^k}{k(k-1)2^{k-1}}(x-2)^k.$

25. $g(x) = x\sin x = x\sum_{k=0}^{\infty}(-1)^k\dfrac{x^{2k+1}}{(2k+1)!} = \sum_{k=0}^{\infty}(-1)^k\dfrac{x^{2k+2}}{(2k+1)!}$

27.
$$\begin{aligned} g(x) &= (1-2x)^{-3} \\ g'(x) &= -2(-3)(1-2x)^{-4} \\ g''(x) &= (-2)^2(4\cdot 3)(1-2x)^{-5} \\ g'''(x) &= (-2)^3(-5\cdot 4\cdot 3)(1-2x)^{-6} \\ &\vdots \\ g^{(k)}(x) &= (-2)^k\left[(-1)^k\frac{(k+2)!}{2}\right](1-2x)^{-k-3}, \quad k \ge 0. \end{aligned}$$

Thus, $\quad g^{(k)}(-2) = (-2)^k\left[(-1)^k\dfrac{(k+2)!}{2}\right]5^{-k-3} = \dfrac{2^{k-1}}{5^{k+3}}(k+2)!$

and $\quad g(x) = \displaystyle\sum_{k=0}^{\infty} (k+2)(k+1)\frac{2^{k-1}}{5^{k+3}}(x-2)^k.$

29. $\quad g(x) = \cos^2 x = \dfrac{1 + \cos 2x}{2} = \dfrac{1}{2} + \dfrac{1}{2}\cos\left[2(x-\pi) + 2\pi\right]$

$\qquad\qquad = \dfrac{1}{2} + \dfrac{1}{2}\cos\left[2(x-\pi)\right] = \dfrac{1}{2} + \dfrac{1}{2}\displaystyle\sum_{k=0}^{\infty}(-1)^k\frac{[2(x-\pi)]^{2k}}{(2k)!}$

$\qquad\qquad = 1 + \displaystyle\sum_{k=1}^{\infty}\frac{(-1)^k 2^{2k-1}}{(2k)!}(x-\pi)^{2k}$

$\qquad\qquad\quad \underset{\big\uparrow}{}$

$\qquad\qquad \rule[0.5ex]{1.2em}{0.4pt}(k=0 \ \text{ term is } \frac{1}{2})$

31.
$$
\begin{aligned}
g(x) &= x^n \\
g'(x) &= nx^{n-1} \\
g''(x) &= n(n-1)x^{n-2} \\
g'''(x) &= n(n-1)(n-2)x^{n-3} \\
&\;\;\vdots \\
g^{(k)}(x) &= n(n-1)\cdots(n-k+1)x^{n-k}, \qquad 0 \le k \le n \\
g^{(k)}(x) &= 0, \qquad k > n.
\end{aligned}
$$

Thus,

$$
g^{(k)}(1) = \begin{cases} \dfrac{n!}{(n-k)!}, & 0 \le k \le n \\[2mm] 0, & k > n \end{cases} \quad \text{and} \quad g(x) = \sum_{k=0}^{n}\frac{n!}{(n-k)!\,k!}(x-1)^k.
$$

33.   **(a)** $\quad \dfrac{e^x}{e^a} = e^{x-a} = \displaystyle\sum_{k=0}^{\infty}\frac{(x-a)^k}{k!}, \quad e^x = e^a\displaystyle\sum_{k=0}^{\infty}\frac{(x-a)^k}{k!}$

      (b) $\quad e^{a+(x-a)} = e^x = e^a\displaystyle\sum_{k=0}^{\infty}\frac{(x-a)^k}{k!}, \quad e^{x_1+x_2} = e^{x_1}\displaystyle\sum_{k=0}^{\infty}\frac{x_2^k}{k!} = e^{x_1}e^{x_2}$

      (c) $\quad e^{-a}\displaystyle\sum_{k=0}^{\infty}(-1)^k\frac{(x-a)^k}{k!}$

35.   **(a)** Let $\ g(x) = \sin x\ $ and $\ a = \pi/6.$ Then

$$
\left|R_{n+1}(x)\right| = \frac{\left|g^{(n+1)}(c)\right|}{(n+1)!}\left|\left(x - \frac{\pi}{6}\right)^{n+1}\right|
$$

$$
\le \frac{\left|\left(x - \dfrac{\pi}{6}\right)\right|^{n+1}}{(n+1)!} \qquad (g^{(n+1)}(c) = \pm\sin c \ \text{or} \ \pm\cos c)
$$

Now, $\ 35° = \dfrac{35\pi}{180}\ $ radians. We want to find the smallest positive integer n such that

$\left|R_{n+1}(35\pi/180)\right| < 0.00005.$

$$
\left|R_{n+1}(35\pi/180)\right| \le \frac{\left(\dfrac{35\pi}{180} - \dfrac{\pi}{6}\right)^{n+1}}{(n+1)!} \cong \frac{(0.087266)^{n+1}}{(n+1)!} < 0.00005 \quad\Longrightarrow\quad n \ge 3
$$

$$g(x) = \sin x; \qquad\qquad g(\pi/6) = \frac{1}{2}$$

$$g'(x) = \cos x; \qquad\qquad g'(\pi/6) = \frac{\sqrt{3}}{2}$$

$$g''(x) = -\sin x; \qquad\qquad g(\pi/6) = -\frac{1}{2}$$

$$g'''(x) = -\cos x; \qquad\qquad g(\pi/6) = -\frac{\sqrt{3}}{2}$$

$$P_3(x) = \frac{1}{2} + \frac{\sqrt{3}}{2}\left(x - \frac{\pi}{6}\right) - \frac{1/2}{2!}\left(x - \frac{\pi}{6}\right)^2 - \frac{\sqrt{3}/2}{3!}\left(x - \frac{\pi}{6}\right)^3$$

$$= \frac{1}{2} + \frac{\sqrt{3}}{2}\left(x - \frac{\pi}{6}\right) - \frac{1}{4}\left(x - \frac{\pi}{6}\right)^2 - \frac{\sqrt{3}}{12}\left(x - \frac{\pi}{6}\right)^3$$

(b) $P_3(35\pi/180) \cong 0.5736$

35.   Let $g(x) = \sqrt{x} = x^{1/2}$ and $a = 36$.

(a) $g(x) = x^{1/2}$ $\qquad\qquad\qquad\qquad g(36) = 6$

$$g'(x) = \frac{1}{2}x^{-1/2} \qquad\qquad\qquad g'(36) = \frac{1}{12}$$

$$g''(x) = -\frac{1}{4}x^{-3/2} \qquad\qquad\qquad g''(36) = -\frac{1}{864}$$

$$g'''(x) = \frac{3}{8}x^{-5/2} \qquad\qquad\qquad g'''(36) = \frac{1}{20,736}$$

We want to find the smallest positive integer n such that $\left|R_{n+1}(38)\right| < 0.0005$:

$$n = 1: \quad \left|R_2(38)\right| = \frac{c^{-3/2}/4}{2!}(38 - 36)^2 = \frac{4}{8c^{3/2}} = \frac{1}{2c^{3/2}}, \quad \text{where} \quad 36 \le c \le 38,$$

and

$$\left|R_2(38)\right| \le \frac{1}{2(36)^{3/2}} = \frac{1}{432} \cong 0.0023.$$

$$n = 2: \quad \left|R_3(38)\right| = \frac{3c^{-5/2}/8}{3!}(38 - 36)^3 = \frac{8}{16c^{5/2}} = \frac{1}{2c^{5/2}}, \quad \text{where} \quad 36 \le c \le 38,$$

and

$$\left|R_3(38)\right| \le \frac{1}{2(36)^{5/2}} = \frac{1}{15,552} \cong 0.000064.$$

Thus, we take $n = 2$:

$$P_2(x) = 6 + \frac{1}{12}(x - 36) - \frac{1}{1728}(x - 36)^2 \quad \text{and} \quad P_2(38) \cong 6.164$$

SECTION 11.8

1. $(-1, 1)$;   ratio test: $\dfrac{b_{k+1}}{b_k} = \dfrac{k+1}{k}|x| \to |x|$, series converges for $|x| < 1$.

   At the endpoints $x = 1$ and $x = -1$ the series diverges since at those points $b_k \not\to 0$.

3. $(-\infty, \infty)$;   ratio test: $\dfrac{b_{k+1}}{b_k} = \dfrac{|x|}{(2k+1)(2k+2)} \to 0$, series converges all x.

5. Converges only at 0;   divergence test: $(-k)^{2k} x^{2k} \to 0$ only if $x = 0$, and series clearly converges at $x = 0$.

7. $[-2, 2)$;   root test: $(b_k)^{1/k} = \dfrac{|x|}{2k^{1/k}} \to \dfrac{|x|}{2}$, series converges for $|x| < 2$.

   At $x = 2$ series becomes $\sum \dfrac{1}{k}$, the divergent harmonic series.

   At $x = -2$ series becomes $\sum (-1)^k \dfrac{1}{k}$, a convergent alternating series.

9. Converges only at 0;   divergence test: $\left(\dfrac{k}{100}\right)^k x^k \to 0$ only if $x = 0$, and series clearly converges at $x = 0$.

11. $\left[-\dfrac{1}{2}, \dfrac{1}{2}\right)$;   root test: $(b_k)^{1/k} = \dfrac{2|x|}{\sqrt{k^{1/k}}} \to 2|x|$, series converges for $|x| < \dfrac{1}{2}$.

   At $x = \dfrac{1}{2}$ series becomes $\sum \dfrac{1}{\sqrt{k}}$, a divergent p-series.

   At $x = -\dfrac{1}{2}$ series becomes $\sum (-1)^k \dfrac{1}{\sqrt{k}}$, a convergent alternating series.

13. $(-1, 1)$;   ratio test: $\dfrac{b_{k+1}}{b_k} = \dfrac{k^2}{(k+1)(k-1)}|x| \to |x|$, series converges for $|x| < 1$.

   At the endpoints $x = 1$ and $x = -1$ the series diverges since there $b_k \not\to 0$.

15. $(-10, 10)$;   root test: $(b_k)^{1/k} = \dfrac{k^{1/k}}{10}|x| \to \dfrac{|x|}{10}$, series converges for $|x| < 10$.

   At the endpoints $x = 10$ and $x = -10$ the series diverges since there $b_k \not\to 0$.

17. $(-\infty, \infty)$;   root test: $(b_k)^{1/k} = \dfrac{|x|}{k} \to 0$, series converges all x.

19. $(-\infty, \infty)$;   root test: $(b_k)^{1/k} = \dfrac{|x-2|}{k} \to 0$, series converges all x.

21. $\left(-\dfrac{3}{2}, \dfrac{3}{2}\right)$;   ratio test: $\dfrac{b_{k+1}}{b_k} = \dfrac{\dfrac{2^{k+1}}{3^{k+2}}|x|}{\dfrac{2^k}{3^{k+1}}} = \dfrac{2}{3}|x|$, series converges for $|x| < \dfrac{3}{2}$.

At the endpoints $x = 3/2$ and $x = -3/2$, the series diverges since there $b_k \not\to 0$.

23. Converges only at $x = 1$;   ratio test: $\dfrac{b_{k+1}}{b_k} = \dfrac{k^3}{(k+1)^2}|x-1| \to \infty$ if $x \neq 1$

The series clearly converges at $x = 1$; otherwise it diverges.

25. $(-4, 0)$;   ratio test: $\dfrac{b_{k+1}}{b_k} = \dfrac{k^2-1}{2k^2}|x+2| \to \dfrac{|x+2|}{2}$, series converges for $|x+2| < 2$.

At the endpoints $x = 0$ and $x = -4$, the series diverges since there $b_k \not\to 0$.

27. $(-\infty, \infty)$;   ratio test: $\dfrac{b_{k+1}}{b_k} = \dfrac{(k+1)^2}{k^2(k+2)}|x+3| \to 0$, series converges for all x.

29. $(-1, 1)$;   root test: $(b_k)^{1/k} = \left(1 + \dfrac{1}{k}\right)|x| \to |x|$, series converges for $|x| < 1$.

At the endpoints $x = 1$ and $x = -1$, the series diverges since there $b_k \not\to 0$

$\left[\text{recall } \left(1 + \dfrac{1}{k}\right)^k \to e\right]$

31. $(0, 4)$;   ratio test: $\dfrac{b_{k+1}}{b_k} = \dfrac{\ln(k+1)}{\ln k}\dfrac{|x-2|}{2} \to \dfrac{|x-2|}{2}$, series converges for $|x-2| < 2$.

At the endpoints $x = 0$ and $x = 4$ the series diverges since there $b_k \not\to 0$.

33. $\left(-\dfrac{5}{2}, \dfrac{1}{2}\right)$;   root test: $(b_k)^{1/k} = \dfrac{2}{3}|x+1| \to \dfrac{2}{3}|x+1|$, series converges for $|x+1| < \dfrac{3}{2}$.

At the endpoints $x = -\dfrac{5}{2}$ and $x = \dfrac{1}{2}$ the series diverges since there $b_k \not\to 0$.

35. $1 - \dfrac{x}{2} + \dfrac{2x^2}{4} - \dfrac{3x^3}{8} + \dfrac{4x^4}{16} - \cdots = 1 + \displaystyle\sum_{k=1}^{\infty} (-1)^k \dfrac{kx^k}{2^k}$

$(-2, 2)$;   ratio test: $\dfrac{b_{k+1}}{b_k} = \dfrac{k+1}{2k}|x| \to \dfrac{|x|}{2}$, series converges for $|x| < 2$.

At the endpoints $x = 2$ and $x = -2$ the series diverges since there $b_k \not\to 0$.

37. $\dfrac{3x^2}{4} + \dfrac{9x^4}{9} + \dfrac{27x^6}{16} + \dfrac{81x^8}{25} + \cdots = \displaystyle\sum_{k=1}^{\infty} \dfrac{3^k}{(k+1)^2} x^{2k}$

$\left[-\dfrac{1}{\sqrt{3}}, \dfrac{1}{\sqrt{3}} \right]$; ratio test: $\dfrac{b_{k+1}}{b_k} = \dfrac{3(k+1)^2}{(k+2)^2} x^2 \to 3x^2$, series converges for $x^2 < \dfrac{1}{3}$

or $|x| < \dfrac{1}{\sqrt{3}}$.

At $x = \pm \dfrac{1}{\sqrt{3}}$, the series becomes $\displaystyle\sum \dfrac{1}{(k+1)^2} \cong \sum \dfrac{1}{n^2}$, a convergent series p-series.

39. Examine the convergence of $\sum |a_k x^k|$; for (a) use the root test and for (b) use the ratio rest.

41. $\displaystyle\sum |a_k r^k| = \sum |a_k (-r)^k|$

SECTION 11.9

1. Use the fact that $\dfrac{d}{dx} \left(\dfrac{1}{1-x} \right) = \dfrac{1}{(1-x)^2}$:

$$\dfrac{1}{(1-x)^2} = \dfrac{d}{dx}(1 + x + x^2 + x^3 + \cdots + x^n + \cdots) = 1 + 2x + 3x^2 + \cdots + nx^{n-1} + \cdots.$$

3. Use the fact that $\dfrac{d^{(k-1)}}{dx^{(k-1)}} \left[\dfrac{1}{1-x} \right] = \dfrac{(k-1)!}{(1-x)^k}$:

$$\dfrac{1}{(1-x)^k} = \dfrac{1}{(k-1)!} \dfrac{d^{(k-1)}}{dx^{(k-1)}} \left[1 + x + \cdots + x^{k-1} + x^k + x^{k+1} + \cdots + x^{n+k-1} + \cdots \right]$$

$$= \dfrac{1}{(k-1)!} \dfrac{d^{(k-1)}}{dx^{(k-1)}} \left[x^{k-1} + x^k + x^{k+1} + \cdots + x^{n+k-1} + \cdots \right]$$

$$= 1 + kx + \dfrac{(k+1)k}{2} x^2 + \cdots + \dfrac{(n+k-1)(n+k-2)\cdots(n+1)}{(k-1)!} x^n + \cdots$$

$$= 1 + kx + \dfrac{(k+1)k}{2!} x^2 + \cdots + \dfrac{(n+k-1)!}{n!(k-1)!} x^n + \cdots.$$

5. Use the fact that $\dfrac{d}{dx}[\ln(1-x^2)] = \dfrac{-2x}{1-x^2}$:

$$\dfrac{1}{1-x^2} = 1 + x^2 + x^4 + \cdots + x^{2n} + \cdots$$

$$\dfrac{-2x}{1-x^2} = -2x - 2x^3 - 2x^5 - \cdots - 2x^{2n+1} - \cdots.$$

By integration

$$\ln\left(1-x^2\right) = \left(-x^2 - \frac{1}{2}x^4 - \frac{1}{3}x^6 - \cdots - \frac{x^{2n+2}}{n+1} - \cdots\right) + C.$$

At $x = 0$, both $\ln\left(1-x^2\right)$ and the series are 0. Thus, $C = 0$ and

$$\ln\left(1-x^2\right) = -x^2 - \frac{1}{2}x^4 - \frac{1}{3}x^6 - \cdots - \frac{1}{n+1}x^{2n+2} - \cdots.$$

7. $\sec^2 x = \dfrac{d}{dx}(\tan x) = \dfrac{d}{dx}\left(x + \dfrac{1}{3}x^3 + \dfrac{2}{15}x^5 + \dfrac{17}{315}x^7 + \cdots\right) = 1 + x^2 + \dfrac{2}{3}x^4 + \dfrac{17}{45}x^6 + \cdots$

9. On its interval of convergence a power series is the Taylor series of its sum. Thus,

$$f(x) = x^2 \sin^2 x = x^2\left(x - \frac{x^3}{3!} + \frac{x^5}{5!} - \frac{x^7}{7!} + \cdots\right)$$

$$= x^3 - \frac{x^5}{3!} + \frac{x^7}{5!} - \frac{x^9}{7!} + \cdots = \sum_{n=0}^{\infty} f^{(n)}(0)\frac{x^n}{n!}$$

implies $\quad f^{(9)}(0) = -9!/7! = -72.$

11. $\sin x^2 = \displaystyle\sum_{k=0}^{\infty}(-1)^k\frac{(x^2)^{2k+1}}{(2k+1)!} = \sum_{k=0}^{\infty}(-1)^k\frac{x^{4k+2}}{(2k+1)!}$

13. $e^{3x^3} = \displaystyle\sum_{k=0}^{\infty}\frac{(3x^3)^k}{k!} = \sum_{k=0}^{\infty}\frac{3^k}{k!}x^{3k}$

15. $\dfrac{2x}{1-x^2} = 2x\left(\dfrac{1}{1-x^2}\right) = 2x\displaystyle\sum_{k=0}^{\infty}(x^2)^k = \sum_{k=0}^{\infty}2x^{2k+1}$

17. $\dfrac{1}{1-x} + e^x = \displaystyle\sum_{k=0}^{\infty}x^k + \sum_{k=0}^{\infty}\frac{x^k}{k!} = \sum_{k=0}^{\infty}\frac{(k!+1)}{k!}x^k$

19. $x\ln\left(1+x^3\right) = x\displaystyle\sum_{k=1}^{\infty}\frac{(-1)^{k+1}}{k}(x^3)^k = \sum_{k=1}^{\infty}\frac{(-1)^{k+1}}{k}x^{3k+1}$

    (11.6.8)

21. $x^3 e^{-x^3} = x^3\displaystyle\sum_{k=0}^{\infty}\frac{(-x^3)^k}{k!} = \sum_{k=0}^{\infty}\frac{(-1)^k}{k!}x^{3k+3}$

23. (a) $\displaystyle\lim_{x\to0}\frac{1-\cos x}{x^2} \overset{\star}{=} \lim_{x\to0}\frac{\sin x}{2x} = \frac{1}{2}$ ($\star$ indicates differentiation of numerator and denominator).

(b) $\lim\limits_{x \to 0} \dfrac{1 - \cos x}{x^2} = \lim\limits_{x \to 0} \dfrac{\dfrac{x^2}{2!} - \dfrac{x^4}{4!} + \dfrac{x^6}{6!} - \cdots}{x^2} = \lim\limits_{x \to 0}\left(\dfrac{1}{2} - \dfrac{x^2}{4!} + \dfrac{x^4}{6!} - \cdots\right) = \dfrac{1}{2}$

25. (a) $\lim\limits_{x \to 0} \dfrac{\cos x - 1}{x \sin x} \overset{\star}{=} \lim\limits_{x \to 0} \dfrac{-\sin x}{\sin x + x \cos x} \overset{\star}{=} \lim\limits_{x \to 0} \dfrac{-\cos x}{2\cos x - x \sin x} = -\dfrac{1}{2}$

(b)

$$\lim_{x \to 0} \frac{\cos x - 1}{x \sin x} = \frac{-\dfrac{x^2}{2!} + \dfrac{x^4}{4!} - \dfrac{x^6}{6!} + \cdots}{x^2 - \dfrac{x^4}{3!} + \dfrac{x^6}{5!} \cdots}$$

$$= \frac{-\dfrac{1}{2} + \dfrac{x^2}{4!} - \dfrac{x^4}{6!} + \cdots}{1 - \dfrac{x^2}{3!} + \dfrac{x^4}{5!} \cdots} = -\dfrac{1}{2}$$

27.

$$\int_0^x \frac{\ln(1+t)}{t}\,dt = \int_0^x \frac{1}{t}\left(\sum_{k=1}^{\infty} \frac{(-1)^{k-1}}{k} t^k\right)dt = \int_0^x \left(\sum_{k=1}^{\infty} \frac{(-1)^{k-1}}{k} t^{k-1}\right)dt$$

$$= \sum_{k=1}^{\infty} \frac{(-1)^{k-1}}{k} \int_0^x t^{k-1}\,dt$$

$$= \sum_{k=1}^{\infty} \frac{(-1)^{k-1}}{k^2} x^k,\ -1 \le x \le 1$$

29.

$$\int_0^x \frac{\tan^{-1} t}{t}\,dt = \int_0^x \frac{1}{t}\left(\sum_{k=0}^{\infty} \frac{(-1)^k}{2k+1} t^{2k+1}\right)dt = \int_0^x \left(\sum_{k=0}^{\infty} \frac{(-1)^k}{2k+1} t^{2k}\right)dt$$

$$= \sum_{k=0}^{\infty} \frac{(-1)^k}{2k+1} \int_0^x t^{2k}\,dt$$

$$= \sum_{k=0}^{\infty} \frac{(-1)^k}{(2k+1)^2} x^{2k+1},\ -1 \le x \le 1$$

31. $0.804 \le I \le 0.808;$ $I = \displaystyle\int_0^1 \left(1 - x^3 + \frac{x^6}{2!} - \frac{x^9}{3!} + \cdots\right)dx$

$$= \left[x - \frac{x^4}{4} + \frac{x^7}{14} - \frac{x^{10}}{60} + \frac{x^{13}}{(13)(24)} - \cdots\right]_0^1$$

$$= 1 - \tfrac{1}{4} + \tfrac{1}{14} - \tfrac{1}{60} + \tfrac{1}{312} - \cdots.$$

Since $\frac{1}{312} < 0.01,$ we can stop there:

$$1 - \frac{1}{4} + \frac{1}{14} - \frac{1}{60} \le I \le 1 - \frac{1}{4} + \frac{1}{14} - \frac{1}{60} + \frac{1}{312} \quad \text{gives} \quad 0.804 \le I \le 0.808.$$

33. $0.600 \leq I \leq 0.603;$

$$I = \int_0^1 \left(x^{1/2} - \frac{x^{3/2}}{3!} + \frac{x^{5/2}}{5!} - \cdots \right) dx$$

$$= \left[\tfrac{2}{3}x^{3/2} - \tfrac{1}{15}x^{5/2} + \tfrac{1}{420}x^{7/2} - \cdots \right]_0^1$$

$$= \tfrac{2}{3} - \tfrac{1}{15} + \tfrac{1}{420} - \cdots .$$

Since $\frac{1}{420} < 0.01,$ we can stop there:

$$\tfrac{2}{3} - \tfrac{1}{15} \leq I \leq \tfrac{2}{3} - \tfrac{1}{15} + \tfrac{1}{420} \quad \text{gives} \quad 0.600 \leq I \leq 0.603.$$

35. $0.294 \leq I \leq 0.304;$

$$I = \int_0^1 \left(x^2 - \frac{x^6}{3} + \frac{x^{10}}{5} - \frac{x^{14}}{7} + \cdots \right) dx$$

$(11.9.7)$

$$= \left[\tfrac{1}{3}x^3 - \tfrac{1}{21}x^7 + \tfrac{1}{55}x^{11} - \tfrac{1}{105}x^{15} + \cdots \right]_0^1$$

$$= \tfrac{1}{3} - \tfrac{1}{21} + \tfrac{1}{55} - \tfrac{1}{105} + \cdots .$$

Since $\frac{1}{105} < 0.01,$ we can stop there:

$$\tfrac{1}{3} - \tfrac{1}{21} + \tfrac{1}{55} - \tfrac{1}{105} \leq I \leq \tfrac{1}{3} - \tfrac{1}{21} + \tfrac{1}{55} \quad \text{gives} \quad 0.294 \leq I \leq 0.304.$$

37. $I \cong 0.9461;$

$$I = \int_0^1 \left(1 - \frac{x^2}{3!} + \frac{x^4}{5!} - \cdots \right) dx$$

$$= \left[x - \frac{x^3}{3 \cdot 3!} + \frac{x^5}{5 \cdot 5!} - \cdots \right]_0^1$$

$$= 1 - \frac{1}{3 \cdot 3!} + \frac{1}{5 \cdot 5!} - \frac{1}{7 \cdot 7!} \cdots .$$

Since $\dfrac{1}{7 \cdot 7!} = \dfrac{1}{35,280} \cong 0.000028 < 0.0001,$ we can stop there:

$$1 - \frac{1}{3 \cdot 3!} + \frac{1}{5 \cdot 5!} - \frac{1}{7 \cdot 7!} < I < 1 - \frac{1}{3 \cdot 3!} + \frac{1}{5 \cdot 5!}; \quad I \cong 0.9461$$

39. $I \cong 0.4485;$

$$I = \int_0^{0.5} \left(1 - \frac{x}{2} + \frac{x^2}{3} - \frac{x^3}{4} + \cdots \right) dx$$

$$= \left[x - \frac{x^2}{2^2} + \frac{x^3}{3^2} - \frac{x^4}{4^2} + \cdots \right]_0^{1/2}$$

$$= \frac{1}{2} - \frac{1}{2^2 \cdot 2^2} + \frac{1}{3^2 \cdot 2^3} - \frac{1}{4^2 \cdot 2^4} + \cdots = \sum_{k=1}^{\infty} \frac{(-1)^{k-1}}{k^2 \cdot 2^k}$$

Now, $\dfrac{1}{8^2 \cdot 2^8} = \dfrac{1}{16,384} \cong 0.000061$ is the first term which is less than 0.0001. Thus

$$\sum_{k=1}^{7} \frac{(-1)^{k-1}}{k^2 \cdot 2^k} < I < \sum_{k=1}^{8} \frac{(-1)^{k-1}}{k^2 \cdot 2^k}; \quad I \cong 0.4485$$

41. $e^{x^3};$ by (11.6.5)

43. $3x^2 e^{x^3} = \dfrac{d}{dx}(e^{x^3})$

45. (a) $f(x) = xe^x = x \sum_{k=0}^{\infty} \frac{x^k}{k!} = \sum_{k=0}^{\infty} \frac{x^{k+1}}{k!}$

(b) Using integration by parts: $\int_0^1 xe^x \, dx = [xe^x - e^x]_0^1 = e - e + 1 = 1.$

Using the power series representation:

$$\int_0^1 xe^x \, dx = \int_0^1 \left(\sum_{k=0}^{\infty} \frac{x^{k+1}}{k!} \right) dx = \sum_{k=0}^{\infty} \int_0^1 \left(\frac{x^{k+1}}{k!} \right) dx$$

$$= \sum_{k=0}^{\infty} \frac{1}{k!} \left[\frac{x^{k+2}}{k+2} \right]_0^1$$

$$= \sum_{k=0}^{\infty} \frac{1}{k!(k+2)}$$

$$= \frac{1}{2} + \sum_{k=1}^{\infty} \frac{1}{k!(k+2)}$$

Thus, $1 = \frac{1}{2} + \sum_{k=1}^{\infty} \frac{1}{k!(k+2)}$ and $\sum_{k=1}^{\infty} \frac{1}{k!(k+2)} = \frac{1}{2}.$

47. Let $f(x)$ be the sum of these series; a_k and b_k are both $\dfrac{f^{(k)}(0)}{k!}.$

49. (a) If f is even, then the odd ordered derivatives $f^{(2k-1)}$, $k = 1, 2, \ldots$ are odd. This implies that $f^{(2k-1)}(0) = 0$ for all k and so $a_{2k-1} = f^{(2k-1)}(0)/(2k-1)! = 0$ for all k.

(b) If f is odd, then all the even ordered derivatives $f^{(2k)}$, $k = 1, 2, \ldots$ are odd. This implies that $f^{(2k)}(0) = 0$ for all k and so $a_{2k} = f^{(2k)}(0)/(2k)! = 0$ for all k.

51. $0.0352 \le I \le 0.0359;$ $I = \int_0^{1/2} \left(x^2 - \frac{x^3}{2} + \frac{x^4}{3} - \frac{x^5}{4} + \cdots \right) dx$

$$= \left[\frac{x^3}{3} - \frac{x^4}{8} + \frac{x^5}{15} - \frac{x^6}{24} + \cdots \right]_0^{1/2}$$

$$= \frac{1}{3(2^3)} - \frac{1}{8(2^4)} + \frac{1}{15(2^5)} - \frac{1}{24(2^6)} + \cdots.$$

Since $\dfrac{1}{24(2^6)} = \dfrac{1}{1536} < 0.001$, we can stop there:

$$\frac{1}{3(2^3)} - \frac{1}{8(2^4)} + \frac{1}{15(2^5)} - \frac{1}{24(2^6)} \le I \le \frac{1}{3(2^3)} - \frac{1}{8(2^4)} + \frac{1}{15(2^5)}$$

gives $0.0352 \le I \le 0.0359.$ Direct integration gives

$$I = \int_0^{1/2} x \ln(1+x) \, dx = \left[\frac{1}{2}(x^2 - 1)\ln(1+x) - \frac{1}{4}x^2 + \frac{1}{2}x \right]_0^{1/2} = \frac{3}{16} - \frac{3}{8}\ln 1.5 \cong 0.0354505.$$

53. $0.2640 \le I \le 0.2643$;

$$I = \int_0^1 \left(x - x^2 + \frac{x^3}{2!} - \frac{x^4}{3!} + \frac{x^5}{4!} - \frac{x^6}{5!} + \frac{x^7}{6!} - \cdots \right) dx$$

$$= \left[\frac{x^2}{2} - \frac{x^3}{3} + \frac{x^4}{4(2!)} - \frac{x^5}{5(3!)} + \frac{x^6}{6(4!)} - \frac{x^7}{7(5!)} + \frac{x^8}{8(6!)} - \cdots \right]_0^1$$

$$= \frac{1}{2} - \frac{1}{3} + \frac{1}{4(2!)} - \frac{1}{5(3!)} + \frac{1}{6(4!)} - \frac{1}{7(5!)} + \frac{1}{8(6!)} - \cdots.$$

Note that $\dfrac{1}{8(6!)} = \dfrac{1}{5760} < 0.001.$ The integral lies between

$$\frac{1}{2} - \frac{1}{3} + \frac{1}{4(2!)} - \frac{1}{5(3!)} + \frac{1}{6(4!)} - \frac{1}{7(5!)}$$

and

$$\frac{1}{2} - \frac{1}{3} + \frac{1}{4(2!)} - \frac{1}{5(3!)} + \frac{1}{6(4!)} - \frac{1}{7(5!)} + \frac{1}{8(6!)}.$$

The first sum is greater than 0.2640 and the second sum is less than 0.2643.

Direct integration gives

$$\int_0^1 xe^{-x} \, dx = \left[-xe^{-x} - e^{-x} \right]_0^1 = 1 - 2/e \cong 0.2642411.$$

SECTION 11.10

1.   Take $\alpha = 1/2$ in (11.10.2) to obtain $1 + \frac{1}{2}x - \frac{1}{8}x^2 + \frac{1}{16}x^3 - \frac{5}{128}x^4$.

3.   In (11.10.2), replace x by x^2 and take $\alpha = 1/2$ to obtain $1 + \frac{1}{2}x^2 - \frac{1}{8}x^4$.

5.   Take $\alpha = -1/2$ in (11.10.2) to obtain $1 - \frac{1}{2}x + \frac{3}{8}x^2 - \frac{5}{16}x^3 + \frac{35}{128}x^4$.

7.   In (11.10.2), replace x by $-x$ and take $\alpha = 1/4$ to obtain $1 - \frac{1}{4}x - \frac{3}{32}x^2 - \frac{7}{128}x^3 - \frac{77}{2048}x^4$.

9. $f(x) = (4 + x)^{3/2} = 8\left(1 + \dfrac{x}{4}\right)^{3/2}$

In 11.10.2, replace x by $x/4$ and take $\alpha = 3/2$ to obtain

$$8\left[1 + \frac{3}{8}\left(\frac{x}{4}\right) + \frac{1}{2!}\left(\frac{3}{2}\right)\left(\frac{1}{2}\right)\left(\frac{x}{4}\right)^2 + \frac{1}{3!}\left(\frac{3}{2}\right)\left(\frac{1}{2}\right)\left(-\frac{1}{2}\right)\left(\frac{x}{4}\right)^3 \right.$$

$$\left. + \frac{1}{4!}\left(\frac{3}{2}\right)\left(\frac{1}{2}\right)\left(-\frac{1}{2}\right)\left(-\frac{3}{2}\right)\left(\frac{x}{4}\right)^4 \right]$$

$$= 8 + 3x + \frac{3}{16}x^2 - \frac{1}{128}x^3 + \frac{3}{4096}x^4$$

11.   (a) $f(x) = \dfrac{1}{\sqrt{1 - x^2}} = \left(1 - x^2\right)^{-1/2}$

In 11.10.2, replace x by x^2 and take $\alpha = -1/2$ to obtain

$$\frac{1}{\sqrt{1-x^2}} = \sum_{k=0}^{\infty} \binom{-1/2}{k}(-1)^k x^{2k}$$

By Problem 2, this series has radius of convergence $R = 1$.

(b)
$$\sin^{-1} x = \int_0^x \frac{1}{\sqrt{1-x^2}}\, dt = \int_0^x \sum_{k=0}^{\infty} \binom{-1/2}{k}(-1)^k t^{2k}\, dt$$

$$= \sum_{k=0}^{\infty} \binom{-1/2}{k}(-1)^k \int_0^x t^{2k}\, dt$$

$$= \sum_{k=0}^{\infty} \binom{-1/2}{k}\frac{(-1)^k}{2k+1} x^{2k+1}$$

By Theorem 11.9.4, the radius of convergence of this series is $R = 1$.

13. 9.8995; $\sqrt{98} = (100-2)^{1/2} = 10\left(1 - \frac{1}{50}\right)^{1/2} \cong 10\left[1 - \frac{1}{100} - \frac{1}{20000}\right] = 9.8995$

15. 2.0799; $\sqrt[3]{9} = (8+1)^{1/3} = 2\left(1 + \frac{1}{8}\right)^{1/3} \cong 2\left[1 + \frac{1}{24} - \frac{1}{576}\right] \cong 2.0799$

17. 0.4925; $17^{-1/4} = (16+1)^{-1/4} = \frac{1}{2}\left(1 + \frac{1}{16}\right)^{-1/4} \cong \frac{1}{2}\left[1 - \frac{1}{64} + \frac{5}{8192}\right] \cong 0.4925$

19.
$$I = \int_0^{1/3} \sqrt{1+x^3}\, dx = \int_0^{1/3} \sum_{k=0}^{\infty} \binom{1/2}{k} x^{3k}\, dx$$

$$= \sum_{k=0}^{\infty} \binom{1/2}{k} \int_0^{1/3} x^{3k}\, dx$$

$$= \sum_{k=0}^{\infty} \binom{1/2}{k} \left[\frac{x^{3k+1}}{3k+1}\right]_0^{1/3}$$

$$= \sum_{k=0}^{\infty} \binom{1/2}{k} \frac{1}{3k+1} \left(\frac{1}{3}\right)^{3k+1}$$

$$= \frac{1}{3} + \left(\frac{1}{2}\right)\left(\frac{1}{4}\right)\left(\frac{1}{3}\right)^4 + \frac{1}{2!}\left(\frac{1}{2}\right)\left(-\frac{1}{2}\right)\left(\frac{1}{7}\right)\left(\frac{1}{3}\right)^7 + \cdots$$

Now, $I - \frac{1}{3} = \left(\frac{1}{2}\right)\left(\frac{1}{4}\right)\left(\frac{1}{3}\right)^4 + \frac{1}{2!}\left(\frac{1}{2}\right)\left(-\frac{1}{2}\right)\left(\frac{1}{7}\right)\left(\frac{1}{3}\right)^7 + \cdots$

is an alternating series and $\frac{1}{2!}\left(\frac{1}{2}\right)\left(-\frac{1}{2}\right)\left(\frac{1}{7}\right)\left(\frac{1}{3}\right)^7 \cong 8.2 \times 10^{-6} < 0.001$

Therefore, $\int_0^{1/3} \sqrt{1+x^3}\, dx \cong \frac{1}{3} + \left(\frac{1}{2}\right)\left(\frac{1}{4}\right)\left(\frac{1}{3}\right)^4 \cong 0.3349$

21.

$$\int_0^{1/2} \frac{1}{\sqrt{1+x^2}}\, dx = \int_0^{1/2} \left(1+x^2\right)^{-1/2} dx = \int_0^{1/2} \sum_{k=0}^{\infty} \binom{-1/2}{k} x^{2k}\, dx$$

$$= \sum_{k=0}^{\infty} \binom{-1/2}{k} \int_0^{1/2} x^{2k}\, dx$$

$$= \sum_{k=0}^{\infty} \binom{-1/2}{k} \left[\frac{1}{2k+1} x^{2k+1}\right]_0^{1/2}$$

$$= \sum_{k=0}^{\infty} \binom{-1/2}{k} \frac{1}{2k+1} \left(\frac{1}{2}\right)^{2k+1}$$

Now

$$\sum_{k=0}^{\infty} \binom{-1/2}{k} \frac{1}{2k+1} \left(\frac{1}{2}\right)^{2k+1} = \frac{1}{2} + \left(-\frac{1}{2}\right)\left(\frac{1}{3}\right)\left(\frac{1}{2}\right)^3 + \frac{1}{2!}\left(-\frac{1}{2}\right)\left(-\frac{3}{2}\right)\left(\frac{1}{5}\right)\left(\frac{1}{2}\right)^5$$

$$+ \frac{1}{3!}\left(-\frac{1}{2}\right)\left(-\frac{3}{2}\right)\left(-\frac{5}{2}\right)\left(\frac{1}{7}\right)\left(\frac{1}{2}\right)^7 + \cdots$$

is an alternating series and

$$\frac{1}{3!}\left(-\frac{1}{2}\right)\left(-\frac{3}{2}\right)\left(-\frac{5}{2}\right)\left(\frac{1}{7}\right)\left(\frac{1}{2}\right)^7 \cong 3.5 \times 10^{-4} < 0.001$$

Therefore,

$$\int_0^{1/2} \frac{1}{\sqrt{1+x^2}}\, dx \cong \frac{1}{2} + \left(-\frac{1}{2}\right)\left(\frac{1}{3}\right)\left(\frac{1}{2}\right)^3 + \frac{1}{2!}\left(-\frac{1}{2}\right)\left(-\frac{3}{2}\right)\left(\frac{1}{5}\right)\left(\frac{1}{2}\right)^5 \cong 0.4815$$

PROJECTS AND EXPLORATIONS

11.1. (a) $j = 3$.

If $f(x) = \dfrac{[x^3 + \ln x]\ln x}{x^{4+\ln x} + 3x^3 + 7}$,

then $f(2) < f(3)$ but the function is not

decreasing on this interval (see the graph

to the right); f is decreasing on $[3, \infty)$.

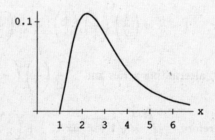

(b) For $x \geq 3$, $\quad f(x) \leq \dfrac{\ln x}{x^{1+\ln x}}$, $\quad$ and

$$\int_k^\infty \frac{\ln x}{x^{1+\ln x}}\,dx = \lim_{b\to\infty} \int_k^b \frac{\ln x}{x^{1+\ln x}}\,dx = \lim_{b\to\infty}\left[-\frac{1}{2}\,x^{-\ln x}\right]_k^b = \frac{1}{(2)k^{\ln k}}, \quad k \geq 3$$

(c) Suppose that f is a continuous, positive, decreasing function on $[a, \infty)$ and let k be a positive integer such that $k \geq a$. Let

$$S_k = \sum_{n=k+1}^\infty f(n) \quad \text{and} \quad I_k = \int_k^\infty f(x)\,dx$$

Then, $\quad I_{k+1} \leq S_k \leq I_k$; $\quad$ see the figures.

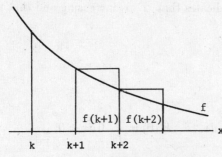 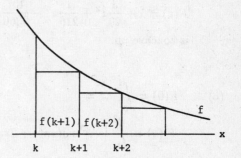

(d) Let $\quad f(n) = \dfrac{[n^3 + \ln(n)]\ln(n)}{n^{4+\ln(n)} + 3n^2 + 7}$, $\quad$ and let

$$S = \sum_{n=1}^\infty f(n), \quad S_k = \sum_{n=k+1}^\infty f(n), \quad \text{and} \quad I_k = \int_k^\infty f(x)\,dx.$$

Then,

$$S = \sum_{n=1}^k f(n) + \sum_{n=k+1}^\infty f(n) \quad \text{and} \quad \left|S - \sum_{n=1}^k f(n)\right| = S_k \leq I_k \leq \frac{1}{(2)k^{\ln k}}.$$

$\displaystyle\sum_{n=1}^k f(n)$ will approximate S with three decimal place accuracy if $I_k < 0.005$.

$$\frac{1}{(2)k^{\ln k}} < 0.005 \quad \Longrightarrow \quad (2)k^{\ln k} > 2000 \quad \Longrightarrow \quad (\ln k)^2 > \ln 1000 \quad \Longrightarrow \quad k \geq 14.$$

11.3. $\quad F(z) = \displaystyle\int_0^1 \left[\cos(zx^2) + \sin(z^2 x^3)\right]\,dx.$

(a)

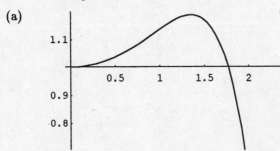

From the graph, F appears to be continuous and increasing on $[0, 1]$, and the graph of F appears to be concave up.

(b) $\cos(zx^2) \cong 1 - \dfrac{1}{2}\left(zx^2\right)^2 + \dfrac{1}{4!}\left(zx^2\right)^4 - \dfrac{1}{6!}\left(zx^2\right)^6$

$\sin(z^2x^3) \cong z^2x^3 - \dfrac{1}{3!}\left(z^2x^3\right)^3 + \dfrac{1}{5!}\left(z^2x^3\right)^5$

$\displaystyle\int_0^1 \cos(zx^2)\,dx \cong \left[x - \dfrac{1}{10}z^2x^5 + \dfrac{1}{216}z^4x^9 - \dfrac{1}{9360}z^6x^13\right]_0^1 \cong 1 - \dfrac{1}{10}z^2 + \dfrac{1}{216}z^4 - \dfrac{1}{9360}z^6$

$\displaystyle\int_0^1 \sin(z^2x^3)\,dx \cong \left[\dfrac{1}{4}z^2x^4 - \dfrac{1}{60}z^6x^{10} + \dfrac{1}{1920}z^{10}x^{16}\right]_0^1 \cong \dfrac{1}{4}z^2 - \dfrac{1}{60}z^6 + \dfrac{1}{1920}z^{10}$

$F(z) \cong 1 + \dfrac{3}{20}z^2 + \dfrac{1}{216}z^4 - \dfrac{157}{9360}z^6$;   this indicates that F is increasing and that its graph is concave up.

(c) $F(0) = \displaystyle\int_0^1 dx = 1$

$F'(z) = \displaystyle\int_0^1 \left[-x^2\sin(zx^2) + 2zx^3\cos(z^2x^3)\right]dx, \qquad F'(0) = \int_0^1 0\,dx = 0;$

$F''(z) = \displaystyle\int_0^1 \left[-x^4\cos(zx^2) + 2x^3\cos(z^2x^3) - 4z^2x^6\sin(z^2x^3)\right]dx,$

$F''(0) = \displaystyle\int_0^1 \left(-x^4 + 2x^3\right)dx = \dfrac{3}{10};$

$F'''(z) = \displaystyle\int_0^1 \left[x^6\sin(zx^2) - 8x^9z^3\cos(z^2x^3) - 12x^6z\sin(z^2x^3)\right]dx,$

$F'''(0) = \displaystyle\int_0^1 0\,dx = 0$

$F^{(4)}(z) = \displaystyle\int_0^1 \left[x^8\cos(zx^2) - 48x^9z^2\cos(z^2x^3) - 12x^6\sin(z^2x^3) + 16x^{12}z^4\sin(x^3z^2)\right]dx,$

$F^{(4)}(0) = \displaystyle\int_0^1 x^8\,dx = \dfrac{1}{9}$

Thus, $F(z) \cong 1 + \dfrac{3}{20}z^2 + \dfrac{1}{216}z^4 + \cdots,$ the same series as before.

CHAPTER 12

SECTION 12.1

1. first order, ordinary   **3.** first order, partial   **5.** second order, ordinary

7. second order, partial

9. $y_1'(x) = \frac{1}{2} e^{x/2}$; $2y_1' - y_1 = 2\left(\frac{1}{2}\right) e^{x/2} - e^{x/2} = 0$; y_1 is a solution.

$y_2'(x) = 2x + e^{x/2}$; $2y_2' - y_2 = 2\left(2x + e^{x/2}\right) - \left(x^2 + 2e^{x/2}\right) = 4x - x^2 \neq 0$;

y_2 is not a solution.

11. $y_1'(x) = \dfrac{-e^x}{(e^x + 1)^2}$; $y_1' + y_1 = \dfrac{-e^x}{(e^x + 1)^2} + \dfrac{1}{e^x + 1} = \dfrac{1}{(e^x + 1)^2} = y_1^2$; y_1 is a solution.

$y_2'(x) = \dfrac{-Ce^x}{(Ce^x + 1)^2}$; $y_2' + y_2 = \dfrac{-Ce^x}{(Ce^x + 1)^2} + \dfrac{1}{Ce^x + 1} = \dfrac{1}{(Ce^x + 1)^2} = y_2^2$;

y_2 is a solution.

13. $y_1'(x) = 2e^{2x}$, $y_1'' = 4e^{2x}$; $y_1'' - 4y_1 = 4e^{2x} - 4e^{2x} = 0$; y_1 is a solution.

$y_2'(x) = 2C \cosh 2x$, $y_2'' = 4C \sinh 2x$; $y_2'' - 4y_2 = 4C \sinh 2x - 4c \sinh 2x = 0$;

y_2 is a solution.

15. $\dfrac{\partial u_1}{\partial x} = -\lambda \sin \lambda x \sin \lambda at$, $\dfrac{\partial^2 u_1}{\partial x^2} = -\lambda^2 \cos \lambda x \sin \lambda at$

$\dfrac{\partial u_1}{\partial t} = -\lambda a \cos \lambda x \cos \lambda at$, $\dfrac{\partial^2 u_1}{\partial t^2} = -(\lambda a)^2 \cos \lambda x \sin \lambda at$

$$a^2 \frac{\partial^2 u_1}{\partial x^2} = -a^2 \lambda^2 \cos \lambda x \sin \lambda at = \frac{\partial^2 u_1}{\partial t^2};   u_1 \text{ is a solution.}$$

$\dfrac{\partial u_2}{\partial x} = \cos(x - at)$, $\dfrac{\partial^2 u_2}{\partial x^2} = -\sin(x - at)$

$\dfrac{\partial u_2}{\partial t} = -a \cos(x - at)$, $\dfrac{\partial^2 u_2}{\partial t^2} = a^2 \sin(x - at)$

$$a^2 \frac{\partial^2 u_2}{\partial x^2} = a^2 \sin(x - at) = \frac{\partial^2 u_2}{\partial t^2};   u_2 \text{ is a solution.}$$

17. $y_1'(x) = -\frac{1}{2} x^{-3/2}$, $y_1'' = \frac{3}{4} x^{-5/2}$;

$$4x^2 y_1'' - 12x y_1' - 9y_1 = 4x^2 \left(\frac{3}{4} x^{-5/2}\right) - 12x \left(-\frac{1}{2} x^{-3/2}\right) - 9x^{-1/2}$$

$$= 3x^{-1/2} + 6x^{-1/2} - 9x^{-1/2} = 0;$$

y_1 is a solution.

$y_2'(x) = -\frac{1}{2} C_1 x^{-3/2} + \frac{9}{2} C_2 x^{7/2}$, $y_2'' = \frac{3}{4} C_1 x^{-5/2} + \frac{63}{4} C_2 x^{5/2}$;

$$4x^2 y_2'' - 12x\, y_2' - 9y_2$$

$$= 4x^2 \left(\tfrac{3}{4} C_1 x^{-5/2} + \tfrac{63}{4} C_2 x^{5/2} \right) - 12x \left(-\tfrac{1}{2} C_1 x^{-3/2} + \tfrac{9}{2} C_2 x^{7/2} \right) - 9 \left(C_1 x^{-1/2} + C_2 x^{9/2} \right)$$

$$= C_1 \left(3x^{-1/2} + 6x^{-1/2} - 9x^{-1/2} \right) + C_2 \left(63x^{9/2} - 54x^{9/2} - 9x^{9/2} \right) = 0;$$

y_2 is a solution.

19. $y'(x) = 5Ce^{5x} = 5y$ Thus, $y = Ce^{5x}$ is a solution.

$y(0) = Ce^{5 \cdot 0} = 2 \implies C = 2;$ $y = 2e^{5x}$ satisfies the side condition.

21. It was shown in Exercise 11 that $y = \dfrac{1}{Ce^x + 1}$ is a one-parameter family of solutions of $y' + y = y^2$.

$y(1) = \dfrac{1}{Ce + 1} = -1 \implies C = -\dfrac{2}{e};$ $y = \dfrac{1}{-2e^{x-1} + 1}$ satisfies the side condition.

23. $y' = C_1 + \tfrac{1}{2} C_2 x^{-1/2},$ $y'' = -\tfrac{1}{4} C_2 x^{-3/2};$

$$2x^2 y'' - xy' + y = 2x^2 \left(-\tfrac{1}{4} C_2 x^{-3/2} \right) - x \left(C_1 + \tfrac{1}{2} C_2 x^{-1/2} \right) + C_1 x + C_2 x^{1/2}$$

$$= -\tfrac{1}{2} C_2 x^{1/2} - C_1 x - \tfrac{1}{2} C_2 x^{1/2} + C_1 x + C_2 x^{1/2} = 0$$

Therefore, $y = C_1 x + C_2 x^{1/2}$ is a two-parameter family of solutions.

$$y(4) = C_1(4) + C_2(4)^{1/2} = 1 = 4C_1 + 2C_2$$

$$y'(4) = C_1 + \tfrac{1}{2} C_2(4)^{-1/2} = -2 = C_1 + \tfrac{1}{4} C_2$$

implies $C_1 = -\tfrac{17}{4},$ $C_2 = 9;$ $y = -\tfrac{17}{4}x + 9x^{1/2}$ satisfies the side conditions.

25. $y' = 2C_1 x + 2C_2 x \ln x + C_2 x,$ $y'' = 2C_1 + 3C_2 + 2C_2 \ln x;$

$x^2 y'' - 3xy' + 4y =$

$x^2 (2C_1 + 3C_2 + 2C_2 \ln x) - 3x (2C_1 x + 2C_2 x \ln x + C_2 x) + 4 \left(C_1 x^2 + C_2 x^2 \ln x \right)$

$= x^2 (2C_1 + 3C_2 - 6C_1 - 3C_2 + 4C_1) + x^2 \ln x (2C_2 - 6C_2 + 4C_2) = 0$

Therefore, $y = C_1 x^2 + C_2 x^2 \ln x$ is a two parameter family of solutions.

$y(1) = C_1 = 0,$ $y'(1) = 2C_1 + C_2 = 1 \implies C_2 = 1;$ $y = x^2 \ln x$ satisfies the side conditions.

27. $y = e^{rx},$ $y' = re^{rx};$ $y' + 3y = re^{rx} + 3e^{rx} = 0 \implies r = -3.$

29. $y = e^{rx},$ $y' = re^{rx},$ $y'' = r^2 e^{rx};$

$$y'' + 6y' + 9y = r^2 e^{rx} + 6re^{rx} + 9e^{rx} = e^{rx}(r+3)^2 = 0 \implies r = -3$$

31. $y = x^r$, $y' = rx^{r-1}$, $y'' = r(r-1)x^{r-2}$;

$$xy'' + y' = x\left[r(r-1)x^{r-2}\right] + rx^{r-1} = \left[r^2 - r + r\right]x^{r-1} = 0 \implies r = 0$$

33. $y = x^r$, $y' = rx^{r-1}$, $y'' = r(r-1)x^{r-2}$;

$$4x^2y'' - 4xy' + 3y = 4x^2r(r-1)x^{r-2} - 4x\,r\,x^{r-1} + 3\,x^r$$

$$= \left(4r^2 - 8r + 3\right)x^r = 0 \implies 4r^2 - 8r + 3 = 0$$

$$\implies (2r-1)(2r-3) = 0 \implies r = \tfrac{1}{2},\ \tfrac{3}{2}$$

35.   (a) $y(0) = C_1 \sin(0) + C_2 \cos(0) = 0 \implies C_2 = 0$;

Since $y = C_1 \sin(4 \cdot \pi/2) = C_1 \sin(2\pi) = 0$, $y = C_1 \sin 4x$ satisfies the boundary conditions

$$y(0) = 0, \qquad y(\pi/2)$$

for all values of C_1.

(b)   As shown above, $y(0) = 0 \implies C_2 = 0$.

Now $y(\pi/8) = C_1 \sin(4 \cdot \pi/8) = C_1 \sin(\pi/2) = 0 \implies C_1 = 0$. Therefore, $y = 0$

is the only member of the family that satisfies the boundary conditions

$$y(0) = 0, \qquad y(\pi/8) = 0.$$

SECTION 12.2

1. $y' - 2y = 1$; $H(x) = \displaystyle\int(-2)\,dx = -2x$,   integrating factor: e^{-2x}

$$e^{-2x}y' - 2e^{-2x}y = e^{-2x}$$

$$\frac{d}{dx}\left[e^{-2x}y\right] = e^{-2x}$$

$$e^{-2x}y = -\tfrac{1}{2}e^{-2x} + C$$

$$y = -\tfrac{1}{2} + Ce^{2x}$$

3. $y' + \dfrac{5}{2}y = 1$; $H(x) = \displaystyle\int\left(\frac{5}{2}\right)dx = \frac{5}{2}x$,   integrating factor: $e^{5x/2}$

$$e^{5x/2}y' + \frac{5}{2}e^{5x/2}y = e^{5x/2}$$

$$\frac{d}{dx}\left[e^{5x/2}y\right] = e^{5x/2}$$

$$e^{5x/2}y = \tfrac{2}{5}e^{5x/2} + C$$

$$y = \tfrac{2}{5} + Ce^{-5x/2}$$

5. $y' - 2y = 1 - 2x$; $H(x) = \displaystyle\int(-2)\,dx = -2x$,   integrating factor: e^{-2x}

$$e^{-2x}y' - 2e^{-2x}y = e^{-2x} - 2xe^{-2x}$$

$$\frac{d}{dx}\left[e^{-2x}y\right] = e^{-2x} - 2xe^{-2x}$$

$$e^{-2x}y = -\tfrac{1}{2}e^{-2x} + xe^{-2x} + \tfrac{1}{2}e^{-2x} + C = xe^{-2x} + C$$

$$y = x + Ce^{2x}$$

7. $y' - \dfrac{4}{x}y = -2n;$ $H(x) = \displaystyle\int\left(-\dfrac{4}{x}\right)dx = -4\ln x = \ln x^{-4},$

integrating factor: $e^{\ln x^{-4}} = x^{-4}$

$$x^{-4}y' - \frac{4}{x}x^{-4}y = -2nx^{-4}$$

$$\frac{d}{dx}\left[x^{-4}y\right] = -2nx^{-4}$$

$$x^{-4}y = \tfrac{2}{3}nx^{-3} + C$$

$$y = \tfrac{2}{3}nx + Cx^{4}$$

9. $y' - e^{x}y = 0;$ $H(x) = \displaystyle\int -e^{x}\,dx = -e^{x},$ integrating factor: $e^{-e^{x}}$

$$e^{-e^{x}}y' - e^{x}e^{-e^{x}}y = 0$$

$$\frac{d}{dx}\left[e^{-e^{x}}y\right] = 0$$

$$e^{-e^{x}}y = C$$

$$y = Ce^{e^{x}}$$

11. $y' + \dfrac{1}{1+e^{x}}y = \dfrac{1}{1+e^{x}};$ $H(x) = \displaystyle\int\dfrac{1}{1+e^{x}}\,dx = \ln\dfrac{e^{x}}{1+e^{x}},$

integrating factor: $e^{H(x)} = \dfrac{e^{x}}{1+e^{x}}$

$$\frac{e^{x}}{1+e^{x}}y' + \frac{1}{1+e^{x}}\cdot\frac{e^{x}}{1+e^{x}}y = \frac{1}{1+e^{x}}\cdot\frac{e^{x}}{1+e^{x}}$$

$$\frac{d}{dx}\left[\frac{e^{x}}{1+e^{x}}y\right] = \frac{e^{x}}{(1+e^{x})^{2}}$$

$$\frac{e^{x}}{1+e^{x}}y = -\frac{1}{1+e^{x}} + C$$

$$y = -e^{-x} + C\left(1+e^{-x}\right)$$

This solution can also be written: $y = 1 + K\left(e^{-x}+1\right),$ where K is an arbitrary constant.

13. $y' + 2xy = xe^{-x^{2}};$ $H(x) = \displaystyle\int 2x\,dx = x^{2},$ integrating factor: $e^{x^{2}}$

$$e^{x^2} y' + 2x e^{x^2} y = x$$
$$\frac{d}{dx}\left[e^{x^2} y\right] = x$$
$$e^{x^2} y = \tfrac{1}{2} x^2 + C$$
$$y = e^{-x^2}\left(\tfrac{1}{2} x^2 + C\right)$$

15. $\quad y' + \dfrac{2}{x+1}\, y = 0; \quad H(x) = \displaystyle\int \frac{2}{x+1}\, dx = 2\ln(x+1) = \ln(x+1)^2,$

integrating factor: $\quad e^{\ln(x+1)^2} = (x+1)^2$

$$(x+1)^2 y' + 2(x+1)\, y = 0$$
$$\frac{d}{dx}\left[(x+1)^2 y\right] = 0$$
$$(x+1)^2 y = C$$
$$y = \frac{C}{(x+1)^2}$$

17. $\quad y' + y = x; \quad H(x) = \displaystyle\int 1\, dx = x, \quad$ integrating factor : $\quad e^x$

$$e^x y' + e^x y = x e^x$$
$$\frac{d}{dx}\left[e^x y\right] = x e^x$$
$$e^x y = x e^x - e^x + C$$
$$y = (x-1) + C e^{-x}$$

$y(0) = -1 + C = 1 \quad \Longrightarrow \quad C = 2.$ Therefore, $\ y = 2e^{-x} + x - 1\ $ is the solution which satisfies the side condition.

19. $\quad y' + y = \dfrac{1}{1+e^x}; \quad H(x) = \displaystyle\int 1\, dx = x, \quad$ integrating factor : $\quad e^x$

$$e^x y' + e^x y = \frac{e^x}{1+e^x}$$
$$\frac{d}{dx}\left[e^x y\right] = \frac{e^x}{1+e^x}$$
$$e^x y = \ln\left(1+e^x\right) + C$$
$$y = e^{-x}\left[\ln\left(1+e^x\right) + C\right]$$

$y(0) = \ln 2 + C = e \quad \Longrightarrow \quad C = e - \ln 2.$ Therefore, $\ y = e^{-x}\left[\ln\left(1+e^x\right) + e - \ln 2\right]\ $ is the solution which satisfies the side condition.

21. $y' - \dfrac{2}{x}y = x^2 e^x$; $H(x) = \displaystyle\int \left(-\dfrac{2}{x}\right) dx = -2 \ln x = \ln x^{-2}$,

integrating factor: $e^{\ln x^{-2}} = x^{-2}$

$$x^{-2} y' - 2x^{-3} y = e^x$$
$$\dfrac{d}{dx}\left[x^{-2} y\right] = e^x$$
$$x^{-2} y = e^x + C$$
$$y = x^2 \left(e^x + C\right)$$

$y(1) = e + C = 0 \implies C = -e$. Therefore, $y = x^2 \left(e^x - e\right)$ is the solution which satisfies the side condition.

23. $y' + 2xy = 0$; $H(x) = \displaystyle\int 2x\,dx = x^2$,   integrating factor : e^{x^2}

$$e^{x^2} y' + 2x\, e^{x^2} y = 0$$
$$\dfrac{d}{dx}\left[e^{x^2} y\right] = 0$$
$$e^{x^2} y = C$$
$$y = Ce^{-x^2}$$

$y(x_0) = Ce^{-x_0^2} = y_0 \implies C = y_0 e^{x_0^2}$.   Therefore, $y = y_0 e^{(x_0^2 - x^2)}$ is the solution which satisfies the side condition.

25. Let $y(t)$ be the amount of water in the tank at time t.

(a) The water drains off at a rate proportional to the amount present:

$$y' = -ky, \quad k > 0 \text{ a constant}, \quad y(0) = 200.$$
$y' + ky = 0$; $H(t) = \displaystyle\int k\,dt = kt$,   integrating factor : e^{kt}

$$e^{kt} y' + ke^{kt} y = 0$$
$$\dfrac{d}{dt}\left[e^{kt} y\right] = 0$$
$$e^{kt} y = C$$
$$y = Ce^{-kt}$$

$$y(0) = 200 \implies C = 200. \text{ Thus } y = 200e^{-kt}$$
$$y(5) = 200e^{-5k} = 0.8(200) = 160 \implies e^{-5k} = \dfrac{4}{5} \implies k = \dfrac{1}{5}\ln\left(\dfrac{4}{5}\right)$$

Therefore, $y = 200 \, e^{(t/5) \ln (4/5)} = 200 \left(\dfrac{4}{5} \right)^{t/5}$.

(b) The water drains off at a rate proportional to the product of the time elapsed and the amount of water present:

$$y' = -k \, t \, y, \qquad k > 0 \ \text{ a constant}, \qquad y(0) = 200.$$

$$y' + kt \, y = 0; \qquad H(t) = \int kt \, dt = \tfrac{1}{2} \, kt^2, \quad \text{integrating factor}: \quad e^{kt^2/2}$$

$$e^{kt^2/2} \, y' + kt e^{kt^2/2} \, y = 0$$

$$\frac{d}{dt} \left[e^{kt^2/2} \, y \right] = 0$$

$$e^{kt^2/2} \, y = C$$

$$y = Ce^{-kt^2/2}$$

$$y(0) = 200 \quad \Longrightarrow \quad C = 200. \quad \text{Thus} \quad y = 200 e^{-kt^2/2}$$

$$y(5) = 200 e^{-25k/2} = 0.8(200) = 160 \quad \Longrightarrow \quad e^{-25k/2} = \frac{4}{5} \quad \Longrightarrow k = \frac{2}{25 \ln \left(\frac{4}{5} \right)}$$

Therefore, $y = 200 \, e^{(t^2/25) \ln (4/5)} = 200 \left(\dfrac{4}{5} \right)^{t^2/25}$.

27. Let $S = S(t)$ be the amount of salt (in pounds) in the tank at time t. Then,

$S(0) = 0.25(100) = 25$ lbs. and

$$\frac{dS}{dt} = (\text{rate in}) - (\text{rate out})$$

$$\frac{dS}{dt} = (0.2)3 - \frac{3}{100} \, S$$

$$\frac{dS}{dt} + 0.03 \, S = 0.6 \qquad (\text{integrating factor}: \quad e^{0.03 \, t})$$

$$e^{0.03 \, t} \frac{dS}{dt} + 0.03 \, e^{0.03 \, t} \, S = 0.6 \, e^{0.03 \, t}$$

$$\frac{d}{dt} \left[e^{0.03 \, t} \, S \right] = 0.6 \, e^{0.03 \, t}$$

$$e^{0.03 \, t} \, S = 20 \, e^{0.03 \, t} + C$$

$$S = 20 + C \, e^{-0.03 \, t}$$

$S(0) = 25 \quad \Longrightarrow \quad C = 5, \quad \text{and} \quad S(t) = 20 + 5 \, e^{-0.03 \, t}$.

29. (a) $\dfrac{dD}{dt} + kT = k\tau; \qquad H(t) = \displaystyle\int k \, dt = kt, \quad \text{integrating factor}: \quad e^{kt}$

$$e^{kt} \frac{dD}{dt} + k e^{kt} \, T = k\tau \, ekt$$

$$\frac{d}{dt} \left[e^{kt} \, T \right] = k\tau \, e^{kt}$$

$$e^{kt} T = \tau \, e^{kt} + C$$

$$T = \tau + Ce^{-kt}$$

(b) $T(0) = \tau + C = T_0 \implies C = T_0 - \tau.$ Thus $T = \tau + (T_0 - \tau) e^{-kt}.$

(c) $k > 0 \implies \lim_{t \to \infty} T(t) = \tau$ for both $T_0 > \tau$ and $T_0 < \tau.$

(d) We have $\tau = 65$ and $T_0 = 185,$ so that $T(t) = 65 + 120e^{-kt}.$ Now

$$T(2) = 65 + 120e^{-2k} = 155 \implies e^{-2k} = \frac{3}{4} \implies k = -\frac{1}{2} \ln\left(\frac{3}{4}\right).$$

Therefore, $T(t) = 65 + 120e^{(t/2)\ln(3/4)} = 65 + 120\left(\frac{3}{4}\right)^{t/2}$

$$T(t) = 65 + 120\left(\frac{3}{4}\right)^{t/2} = 105 \implies t = \frac{-2\ln 3}{\ln(3/4)} \cong 7.64 \text{ min.}$$

Therefore, you should wait another 5.64 minutes.

31. (a) $\dfrac{dP}{dt} = k(M - P)$

(b) $\dfrac{dP}{dt} + kP = kM;$ $H(t) = \displaystyle\int k\,dt = kt,$ integrating factor : e^{kt}

$$e^{kt}\frac{dP}{dt} + ke^{kt}P = kM\,ekt$$

$$\frac{d}{dt}\left[e^{kt}P\right] = kM\,e^{kt}$$

$$e^{kt}P = Me^{kt} + C$$

$$P = M + Ce^{-kt}$$

(b) $P(0) = M + C = 0 \implies C = -M$ and $P(t) = M\left(1 - e^{-kt}\right)$

$P(10) = M\left(1 - e^{-10k}\right) = 0.3M \implies k \cong 0.0357$ and $P(t) = M\left(1 - e^{-0.0357t}\right)$

(c) $P(t) = M\left(1 - e^{-0.0357t}\right) = 0.9M \implies e^{-0.0357t} = 0.1 \implies t \cong 65$

Therefore, it will take approximately 65 days for 90 % of the population to be aware of the product.

33. This is a Bernoulli equation with $a(x) = 1,$ $b(x) = 3e^x,$ $k = 3.$ Using Exercise 32,

$$y' + y = 3e^x y^3 \quad \text{ is transformed into } \quad v' - 2v = -6e^x; \quad v = y^{-2}$$

$$H(x) = \int (-2)\,dx = -2x, \quad \text{ integrating factor: } \quad e^{-2x}$$

$$e^{-2x}v' - 2e^{-2x}v = -6e^{-x}$$

$$\frac{d}{dx}\left[e^{-2x}v\right] = -6e^{-x}$$

$$e^{-2x}v = 6e^{-x} + C$$

$$v = 6e^x + Ce^{2x}$$

Therefore, $y^2 = \dfrac{1}{v} = \dfrac{1}{6e^x + Ce^{2x}}$

35. This is a Bernoulli equation with $a(x) = \dfrac{1}{x}$, $b(x) = x^3$, $k = 2$. Using Exercise 32,

$$y' + \frac{1}{x}y = x^3 y^2 \quad \text{is transformed into} \quad v' - \frac{1}{x}v = -x^3; \quad v = \frac{1}{y}$$

$H(x) = \displaystyle\int -\frac{1}{x}\,dx = -\ln x,$ integrating factor: $e^{-\ln x} = \dfrac{1}{x}$

$$\frac{1}{x}v' - \frac{1}{x^2}v = -x^2$$

$$\frac{d}{dx}\left[\frac{1}{x}v\right] = -x^2$$

$$\frac{1}{x}v = -\tfrac{1}{3}x^3 + C$$

$$v = -\tfrac{1}{3}x^4 + Cx$$

Therefore, $y = \dfrac{1}{v} = \dfrac{1}{Cx - \frac{1}{3}x^4}$

37. If $y = y(x)$ is a solution of (12.2.1), then

$$y'(x) + p(x)y(x) = q(x)$$

and

$$e^{h(x)}y'(x) + e^{h(x)}p(x)y(x) = e^{h(x)}q(x) \quad \text{where} \quad h(x) = \int p(x)\,dx$$

This implies that

$$\frac{d}{dx}\left[e^{h(x)}y(x)\right] = e^{h(x)}q(x) \quad \text{and} \quad e^{h(x)}y(x) = \int e^{h(x)}q(x)\,dx + C$$

It now follows that $y(x)$ is a member of the one-parameter family (12.2.2).

39. (a) Let y_1 and y_2 be solutions of (12.2.3), and let $z = y_1 + y_2$. Then

$$z' = y_1' + y_2' = y_1' + y_2' = -py_1 - py_2 = -p(y_1 + y_2) = -pz$$

Thus, $z' + pz = 0$ and z is a solution of (12.2.3).

(b) Let $z = cy_1$, c a constant. Then

$$z' = cy_1' = -cpy_1 = -pz$$

Thus, $z' + pz = 0$ and z is a solution of (12.2.3).

41. (a) From Exercise 38, $y(x) = Ce^{-h(x)}$. If $y(a) = 0$, then $Ce^{h(a)} = 0 \implies C = 0$ since $e^{-h(a)} \neq 0$. Thus $y(x) = 0$ for all $x \in I$.

(b) Put $u(x) = y_1(x) - y_2(x)$. Then u is a solution of (12.2.3) by Exercise 39. Since $u(a) = y_1(a) - y_2(a) = 0$, $u(x) = 0$ for all $x \in I$ by part (a). Thus $y_1(x) = y_2(x)$ for all $x \in I$.

43. The general solution of $y' + ry = 0$ is $y = Ce^{-rx}$.

(a) Suppose $r > 0$. Then $\lim_{x \to \infty} y(x) = \lim_{x \to \infty} e^{-rx} = 0$.

(b) Suppose $r < 0$. Then $\lim_{x \to \infty} y(x) = \lim_{x \to \infty} e^{-rx} = \infty$. Thus, y is unbounded.

SECTION 12.3

1.
$$y' = y \sin(2x + 3)$$
$$\frac{1}{y} \, dy = \sin(2x + 3) \, dx$$
$$\int \frac{1}{y} \, dy = \int \sin(2x + 3) \, dx$$
$$\ln |y| = -\frac{1}{2} \cos(2x + 3) + C$$

This solution can also be written: $y = Ce^{-(1/2)\cos(2x+3)}$.

3.
$$y' = (xy)^3$$
$$\frac{1}{y^3} \, dy = x^3 \, dx, \qquad y \neq 0$$
$$\int \frac{1}{y^3} \, dy = \int x^3 \, dx$$
$$-\frac{1}{2} y^{-2} = \frac{1}{4} x^4 + C$$

This solution can also be written: $x^4 + \dfrac{2}{y^2} = C$, or $y^2 = \dfrac{2}{C - x^4}$;

$y = 0$ is a singular solution.

5.
$$y' = -\frac{\sin(1/x)}{x^2 y \cos y}$$
$$y \cos y \, dy = -\frac{1}{x^2} \sin(1/x) \, dx$$
$$\int y \cos y \, dy = \int -\frac{1}{x^2} \sin(1/x) \, dx$$
$$y \sin y + \cos y = -\cos(1/x) + C$$

7.
$$y' = x \, e^{y-x}$$
$$e^{-y} \, dy = x e^{-x} \, dx$$
$$\int e^{-y} \, dy = \int x e^{-x} \, dx$$
$$e^{-y} = x e^{-x} + e^{-x} + C$$

9.

$$(y \ln x)y' = \frac{(y+1)^2}{x}$$

$$\frac{y}{(y+1)^2}\,dy = \frac{1}{x \ln x}\,dx$$

$$\int \frac{y}{(y+1)^2}\,dy = \int \frac{1}{x \ln x}\,dx$$

$$\ln |y+1| + \frac{1}{y+1} = \ln |\ln x| + C$$

11.

$$y' = x\sqrt{\frac{1-y^2}{1-x^2}}, \qquad y(0) = 0$$

$$\frac{1}{\sqrt{1-y^2}}\,dy = \frac{x}{\sqrt{1-x^2}}\,dx$$

$$\int \frac{1}{\sqrt{1-y^2}}\,dy = \int \frac{x}{\sqrt{1-x^2}}\,dx$$

$$\sin^{-1} y = -\sqrt{1-x^2} + C$$

$$y(0) = 0 \quad\Longrightarrow\quad \sin^{-1} 0 = -1 + C \quad\Longrightarrow\quad C = 1$$

Thus, $\quad \sin^{-1} y = 1 - \sqrt{1-x^2}.$

13.

$$y' = \frac{x^2 y - y}{y+1}, \qquad y(3) = 1$$

$$\frac{y+1}{y}\,dy = (x^2 - 1)\,dx, \qquad y \neq 0$$

$$\int \frac{y+1}{y}\,dy = \int (x^2 - 1)\,dx$$

$$y + \ln |y| = \frac{1}{3}x^3 - x + C$$

$$y(3) = 1 \quad\Longrightarrow\quad 1 + \ln 1 = \frac{1}{3}(3)^3 - 3 + C \quad\Longrightarrow\quad C = -5.$$

Thus, $\quad y + \ln |y| = \frac{1}{3}x^3 - x - 5.$

15. $\qquad (xy^2 + y^2 + x + 1)\,dx + (y - 1)\,dy = 0, \qquad y(2) = 0$

$$(x+1)(y^2+1)\,dx + (y-1)\,dy = 0$$

$$(x+1)\,dx + \frac{y-1}{y^2+1}\,dy = 0$$

$$\int (x+1)\,dx + \int \frac{y-1}{y^2+1}\,dy = 0$$

$$\frac{x^2}{2} + x + \frac{1}{2}\ln (y^2+1) - \tan^{-1} y = C$$

$y(2) = 0 \quad\Longrightarrow\quad C = 4.$ Thus, $\frac{1}{2}x^2 + x + \frac{1}{2}\ln (y^2+1) - \tan^{-1} y = 4$

17. $f(x,y) = \dfrac{x^2+y^2}{2xy};$ $f(tx,ty) = \dfrac{(tx)^2+(ty)^2}{2(tx)(ty)} = \dfrac{t^2(x^2+y^2)}{t^2(2xy)} = \dfrac{x^2+y^2}{2xy} = f(x,y)$

Set $vx = y.$ Then, $v + xv' = y'$ and

$$v + xv' = \frac{x^2 + v^2x^2}{2vx^2} = \frac{1+v^2}{2v}$$

$$v - \frac{1+v^2}{2v} + xv' = 0$$

$$v^2 - 1 + 2xvv' = 0$$

$$\frac{1}{x}\,dx + \frac{2v}{v^2-1}\,dv = 0$$

$$\int \frac{1}{x}\,dx + \int \frac{2v}{v^2-1}\,dv = 0$$

$$\ln|x| + \ln|v^2 - 1| = K \qquad \text{or} \qquad x(v^2 - 1) = C$$

Replacing v by y/x, we get

$$x\left(\frac{y^2}{x^2} - 1\right) = C \qquad \text{or} \qquad y^2 - x^2 = Cx$$

19. $f(x,y) = \dfrac{x-y}{x+y};$ $f(tx,ty) = \dfrac{(tx)-(ty)}{tx+ty} = \dfrac{t(x-y)}{t(x+y)} = \dfrac{x-y}{x+y} = f(x,y)$

Set $vx = y.$ Then, $v + xv' = y'$ and

$$v + xv' = \frac{x - vx}{x + vx} = \frac{1-v}{1+v}$$

$$v^2 + 2v - 1 + x(1+v)v' = 0$$

$$\frac{1}{x}\,dx + \frac{1+v}{v^2+2v-1}\,dv = 0$$

$$\int \frac{1}{x}\,dx + \int \frac{1+v}{v^2+2v-1}\,dv = 0$$

$$\ln|x| + \tfrac{1}{2}\ln|v^2 + 2v - 1| = K \qquad \text{or} \qquad x\sqrt{v^2 + 2v - 1} = M$$

Replacing v by y/x, we get

$$x\sqrt{\frac{y^2}{x^2} + 2\frac{y}{x} - 1} = M \qquad \text{or} \qquad y^2 + 2xy - x^2 = C$$

21. $f(x,y) = \dfrac{x^2 e^{y/x} + y^2}{xy};$ $f(tx,ty) = \dfrac{(tx)^2 - e^{(ty)/(tx)} + (ty)^2}{(tx)(ty)} = \dfrac{t^2\left(x^2 e^{y/x} + y^2\right)}{t^2(xy)} = f(x,y)$

Set $vx = y.$ Then, $v + xv' = y'$ and

$$v + xv' = \frac{x^2 e^v + v^2 x^2}{vx^2} = \frac{e^v + v^2}{v}$$

$$v^2 + 2v - 1 + x(1+v)v' = 0$$

$$v^2 + xvv' = e^v + v^2$$

$$-e^v + xvv' = 0$$

$$\frac{1}{x}\,dx = ve^{-v}\,dv$$

$$\int \frac{1}{x}\,dx = \int ve^{-v}\,dv$$

$$\ln|x| = -ve^{-v} - e^{-v} + C$$

Replacing v by y/x, and simplifying, we get

$$y + x = xe^{y/x}(C - \ln|x|)$$

23. $f(x,y) = \dfrac{y}{x} + \sin(y/x); \qquad f(tx,ty) = \dfrac{(ty)}{tx} + \sin[(ty/tx)] = \dfrac{y}{x} + \sin(y/x) = f(x,y)$

Set $vx = y$. Then, $v + xv' = y'$ and

$$v + xv' = \frac{vx}{x} + \sin[(vx)/x] = v + \sin v$$

$$xv' = \sin v$$

$$\csc v\,dv = \frac{1}{x}\,dx$$

$$\int \csc v\,dv = \int \frac{1}{x}\,dx$$

$$\ln|\csc v - \cot v| = \ln|x| + K \qquad \text{or} \qquad \csc v - \cot v = Cx$$

Replacing v by y/x, and simplifying, we get

$$1 - \cos(y/x) = Cx\,\sin(y/x) \quad \text{or} \quad M\sin(y/x) = x[1 + \cos(y/x)]$$

25. The differential equation is homogeneous since

$$f(x,y) = \frac{y^3 - x^3}{xy^2}; \qquad f(tx,ty) = \frac{(ty)^3 - (tx)^3}{(tx)(ty)^2} = \frac{t^3(y^3 - x^3)}{t^3(xy^2)} = \frac{y^3 - x^3}{xy^2} = f(x,y)$$

Set $vx = y$. Then, $v + xv' = y'$ and

$$v + xv' = \frac{(vx)^3 - x^3}{v^2 x^3} = \frac{v^3 - 1}{v^2}$$

$$1 + xv^2 v' = 0$$

$$\frac{1}{x}\,dx + v^2\,dv = 0$$

$$\int \frac{1}{x}\,dx + \int v^2\,dv = 0$$

$$\ln|x| + \frac{1}{3}v^3 = C$$

Replacing v by y/x, we get

$$y^3 + 3x^3 \ln|x| = Cx^3$$

Applying the side condition $y(1) = 2$, we have

$$8 + 3\ln 1 = C \implies C = 8 \quad \text{and} \quad y^3 + 3x^3 \ln|x| = 8x^3$$

27. (a) $\dfrac{dy}{dx} = \dfrac{y'(t)}{x'(t)} = \dfrac{y(a_4 x - a_3)}{x(a_1 - a_2 y)}$

(b) This equation is separable.

$$\frac{a_1 - a_2 y}{y}\, dy = \frac{a_4 x - a_3}{x}\, dx$$

$$\int \frac{a_1 - a_2 y}{y}\, dy = \int \frac{a_4 x - a_3}{x}\, dx$$

$$a_1 \ln|y| - a_2 y = a_4 x - a_3 \ln|x| + K$$

$$x^{a_3} y^{a_1} = C e^{a_4 x + a_2 y}$$

29. $2x + 3y = C \implies 2 + 3y' = 0 \implies y' = -\dfrac{2}{3}$

The orthogonal trajectories are the solutions of:

$$y' = \frac{3}{2}.$$

$$y' = \tfrac{3}{2} \implies y = \tfrac{3}{2} x + C$$

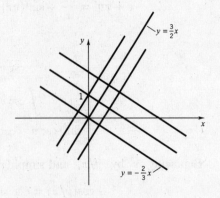

31. $xy = C \implies y + xy' = 0 \implies y' = -\dfrac{y}{x}$

The orthogonal trajectories are the solutions of:

$$y' = \frac{x}{y}.$$

$$y' = \frac{x}{y}$$

$$x\, dx = y\, dy$$

$$\tfrac{1}{2} x^2 = \tfrac{1}{2} y^2 + C$$

$$\text{or} \quad x^2 - y^2 = C$$

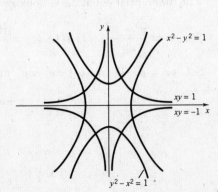

33. $y = Ce^x \implies y' = Ce^x = y$

The orthogonal trajectories are the solutions of:

$$y' = -\frac{1}{y}.$$

$$y' = -\frac{1}{y}$$

$$y\,dy = -dx$$

$$\tfrac{1}{2}y^2 = -x + K$$

or $y^2 = -2x + C$

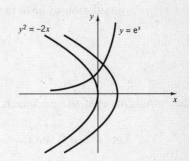

35.   A differential equation for the given family is:

$$y^2 = 2xyy' + y^2(y')^2$$

A differential equation for the family of orthogonal trajectories is found by replacing y' by $-1/y'$. The result is:

$$y^2 = -\frac{2xy}{y'} + \frac{y^2}{(y')^2} \quad \text{which simplifies to} \quad y^2 = 2xyy' + y^2(y')^2,$$

the original equation. Thus, the given family is self-orthogonal.

37.   (a) Let $P = P(t)$ denote the number of people who have the disease at time t. Then

$$\frac{dP}{dt} = kP(25,000 - P) \quad k > 0 \text{ constant}$$

$$\frac{dP}{P(25,000 - P)} = k\,dt$$

$$\int \frac{1}{P(25,000 - P)}\,dP = \int k\,dt$$

$$\frac{1}{25,000}\ln\left|\frac{P}{25,000 - P}\right| = kt + M$$

Solving for P, we get

$$P(t) = \frac{25,000}{1 + Ce^{25,000\,kt}}$$

Now, $P(0) = \dfrac{25,000}{1 + C} = 100 \implies C = 249.$

Also, $P(10) = \dfrac{25,000}{1 + 249e^{25,000\,(10k)}} = 400 \implies 25,000k \cong -0.1382.$

Therefore, $P(t) = \dfrac{25,000}{1 + 249e^{-0.1382\,t}}.$

$P(20) = \dfrac{25,000}{1 + 249e^{-0.1382\,(20)}} \cong 1498;$ 1498 people will have the disease after 20 days.

(b) $\dfrac{25,000}{1+249e^{-0.1382\,t}} = 12,500 \implies t \cong 40;$ (c)

It will take 40 days for half the
population to have the disease.

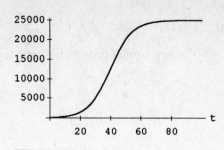

39.   Assume that the package is dropped from rest.

(a) Let $v = v(t)$ be the velocity at time t. Then

$$100\frac{dv}{dt} = 100g - 2v \quad \text{or} \quad \frac{dv}{dt} + \frac{1}{50}v = g \quad (g = 9.8\text{m/sec}^2)$$

This is a linear differential equation; $e^{t/50}$ is an integrating factor.

$$e^{t/50}\frac{dv}{dt} + \frac{1}{50}e^{t/50}v = g\,e^{t/50}g$$

$$\frac{d}{dt}\left[e^{t/50}v\right] = g\,e^{t/50}$$

$$e^{t/50}v = 50g\,e^{t/50} + C$$

$$v = 50g + Ce^{-t/50}$$

Now, $v(0) = 0 \implies C = -50g$ and $v(t) = 50g\left(1 - e^{-t/50}\right)$.

At the instant the parachute opens, $v(10) = 50g\left(1 - e^{-1/5}\right) \cong 50g(0.1813) \cong 88.82$ m/sec.

(b) Now let $v = v(t)$ denote the velocity of the package t seconds after the parachute opens. Then

$$100\frac{dv}{dt} = 100g - 4v^2 \quad \text{or} \quad \frac{dv}{dt} = g - \frac{1}{25}v^2$$

This is a separable differential equation:

$$\frac{dv}{dt} = g - \frac{1}{25}v^2 \quad \text{set } u = v/5, \ du = (1/5)dv$$

$$\frac{du}{g - u^2} = \frac{1}{5}\,dt$$

$$\frac{1}{2\sqrt{g}}\ln\left|\frac{u + \sqrt{g}}{u - \sqrt{g}}\right| = \frac{t}{5} + K$$

$$\ln\left|\frac{u + \sqrt{g}}{u - \sqrt{g}}\right| = \frac{2\sqrt{g}}{5}t + M$$

$$\frac{u + \sqrt{g}}{u - \sqrt{g}} = Ce^{2\sqrt{g}t/5} \cong Ce^{1.25\,t}$$

$$u = \sqrt{g}\,\frac{Ce^{1.25\,t} + 1}{Ce^{1.25t} - 1}$$

$$v = 5\sqrt{g}\,\frac{Ce^{1.25\,t} + 1}{Ce^{1.25t} - 1}$$

Now, $v(0) = 88.82 \quad \Longrightarrow \quad 5\sqrt{g}\,\dfrac{C+1}{C-1} = 88.82 \quad \Longrightarrow \quad C \cong 1.43.$

Therefore, $\quad v(t) = 5\sqrt{g}\,\dfrac{1.43e^{1.25\,t}+1}{1.43e^{1.25t}-1} = \dfrac{15.65\left(1+0.70e^{-1.25\,t}\right)}{1-0.70e^{-1.25\,t}}$

(c)   From part (b), $\quad \lim\limits_{t\to\infty} v(t) = 15.65\text{m/sec}.$

SECTION 12.4

1.   The characteristic equation is:

$$r^2 + 2r - 15 = 0 \quad \text{ or } \quad (r+5)(r-3) = 0.$$

The roots are: $\;r = 3,\,-5.$ The general solution is:

$$y = C_1 e^{3x} + C_2 e^{-5x}.$$

3.   The characteristic equation is:

$$r^2 + 8r + 16 = 0 \quad \text{ or } \quad (r+4)^2 = 0.$$

There is only one root: $\;r = -4.$ By Theorem 12.4.6 II, the general solution is:

$$y = C_1 e^{-4x} + C_2 x e^{-4x}.$$

5.   The characteristic equation is:

$$r^2 + 2r + 5 = 0.$$

The roots are complex: $\;r = -1 \pm 2i.$ By Theorem 12.4.6 III, the general solution is:

$$y = e^{-x}\left(C_1 \cos 2x + C_2 \sin 2x\right).$$

7.   The characteristic equation is:

$$2r^2 + 5r - 3 = 0 \quad \text{ or } \quad (2r-1)(r+3) = 0.$$

The roots are: $\;r = \tfrac{1}{2},\,-3.$ The general solution is:

$$y = C_1 e^{x/2} + C_2 e^{-3x}.$$

9.   The characteristic equation is:

$$r^2 + 12 = 0.$$

The roots are complex: $\;r = \pm 2\sqrt{3}\,i.$ The general solution is:

$$y = C_1 \cos 2\sqrt{3}\,x + C_2 \sin 2\sqrt{3}\,x.$$

11.   The characteristic equation is:

$$5r^2 + \tfrac{11}{4}r - \tfrac{3}{4} = 0 \quad \text{or} \quad 20r^2 + 11r - 3 = (5r - 1)(4r + 3) = 0.$$

The roots are: $r = \tfrac{1}{5}, \, -\tfrac{3}{4}$.   The general solution is:

$$y = C_1 e^{x/5} + C_2 e^{-3x/4}.$$

13.   The characteristic equation is:

$$r^2 + 9 = 0.$$

The roots are complex: $r = \pm 3i$.   The general solution is:

$$y = C_1 \cos 3x + C_2 \sin 3x.$$

15.   The characteristic equation is:

$$2r^2 + 2r + 1 = 0.$$

The roots are complex: $r = -\tfrac{1}{2} \pm \tfrac{1}{2} i$.   The general solution is:

$$y = e^{-x/2} \left[C_1 \cos(x/2) + C_2 \sin(x/2) \right].$$

17.   The characteristic equation is:

$$8r^2 + 2r - 1 = 0 \quad \text{or} \quad (4r - 1)(2r + 1) = 0.$$

The roots are: $r = \tfrac{1}{4}, \, -\tfrac{1}{2}$.   The general solution is:

$$y = C_1 e^{x/4} + C_2 e^{-x/2}.$$

19.   The characteristic equation is:

$$r^2 - 5r + 6 = 0 \quad \text{or} \quad (r - 3)(r - 2) = 0.$$

The roots are: $r = 3, \, 2$.   The general solution and its derivative are:

$$y = C_1 e^{3x} + C_2 e^{2x}, \qquad y' = 3C_1 e^{3x} + 2C_2 e^{2x}.$$

The conditions: $y(0) = 1$, $y'(0) = 1$ require that

$$C_1 + C_2 = 1 \quad \text{and} \quad 3C_1 + 2C_2 = 1.$$

Solving these equations simultaneously gives $C_1 = -1$, $C_2 = 2$.

The solution of the initial value problem is: $y = 2e^{2x} - e^{3x}$.

21. The characteristic equation is:

$$r^2 + \tfrac{1}{4} = 0.$$

The roots are: $r = \pm\tfrac{1}{2}i.$ The general solution and its derivative are:

$$y = C_1 \cos(x/2) + C_2 \sin(x/2) \qquad y' = -\tfrac{1}{2}C_1 \sin(x/2) + \tfrac{1}{2}C_2 \cos(x/2).$$

The conditions: $y(\pi) = 1, \quad y'(\pi) = -1$ require that

$$C_2 = 1 \quad \text{and} \quad C_1 = 2.$$

The solution of the initial value problem is: $y = 2\cos(x/2) + \sin(x/2).$

23. The characteristic equation is:

$$r^2 + 4r + 4 = 0 \quad \text{or} \quad (r+2)^2 = 0.$$

There is only one root: $r = -2.$ The general solution and its derivative are:

$$y = C_1 e^{-2x} + C_2 x e^{-2x} \qquad y' = -2C_1 e^{-2x} + C_2 e^{-2x} - 2C_2 x e^{-2x}.$$

The conditions: $y(-1) = 2, \quad y'(-1) = 1$ require that

$$C_1 e^2 - C_2 e^2 = 2 \quad \text{and} \quad -2C_1 e^2 + 3C_2 e^2 = 1.$$

Solving these equations simultaneously gives $C_1 = 7e^{-2}, \; C_2 = 5e^{-2}.$

The solution of the initial value problem is: $y = 7e^{-2}e^{-2x} + 5e^{-2}x e^{-2x} = 7e^{-2(x+1)} + 5xe^{-2(x+1)}.$

25. The characteristic equation is:

$$r^2 - r - 2 = 0 \quad \text{or} \quad (r-2)(r+1) = 0.$$

The roots are: $r = 2, -1.$ The general solution and its derivative are:

$$y = C_1 e^{2x} + C_2 e^{-x} \qquad y' = 2C_1 e^{2x} - C_2 e^{-x}.$$

(a) $y(0) = 1 \implies C_1 + C_2 = 1 \implies C_2 = 1 - C_1.$

Thus, the solutions that satisfy $y(0) = 1$ are: $y = Ce^{2x} + (1 - C)e^{-x}.$

(b) $y'(0) = 1 \implies 2C_1 - C_2 = 1 \implies C_2 = 2C_1 - 1.$

Thus, the solutions that satisfy $y'(0) = 1$ are: $y = Ce^{2x} + (2C - 1)e^{-x}.$

(c) To satisfy both conditions, we must have $2C - 1 = 1 - C \implies C = \tfrac{2}{3}.$

The solution that satisfies $y(0) = 1, \; y'(0) = 1$ is: $y = \tfrac{2}{3}e^{2x} + \tfrac{1}{3}e^{-x}.$

27. $\quad \alpha = \dfrac{r_1 + r_2}{2}, \qquad \beta = \dfrac{r_1 - r_2}{2};$

$\quad y = k_1 e^{r_1 x} + k_2 e^{r_2 x} = e^{\alpha x} \left(C_1 \cosh \beta x + C_2 \sinh \beta x \right), \quad \text{where} \quad k_1 = \dfrac{C_1 + C_2}{2}, \quad k_2 = \dfrac{C_1 - C_2}{2}.$

29. (a) Let $\ y_1 = e^{rx}, \quad y_2 = xe^{rx}.\quad$ Then

$$W(x) = y_1 y_2' - y_2 y_1' = e^{rx} \left[e^{rx} + rxe^{rx} \right] - xe^{rx} \left[re^{rx} \right] = e^{2rx} \neq 0$$

(b) Let $\ y_1 = e^{\alpha x} \cos \beta x, \quad y_2 = e^{\alpha x} \sin \beta x, \quad \beta \neq 0.\quad$ Then

$W(x) = y_1 y_2' - y_2 y_1'$

$\qquad = e^{\alpha x} \cos \beta x \left[\alpha e^{\alpha x} \sin \beta x + \beta e^{\alpha x} \cos \beta x \right] - e^{\alpha x} \sin \beta x \left[\alpha e^{\alpha x} \cos \beta x - \beta e^{\alpha x} \sin \beta x \right]$

$\qquad = \beta e^{2\alpha x} \neq 0$

31. (a) The solutions $\ y_1 = e^{2x}, \quad y_2 = e^{-4x}\ $ imply that the roots of the characteristic equation are $r_1 = 2 \ \ r_2 = -4$. Therefore, the characteristic equation is:

$$(r - 2)(r + 4) = r^2 + 2r - 8 = 0$$

and the differential equation is: $\quad y'' + 2y' - 8y = 0.$

(b) The solutions $\ y_1 = 3e^{-x}, \quad y_2 = 4e^{5x}\ $ imply that the roots of the characteristic equation are $r_1 = -1 \ \ r_2 = 5$. Therefore, the characteristic equation is

$$(r + 1)(r - 5) = r^2 - 4r - 5 = 0$$

and the differential equation is: $\quad y'' - 4y' - 5y = 0.$

(c) The solutions $\ y_1 = 2e^{3x}, \quad y_2 = xe^{3x}\ $ imply that 3 is the only root of the characteristic equation. Therefore, the characteristic equation is

$$(r - 3)^2 = r^2 - 6r + 9 = 0$$

and the differential equation is: $\quad y'' - 6y' + 9y = 0.$

33. Suppose that $\quad y_1(x) = ky_2(x),\ $ where $\ k\ $ is a constant. Then

$$W(x) = y_1 y_2' - y_2 y_1' = ky_2 y_2' - ky_2 y_2' = 0.$$

Now suppose that $\ y_1 y_2' - y_2 y_1' = 0.\quad$ Then

$$\left[\frac{y_1(x)}{y_2(x)} \right]' = \frac{y_2 y_1' - y_1 y_2'}{y_2^2} = -\frac{y_1 y_2' - y_2 y_1'}{y_2^2} = 0.$$

This implies that $\ \dfrac{y_1}{y_2} = k,\ $ for some constant k, that is $\ y_1 = ky_2.$

35. (a) If $a = 0$, $b > 0$, then the general solution of the differential equation is:

$$y = C_1 \cos \sqrt{b}\, x + C_2 \sin \sqrt{b}\, x = A \cos \left(\sqrt{b}\, x + \phi \right)$$

where A and ϕ are constants. Clearly $|y(x)| \leq |A|$ for all x.

(b) If $a > 0$, $b = 0$, then the general solution of the differential equation is:

$$y = C_1 + C_2 e^{-ax} \qquad \text{and} \qquad \lim_{x \to \infty} y(x) = C_1.$$

The solution which satisfies the conditions: $y(0) = y_0$, $y'(0) = y_1$ is:

$$y = y_0 + \frac{y_1}{a} - \frac{y_1}{a} e^{-ax} \quad \text{and} \quad \lim_{x \to \infty} y(x) = y_0 + \frac{y_1}{a}; \qquad k = y_0 + \frac{y_1}{a}.$$

37. From Exercise 36, the change of variable $z = \ln x$ transforms the equation

$$x^2 y'' - xy' - 8y = 0$$

into the differential equation with constant coefficients

$$\frac{d^2 y}{dz^2} - 2\frac{dy}{dz} - 8y = 0.$$

The characteristic equation is:

$$r^2 - 2r - 8 = 0 \qquad \text{or} \qquad (r - 4)(r + 2) = 0$$

The roots are: $r = 4$, $r = -2$, and the general solution (in terms of z) is:

$$y = C_1 e^{4z} + C_2 e^{-2z}.$$

Replacing z by $\ln x$ we get

$$y = C_1 e^{4 \ln x} + C_2 e^{-2 \ln x} = C_1 x^4 + C_2 x^{-2}.$$

39. From Exercise 36, the change of variable $z = \ln x$ transforms the equation

$$x^2 y'' - 3xy' + 4y = 0$$

into the differential equation with constant coefficients

$$\frac{d^2 y}{dz^2} - 4\frac{dy}{dz} + 4y = 0.$$

The characteristic equation is:

$$r^2 - 4r + 4 = 0 \qquad \text{or} \qquad (r - 2)^2 = 0.$$

The only root is: $r = 2$, and the general solution (in terms of z) is:

$$y = C_1 e^{2z} + C_2 z e^{-2z}.$$

Replacing z by $\ln x$ we get: $y = C_1 e^{2 \ln x} + C_2 \ln x\, e^{2 \ln x} = C_1 x^2 + C_2 x^2 \ln x.$

41. (a) $e^x = 1 + x + \dfrac{x^2}{2!} + \dfrac{x^3}{3!} + \cdots + \dfrac{x^n}{n!} \cdots = \displaystyle\sum_{k=0}^{\inf} \dfrac{1}{k!} x^k$

(b) $e^{i\theta} = 1 + (i\theta) + \dfrac{(i\theta)^2}{2!} + \dfrac{(i\theta)^3}{3!} + \cdots + \dfrac{(i\theta)^n}{n!} + \cdots$

$\qquad = 1 + i\theta - \dfrac{1}{2!}\theta^2 - i\dfrac{1}{3!}\theta^3 + \cdots + (i)^n \dfrac{1}{n!}\theta^n + \cdots = \cos\theta + i\sin\theta$

(c) $e^{-i\theta} = 1 + (-i\theta) + \dfrac{(-i\theta)^2}{2!} + \dfrac{(-i\theta)^3}{3!} + \cdots + \dfrac{(-i\theta)^n}{n!} + \cdots$

$\qquad = 1 - i\theta - \dfrac{1}{2!}\theta^2 + i\dfrac{1}{3!}\theta^3 + \cdots + (-i)^n \dfrac{1}{n!}\theta^n + \cdots = \cos\theta - i\sin\theta$

SECTION 12.5

1. First consider the reduced equation. The characteristic equation is:

$$r^2 + 5r + 6 = (r+2)(r+3) = 0$$

and $\quad u_1(x) = e^{-2x}, \quad u_2(x) = e^{-3x} \quad$ are fundamental solutions. Therefore, a particular solution of the given equation has the form

$$y = Ax + B.$$

The derivatives of y are: $\quad y' = A, \quad y'' = 0.$

Substituting y and its derivatives into the given equation gives

$$0 + 5A + 6(Ax + B) = 3x + 4.$$

Thus,

$$6A = 3$$

$$5A + 6B = 4$$

The solution of this pair of equations is: $\quad A = \frac{1}{2}, \quad B = \frac{1}{4}, \quad$ and $\quad y = \frac{1}{2}x + \frac{1}{4}.$

3. First consider the reduced equation. The characteristic equation is:

$$r^2 + 2r + 5 = 0$$

and $\quad u_1(x) = e^{-x}\cos 2x, \quad u_2(x) = e^{-x}\sin 2x \quad$ are fundamental solutions. Therefore, a particular solution of the given equation has the form

$$y = Ax^2 + Bx + C.$$

The derivatives of y are: $\quad y' = 2Ax + B, \quad y'' = 2A.$

Substituting y and its derivatives into the given equation gives

$$2A + 2(2Ax + B) + 5(Ax^2 + Bx + C) = x^2 - 1.$$

Thus,
$$5A = 1$$
$$4A + 5B = 0$$
$$2A + 2B + 5C = -1$$

The solution of this system of equations is: $A = \frac{1}{5}$, $B = -\frac{4}{25}$, $C = -\frac{27}{125}$, and

$$y = \frac{1}{5}x^2 - \frac{4}{25}x - \frac{27}{125}.$$

5. First consider the reduced equation. The characteristic equation is:

$$r^2 + 6r + 9 = (r+3)^2 = 0$$

and $u_1(x) = e^{-3x}$, $u_2(x) = xe^{-3x}$ are fundamental solutions. Therefore, a particular solution of the given equation has the form

$$y = Ae^{3x}.$$

The derivatives of y are: $y' = 3Ae^{3x}$, $y'' = 9Ae^{3x}0$.

Substituting y and its derivatives into the given equation gives

$$9Ae^{3x} + 18Ae^{3x} + 9Ae^{3x} = e^{3x}.$$

Thus, $36A = 1 \implies A = \frac{1}{36}$, and $y = \frac{1}{36}e^{3x}.$

7. First consider the reduced equation. The characteristic equation is:

$$r^2 + 2r + 2 = 0$$

and $u_1(x) = e^{-x}\cos x$, $u_2(x) = e^{-x}\sin x$ are fundamental solutions. Therefore, a particular solution of the given equation has the form

$$y = Ae^x.$$

The derivatives of y are: $y' = Ae^x$, $y'' = Ae^x$.

Substituting y and its derivatives into the given equation gives

$$Ae^x + 2Ae^x + 2Ae^x = e^x.$$

Thus, $5A = 1 \implies A = \frac{1}{5}$ and $y = \frac{1}{5}e^x.$

9. First consider the reduced equation. The characteristic equation is:

$$r^2 - r - 12 = (r-4)(r+3) = 0$$

and $u_1(x) = e^{4x}$, $u_2(x) = e^{-3x}$ are fundamental solutions. Therefore, a particular solution of the given equation has the form

$$y = A\cos x + B\sin x.$$

The derivatives of y are: $y' = -A\sin x + B\cos x,$ $y'' = -A\cos x - B\sin x.$

Substituting y and its derivatives into the given equation gives

$$-A\cos x - B\sin x - (-A\sin x + B\cos x) - 12(A\cos x + B\sin x) = \cos x.$$

Thus,

$$-13A - B = 1$$

$$A - 13B = 0$$

The solution of this system of equations is: $A = -\frac{13}{170},$ $B = -\frac{1}{170},$ and

$$y = -\frac{13}{170}\cos x - \frac{1}{170}\sin x.$$

11.   First consider the reduced equation. The characteristic equation is:

$$r^2 + 7r + 6 = (r+6)(r+1) = 0$$

and $u_1(x) = e^{-6x},$ $u_2(x) = e^{-x}$ are fundamental solutions. Therefore, a particular solution of the given equation has the form

$$y = A\cos 2x + B\sin 2x.$$

The derivatives of y are: $y' = -2A\sin 2x + 2B\cos 2x,$ $y'' = -4A\cos 2x - 4B\sin 2x.$

Substituting y and its derivatives into the given equation gives

$$-4A\cos 2x - 4B\sin 2x + 7(-2A\sin 2x + 2B\cos 2x) + 6(A\cos 2x + B\sin 2x) = 3\cos 2x.$$

Thus,

$$2A + 14B = 3$$

$$-14A + 2B = 0$$

The solution of this system of equations is: $A = \frac{3}{100},$ $B = \frac{21}{100}$ and

$$y = \frac{3}{100}\cos 2x + \frac{21}{100}\sin 2x.$$

13.   First consider the reduced equation. The characteristic equation is:

$$r^2 - 2r + 5 = 0$$

and $u_1(x) = e^x\cos 2x,$ $u_2(x) = e^x\sin 2x$ are fundamental solutions. Therefore, a particular solution of the given equation has the form

$$y = Ae^{-x} \cos 2x + Be^{-x} \sin 2x$$

The derivatives of y arc: $y' = -Ae^{-x} \cos 2x - 2Ae^{-x} \sin 2x - Be^{-x} \sin 2x + 2be^{-x} \cos 2x$,

$y'' = 4Ae^{-x} \sin -3Ae^{-x} \cos 2x - 4be^{-x} \cos 2x - 3Be^{-x} \sin 2x$.

Substituting y and its derivatives into the given equation gives

$$4Ae^{-x} \sin -3Ae^{-x} \cos 2x - 4be^{-x} \cos 2x - 3Be^{-x} \sin 2x-$$

$$2\left(-Ae^{-x} \cos 2x - 2Ae^{-x} \sin 2x - Be^{-x} \sin 2x + 2be^{-x} \cos 2x\right)+$$

$$5\left(Ae^{-x} \cos 2x + Be^{-x} \sin 2x\right) = e^{-x} \sin 2x.$$

Equating the coefficients of $e^{-x} \cos 2x$ and $e^{-x} \sin 2x$ we get,

$$8A + 4B = 1$$

$$4A - 8B = 0$$

The solution of this system of equations is: $A = \frac{1}{10}$, $B = \frac{1}{20}$ and

$$y = \frac{1}{10} e^{-x} \cos 2x + \frac{1}{20} e^{-x} \sin 2x.$$

15.   First consider the reduced equation. The characteristic equation is:

$$r^2 + 6r + 8 = (r + 4)(r + 2) = 0$$

and $u_1(x) = e^{-4x}$, $u_2(x) = e^{-2x}$ are fundamental solutions. Therefore, a particular solution of the given equation has the form

$$y = Axe^{-2x}.$$

The derivatives of y are: $y' = Ae^{-2x} - 2Axe^{-2x}$, $y'' = -4Ae^{-2x} + 4Axe^{-2x}$.

Substituting y and its derivatives into the given equation gives

$$-4Ae^{-2x} + 4Axe^{-2x} + 6\left(Ae^{-2x} - 2Axe^{-2x}\right) + 8Axe^{-2x} = 3e^{-2x}$$

Thus, $2A = 3 \implies A = \frac{3}{2}$ and $y = \frac{3}{2} xe^{-2x}$.

17.   First consider the reduced equation: $y'' + y = 0$.   The characteristic equation is:

$$r^2 + 1 = 0$$

and $u_1(x) = \cos x$, $u_2(x) = \sin x$ are fundamental solutions. A particular solution of the given equation has the form

$$y = Ae^x.$$

The derivatives of y are: $y' = y'' = Ae^x$.

Substitute y and its derivatives into the given equation:

$$Ae^x + Ae^x = e^x \implies A = \tfrac{1}{2} \text{ and } y = \tfrac{1}{2}e^x.$$

The general solution of the given equation is: $y = C_1 \cos x + C_2 \sin x + \tfrac{1}{2}e^x.$

19.   First consider the reduced equation: $y'' - 3y' - 10y = 0.$ The characteristic equation is:

$$r^2 - 3r - 10 = (r - 5)(r + 2) = 0$$

and $u_1(x) = e^{5x},$ $u_2(x) = e^{-2x}$ are fundamental solutions. A particular solution of the given equation has the form

$$y = Ax + B.$$

The derivatives of y are: $y' = A,$ $y'' = 0.$

Substitute y and its derivatives into the given equation:

$$-3A - 10(Ax + B) = -x - 1 \implies A = \tfrac{1}{10}, \ B = \tfrac{7}{100} \text{ and } y = \tfrac{1}{10}x + \tfrac{7}{100}$$

The general solution of the given equation is:

$$y = C_1 e^{5x} + C_2 e^{-2x} + \cos x + C_2 \sin x + \tfrac{1}{10}x + \tfrac{7}{100}$$

21.   First consider the reduced equation: $y'' + 3y' - 4y = 0.$ The characteristic equation is:

$$r^2 + 3r - 4 = (r + 4)(r - 1) = 0$$

and $u_1(x) = e^x,$ $u_2(x) = e^{-4x}$ are fundamental solutions. A particular solution of the given equation has the form

$$y = Axe^{-4x}.$$

The derivatives of y are: $y' = Ae^{-4x} - 4Axe^{-4x},$ $y'' = -8Ae^{-4x} + 16Axe^{-4x}.$

Substitute y and its derivatives into the given equation:

$$-8Ae^{-4x} + 16Axe^{-4x} + 3\left(Ae^{-4x} - 4Axe^{-4x}\right) - 4Axe^{-4x} = e^{-4x}.$$

This implies $-5A = 1$, so $A = -\tfrac{1}{5}$ and $y = -\tfrac{1}{5}xe^{-4x}.$

The general solution of the given equation is: $y = C_1 e^x + C_2 e^{-4x} - \tfrac{1}{5}xe^{-4x}.$

23.   First consider the reduced equation: $y'' + y' - 2y = 0.$ The characteristic equation is:

$$r^2 + r - 2 = (r + 2)(r - 1) = 0$$

and $u_1(x) = e^{-2x},$ $u_2(x) = e^x$ are fundamental solutions. A particular solution of the given equation has the form

$$y = (Ax^2 + Bx)e^x = Ax^2e^x + Bxe^x.$$

The derivatives of y are:

$$y' = 2Axe^x + Ax^2e^x + Be^x + bxe^x, \quad y'' = 2Ae^x + 4Axe^x + Ax^2e^x + 2Be^x + Bxe^x.$$

Substitute y and its derivatives into the given equation:

$$2Ae^x + 4Axe^x + Ax^2e^x + 2Be^x + Bxe^x +$$

$$\left(2Axe^x + Ax^2e^x + Be^x + Bxe^x\right) - 2\left(Ax^2e^x + Bxe^x\right) = 3xe^x.$$

Equating coefficients, we get

$$6A = 3$$

$$2A + 3B = 0$$

The solution of this system of equations is: $A = \frac{1}{2}$, $B = -\frac{1}{3}$ and $y = \frac{1}{2}x^2e^x - \frac{1}{3}xe^x$.

The general solution of the given equation is: $y = C_1e^{-2x}x + C_2e^x + \frac{1}{2}x^2e^x - \frac{1}{3}xe^x$.

25. Let $y_1(x)$ be a solution of $y'' + ay' + by = \phi_1(x)$, let $y_2(x)$ be a solution of $y'' + ay' + by = \phi_2(x)$, and let $z = y_1 + y_2$. Then

$$z'' + az' + bz = (y_1'' + y_2'') + a(y_1' + y_2') + b(y_1 + y_2)$$

$$= (y_1'' + ay_1' + by_1) + (y_2'' + ay_2' + y_2) = \phi_1 + \phi_2.$$

27. First consider the reduced equation: $y'' + 4y' + 3y = 0$. The characteristic equation is:

$$r^2 + 4r + 3 = (r + 3)(r + 1) = 0$$

and $u_1(x) = e^{-3x}$, $u_2(x) = e^{-x}$ are fundamental solutions. Since $\cosh x = \frac{1}{2}(e^x + e^{-x})$, a particular solution of the given equation has the form

$$y = Ae^x + Bxe^{-x}$$

The derivatives of y are: $y' = Ae^x + Be^{-x} - Bxe^{-x} \quad y'' = Ae^x - 2Be^{-x} + Bxe^{-x}$.

Substitute y and its derivatives into the given equation:

$$Ae^x - 2Be^{-x} + Bxe^{-x} + 4\left(Ae^x + Be^{-x} - Bxe^{-x}\right) + 3\left(Ae^x + Bxe^{-x}\right) = \frac{1}{2}\left(e^x + e^{-x}\right).$$

Equating coefficients, we get $A = \frac{1}{16}$, $B = \frac{1}{4}$, and so $y = \frac{1}{16}e^x + \frac{1}{4}xe^{-x}$.

The general solution of the given equation is:

$$y = C_1e^{-3x} + C_2e^{-x} + \frac{1}{16}e^x + \frac{1}{4}xe^{-x}.$$

29. First consider the reduced equation $y'' - 2y' + y = 0$.   The characteristic equation is:

$$r^2 - 2r + 1 = (r-1)^2 = 0$$

and $u_1(x) = e^x$, $u_2(x) = xe^x$ are fundamental solutions. Their Wronskian is given by

$$W = u_1 u_2' - u_2 u_1' = e^x(e^x + xe^x) - xe^x(e^x) = e^{2x}$$

Using variation of parameters, a particular solution of the given equation will have the form

$$y = u_1 z_1 + u_2 z_2,$$

where

$$z_1 = -\int \frac{xe^x(xe^x \cos x)}{e^{2x}}\, dx = -\int x^2 \cos x\, dx = -x^2 \sin x - 2x \cos x + 2\sin x,$$

$$z_2 = \int \frac{e^x(xe^x \cos x)}{e^{2x}}\, dx = \int x \cos x\, dx = x \sin x + \cos x$$

Therefore,

$$y = e^x\left(-x^2 \sin x - 2x \cos x + 2\sin x\right) + xe^x\left(x \sin x + \cos x\right) = 2e^x \sin x - xe^x \cos x.$$

31. First consider the reduced equation $y'' - 4y' + 4y = 0$.   The characteristic equation is:

$$r^2 - 4r + 4 = (r-2)^2 = 0$$

and $u_1(x) = e^{2x}$, $u_2(x) = xe^{2x}$ are fundamental solutions. Their Wronskian is given by

$$W = u_1 u_2' - u_2 u_1' = e^{2x}\left(e^{2x} + 2xe^{2x}\right) - xe^{2x}(2e^{2x}) = e^{4x}.$$

Using variation of parameters, a particular solution of the given equation will have the form

$$y = u_1 z_1 + u_2 z_2,$$

where

$$z_1 = -\int \frac{xe^{2x}\left(\frac{1}{3}x^{-1}e^{2x}\right)}{e^{4x}}\, dx = -\frac{1}{3}\int dx = -\frac{1}{3}x,$$

$$z_2 = \int \frac{e^{2x}\left(\frac{1}{3}x^{-1}e^{2x}\right)}{e^{4x}}\, dx = -\frac{1}{3}\int \frac{1}{x}\, dx = \frac{1}{3}\ln|x|.$$

Therefore,

$$y = e^{2x}\left(-\frac{1}{3}x\right) + xe^{2x}\left(\frac{1}{3}\ln|x|\right) = -\frac{1}{3}xe^{2x} + \frac{1}{3}x\ln|x|\,e^{2x}.$$

Note: Since $u = -\frac{1}{3}xe^{2x}$ is a solution of the reduced equation,

$$y = \frac{1}{3}x\ln|x|\,e^{2x}$$

is also a particular solution of the given equation.

33. First consider the reduced equation $y'' + 4y' + 4y = 0$. The characteristic equation is:

$$r^2 + 4r + 4 = (r + 2)^2 = 0$$

and $u_1(x) = e^{-2x}$, $u_2(x) = xe^{-2x}$ are fundamental solutions. Their Wronskian is given by

$$W = u_1 u_2' - u_2 u_1' = e^{-2x} \left(e^{-2x} - 2xe^{2x} \right) - xe^{-2x}(-2e^{-2x}) = e^{-4x}.$$

Using variation of parameters, a particular solution of the given equation will have the form

$$y = u_1 z_1 + u_2 z_2,$$

where

$$z_1 = -\int \frac{xe^{-2x} \left(x^{-2}e^{-2x} \right)}{e^{-4x}} \, dx = -\int \frac{1}{x} \, dx = -\ln|x|$$

$$z_2 = \int \frac{e^{-2x} \left(x^{-2}e^{-2x} \right)}{e^{-4x}} \, dx = \int \frac{1}{x^2} \, dx = -\frac{1}{x}$$

Therefore,

$$y = e^{-2x} \left(-\ln|x| \right) + xe^{-2x} \left(-\frac{1}{x} \right) = -e^{-2x} \ln|x| - e^{-2x}.$$

Note: Since $u = -e^{-2x}$ is a solution of the reduced equation, we can take

$$y = -e^{-2x} \ln|x|$$

35. First consider the reduced equation $y'' - 2y' + 2y = 0$. The characteristic equation is:

$$r^2 - 2r + 2 = 0$$

and $u_1(x) = e^x \cos x$, $u_2(x) = e^x \sin x$ are fundamental solutions. Their Wronskian is given by

$$W = e^x \cos x \left[e^x \sin x + e^x \cos x \right] - e^x \sin x \left[e^x \cos x - e^x \sin x \right] = e^{2x}$$

Using variation of parameters, a particular solution of the given equation will have the form

$$y = u_1 z_1 + u_2 z_2,$$

where

$$z_1 = -\int \frac{e^x \sin x \cdot e^x \sec x}{e^{2x}} \, dx = -\int \tan x \, dx = -\ln|\sec x| = \ln|\cos x|$$

$$z_2 = \int \frac{e^x \cos x \cdot e^x \sec x}{e^{2x}} \, dx = \int dx = x$$

Therefore,

$$y = e^x \cos x \left(\ln|\cos x| \right) + e^x \sin x(x) = e^x \cos x \ln|\cos x| + xe^x \sin x.$$

37. Assume that the forcing function $F(t) = F_0$ (constant). Then the differential equation has a particular solution of the form $i = A$. The derivatives of i are: $i' = i'' = 0$. Substituting i and its derivatives into the equation, we get

$$\frac{1}{C}A = F_0 \quad \Longrightarrow \quad A = CF_0 \quad \Longrightarrow \quad i = CF_0.$$

The characteristic equation for the reduced equation is:

$$Lr^2 + Rr + \frac{1}{C} = 0 \quad \Longrightarrow \quad r_1, r_2 = \frac{-R \pm \sqrt{R^2 - 4L/C}}{2L} = \frac{-R \pm \sqrt{CR^2 - 4L}}{2L\sqrt{C}}$$

(a) If $CR^2 = 4L$, then the characteristic equation has only one root: $r = -R/2L$,

and $u_1 = e^{-(R/2L)t}$, $u_2 = t\,e^{-(R/2L)t}$ are fundamental solutions.

The general solution of the given equation is:

$$i(t) = C_1 e^{-(R/2L)t} + C_2 t\,e^{-(R/2L)t} + CF_0$$

and its derivative is:

$$i'(t) = -C_1 (R/2L) e^{-(R/2L)t} + C_2 e^{-(R/2L)t} - C_2 (R/2L) t\,e^{-(R/2L)t}.$$

Applying the side conditions $i(0) = 0$, $i'(0) = F_0/L$, we get

$$C_1 + CF_0 = 0$$

$$(-R/2L)C_1 + C_2 = F_0/L$$

The solution is $C_1 = -CF_0$, $C_2 = \frac{F_0}{2L}(2 - RC)$.

The current in this case is:

$$i(t) = -CF_0 e^{-(R/2L)t} + \frac{F_0}{2L}(2 - RC)\,t\,e^{-(R/2L)t} + CF_0.$$

(b) If $CR^2 - 4L < 0$ then the characteristic equation has complex roots:

$$r_1 = -R/2L \pm i\beta, \quad \text{where} \quad \beta = \sqrt{\frac{4L - CR^2}{4CL^2}} \quad (\text{here } i^2 = -1)$$

and fundamental solutions are: $u_1 = e^{-(R/2L)t} \cos\beta t$, $u_2 = e^{-(R/2L)t} \sin\beta t$.

The general solution of the given differential equation is:

$$i(t) = e^{-(R/2L)t}\left(C_1 \cos\beta t + C_2 \sin\beta t\right) + CF_0$$

and its derivative is:

$$i'(t) = (-R/2L)e^{-(R/2L)t}\left(C_1 \cos\beta t + C_2 \sin\beta t\right) + \beta e^{-(R/2L)t}\left(-C_1 \sin\beta t + C_2 \cos\beta t\right).$$

Applying the side conditions $i(0) = 0$, $i'(0) = F_0/L$, we get

$$C_1 + CF_0 = 0$$

$$(-R/2L)C_1 + \beta C_2 = F_0/L$$

The solution is $C_1 = -CF_0$, $C_2 = \frac{F_0}{2L\beta}(2 - RC)$.

The current in this case is:

$$i(t) = e^{-(R/2L)t}\left(\frac{F_0}{2L\beta}(2 - RC) \sin\beta t - CF_0 \cos\beta t\right) + CF_0.$$

39. (a) Let $y_1(x) = \sin\left(\ln x^2\right)$. Then

$$y_1' = \left(\frac{2}{x}\right)\cos\left(\ln x^2\right) \quad \text{and} \quad y_1'' = -\left(\frac{4}{x^2}\right)\sin\left(\ln x^2\right) - \left(\frac{2}{x^2}\right)\cos\left(\ln x^2\right)$$

Substituting y_1 and its derivatives into the differential equation, we have

$$x^2\left[-\left(\frac{4}{x^2}\right)\sin\left(\ln x^2\right) - \left(\frac{2}{x^2}\right)\cos\left(\ln x^2\right)\right] - x\left[\left(\frac{2}{x}\right)\cos\left(\ln x^2\right)\right] + 4\sin\left(\ln x^2\right) = 0$$

The verification that y_2 is a solution is done in exactly the same way.

The Wronskian of y_1 and y_2 is:

$$\begin{aligned}
W(x) &= y_1 y_2' - y_2 y_1' \\
&= \sin\left(\ln x^2\right)\left[-\left(\frac{2}{x}\right)\sin\left(\ln x^2\right)\right] - \cos\left(\ln x^2\right)\left[\left(\frac{2}{x}\right)\cos\left(\ln x^2\right)\right] \\
&= -\frac{2}{x} \neq 0 \text{ on } (0, \infty)
\end{aligned}$$

(b) To use the method of variation of parameters as described in the text, we first re-write the equation in the form

$$y'' + x^{-1}y' + 4x^{-2}y = x^{-2}\sin(\ln x).$$

Then, a particular solution of the equation will have the form $y = y_1 z_1 + y_2 z_2$, where

$$\begin{aligned}
z_1 &= -\int \frac{\cos(\ln x^2)x^{-2}\sin(\ln x)}{-2/x}\,dx \\
&= \tfrac{1}{2}\int \cos(2\ln x)x^{-1}\sin(\ln x)\,dx \\
&= \tfrac{1}{2}\int \cos 2u \sin u\,du \qquad (u = \ln x) \\
&= \tfrac{1}{2}\int (2\cos^2 u - 1)\sin u\,du \\
&= -\tfrac{1}{3}\cos^3 u + \tfrac{1}{2}\sin u
\end{aligned}$$

and

$$\begin{aligned}
z_2 &= \int \frac{\sin(\ln x^2)x^{-2}\sin(\ln x)}{-2/x}\,dx \\
&= -\tfrac{1}{2}\int \sin(2\ln x)x^{-1}\sin(\ln x)\,dx \\
&= -\tfrac{1}{2}\int \sin 2u \sin u\,du \qquad (u = \ln x) \\
&= -\int \sin^2 u \cos u\,du \\
&= -\tfrac{1}{3}\sin^3 u
\end{aligned}$$

Thus, $y = \sin 2u\left(-\tfrac{1}{3}\cos^3 u + \tfrac{1}{2}\sin u\right) - \cos 2u\left(\tfrac{1}{3}\sin^3 u\right)$ which simplifies to:

$$y = \tfrac{1}{3}\sin u = \tfrac{1}{3}\sin(\ln x).$$

SECTION 12.6

CTION 12.6

1. The equation of motion is of the form

$$x(t) = A \sin(\omega t + \phi_0).$$

The period is $T = 2\pi/\omega = \pi/4$. Therefore $\omega = 8$. Thus

$$x(t) = A \sin(8t + \phi_0) \quad \text{and} \quad v(t) = 8A \cos(8t + \phi_0).$$

Since $x(0) = 1$ and $v(0) = 0$, we have

$$1 = A \sin \phi_0 \quad \text{and} \quad 0 = 8A \cos \phi_0.$$

These equations are satisfied by taking $A = 1$ and $\phi_0 = \pi/2$.

Therefore the equation of motion reads

$$x(t) = \sin\left(8t + \tfrac{1}{2}\pi\right).$$

The amplitude is 1 and the frequency is $8/2\pi = 4/\pi$.

3. We can write the equation of motion as

$$x(t) = A \sin\left(\frac{2\pi}{T}t\right).$$

Differentiation gives

$$v(t) = \frac{2\pi A}{T} \cos\left(\frac{2\pi}{T}t\right).$$

The object passes through the origin whenever $\sin[(2\pi/T)] = 0$.

Then $\cos[(2\pi/T)t] = \pm 1$ and $v = \pm 2\pi A/T$.

5. In this case $\phi_0 = 0$ and, measuring t_2 in seconds, $T = 6$.

Therefore $\omega = 2\pi/6 = \pi/3$ and we have

$$x(t) = A \sin\left(\frac{\pi}{3}t\right), \quad v(t) = \frac{\pi A}{3} \cos\left(\frac{\pi}{3}t\right).$$

Since $v(0) = 5$, we have $\pi A/3 = 5$ and therefore $A = 15/\pi$.

The equation of motion can be written

$$x(t) = (15/\pi) \sin\left(\tfrac{1}{3}\pi t\right)$$

7. $x(t) = x_0 \sin\left(\sqrt{k/m}\, t + \tfrac{1}{2}\pi\right)$

9. The equation of motion for the bob reads

$$x(t) = x_0 \sin\left(\sqrt{k/m}\, t + \tfrac{1}{2}\pi\right). \qquad \text{(Exercise 7)}$$

Since $v(t) = \sqrt{k/m}\, x_0 \cos\left(\sqrt{k/m}\, t + \tfrac{1}{2}\pi\right)$, the maximum speed is $\sqrt{k/m}\, x_0$.

The bob takes on half of that speed where $\left|\cos\left(\sqrt{k/m}\, t + \tfrac{1}{2}\pi\right)\right| = \tfrac{1}{2}$. Therefore

$$\left|\sin\left(\sqrt{k/m}\,t+\tfrac{1}{2}\pi\right)\right|=\sqrt{1-\tfrac{1}{4}}=\tfrac{1}{2}\sqrt{3}\quad\text{and}\quad x\left(t\right)=\pm\tfrac{1}{2}\sqrt{3}\,x_0.$$

11.
$$\text{KE}=\tfrac{1}{2}m[v(t)]^2=\tfrac{1}{2}m(k/m)x_0{}^2\cos^2\left(\sqrt{k/m}\,t+\tfrac{1}{2}\pi\right)$$

$$=\tfrac{1}{4}kx_0{}^2\left[1+\cos\left(2\sqrt{k/m}\,t+\tfrac{1}{2}\pi\right)\right].$$

$$\text{Average KE}=\frac{1}{2\pi\sqrt{m/k}}\int_0^{2\pi\sqrt{m/k}}\tfrac{1}{4}kx_0{}^2\left[1+\cos\left(2\sqrt{k/m}\,t+\tfrac{1}{2}\pi\right)\right]\,dt$$

$$=\tfrac{1}{4}kx_0{}^2.$$

13. Setting $\quad y(t)=x(t)-2,\quad$ we can write $\quad x''(t)=8-4x(t)\quad$ as $y''(t)+4y(t)=0.$

This is simple harmonic motion about the point $y=0$; that is, about the point $x=2$. The equation of motion is of the form
$$y(t)=A\sin\left(2t+\phi_0\right).$$

Since $y(0)=x(0)-2=-2$, the amplitude A is 2. Since $\omega=2$, the period T is $2\pi/2=\pi$.

15. (a) Take the downward direction as positive. We begin by analyzing the forces on the buoy at a general position x cm beyond equilibrium. First there is the weight of the buoy: $F_1=mg$. This is a downward force. Next there is the buoyancy force equal to the weight of the fluid displaced; this force is in the opposite direction: $F_2=-\pi r^2\left(L+x\right)\rho$. We are neglecting friction so the total force is

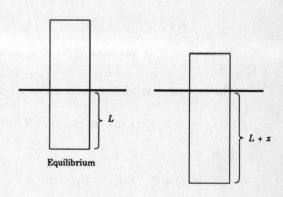

Equilibrium

$$F=F_1+F_2=mg-\pi r^2\left(L+x\right)\rho=\left(mg-\pi r^2L\rho\right)-\pi r^2x\rho.$$

We are assuming at the equilibrium point that the forces (weight of buoy and buoyant force of fluid) are in balance:
$$mg-\pi r^2L\rho=0.$$

Thus,
$$F=-\pi r^2x\rho.$$

By Newton's
$$F=ma\qquad\qquad(\text{force}=\text{mass}\times\text{acceleration})$$

we have

$$ma = -\pi r^2 x \rho \qquad \text{and thus} \qquad a + \frac{\pi r^2 \rho}{m} x = 0.$$

Thus, at each time t,

$$x''(t) + \frac{\pi r^2 \rho}{m} x(t) = 0.$$

(b) The usual procedure shows that

$$x(t) = x_0 \sin\left(r\sqrt{\pi\rho/m}\, t + \tfrac{1}{2}\pi\right).$$

The amplitude A is x_0 and the period T is $(2/r)\sqrt{m\pi/\rho}$.

17. From (12.6.4), we have

$$x(t) = Ae^{(-c/2m)t}\sin(\omega t + \phi_0) = \frac{A}{e^{(c/2m)t}}\sin(\omega t + \phi_0) \quad \text{where} \quad \omega = \frac{\sqrt{4km - c^2}}{2m}$$

If c increases, then both the amplitude, $\left|\dfrac{A}{e^{(c/2m)t}}\right|$ and the frequency $\dfrac{\omega}{2\pi}$ decrease.

19. Set $x(t) = 0$ in (12.6.6). The result is:

$$C_1 e^{(-c/2m)t} + C_2 t e^{(-c/2m)t} = 0 \implies C_1 + C_2 t = 0 \implies t = -C_1/C_2$$

Thus, there is at most one value of t at which $x(t) = 0$.

The motion changes directions when $x'(t) = 0$:

$$x'(t) = -C_1(c/2m)e^{(-c/2m)t} + C_2 e^{(-c/2m)t} - C_2(c/2m)t e^{(-c/2m)t}.$$

Now,

$$x'(t) = 0 \implies -C_1(c/2m) + C_2 - C_2 t(c/2m) = 0 \implies t = \frac{C_2 - C_1(c/2m)}{C_2(c/2m)}$$

and again we conclude that there is at most one value of t at which $x'(t) = 0$.

21. $\qquad x(t) = A\sin(\omega t + \phi_0) + \dfrac{F_0/m}{\omega^2 - \gamma^2}\cos(\gamma t)$

If $\omega/\gamma = m/n$ is rational, then $m/\omega = n/\gamma$ is a period.

23. The characteristic equation is

$$r^2 + 2\alpha r + \omega^2 = 0; \qquad \text{the roots are} \quad r_1, r_2 = -\alpha \pm \sqrt{\alpha^2 - \omega^2}$$

Since $0 < \alpha < \omega$, $\alpha^2 < \omega^2$ and the roots are complex. Thus, $u_1(t) = e^{-\alpha t}\cos\beta t$, $u_2(t) = e^{-\alpha t}\sin\beta t$, where $\beta = \sqrt{\alpha^2 - \omega^2}$ are fundamental solutions, and the general solution is:

$$x(t) = e^{-\alpha t}(C_1\cos\beta t + C_2\sin\beta t) = Ae^{-\alpha t}\sin(\beta t + \phi_0); \qquad \beta = \sqrt{\alpha^2 - \omega^2}$$

25.   Set $\omega = \gamma$ in the particular solution x_p given in Exercise 24. Then we have

$$x_p = \frac{F_0}{2\alpha\gamma m} \sin \gamma t$$

As $c = 2\alpha m \to 0^+$,   the amplitude $\left| \dfrac{F_0}{2\alpha\gamma m} \right| \to \infty$

27. $\left(\omega^2 - \gamma^2\right)^2 + 4\alpha^2\gamma^2 = \omega^4 + \gamma^4 + 2\gamma^2(2\alpha^2 - \omega^2)$ increases as γ increases.

PROJECTS AND EXPLORATIONS

18.1.   (b) The exact solutions are:

$$(i) \quad y = e^x \qquad (ii) \quad y = \tfrac{5}{4} e^{2x} - \tfrac{1}{2} x - \tfrac{1}{4} \qquad (iii) \quad y = e^{x^2 - 1}$$

Numerical vs Exact Solutions $(n = 4)$:

(i)

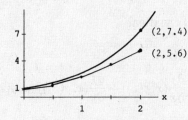

(ii)

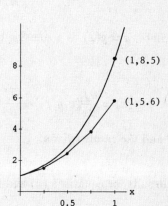

(iii)

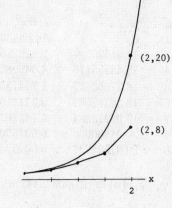

(c)

(i) $\quad y' = y$ (ii) $\quad y' = x + 2y$

n	y	error	n	y	error
1	3	4.38905610	1	3	5.48632012
2	4	3.38905610	2	4.25	4.23632012
4	5.0625	2.32655610	4	5.578125	2.90819512
8	5.96046448	1.42859162	8	6.70058060	1.78573953
16	6.58325017	0.80580593	16	7.47906272	1.00725749
32	6.95866676	0.43038934	32	7.94833345	0.53798668
64	7.16627615	0.22277995	64	8.20784519	0.20553560
128	7.27566979	0.11338631	128	8.34458724	0.14173288

SECTION 12.6

Using the TI-85 Stat menu to fit a curve:

$$err(n) \cong (5.83373529)n^{-0.77009098} \qquad err(n) \cong (7.32113942)n^{-0.77417682}$$

(iii) $y' = 2xy$

n	y	error
1	3	17.08553692
2	5	15.08553692
4	7.98046875	12.08749005
8	11.53141620	8.55412072
16	14.72391360	5.38309872
32	17.02438201	3.06115491
64	18.43970696	1.64582997
128	19.23067287	0.85486405

$$err(n) \cong (24.35752860)n^{-0.62905041}$$

(d) Since $g = g(x)$ is a solution of the differential equation,

$$g'(x) = f(x, g(x)),$$

$$g''(x) = \frac{\partial f}{\partial x}(x, g(x)) + \frac{\partial f}{\partial y}(x, g(x))g'(x) = \frac{\partial f}{\partial x}(x, g(x)) + \frac{\partial f}{\partial y}(x, g(x))f(x, g(x)),$$

and the result follows.

(e) Results for equations (ii) and (iii) are:

n	y	error	n	y	error
1	5.5	2.89632012	1	6	14.08553692
2	7.0625	1.42382012	2	10.65625	9.42928692
4	7.96612549	0.52019463	4	15.43555176	4.64998516
8	8.32780899	0.15851114	8	18.37080327	1.71473365
16	8.44260363	0.04371649	16	19.56382662	0.52171030
32	8.47485046	0.01146967	32	19.94213324	0.14340368
64	8.48338356	0.00293656	64	20.04800668	0.03753024
128	8.48557725	0.00074270	128	20.07594231	0.00959461

$$err(n) \cong (4.61414932)n^{-1.74744901} \qquad err(n) \cong (27.60325443)n^{-1.55091132}$$

APPENDIX A

SECTION A.1

1. $\Delta = 1^2 - 4(0)(0) = 1;$

hyperbola or intersecting lines

$\alpha = \frac{1}{4}\pi, \quad \frac{1}{2}(X^2 - Y^2) = 1$

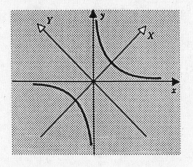

3. $\Delta = (10\sqrt{3})^2 - 4(11)(1) = 256;$

hyperbola or intersecting lines

$\alpha = \frac{1}{6}\pi, \quad 4X^2 - Y^2 = 1$

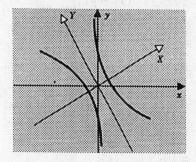

5. $\Delta = (-2)^2 - 4(1)(1) = 0;$

parabola or line(s)

$\alpha = \frac{1}{4}\pi, \quad 2Y^2 + \sqrt{2}X = 0$

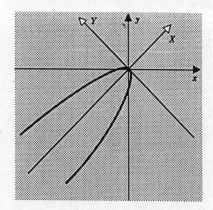

7. $\Delta = (2\sqrt{3})^2 - 4(1)(3) = 0;$

parabola or line(s)

$\alpha = \frac{1}{3}\pi, \quad Y = X^2$

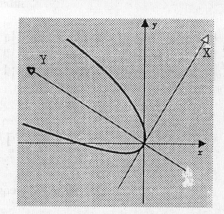

9. $\alpha = \frac{1}{8}\pi, \quad \cos\alpha = \frac{1}{2}\sqrt{2+\sqrt{2}}, \quad \sin\alpha = \frac{1}{2}\sqrt{2-\sqrt{2}}$

11. Set $x = (\cos\alpha)X - (\sin\alpha)Y, \; y = (\sin\alpha)X - (\cos\alpha)Y$ in the equation

$$ax^2 + bxy + cy^2 + dx + ey + f = 0$$

and expand to obtain the coefficients of X^2, XY, Y^2, etc.

TION A.2

1.   −2   **3.**   0   **5.**   5   **7.**   −6

9.   (a) $\begin{vmatrix} a_1 & a_2 \\ b_1 & b_2 \end{vmatrix} = a_1 b_2 - b_1 a_2 = \begin{vmatrix} a_1 & b_1 \\ a_2 & b_2 \end{vmatrix}$

      (b) $\begin{vmatrix} a_1 & b_1 & c_1 \\ a_2 & b_2 & c_2 \\ a_3 & b_3 & c_3 \end{vmatrix} = a_1 \begin{vmatrix} b_2 & c_2 \\ b_3 & c_3 \end{vmatrix} - b_1 \begin{vmatrix} a_2 & c_2 \\ a_3 & c_3 \end{vmatrix} + c_1 \begin{vmatrix} a_2 & b_2 \\ a_3 & b_3 \end{vmatrix}$

$$= a_1 b_2 c_3 - a_1 c_2 b_3 - b_1 a_2 c_3 + b_1 c_2 a_3 + c_1 a_2 b_3 - c_1 b_2 a_3$$

$$= a_1(b_2 c_3 - b_3 c_2) - a_2(b_1 c_3 - b_3 + c_1) + a_3(b_1 c_2 - b_2 c_1)$$

$$= a_1 \begin{vmatrix} b_2 & b_3 \\ c_2 & c_3 \end{vmatrix} - a_2 \begin{vmatrix} b_1 & b_3 \\ c_1 & c_3 \end{vmatrix} + a_3 \begin{vmatrix} b_1 & b_2 \\ c_1 & c_2 \end{vmatrix}$$

$$= \begin{vmatrix} a_1 & a_2 & a_3 \\ b_1 & b_2 & b_3 \\ c_1 & c_2 & c_3 \end{vmatrix}$$

11. $\begin{vmatrix} 1 & 2 & 3 \\ 4 & 5 & 6 \\ 7 & 8 & 9 \end{vmatrix} = - \begin{vmatrix} 4 & 5 & 6 \\ 1 & 2 & 3 \\ 7 & 8 & 9 \end{vmatrix} = \begin{vmatrix} 4 & 5 & 6 \\ 7 & 8 & 9 \\ 1 & 2 & 3 \end{vmatrix}$

                                  interchange two rows

13. $\dfrac{1}{2} \begin{vmatrix} 1 & 0 & 7 \\ 3 & 4 & 5 \\ 2 & 4 & 6 \end{vmatrix} = \begin{vmatrix} 1 & 0 & 7 \\ 3 & 4 & 5 \\ \frac{1}{2}(2) & \frac{1}{2}(4) & \frac{1}{2}(6) \end{vmatrix} = \begin{vmatrix} 1 & 0 & 7 \\ 3 & 4 & 5 \\ 1 & 2 & 3 \end{vmatrix}$

15.   (a) $\dfrac{\begin{vmatrix} 6 & 4 \\ 7 & -3 \end{vmatrix}}{\begin{vmatrix} 3 & 4 \\ 2 & -3 \end{vmatrix}} = \dfrac{46}{17}, \qquad \dfrac{\begin{vmatrix} 3 & 6 \\ 2 & 7 \end{vmatrix}}{\begin{vmatrix} 3 & 4 \\ 2 & -3 \end{vmatrix}} = -\dfrac{9}{17}$

           $x = \dfrac{46}{17}, \quad y = -\dfrac{9}{17}$ satisfy the equations:

$$3x + 4y = 3\left(\tfrac{46}{17}\right) + 4\left(-\tfrac{9}{17}\right) = \tfrac{102}{17} = 6$$

$$2x - 3y = 2\left(\tfrac{46}{17}\right) - 3\left(-\tfrac{9}{17}\right) = \tfrac{119}{17} = 7$$

      (b)   first quotient $= \dfrac{db_2 - a_2 e}{a_1 b_2 - a_2 b_1}$,   second quotient $= \dfrac{a_1 e - db_1}{a_1 b_2 - a_2 b_1}$

$$a_1 \left(\frac{db_2 - a_2 e}{a_1 b_2 - a_2 b_1} \right) + a_2 \left(\frac{a_1 e - db_1}{a_1 b_2 - a_2 b_1} \right) = \frac{a_1 db_2 e - a_2 db_1}{a_1 b_2 - a_2 b_1} = d$$

$$b_1 \left(\frac{db_2 - a_2 e}{a_1 b_2 - a_2 b_1} \right) + b_2 \left(\frac{a_1 e - db_1}{a_1 b_2 - a_2 b_1} \right) = \frac{-b_1 a_2 e + b_2 a_1 e}{a_1 b_2 - a_2 b_1} = e$$

(c) The equations

$$a_1 x + a_2 y + a_3 z = e$$

$$b_1 x + b_2 y + b_3 z = f$$

$$c_1 x + c_2 y + c_3 z = g$$

are solved by

$$x = \frac{1}{D} \begin{vmatrix} e & a_2 & a_3 \\ f & b_2 & b_3 \\ g & c_2 & c_3 \end{vmatrix}, \quad y = \frac{1}{D} \begin{vmatrix} a_1 & e & a_3 \\ b_1 & f & b_3 \\ c_1 & g & c_3 \end{vmatrix}, \quad z = \frac{1}{D} \begin{vmatrix} a_1 & a_2 & e \\ b_1 & b_2 & f \\ c_1 & c_2 & g \end{vmatrix},$$

where

$$D = \begin{vmatrix} a_1 & a_2 & a_3 \\ b_1 & b_2 & b_3 \\ c_1 & c_2 & c_3 \end{vmatrix}$$